Advances in Material Research and Technology

Series Editor

Shadia Jamil Ikhmayies, Physics Department, Isra University, Amman, Jordan

This Series covers the advances and developments in a wide range of materials such as energy materials, optoelectronic materials, minerals, composites, alloys and compounds, polymers, green materials, semiconductors, polymers, glasses, nanomaterials, magnetic materials, superconducting materials, high temperature materials, environmental materials, Piezoelectric Materials, ceramics, and fibers.

Shadia Jamil Ikhmayies
Editor

Advances in Catalysts Research

Springer

Editor
Shadia Jamil Ikhmayies
Department of Physics
School of Science
University of Jordan
Amman, Jordan

ISSN 2662-4761 ISSN 2662-477X (electronic)
Advances in Material Research and Technology
ISBN 978-3-031-49107-8 ISBN 978-3-031-49108-5 (eBook)
https://doi.org/10.1007/978-3-031-49108-5

This Springer imprint is published by the registered company Springer Nature Switzerland AG
The registered company address is: Gewerbestrasse 11, 6330 Cham, Switzerland

Paper in this product is recyclable.

Preface

Catalysis is a process in which a substance known as a catalyst is used to increase the rate of a chemical reaction. The catalyst changes the path of the chemical reaction but is not consumed by the reaction. Selectivity is the fundamental property of the catalyst, by which it can direct the reaction to increase the amount of desired product and reduce the amount of unwanted byproducts. Catalysts can produce entirely new materials with entirely new potential uses. Catalysts can be classified into four types; homogeneous catalysts, heterogeneous (solid) catalysts, heterogenized homogeneous catalysts, and biocatalysts. Some of the applications of catalysts are: they can enable the synthesis of complex molecules in fewer steps, can be successfully used to produce low-sulfur fuels in refineries, and they play an effective role in reducing emissions of carbon dioxide, nitrogen oxides, and unburned hydrocarbons from petrol, diesel, and jet fuel combustion vehicles. Despite these wonderful applications, there is still a need to develop new concepts in catalysis, and there are many issues associated with cost, availability, and toxicity of many of the precious metals used as catalysts, and the need for expensive and complex ligands. Also, there is still a need to design new catalysts with high selectivity, reactivity, stability, and low catalyst loading with high turnover number (TON). So, the door is open for more developments and more innovations in this field of research.

Eleven chapters in this book shine light on a wide set of diverse catalysts of the above, along with their synthesis methods, properties, applications, and challenges. The book provides beginners with introductory material on techniques used and emerging technologies in catalysts research and concepts, synthesis and characterization techniques for photocatalysis, electrocatalysis, and biocatalysis. It reviews the use of nanostructured materials, nanoparticles, and nanocomposites in heterogeneous catalysts, photocatalysts, semiconductor photocatalysts and presents their use in pollutant degradation, environmental remediation, water splitting, and removal of toxic and harmful substances from water. Developments and improvements in advanced catalysts for chemical, petrochemical, environmental, energy storage, and energy conversion are discussed in this book. The electrochemical approach to hydrogen technologies, converting waste frying oil into biodiesel using acidic and basic catalysts, is presented. The following paragraphs provide a brief description of chapters included in this book.

Chapter one, "Emerging Technologies in Catalyst Research" by Aisha Khalid et al. is an introductory chapter that deals with the barriers that were being faced during the advancing of different catalysts and their related technologies. It also focuses on recent advances in photocatalysis, electrocatalysis, and biocatalysis. Biocatalysis was described in detail due to its green nature as emerging science that is stressing upon green synthesis and catalysts. Chapter two "Advanced Nanostructured Materials for Heterogeneous Catalysis—Past, Present and Future" by Agnieszka Feliczak-Guzik et al. provides a review of the fundamental information on heterogeneous catalysis with its advantages and disadvantages together with its use in different catalytic reactions. Chapter three "Nanoarchitectured Ferrites, Graphene Oxide, and Their Composites as Enzyme Mimics and Photocatalysts" by Vibha Verma et al. provides an overview of the recent strategies for the synthesis methods and properties of ferrites nanoparticles (NPs), graphene oxide (GO), and their nanocomposites followed by their applications in the fields of photocatalysis and enzyme mimics. Chapter four "Advances in Polyphenol Oxidase Mimics as Catalyst" by Harmilan Kaur et al. summarizes the latest progress in the field of polyphenol oxidase mimics and highlights the factors affecting the polyphenol oxidase mimic activity. They also reported biosensing techniques to evaluate the polyphenol oxidase and mimic activity of synthesized materials and summarized the future research challenges and opportunities to enhance the polyphenol oxidase mimic activity. Chapter five, "Ru-Doped ZnS as an Enhanced Visible Light-Driven Photocatalyst" by Sujoy Kumar Mandal et al. presents experimental and computational works on sphalerite ZnS nanoparticles (NPs) doped with various ruthenium (Ru) concentrations synthesized via solvothermal method. The feasible doping of Ru into the ZnS matrix has been confirmed by structural characterizations. The electronic, photocatalytic, and optical properties have been investigated both experimentally as well as theoretically by first-principle calculations. Chapter six, "Recent Advances and Applications of Modified-Semiconductor Photocatalyst in Pollutant Degradation" by Pin Chen et al. presents the fundamentals of photocatalytic overall pollutant degradation, focusing on the design and synthesis of photocatalysts. Moreover, the chapter shines light on the recent advances in photocatalytic degradation to illustrate several general methods that are expected to mitigate several distinct obstacles concerned with achieving environmental remediation.

Chapter seven, "Past, Present, and Future in the Development of Medium and High-Temperature Catalytic Processes for N_2O Decomposition" by Yihao Wu et al. is a review that discusses different theoretical and experimental approaches for the preparation of more stable and selective catalysts. It focuses on medium and high temperature (350–900 °C) application. In chapter eight, "Nitrite Removal from Water: New Support Materials for Pd-Based Catalysts Aiming for a Low Ammonium Production" by F. M. Zoppas et al., the authors overview the current state of catalytic nitrite reduction and the use of catalysts supported on CeO_2, Nb_2O_5, ZrO_2, TiO_2, γ-Al_2O_3, SiO_2, and ZSM-5 for the removal of high concentrations of nitrite from water. The total conversion was achieved by catalysts supported in

γAl_2O_3, ZSM-5 (Si: Al = 30), SiO_2, and TiO_2. Chapter nine, "Noble-Metal-Free Bifunctional Electrocatalysts for Overall Water Splitting in Alkaline Medium" by Subhasis Shit et al. provides basic ideas about the bifunctional electrocatalysts for overall water splitting and summarizes some recent advancement in that field. Chapter ten, "Electrochemical Approach for Hydrogen Technology: Fundamental Concepts and Materials" by Victor Márquez et al. presents the aspects related to the electrochemical approach to hydrogen technologies, considering key concepts that drive both the thermodynamic and kinetic phenomena of the redox processes involved. Strategies for optimizing surface processes on different electrode materials were considered. Chapter eleven "Modification of TiO_2 as SO_4/TiO_2 Acid and CaO/TiO_2 Base Catalysts and Their Applications in Conversion of Waste Frying Oil (WFO) into Biodiesel" by Karna Wijaya et al. presents an experimental work about the conversion of waste frying oil into biodiesel using modified oxides (TiO_2, to SO_4/TiO_2) and CaO/TiO_2 to acidic and basic catalysts, respectively.

Amman, Jordan Shadia Jamil Ikhmayies

Contents

Emerging Technologies in Catalyst Research

Aisha Khalid, Maria Batool, Maryam Saghir, Tahoor Khalid, and Muhammad Faizan Nazar

Abstract The discovery of new technologies to potentially improve catalyst efficiency is seen as an emerging approach. The current chapter deals with the barriers that are being faced during advancing of different catalysts. Various strategies have been developed over the past few decades to overcome shortcomings such as cost inefficacy, instability due to moisture, salty environment and reduced work efficiency in heterogeneous condition. Extensive research is being carried out to find cheap and environmentally friendly renewable resources for the development of catalysts for different reactions. This chapter focuses on photocatalysis, electrocatalysis, and recent advances in biocatalysis. These techniques are employed due to their cost effectiveness and efficiency. So, we have detailed here working condition and mechanism of these techniques. Biocatalysis is described here in detail due to its green nature as emerging science is stressing upon green synthesis and catalysts.

Keywords Catalyst · Photocatalysis · Electrocatalysis · Biocatalysis

---

A. Khalid
Department of Biology, Lahore Garrison University, Lahore 54000, Pakistan

M. Batool
Department of Chemistry, University of Gujrat, Gujrat 50700, Pakistan

M. Saghir · T. Khalid
Department of Biotechnology, University of Management and Technology, Sialkot Campus, Sialkot 51310, Pakistan

M. F. Nazar (✉)
Department of Chemistry, University of Education Lahore, Multan Campus, Lahore 60700, Pakistan
e-mail: faizan_qau@yahoo.com

S. J. Ikhmayies (ed.), *Advances in Catalysts Research*, Advances in Material Research and Technology, https://doi.org/10.1007/978-3-031-49108-5_1

1 Introduction

Catalysts with increased reaction efficiency enable the industry to provide high quality, stable and environment friendly products recently. The highly reaction-specific nature of the catalyst reduces the production of hazardous products by avoiding undesirable side reactions. The efficiency of industrial catalysts has been scaled through the use of advanced technologies to improve their functionality and stability in harsh environmental conditions [1, 2].

The catalyzed mechanism of homogeneous catalysts has the advantage of better understanding, but their use on an industrial scale is precluded due to their reliance on expensive metal ions, which are later difficult to recover, by the process. However, the processes catalyzed by heterogeneous catalysts dictate the problems associated with homogeneous catalysis and the reuse of the metal catalyst [3].

In industries more than 90% of the chemical processes require homogeneous and heterogeneous catalyst for facilitation of reactions. However, the processes catalyzed by heterogeneous and homogeneous catalysts dictate the problems associated with catalysis and the reuse of the metal catalyst. Homogeneous catalysts at industrial scale are not efficient or economically viable because they are irrecoverable after completion of reaction while on the other hand major limitation for the introduction of heterogeneous catalyst at industrial scale are an expensive support material and adsorption step is the rate limiting step as the reaction cannot proceed until few of the product molecules leave and allow reactant molecules to attach for the reaction [4].

One of the founding father of the chemical thermodynamics Ostwald defined catalyst as these are the substances will leave the equilibrium of the reaction unchanged. Currently, the chemical transformation with increased rate is carried out due to the central role of the catalyst. Countless chemical approaches at academic research and chemical industry level seem to be impossible without use of catalyst. The other advantages of using catalysts include lowering the requirement of increase temperature, less byproduct formation, reaction specificity and reduced use of chemical reagent etc. [5].

Typical industrial catalysts include unsupported catalyst, supported catalyst, confined and hybrid catalyst etc. Unsupported catalysts have simple composition such as metal oxide but catalytically active application in the chemical industry. The work efficiency of these catalysts depends on the bonding efficiency of electronegative oxygen atom with metal. Support material on the other hand play significant role not only just in providing the high surface area but also stabilizing the active component such as Al_2O_3, TiO_2, SiO_2, MgO etc. binary oxide catalysts. However in case of support catalyst, inert material is usually recommended in order to avoid the interference of the support material in catalytic processes [5]. In various industrial important reactions confined catalysis strategy is efficiently utilized such as metal organic framework (MOFs). In this approach a unique nanoscale chemical environment energetically and kinetically modulates the reaction [6]. Due to increase demand to save energy the scientists are motivated to develop a hybrid catalyst that can draw two to three field of catalysis in a single reaction and maintain the boundaries between them such as Fe-ZSM-5 [7].

In the future, researchers could face many exciting challenges and opportunities for developing new catalytic techniques with industrial application. Human/public concerns about the environment and industrial workers' health forced scientists to discover new technologies that nearly eliminate the possibility of hazardous emissions. Human/public major concern about the environment and health status of the industrial workers due to use of hazardous chemicals for catalysis of chemical reactions at industrial scale forced scientists to design an efficient catalyst such as photocatalyst, electro-catalyst and biocatalyst via adopting new technologies like doping, oxygen evolution reaction and enzyme engineering that nearly eliminate the possibility of hazardous emissions also. Another particular interest in advancing various catalytic techniques such as photo-catalysis, electro-catalysis and enzyme catalysis is to use cheap alternative resources. The pressures of badly affected resources are forcing scientists to discover and develop the novel catalytic technologies.

There are several types of catalytic research referred to as;

- Photocatalytic research
- Electrocatalytic research
- Environmental catalytic research
- Biocatalytic research (enzymes, enzyme catalysis)
- Catalysis for biomass conversion
- Organocatalytic based research (catalysis in organic and polymer chemistry)
- Nanostructured based catalytic research
- Computational catalytic research

In this chapter our main focus is to address the recent advances of photocatalysis, electrocatalysis, and biocatalysis in detail among various catalytic research that confronts developing and emerging economies. Moreover, these technologies have promising potential application at the industrial scale. These three catalysis reactions are focus of attention due to ongoing immense work owing to their excellent merits such as light which is the free source of energy so using it as a catalyst is economic friendly. Electro catalysis is selected due to its high use for energy production and a lot researchers have keen interest in it. Bio catalysis is selected due to its environment friendly nature. Scientific research is shifting toward green approaches due to increasing rise in environmental pollution so biocatalysis is gaining a lot of popularity among researchers Table 1.

Table 1 Comparison of cost effectiveness, efficiency and recyclability rate of the described catalyst

Catalyst	Cost effective b	Efficient	Recyclability rate
Photocatalyst	✓	✓	✓
Electrocatalyst	×	✓	×
Biocatalyst	✓	✓	×

1.1 Photocatalytic Research

Photo-catalysis is photosynthesis inspired. So the thermodynamically uphill reactions are favoured in the presence of light. However, the thermodynamically uphill reaction creates confusion as these reaction are not spontaneous. As use of the canonical catalyst doesn't enable the non-spontaneous reaction. That's why, to combat the confusion, light is regarded as the reactant which will alter the thermodynamic of system thus allowing broader application of photo-catalysis. As a result, such reactions are described ideally by the terminology of photocatalysis.

Kinetic studies of photo catalyzed reactions at the interface is too complex to study. Following there are five lessons learnt regarding photo catalytic kinetics.

1. Presumed reaction steps leads to a ingle kinetic model. However, Langmuir Hinshelwood rate equation, $\mathbf{r = k_{cat}\ K\ C/[1 + KC]}$ resulted from multiple mechanisms, henceforth models alone are not enough to observe reaction mechanisms.
2. k_{cat} in Langmuir Hinshelwood equation signifies the slow step at a catalyst surface. It depends upon the reactant surface in thermal catalysis. However, Single valued k_{cat} in chlorination of the hydrocarbon didn't seem to depend upon the reactant structure.
3. k_{cat} and K in Langmuir Hinshelwood depend upon the intensity of light thus indicting that the pseudo-steady state approach is more important than the pre-expected equilibrated adsorption of the Langmuir Hinshelwood model.
4. Commonly studied cases such as those related to dyes and phenols are claimed as first order reactions but decrease in the rate constant is observed by increasing concentration of these pollutants. This effect is due to plotting of data of intrinsic zero order on a semi-log graph, and include zero order rate restrictions due to reactant saturation, electron transfer to oxygen, oxygen mass transfer, or light supply.
5. Configuration of the reactant, catalyst porosity and light absorption by the reactant affects the kinetics related to the removal of the contaminant from the self-cleaning photocatalyst surface. Rate equation for the self-cleaning photocatalyst surface is $\mathbf{rate = k\ C^n}$ **(1)**.

It is well known that the main impact of human health at present is pollution. The use of photocatalytic techniques, which are an effective and environmentally sound technology, could overcome this current pollution problem. In general, a photocatalytic reaction takes place in three steps (Fig. 1) [3, 8].

In the first step, a photoexcitation approach is used to generate electrons (e^-) and holes (h^+), and then these electrons and holes interact with the electron acceptors and donors accordingly after migrating to the surface of the photocatalysts. The semiconductor with the appropriate band gap is required to collect light, face separation is required, charge carriers should be considered, and valence band (VB) and conduction band (CB) edge potential should be considered depending on the requirement.

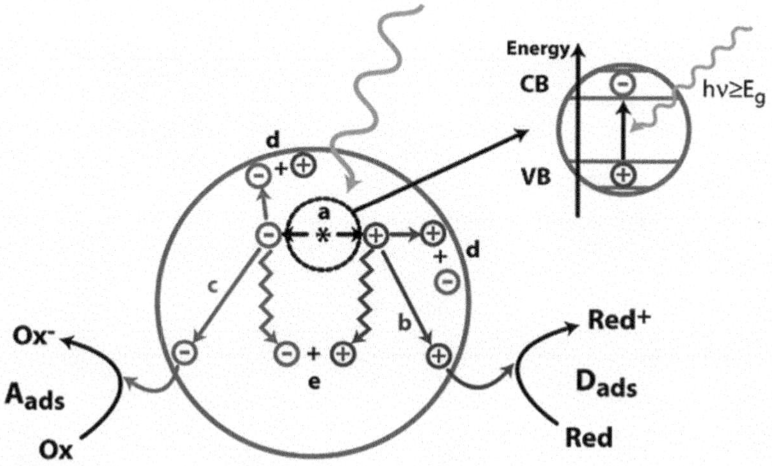

Fig. 1 Light absorption by photocatalyst **a** showing electron-pair formation **b** donor oxidation by hole **c** representing electronic reduction **d, e** electron and hole combination at the surface and interior. Adapted with permission from Hoffmann [9]. Copyright (1995). American Chemical Society

The performance of the photocatalytic system is improved by improving the light absorption gain, proper charge separation and surface reactivity capacity. For this purpose, certain methods are recommended, such as doping, modification with graphene, with metal particles and synthesis via crystal growth design and hetero-structuring.

In recent years, the TiO_2 nanomaterial has attracted notable attention in the field of pollution elimination and photocatalytic hydrogen generation. Due to the large electronic band gaps, less than 5% of the total solar energy is consumed. Therefore, much emphasis was placed on the absorption of titina in the range of visible light, which further increased the photocatalytic activity [4]. The development of this novel photocatalytic strategy showed that orbitals of some p-block metals with full d-subshell configuration, such as Ag, can hybridize with O-subshell of 2p to form a hybridized valence band, thereby reducing the band gap to use visible light. Heterostructures obtained after hybridization of two or more semiconductor systems have the potential advantage of using sunlight more efficiently than the single-phase semiconductor photocatalyst. Heterojunction of semiconductor based photocatalysts were developed that includes type I and type II heterojunctions, p-n hetero junction and Z scheme etc. (Fig. 2) [3, 10].

1.1.1 TiO_2-Based Photocatalysts

In this section, strategies for the modulation of TiO_2 energy band structure are reviewed in detail. TiO_2 is used as an example to understand how photocatalyst work with some latest approaches. Recently, the latest approaches including doping,

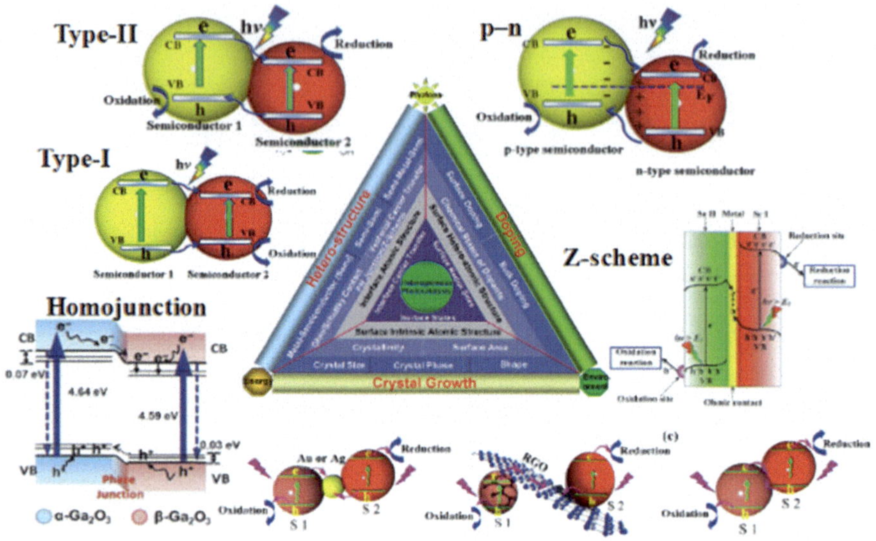

Fig. 2 Hetrojunction based photocatalyst with band alignments. Reproduced from Wang et al. [3], under the terms of the Creative Commons Atribution lisence CC 3.0. https://creativecommons.org/licenses/by/3.0/

coupling with graphene, precipitating with metal particles, crystal growth design, and heterostructuring have been developed. Here we will discuss the one related to the modification of the existing photocatalyst by doping, coupling with graphene and plasmonic resonance [5].

1.1.2 Doping

This is the most common approach used currently to tune the absorption band of wide band photocatalysts. Studies have been investigated about non-metal doping. Co-doping by combining the two suitable heteroatoms are also developed due to the advancement of technology for example nitrogen, fluorine, carbon, sulphur, boron, bromine, iodine and phosphorous. Scientists successfully prepared the nitrogen doped titania nanoparticles (NPs) in short time period via the microwave assisted solvothermal process which showed visible light absorption in the range of 400–550 nm. The resulted visible light induced photocatalysts have broad spectrum of potent application [5].

1.1.3 TiO_2-Graphene Composite

The two-dimensional carbonaceous material, graphene has broad application in various techniques due to unique feature of chemical stability, high conductivity and has large surface area. Graphene semiconductor improves the performance of the photocatalyst via aid to charge separation and its migration. Thus grapheme is integrated with TiO_2 for the development of the nanocomposite, which further improves the photocatalytic activity of TiO_2. Recently, many scientists synthesized the Graphene-TiO_2 nanostructure and reported the higher photocatalytic activity [6].

1.1.4 TiO_2-Based Plasmonic Photocatalysts

The efficiency of photocatalysts is significantly enhanced via introducing the new emerging technique of plasmonic photocatalysis. The coupling of nanostructured plasmonic metals with TiO_2 increases the local surface resonance (LSPR) effect, which further improves the photocatalytic activity. This LSPR effect allows the metal non crystal to absorb the large spectrum and scattering cross section [7].

2 Electrocatalytic Research

Electrocatalysis is the field of electrochemistry which has been gaining a lot of attention of both chemists and engineers for industrial application of electrocatalysis. Early literature was only about chemical engineering with minute reference of electrochemical explanation. Emerging research is about designing of the electrocatalyst in reference of surface chemistry, atomic topographic profiles, catalytic sites, phase transition and atomic rearrangement. Different crystallographic orientations at the electrode surface were observed to create different electrode kinetics. Teliz et al. explained electrocatalytic oxidation of methanol by using (111) stepped planes generated via cathodic treatment [11]. Electrocatalytic reactions involving amino acids have been observed to show changes in configuration and responses due to change of the electrode surface. Till now a major discussion about electrocatalysts has been seen to focus upon ultra-high vacuumed electrochemical techniques and surface chemistry. Celorrio et al. used differential electrochemical mass spectrometry for carbon dioxide reduction [12]. Various applications of electrocatalysis includes organic electrosynthesis, electrode sensors, fuel cells and galvanoplasty.

Water electrolysis can be summarized into two parts involving two half-cell reaction: one is hydrogen evolution reaction (HER) and second is the oxygen evolution reaction (OER). In HER, water gets reduced at the cathode and produced hydrogen and in OER, water get oxidized at the anode and produced oxygen. The major barrier which prevents application of water electrocalysis at the commercial level is high over the potential (measure of kinetic energy barrier). So catalysts help lowering this potential barrier during HER and OER. There is a need for favorable designing to make possible commercial implication of the electro-catalyst. This designing depends upon the working condition of the electrochemical cell. There are three approaches for accomplishing this designing.

(1) Proton exchange membrane electrolysis
(2) Alkaline electrolysis
(3) High temperature solid oxide electrolysis

There has been excellent review articles comprising water electrolysis due to rising interest of hydrogen production from water electrolysis [13–15]. Much literature has majorly been focused on the mechanism of the half-cell reaction in alkaline and preparation of different noble metal free electro catalyst in relation to better performance and durability during HER and OER.

Very few literature has been found which covers a mechanistic approach of the catalyst by accounting structure, composition, active sites and their stability in relation to HER and OER. Currently, hydrogen as an alternative source of energy has been placed in high hopes due to gradual depletion of natural energy sources such as coal, oil and natural gas [16]. Investigation showed that the current consumption of hydrogen is- billion Nm^3 range and will share 34% of total energy need in 2050 [17, 18]. This predictability is due to the high calorific value, non-toxicity and prospective economic fact of the hydrogen fuel [19].

The current challenge worldwide is to discover the new efficient approaches for hydrogen production. Electrochemical strategy is gaining more attraction rather than methane reforming, coal gasification and biomass conversion technologies due to the use of not only water as the natural electron carrier but also use electricity for generation of carbon free energy source. This is the beginning step of developing an environmental friendly methodology. This strategy also contributes to the upgradation of the biomass, efficient transportation of the fuel cell powder and better energy storage and conversion. Moreover, electrochemical water splitting is rapidly emerging as promising and feasible technology for efficient transformation of solar and wind power energy to clean the hydrogen fuel.

Generally, hydrogen evolution reaction [HER] and oxygen evolution reaction [OER] approaches used in electrochemical water splitting for the production of pure hydrogen and oxygen respectively [20]. The poor efficiency of both HER and OER caused only 4% of hydrogen product via electrochemical splitting of water [21].

However, both HER and OER are catalyst dependent and used as an efficient catalyst of the platinum-group metals and IrO_2 or RuO_2 respectively. High cost of these noble catalysts hamper the application electrochemical water splitting on large scale [22]. In order to address this prohibitively high cost issue it is very important to explore alternative comparable stable and low cost catalyst for water splitting reactions. Both HER and OER have the capability to be carried out in acidic, alkaline and neutral environment.

Preferably, HER types of reactions are carried out in acidic condition and will get the benefit of sufficient proton [H^+] adsorption on the electrode surface but not in alkaline condition while in case of OER basic conditions are more suitable rather than acidic condition [23]. This phenomenon decreased the possibility of simultaneous generation of H_2 and O_2. Consequently, this challenge encourages the studies of water splitting electrolysis under neutral conditions but unfortunately no method is still developed for neutral water electrolysis [24].

In near future, it is highly paramount to develop a methodology for water splitting under neutral condition. Recently, investigations are carried out in search of a cheap and efficient electrocatalyst including transition metal based catalyst, such as alloys, carbides, oxides, (oxy) hydroxide, sulfides, selenides, nitrides, phosphides, and dichalcogenides, as well as carbon-based catalysts, such as heteroatom-doped carbon and metal-embedded carbon.

Unique physical and chemical features and adjustable electronic configuration structure of these transition metals electrocatalyst would enable the balanced performance of electrocatalysis. The combination of different transition metals or their proportion would create a different type of material one dimensional (MnO_2 nanowires), two dimensional (MoS_2, $MoTe_2$, $NiSe_2$, and Mxene-group), and three dimensional (transition metal dichalcogenid, metal–organic framework, or other porous structure based on carbonaceous materials as matrix usually) [25].

2.1 Design Strategies for Electrocatalysts

Generally in neutral media, two approaches such as enhancing the intrinsic performance of each active site through the formation of polycrystalline, core–shell assembly and alloy structures, etc., and increasing the number of active sites through increased loadings or regulated structures of the catalyst that improve the catalytic efficiency of electrocatalysis [26].

For this reason, electrocatalyst design strategies are based on the structure-activity relationship and design mechanism of HER/OER. HER is a two-electron transfer mechanism strategy for electrocatalysts. It is considered an intermediate adsorption-desorption phenomenon. It can happen through the Volmer-Heyrovsky or Volmer-Tafel mechanism [27].

In an acidic medium

$$H^+ + e^- + * \rightarrow H^* \qquad Volmer\ step$$

$$H^+ + e^- + H^* \rightarrow H_2 + * \qquad Heyrovsky\ step$$

$$H^* + H^* \rightarrow H_2 + 2^* \qquad Tafel\ step$$

In an alkaline medium

$$H_2O + e^- + * \rightarrow H^* + OH^- \qquad Volmer\ step$$

$$H_2O + e^- + H^* \rightarrow H_2 + OH^- + * \quad Heyrovsky\ step$$

$$H^* + H^* \rightarrow H_2 + 2^* \quad Tafel\ step$$

Here * represents active adsorption areas on the surface of the catalyst. H* is seen here involved in every step in acidic medium so HER is described via free energy of H*[28]. According to the Stabalier principle catalytic optimum activity is achieved when the equilibrium is maintained between adsorption and desorption followed by the formation of bond with intermediate strength on the surface of the electrocatalyst [29]. At equilibrium, free energy change of H* should be equal to zero in neutral medium, HER is followed with the help of two intermediate absorbed H* and OH* and here only H_2O provides hydrogen atoms to form H^+ thus leading to low conductivity. [30]. Thus, its concluded that the energy barrier happened to be due to dissociation of extra water which is the reason for delayed kinetics [31]. So, under these circumstances, electrocatalysts with moderate energies of OH* and H* were chosen for water splitting.

2.2 Oxygen Evolution Reaction

Oxygen evolution reaction (OER) based electrocatalyst has been very challenging to deal with due to complex kinetic barrier as it has high-energy barrier with four electron processes. The mechanism of these steps at different pH is as follows [32].

In an acidic medium

$$2H_2O + * \rightarrow OH^* + H_2O + H^+ + e^- \tag{1}$$

$$OH^* + H_2O \rightarrow O^* + H_2O + H^+ + e^- \tag{2}$$

$$OH^* + H_2O \rightarrow OOH^* + H^+ + e^- \tag{3}$$

$$OOH^* \rightarrow O_2 + * + H^+ + e^- \tag{4}$$

In an alkaline medium

$$4OH^- + * \rightarrow OH^* + 3OH^- + e^- \tag{5}$$

$$OH^* + 3OH^- \rightarrow O^* + 2OH^- + H_2O + e^- \tag{6}$$

$$O^* + 2OH^- + H_2O \rightarrow OOH^* + OH^- + H_2O + e^- \tag{7}$$

$$OOH^* + OH^- + H_2O \rightarrow O_2 + 2H_2O + e^- \quad (8)$$

In comparison to HER, OER have many intermediates such as OOH*, OH* and O* thus making the catalytic process more complex where catalysts undergo many processes; adsorption, desorption, dissociation, bond formation, bond breakage and oxygen desorption. Here the associated Gibbs free energy change due to the formation of peroxide which less favored in terms of energy [33].

3 Enzymatic Catalysis

Enzymes are considered as potential biocatalysts because of various distinct advantages and successfully deployed at industrial scale catalysis. They have an efficient capability to work in a friendly environment and limited the environmental and physiological toxicity factors at the commercial scale. Furthermore, these biocatalysts reduce the operational cost indicating that these are also economically competitive in other industrial sectors such as biofuel production [34]. These biocatalysts followed a different mechanism such as acid-base, covalent and metal ion catalysis to go through an uninterrupted pathway. In the first step of catalysis enzymes help to transfer the proton or electron and in the second step amino acid residues in the active site are regenerated to its original form for reuse. Moreover, we know that enzymes are simply the substances that speed up the chemical reaction without itself being consumed in the reaction procedure. The activity of the enzyme has greatly influenced on the rate of the reaction. The rate of the reaction is very high in case of a highly active catalyst. The performance of the enzyme is decreased progressively due to various factors such as denaturation and chemical positioning [35]. This scenario will lead to replace or regenerate the catalyst for further experiment.

However, the technical barrier associated with enzymes such as un-stability at high temperature, pH change and solubility in toxic solvents will limit their usage [36]. So, various approaches are analyzed for the production of robust and novel biocatalyst that will indicate them suitable candidates to work in the broad range of industrial setting [37].

Looking into the future, this chapter will also discuss in depth about the discovery of latest surprising approaches that will scale up the utilization of enzymes at various industrial scales.

3.1 Engineered Enzymes

The artificial construction of the bio inspired catalyst needs a well depth level of understanding about the structure and functional mechanism of natural enzymes. This knowledge must be augmented with practical sense of the engineered tolerance mechanism. The basic understanding about the natural protein complexity should be useful to cope the significant barriers that arise usually in the experimental work [38]. Current research methodology focused on creation of an artificial functional protein armed with full conceptual understanding about the mechanism pathway. Enzyme engineering tool can be used to increase the stability and functionality of natural enzymes at elevated level of temperature, harsh pH condition (alkaline & acidic) and in specific organic solvents [39].

In today's world due to increasing public concern about the health and environment demands the use of cheap, reusable and readily available material at the commercial scale for the production of products is adopted. Microbial enzymes are used at commercial scale for the production of targeted products. The operationally thermostable enzymes are required in these industries to support better mixing and solubility rate, increasing reaction efficiency and lowering the contamination risk etc. Enzymes showing stability at high temperature are being encouraged or proposed to use at the industrial scale. Thermophile microorganisms preferably grow above at 55 °C while the hyperthermophile growth range is beyond 80 °C. Archean mostly dominate in the hyperthermophile category as compared to bacteria with some exceptional cases such as *Thermotoga.*

From last two decades, research has been more focused on the production of thermostable enzymes from microorganisms. In early 1960s Brock and his coworkers started to work on the stability and functionality of enzymes at elevated temperature. In 1990s gene transfer recombinant DNA technology dramatically increases the production of a remarkable list of thermostable enzymes has the potential application at commercial scale. This development of molecular biology stimulates the interest for screening of thermophile strains and isolation of the gene of interest from the thermal environment. The potential example of thermostable enzyme in the current era is commercial Taq-polymerase and various other thermostable enzymes that are also being commercialized. Engineered thermostable enzymes withstand harsh environmental conditions and offer as being an alternative resource for chemical catalysts. Discovery and development of site directed technology removed the significant barrier identified earlier. Today industries are well able to adapt the usage of genetically engineered enzymes at higher volumes in various processes under harsh environmental conditions [40].

3.2 *Enzyme Immobilization*

Enzymes are the biocatalysts of biochemical reactions. They are proteinaceous molecules that promote several chemical reactions in the human body as well as in research laboratories. Specific arrangement of atoms and amino acids make them uniquely designed species. Enzymes being biocatalysts have broad application at the commercial scale such as in industries, health, disease, pharmacy, textile, food processing and so many other areas. Enzymes are highly specific in their functionality and structures.

In twentieth century, during the dawn of microbial fermentation and enzyme technology, the importance of these biocatalysts was highly acknowledged. The detailed research investigation revealed that the handling of highly unstable enzymes during the biochemical reactions raised the issue of enzyme isolation and recovery. So, the whole process becomes complicated and costly. This issue is recently addressed via the invention of a biotechnological strategy "enzyme immobilization" [41].

Enzyme immobilization is an area of deep research and interest nowadays. Several studies have been conducted worldwide for the exploitation of best techniques through this method. Current advancements in the biotechnological world are based on enzyme production and exploring their different properties. Enzyme immobilization is yet another powerful tool that is used in industries and laboratories for achieving multiple tasks.

Enzymes catalyze the reactions in various states, whether being independently in a solution or being attached to some sort of species or surface called "immobilized enzymes", and the technique through which biocatalysts are immobilized is termed as Enzyme Immobilization. In this technique enzymes are made physically or locally confined to some surface or matrix, in order to retain their catalytic activities and recover them at the end of the reactions. The purpose of adopting this procedure is to use them repeatedly and lower the recovery costs especially at the industrial scale [42].

In the immobilization process the enzyme is generally attached to an inert insoluble material for the specificity of the reaction. Enzymes are also imprisoned in the cell or in certain support or matrix, and to restrict enzyme activity in a fixed space/region. Their reusable property is what makes them unique in laboratory experiments and also make them economical ideal candidate to introduce at the commercial scale. The main objective of this strategy is to keep enzyme in its stable conformation structurally and functionally during various biochemical reactions and improving the recovery and isolation phase economical. The key components for the enzyme immobilization are an enzyme, a matrix/support and method of attachment (Fig. 3).

3.3 Enzyme Immobilization: Factors to Be Considered

Enzyme immobilization is a very peculiar technique which requires a lot of background research and considerations before performing it. There are certain factors that need to be considered before immobilizing an enzyme. For example; the microenvironment provided during the experimentation decides the stability and kinetic properties of enzymes. Secondly, a particular immobilization method will impart certain physical and chemical changes in enzymes and these too are to be kept in mind before performing this technique. Thirdly, the surface of support is significant in the whole procedure, because the matrix surface makes different kinds of bonding with enzymes that ultimately have an important role in maintaining the structural properties of enzymes.

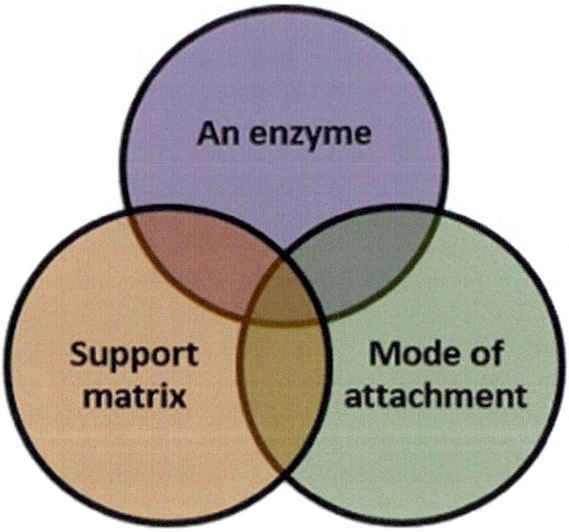

Fig. 3 Indicating the three key components for Enzyme Immobilization

Keeping in view the above discussion, the three basic factors for immobilizing any biocatalyst are;

(1) What support or matrix one is going to use?
(2) What conditions are given?
(3) Which method of immobilization is being used?

The key factor for the immobilization technology is to stabilize the tertiary structure of the protein for proper functioning (Fig. 4).

3.4 Classification of Supports/Support or Matrix Material

The most important component of the whole immobilization technology is the selection of the support to be used for immobilizing the particular enzyme. There are no specific properties dictated for selecting the matrix support but few basic things need to be considered for an ideal situation. The performance of an enzyme-immobilized system depends on the hydrophilicity, biocompatibility, physical resistance to compression, low cost availability, resistance to microbial attack and enzyme inertness nature.

Moreover some general physical features such as particle diameter, mechanical strength, and compression are also of significant considerations.

In immobilization technology, the organic support material can either be a natural polymer such as; agar, cellulose, chitin, dextrans, and proteins like collagen, albumin etc. or synthetic polymers includeing Polystyrene, polyacrylate, polyamide, allyl polymers, and vinyl polymers while on the other hand silica, bauxite, clay minerals

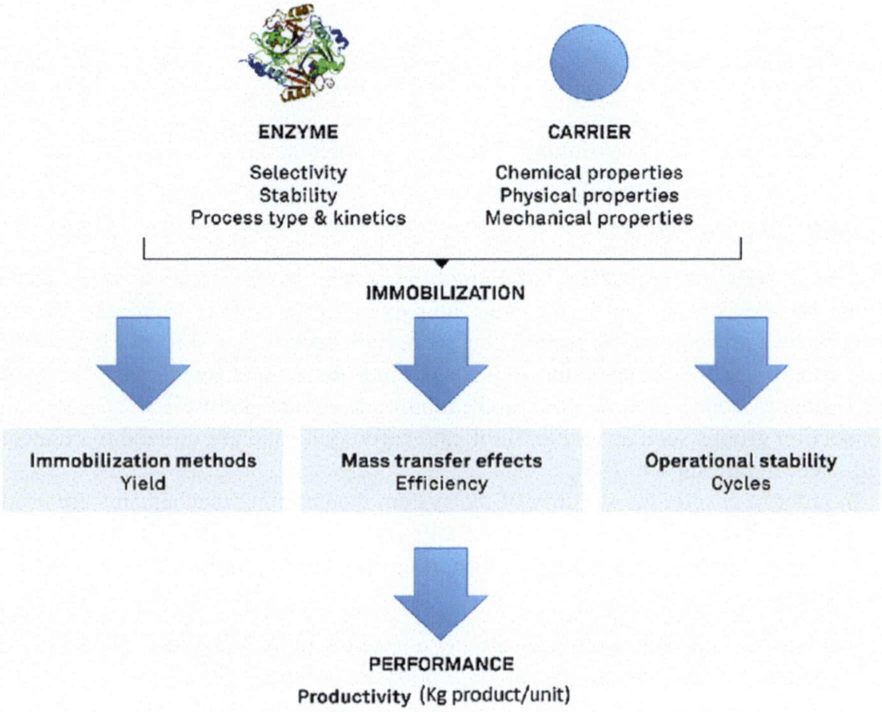

Fig. 4 Enzyme immobilization and factors affecting its performance. Reproduced from Basso and Serban [43] under the terms of the Creative Commons CC. https://creativecommons.org/licenses/

pore metal oxides, and glass constitute the inorganic matrix. An organic matrix such as agarose is used extensively in industrial applications due to their good hydrophilic nature, good commercial availability, and excellent property of derivitization [44].

3.4.1 Methods of Immobilization

Enzymes are immobilized by employing various methods. In order to get best possible result, more than two to three immobilized approaches have been employed during a synthesis process. Moreover the methodology may be according to the physical and chemical nature of the enzyme.

There are two categories in which the methods of enzyme immobilization are divided;

(1) Reversible Methods
(2) Irreversible Methods

Irreversible Methods

In this methodology, the enzyme is strongly linked with the matrix region and could not be detached until damage or destroyed. Irreversible linkage may occur by covalent bond, entrapping and cross-linking. The detail of such methods is given below;

Covalent Bond Formation

The most often employed way of immobilization is by the creation of covalent bonds between enzyme molecules and the support surface. This approach ensures that amino acids required for the enzyme's catalytic activity do not form a bond with the surface while the connection is formed using the unique functional groups of the lysine, cysteine, aspartic acid, and glutamic acid side chains. The nucleophile connection groups such as amino, thiol, and carboxylic acid are utilized to connect with electrophile support [45].

In order to ensure, the stability of the system the coupling mechanisms are used in order to "couple the enzyme" to the support, in a manner that allows efficient and accurate bonding. Activation or coupling may be accomplished in one of the following two ways:

i. Support or matrix activation by adding a function to the polymer
ii. Production of an activated group by modifying polymer

These activation methods are done to produce the proper electrophilic groups on support for strong and precise bonding.

Entrapment

This irreversible immobilization methodology involves the entrapping of enzymes within the fibers of support and allow the substrate and product molecules to pass through the network of matrix. There are different methods used for entrapment such as gel entrapment, fiber entrapment and micro encapsulation [46].

Cross Linking

This method is also called, as "carrier free immobilization" because, in this method the enzyme has its own carrier and recovery of enzyme at the end of the biochemical reaction that is possible just by the technique of cross-linking.

The two latest methods of cross-linking are:

i. Cross Linking Enzyme Crystals (CLEC)
ii. Cross Linking Enzyme Aggregate (CLEA)

In these both procedures, we use some sort of linking agents like; glutaraldehyde. The agent is poured into the reaction mixture and the pure enzyme is isolated at the end of the reaction and unlike other methods enzyme stays in the mixture independently

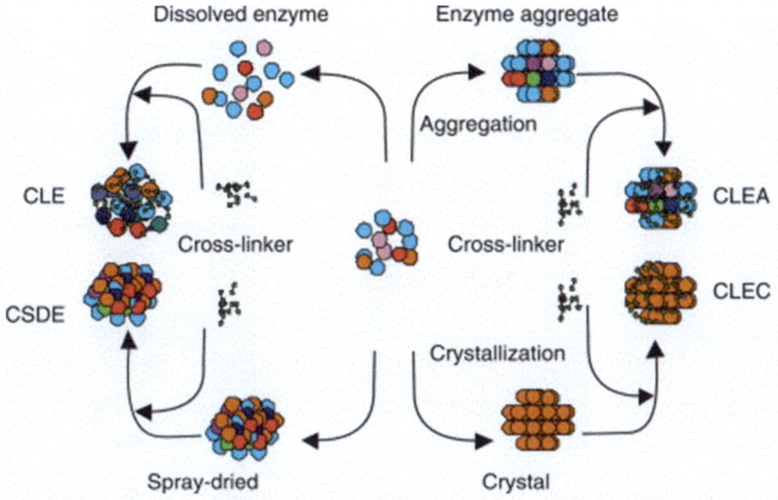

Fig. 5 Schematic representation of enzyme working with dissolved enzyme and enzyme aggregate. Reproduced from Sheldon [47] under the terms of the Creative Commons Attribution (CC BY). https://creativecommons.org/licenses/by/4.0/

without binding to matrix. The purpose of immobilization is ultimately achieved this way.

The difference between CLEC and CLEA is that in the former enzyme crystals are formed, but in CLEA, which is an improved version of CLEA aggregates of enzymes, they are propagated and hence this method can be used in aqueous environments as well. Hence, CLEA is just an improved and better technique to do cross-linking of enzyme and achieve immobilization without any support medium (Fig. 5) [47].

Reversible Methods

The reversible methods of immobilization are those in which the enzymes are able to detach themselves from the support easily after the biochemical reaction. The reversible methods are of quite good economic importance due to the reusable benefit of support after the decay of enzymatic activity. These methods find great applications in bio analytical systems and especially for enzymes which are easily altered [48]. Some of the important reversible techniques of enzyme immobilization are;

- Adsorption
- Disulfide Bonding
- Chelation
- Ionic bonding
- Affinity binding

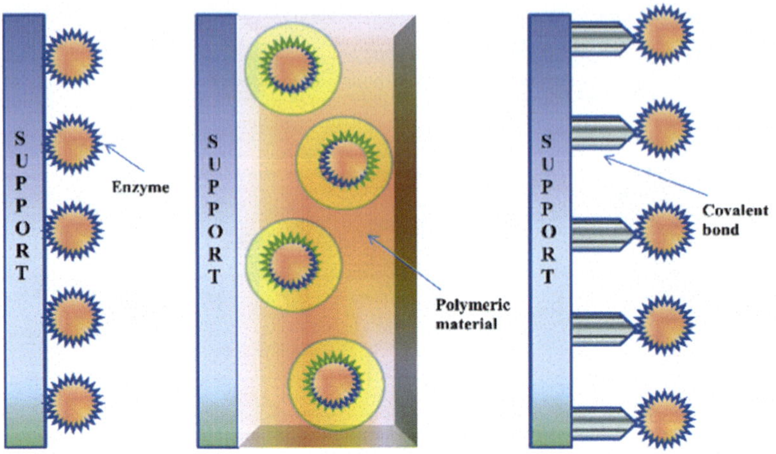

Fig. 6 Immobilization of enzyme and polymeric material at support. Reproduced from Mohamad et al. [51] under the terms of the Creative Commons (CC BY) license. https://creativecommons.org/licenses/by/4.0/

Adsorption

It is a physical reversible method of immobilization in which enzymes are simple adsorbed onto the support surface. It is the oldest and simplest method of making enzyme immobilized. In fact, the first ever-immobilized enzyme aminocyclase was made by this same technique in which a surface DEAE-Sephadex was used for its adsorption. The method of adsorption is used for large industrial scales [49].

Adsorption involves the reversible non-covalent relatively weak binding of the enzyme with a support matrix. Hydrogen bonds, van der waal forces, or hydrophobic linkages are the weak bonds that are used in adsorption. Ionic or metallic bonds are relatively stronger than others (Fig. 6) [50]. A major drawback of using the adsorption technique is that enzymes can detach and get leaked from the matrix easily. The reversal of bonding depends on;

- pH
- Temperature
- Polarity
- Ionic interactions
- Force of attractions

Disulfide Bonding

In this approach, the non-essential thiol group of the enzymes is immobilized on the reactive support under mild conditions. The remarkable feature of this approach is the reversibility of the disulfide bond formed between the thiol enzyme and reactive solid support via using excess amount of DTT (dithiothreitol). Another potential advantage of this strategy over the adsorbent is that the denatured enzyme can be easily removed from the solid support. So this is a point of particular interest especially at the industrial scale where the cost of the solid polymeric surface is hampered [51].

Chelation

It is also defined as the metal binding or metal link immobilization strategy. In this remarkable approach, the transition metal ions salt such as titanium and zirconium have been bounded with some nucleophilic group present on the organic matrix. The metal salt is simply precipitated via heating on to the support material such as chitin, cellulose and silica based support. Then the free available spaces of metals coordinated with the enzymes. In order to avoid the significant leakage of the metal ions from support ligands are immobilized on matrix via covalent linkage [52].

Ionic Bonding

In this a straightforward approach Enzyme is reversibly immobilized on the ligand. This obvious approach is based on the simple chromatography principle of protein ligand interaction. The challenging task in this strategy is to optimize the condition under which enzyme in a fully active condition remains bound. Currently the use of polymeric ionic ligands as immobilized allowed the strong interaction of the enzyme with matrix [53].

Affinity Binding

In this approach, the affinity principle between the immobilized enzyme and the complementary support molecule is applied. However, the cost of this technique is very high due to the requirement of covalent binding of antibody and lectin types of ligand on the matrix [54].

4 Conclusion

In recent era, advancement in catalysts via novel emerging technologies enabled them as a powerful tool to work efficiently in various processes especially at the industrial scale. They also have capability to work potentially in a wide range of conditions and produce various industrial products. Employment of modern catalysts offers advantages of high catalytic efficiency under mild conditions. In this chapter our main focus is to address the recent advances of photocatalysis, electrocatalysis, and biocatalysis in detail among various catalytic research that confronts developing and emerging economies. Moreover, these technologies have promising potential application at the industrial scale. These three branches of catalysis attracted attention due to ongoing immense work owing to their excellent merits such as light which is the free source of energy so using it as catalyst is economically friendly. Electro catalysis is selected due to its high use for energy production, and a lot of researchers have keen interest in it. Bio catalysis is selected due to its eco-friendly nature. Scientific research is shifting toward green approaches due to the increasing rise in environmental pollution so bio catalysis is gaining a lot of popularity among researchers.

References

1. D.F. Ollis, Fronti. Chem. **11**, 378 (2018)
2. T. Jesionowski, J. Zdarta, B. Krajewska, Adsorption **20**, 801 (2014)
3. Y. Wang, X. Ma, H. Li, B. Liu, H. Li, S. Yin,T. Sato, in *Advanced catalytic materials: photocatalysis and other current trends*, ed. by N. Luis (Intechopen, 2016), vol. 12, pp. 337. https://doi.org/10.5772/61864
4. A. Nicosia, F. Vento, G.M. Di Mari, L. D'Urso, P.G. Mineo, Nanomaterials **11**, 400 (2021)
5. R. Verma, J. Gangwar, A.K. Srivastava, RSC Adv. **7**, 44199 (2017)
6. D.M. Tobaldi, D. Dvoranová, L. Lajaunie, N. Rozman, B. Figueiredo, M.P. Seabra, A.S. Škapin, J.J. Calvino, V. Brezová, J.A. Labrincha, Chem. Eng. J. **405**, 126651 (2021)
7. M. Sakar, R. Mithun Prakash, T.-O. Do, Cataly. **9**, 680 (2019)
8. J.M. Guisán, G. Alvaro, C.M. Rosell, Fernandez-Lafuente R. Biotechnol. Appl. Biochem. **20**, 357 (1994)
9. M.R. Hoffmann, S.T. Martin, W. Choi, D.W. Bahnemann, Chemi. revi. **95**, 76 (1995)
10. G. Liu, C.Y. Jimmy, G.Q.M. Lu, H.M. Cheng, Chemi. Communicati. **47**, 79 (2011)
11. E. Teliz, V. Díaz, R. Faccio, A. Mombrú, C. Zinola, Int. J. Electrochem. **2011**(2011)
12. V. Celorrio, L. Calvillo, R. Moliner, E. Pastor, M. Lázaro, J. Po. Sour. **239**, 72 (2013)
13. J. Song, C. Wei, ZF. Huang, C. Liu, L. Zeng, X. Wang X, Chem. Soc. Rev. **49**, 7(2020)
14. C. Hu, L. Zhang, J. Gong, Enemy Environ. Sci. **12**, 2620 (2019)
15. M. Huynh, T. Ozel, C. Liu, E.C. Lau, D.G. Nocera, Chem. Sci. **8**, 4779 (2017)
16. J. M. Ogden. Phys. To. **55**, 69 (2002)
17. A. Demirbas, Biohydro. Fut. Eng. F. Dema. **11**, 163 (2009)
18. A. Demirbas, Energy sources part B. Econ. Plan. Policy **12**, 172 (2017)
19. G.W. Crabtree, M.S. Dresselhaus, M.V. Buchanan, Phys. Today **57**, 39 (2004)
20. N.P. Brandon, Z. Kurban, Philos. Trans. R. Soc. Math. Phys. Eng. Sci. **375**, 20160400 (2017)
21. S. Dunn, Int. J. Hydrog. Energy **27**, 235 (2002)
22. S. Trasatti, J. Electroanal. Chem. Interfacial Electrochem. **39**, 163 (1972)
23. N. Han, K.R. Yang, Z. Lu, Y. Li, W. Xu, T. Gao, Z. Cai, Y. Zhang, V.S. Batista, W. Liu, X. Sun, Nat. Commun. **9**, 924 (2018)
24. K. Qu, Y. Zheng, Y. Jiao, X. Zhang, S. Dai, S.-Z. Qiao, Adv. Energy Mater. **7**, 1602068 (2017)
25. S. Anantharaj, V. Aravindan, Adv. Energy Mater. **10**, 1902666 (2020)
26. J.D. Benck, T.R. Hellstern, J. Kibsgaard, P. Chakthranont, T.F. Jaramillo, ACS Catal. **4**, 3957 (2014)
27. E.B. Carneiro-Neto, M.C. Lopes, E.C. Pereira, J. Electroanal. Chem. **765**, 92 (2016)
28. X. Zou, Y. Zhang, Chem. Soc. Rev. **44**, 5148 (2015)
29. R. Parsons, Trans. Faraday Soc. **54**, 1053 (1958)
30. T. Shinagawa, K. Takanabe, J. Phys. Chem. C **119**, 20453 (2015)
31. X. Wang, Y. Zheng, W. Sheng, Z. Xu, M. Jaroniec, S.-Z. Qiao, Mater. Today **36**, 125 (2020)
32. M. Bajdich, M. García-Mota, A. Vojvodic, J.K. Nørskov, A.T. Bell, J. Am. Chem. Soc. **135**, 13521 (2013)
33. I.C. Man, H.-Y. Su, F. Calle-Vallejo, H.A. Hansen, J.I. Martínez, N.G. Inoglu, J. Kitchin, T.F. Jaramillo, J.K. Nørskov, J. Rossmeisl, ChemCatChem **3**, 1159 (2011)
34. A.S. Bommarius, M.F. Paye, Chem. Soc. Rev. **42**, 6534 (2013)
35. A.G. Grigoras, Biochem. Eng. J. **117**, 1 (2017)

36. C.S. Bezerra, C.M.G. de Farias Lemos, M. de Sousa, L.R.B. Gonçalves, J. Appl. Polym. Sci. **132**, 26 (2015)
37. F.A. Chao, A. Morelli, J.C. Haugner, L. Churchfield, L.N. Hagmann, L. Shi, L.R. Masterson, R. Sarangi, G. Veglia, B. Seelig, Nat. Chem. Biol. **9**, 81 (2013)
38. T.D. Brock, H. Freeze, J. Bacteriol. **98**, 289 (1969)
39. T.D. Brock, Sprin. Scie. & Busin. Me. **1986**, 1 (2012)
40. K.O. Stetter, F.E.M.S. Microbiol, Rev. **18**, 149 (1996)
41. J.M. Berg, J.L. Tymoczko, L. Stryer, L. Stryer (eds.), *Biochemis* (W.H. Freeman and Company, New York and Basingstoke, 2007)
42. E. Katchalski-Katzir, Trends Biotechnol. **11**, 471 (1993)
43. A. Basso, S. Serban, Mol. Catal. **479**, 110607 (2019)
44. J.M. Cabral, J.F. Kennedy, Bioprocess Technol. **14**, 73 (1991)
45. J.S. Weltz, D.F. Kienle, D.K. Schwartz, J.L. Kaar, J. Am. Chem. Soc. **142**, 3463 (2020)
46. E. Górecka, M. Jastrzębska, Biotechnol. Food Sci. **75**, 65 (2011)
47. A.A. Homaei, R. Sariri, F. Vianello, R. Stevanato, Chemi. Boil. **6**, 185 (2013)
48. R.A. Sheldon, Biochem. Soc. Trans. **35**, 1583 (2007)
49. M.N. Gupta, B. Mattiasson, Meth. Biochem. Anal. **36**, 1 (1992)
50. F. Studt, Front. Catal. **1**, 658965 (2021)
51. N. Mohamad, N. Marzuki, H.C. Buang, F. Huyop, R.A. Wahab, Biotechnol. Biotechnol. Equipm. **29**(2), 205 (2015)
52. M. Salazar, R. Becker, W. Grünert, Appl. Catal. B Environ. **165**, 316 (2015)
53. K. Ovsejevi, C. Manta, F. Batista-Viera, Meth. Mol. Biol. Clifton NJ **1051**, 89 (2013)
54. P.R. Coulet, J. Carlsson, J. Porath, Biotechnol. Bioeng. **23**, 663 (1981)

Advanced Nanostructured Materials for Heterogeneous Catalysis—Past, Present and Future

Agnieszka Feliczak-Guzik, Paulina Szczyglewska, and Izabela Nowak

Abstract Heterogeneous catalysis has become pivotal to the world chemical and petrochemical industry as nearly 90% of all chemical products are produced in the processes in which at least one stage involves the use of catalysts. In this chapter fundamental information on heterogeneous catalysis with its advantages and disadvantages is provided together with its use in different catalytic reactions.

Keywords Heterogeneous catalysis · Heterogeneous photocatalysis nanomaterials · Catalytic reactions

1 Introduction

The definition of catalysis formulated by the International Union of Pure and Applied Chemistry (IUPAC) postulates that it is a phenomenon in which the rate of the chemical reaction is increased as a result of addition of a small amount of a substance called a catalyst. Thanks to their specific chemical interactions with the reaction substrates, catalysts form with them the transition systems characterized by a reduced activation energy. The catalysts are not converted in the reactions and thus are not included in the stoichiometric equation of the reaction [1]. According to the catalyst phase, the catalytic reactions can be divided into two types [2]: (i) heterogeneous catalysis—when the catalyst and the reagents are in separate phases and the process takes place at the catalyst-substrates interface, (ii) homogeneous catalysis—when the catalyst and the substrates are in the same phase, there is no interface.

The first heterogeneous catalytic reaction, the dehydration of ethanol on activated clay, was studied by Joseph Priestley in 1778. Later, in 1796, the Dutch chemist Martinus van Marum was the first to use metal catalysts to dehydrogenate ethanol, while Louis Jacques Thénard in 1813 revealed that ammonia decomposes into nitrogen and hydrogen when passed through various metals at high temperatures. Ten years later,

A. Feliczak-Guzik · P. Szczyglewska · I. Nowak (✉)
Faculty of Chemistry, Adam Mickiewicz University, Uniwersytetu Poznańskiego 8, 61-614 Poznań, Poland
e-mail: nowakiza@amu.edu.pl

S. J. Ikhmayies (ed.), *Advances in Catalysts Research*, Advances in Material Research and Technology, https://doi.org/10.1007/978-3-031-49108-5_2

Pierre Dulong concluded that the activity of various metals, i.e., Fe, Cu, Ag, Au, Pt in the decomposition of ammonia decreases in the order given. In 1814 Kirchoff stated that acids assist in the hydrolysis of starch to glucose and 3 years later (1817) cousins Davy (H. Davy and E. Davy) described in their paper the oxidation of hydrogen by air in the presence of finely-divided platinum at room temperature [3].

The term *catalysis* (from the Greek kata - "down" and lyein - "loosen") was first proposed by the Swedish chemist Jöns Jacob Berzelius in 1836 and it is still one of its most important fields today [4]. The presence of more or less stable intermediate compounds was first suggested and described by the 1912 Nobel Prize winner Paul Sabatier. He introduced a concept referred to as "Sabatier's principle", according to which, when a reactant is adsorbed on the surface of a catalyst, its interaction energy should be strong enough for activation but not too strong to allow desorption of the products [5]. The concept was elaborated by studying the mechanism of hydrogenation of ethylene and CO over Ni and Co catalysts. On the other hand, much of the progress in the field of catalysis is related to the research by Nernst, Kirchoff, Ostwald, Langmuir and many other renowned scientists [3].

These and many other developments initiated the development of the immensely important field of chemistry that is heterogeneous catalysis today. Moreover, around 85% of catalytic processes involve heterogeneous catalysis, making it a thriving branch of chemistry (this is shown in Fig. 1) [6]. Thcrcforc, The use of catalysts makes production protocols more economical, ecological and sustainable [7].

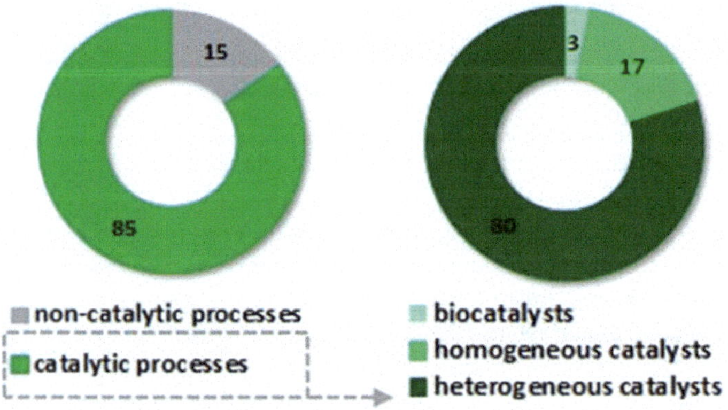

Fig. 1 Diagram presenting: **a** the contribution of catalytic processes to the chemical industry and **b** the contribution of heterogeneous catalysis in comparison to other catalytic processes. Redrawn after Wacławek et al. [6]

1.1 Parameters Characterizing Catalyst Performance

The catalyst performance in a certain chemical reaction is described by a number of parameters, among which the most important are activity, selectivity and lifetime [1, 8]. The catalyst activity is defined as the increase in the rate of a certain reaction in the presence of the catalyst relative to the rate of the same reaction without it.

Catalyst activity is related to the nature, number, strength and spatial distribution of the chemical bonds that are temporarily formed between the reactants and the catalyst surface, which in turn depend on the composition, structure and morphology of the catalytic system. Thus, the activity of the catalyst depends to the greatest extent on the active component and to a lesser extent on the other components of the final catalyst (the components of the catalyst and their functions are described later in this article) [3].

The rate of a catalytic reaction can be referred to as a unit of mass, volume, catalyst area, or number of active sites. Depending on the reference, it is expressed as:

- TOF—turnover frequency (frequency of catalytic cycles)—defined as the number of moles of substrate reacting per unit time per mole of catalyst
- TON—turnover number (number of catalytic cycles)—number of moles of reacting substrate per mole of catalyst [8].

The catalyst selectivity is its ability to direct the reaction towards formation of specific products [1, 8]. Consequently, high selectivity indicates a high yield of the desired product, while blocking undesirable, competing and successive reactions, giving different specific products, which react further in subsequent reactions to give different types of secondary products. When catalyst activity and/or selectivity is markedly reduced during operation, usually a given treatment allows the catalyst to be restored to its original properties [3]. Activity and selectivity in heterogeneous catalysis have been found to depend on many factors, for example, catalyst preparation, activation, chemical composition, as well as volumetric and/or surface crystal structures, electronic effects, interaction between the active phase and its support, etc. [5].

Another important catalyst feature is its lifetime defined as the time in which a given catalyst shows unchanged activity and selectivity, of course the best catalysts should have the longest possible lifetimes [1, 8]. The main characteristics of a good catalyst, in addition to the parameters described above, are: reproducibility, thermal stability, mechanical stability and the ability to be easily regenerated [3].

1.2 Homogeneous Versus Heterogeneous Catalysts

It is known that close to 90% of industrial processes involve at least one catalytic stage. Thus, it is beyond doubt that catalysis is of profound importance in our civilization [9, 10]. From among the two types of catalysis, the heterogeneous one is more attractive in the development of sustainable processes because of a number of its advantages of which one of the most important is the possibility of easy separation of the catalyst from the post-reaction system. Thus, it can be recovered and used again. Heterogeneous catalysis, whether carried out in a gas-solid or liquid-solid system, appears to be more environmentally friendly than homogeneous catalysis as it avoids the use of organic solvents [5]. Moreover these catalysts are highly stable and can be used at high temperatures. It is worth mentioning that heterogeneous catalysis can operate over a wide pH spectrum, where homogeneous catalysis is often pH dependent and at the same time can be less efficient and less environmentally friendly [6].

Thanks to the use of heterogeneous catalysts it is possible to achieve the industrial objectives meeting the principles of "green chemistry" [11]. The disadvantages of heterogeneous catalysts are related to the fact that they are inhomogeneous systems and thus can show lower selectivity and activity so also lower efficiency. In view of the above, much effort is directed to designing and obtaining heterogeneous systems that would be characterized by the activity and selectivity competitive to those of homogeneous catalysts, as the heterogeneous catalysis is the driving force of industry development [2].

Heterogeneous catalysis is applied in:

- refinement of crude oil, energy production, transport, petrochemistry,
- production of bulk chemicals, fine chemicals,
- production of polymers, materials, detergents, textiles,
- production of pharmaceutical and medical chemicals
- production of food,
- monitoring and control of chemical reactions,
- technologies aimed at restriction of pollutants emission [9, 10].

2 Heterogeneous Catalysis

2.1 Stages of Catalytic Reactions

Heterogeneous catalytic reactions in the presence of catalyst in solid phase are realized in a few subsequent stages. A catalytic reaction in the system of gas or liquid reagents and a solid catalyst includes seven stages [9, 12].

Gas phase reaction:

(1) external diffusion of substrates in which reagents are transferred from the internal gas or liquid part to the external part of the catalyst,
(2) internal diffusion of substrates in which reagents are transferred in the catalyst pores towards its internal surface,
(3) chemisorption—chemical adsorption of at least one substrate on the catalyst surface,

(4) surface reaction—transformation of adsorbed reagents and formation of reaction products,
(5) desorption—separation of products from the catalyst surface,
(6) internal diffusion of products—transport of reaction products from the internal surface of the catalyst towards its external surface,
(7) external diffusion of the products—transport of reaction products from the external surface of the catalyst to the gas or liquid phase.

Liquid bulk phase:

(1) internal diffusion of substrates in which reagents are transferred in the catalyst pores towards its internal surface,
(2) chemisorption—chemical adsorption of at least one substrate on the catalyst surface,
(3) surface reaction—transformation of adsorbed reagents and formation of reaction products,
(4) desorption—separation of products from the catalyst surface,
(5) internal diffusion of products—transport of reaction products from the internal surface of the catalyst towards its external surface.

If all the steps involved in the catalytic process are analyzed, the end result is very complicated. The processes of diffusion, adsorption and desorption occur very fast compared to a chemical reaction and therefore the adsorption process is in equilibrium during a catalytic reaction because it is a fast process and therefore the adsorption isotherm (e.g. Langmuir—if a catalyzed reaction takes place between gaseous reactants; e.g. Freundlich—if the catalyzed reaction takes place in solution) can be used to calculate the amount of reactant on the surface. There are transport models based on film and pore diffusion control, however that will be not discussed here. A chemical reaction usually consists of various intermediate steps and various rate laws and reaction mechanisms are relevant to surface catalyzed reactions. However, if one makes the simple assumption that a chemical reaction consists of a single one- or two-molecule elementary reaction or a simple rate-determining reaction followed by one or more rapid steps, then the reaction kinetics can be treated mathematically [3].

It is well known that heterogeneous catalytic reactions are most often explained by two mechanisms (Fig. 2):

(i) the Langmuir–Hinshelwood (L–H, later known as Langmuir–Hinshelwood-Hougen-Watson) mechanism in which product generation occurs through the reaction of two or more adsorbed species followed by product desorption, where the adsorbed species are in thermal equilibrium with the catalyst surface. Consequently, the reaction is initiated by thermal energy supplied by the surface [3],
(ii) the Eley–Rideal (E–R) mechanism according to which a new chemical bond is formed by a simple collision between a molecule or a gas phase atom and an adsorbed substance. Then, once formed , the product immediately desorbs. In

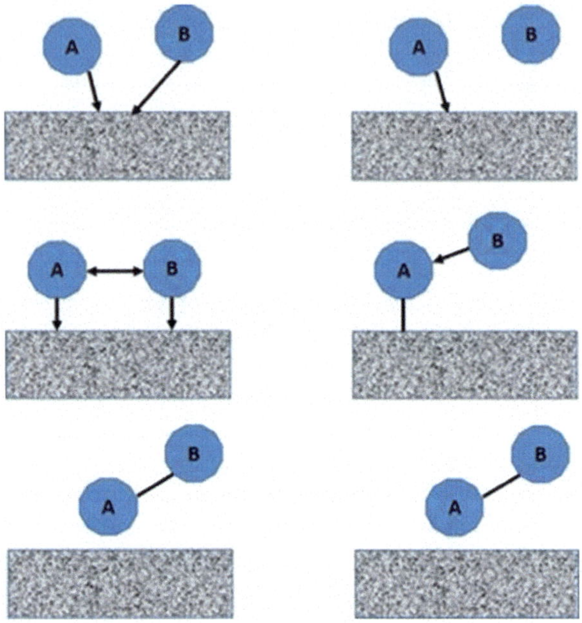

Fig. 2 Illustrations of the Langmuir–Hinshelwood (left) and Eley–Rideal (right) mechanisms for heterogeneous catalysis of bimolecular gas-phase reactions. Redrawn after Malherbe [3]

this case, if the reaction is activated, the energy required to overcome the barrier comes from the translational energy and/or the internal energy of the impacting molecule/atom [3].

2.2 Components of a Heterogeneous Catalyst and Their Functions

In 1925, Hugh Stott Taylor suggested that a catalyzed chemical reaction is not catalyzed over the entire solid surface of the catalyst, but only at certain "active sites" or centers. He also suggested that chemisorption may be an activated process and may occur slowly. Furthermore, he came up with the idea that chemically active sites may consist of an atom or set of atoms and may be few in number on the catalyst surface, and therefore may be inhibited/poisoned by relatively few molecules. The idea that adsorbed species and intermediates in catalysis by metals may require assemblies of several adjacent atoms on the free surface was also expressed during the same period by Aleksey Aleksandrovich Balandin, who suggested that a reacting molecule may be simultaneously adsorbed on several atoms (multiplet theory) [5].

Heterogeneous catalysis is the catalytic process in which the catalyst and the reagents are in separate different phases. The heterogeneous catalyst is composed of the active ingredients, support and promoters [8]. The active ingredients are the most important components of the catalyst as they are responsible for the main chemical reaction. They are the source of active centers at which transient complexes are formed. The active centers can be divided into two groups depending on their chemical character: acidic and basic ones (of Lewis or Brønsted type), redox centers (containing ions of different valency or those capable of formation of charge transfer complexes).

The catalysts are deposited on the porous and fibrous structures called the supports for a number of reasons. The most important benefit obtained from this deposition is greater diffusion of the catalyst, which ensures a desirable distribution of the active centers and increases the active phase area. Moreover, such a supported system has higher mechanical strength and is resistant to elevated temperatures. Certain supports can enhance the catalytic process and can even change the direction of the reaction, but they should not promote undesirable side effects. Thanks to the catalysts deposition on the supports the resistance of the former to poisoning increases. The supports stabilize the active ingredient and thus prevent agglomeration of crystallites. Promoters, also referred to as activators, are the substances whose presence in the reaction system, even in a small amount, increases the catalyst activity, selectivity and stability. Promoters can accompany the support or the active ingredient, their role includes enhancement of preferred activity and inhibition of the undesirable ones.

3 Synthesis of New Materials Used in Heterogeneous Catalysis

Both homogeneous and heterogeneous catalysis have their advantages and disadvantages (Fig. 3), so there is a critical need for a different synergistic structure, which should be dynamic like homogeneous catalysis and should also be efficiently recoverable which is the case in heterogeneous catalysis. The nanocatalyst combines the advantages of both homogeneous and heterogeneous catalytic systems and allows fast, specific chemical conversion with higher yields combined with the simplicity of catalyst separation and recovery. Catalyst recovery is the most essential feature of any catalyst before it is suitable for industrial-level green synthesis processes. Due to the nanoscale, the contact between the reactants and catalyst is greatly increased (this is close to homogeneous catalysis), while the insolubility in reactive solvents makes the catalyst heterogeneous and therefore, it can be effectively isolated from the reaction mixture (this is close to heterogeneous catalysis) [13].

In summary, the most important features of NPs that make them so attractive in materials chemistry and chemical engineering are generally related to the following properties: (i) high volume to surface ratio, (ii) presence of surface plasmon resonance effect, (iii) different physical properties with respect to the starting metal, (iv) large number of low coordination sites on the surface with respect to the starting material, with remarkable effects on chemical reactivity and catalytic properties, (v) easy surface functionalization.

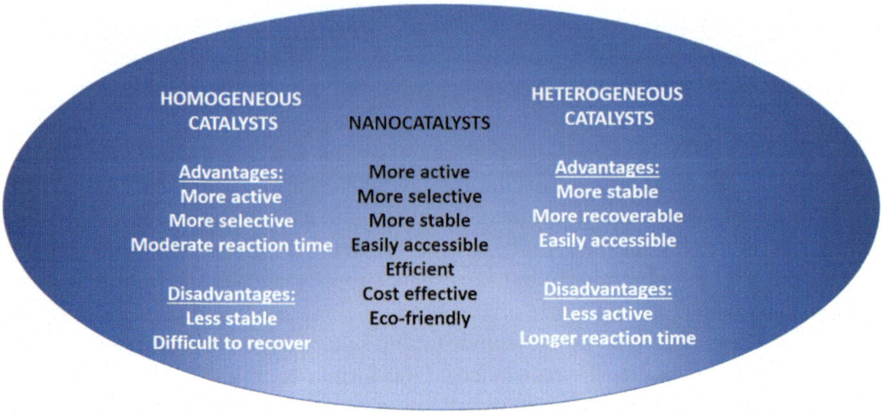

Fig. 3 Differences between homogeneous, heterogeneous and nanocatalysts. Redrawn after Somwanshi et al. [13]

Traditional heterogeneous catalytic systems have two major disadvantages compared to their homogeneous counterparts are (i) the reduced surface area available to the reactant molecules, thus limiting their catalytic activity and (ii) leading to unnecessarily high consumption of expensive catalytic materials. Nanoscaled catalytic materials can solve these problems by increasing the surface area to volume ratio. These specifications form the basis of the current state of nanocatalysis [14].

The use of nanomaterials in heterogeneous catalysis has been known for several decades, although it was the beginning of the twenty-first century that brought the real revolution in the above mentioned catalytic systems. Nanomaterials have specified components with at least one dimension smaller than 100 nm. Due to their much larger specific surface area than that of macroscale materials, the efficiency of nanocatalysis is much higher. Interestingly, the catalysts with nanometer dimensions were already in use long before the term "nanotechnology" was introduced to the world by R. Feynman [6].

The most important fundamentals of nanocatalysis include:

(i) the physical and chemical properties of nanocatalysts can be tailored,
(ii) nanocatalysts are often easily removable from the reaction solvent,
(iii) nanocatalysts have a high surface area to volume ratio,
(iv) nanocatalysts form colloidal suspensions and mimic homogeneous catalysis.

By demonstrating how the physical properties and preparation parameters of nanoparticles affect their catalytic characteristics, nanocatalysts with high activity, selectivity and resistance can be designed. All these advantages lead to enabling industrial chemical reactions to produce: less waste, use less energy and use resources more efficiently, reducing the environmental effect caused by the use of chemical processes. Nanoparticles are one of the most important catalysts for industrial purposes with many applications especially in chemical production, energy storage and conversion. The heterogeneity and differences in the shape and size of nanoparticles lead to their special catalytic performance.

More efficient catalytic processes require improvements in catalytic activity and selectivity. Both of these aspects can be improved by designing catalytic materials with the desired structure and the desired dispersion of active sites. Materials such as zeolites, AlPO (aluminophosphate), SAPO (silicoaluminophosphate), mesoporous materials, activated carbons, PILC (pillared inter-layered *clays),* MOFs (metal–organic frameworks), etc., offer a wide range of possibilities in this regard [15]. Moreover, one of the most important applications of porous solid catalysts is the replacement of environmentally hazardous, corrosive, difficult to separate and remove homogeneous catalysts (i.e. Lewis substances, organic and inorganic bases and toxic metallic compounds) used in the synthesis of various chemicals and high-value substances. There are many families of solid catalysts such as metals, oxides, sulfides, carbons, etc. as bulk or support materials on more or less catalytically active supports such as silica, alumina, zirconia, titanium, ceria, carbon, etc. These materials can have specific chemical properties such as acid-base, redox, dehydrogenation, hydrogenation or oxidation, and physical properties such as porosity, large surface area, abrasion resistance, thermal and/or electrical conductivity, etc. The largest family consists of oxides, alumina, mesoporous materials, mixed oxides—as catalysts or supports. Among these, porous materials exhibit some fascinating properties [15]. Although their common features endow these porous materials with many beneficial properties, the differences between them determine the advantages and disadvantages for particular applications, especially catalytic ones [16].

As mentioned in the subsection on "Components of a heterogeneous catalyst and their functions", the majority of traditional heterogeneous catalysts are composed of the active ingredient on an appropriate support. According to literature, the most often used supports are nanomaterials (activated carbons, mesoporous material, zeolites and metal–organic frameworks) [17]. The use of materials of atomic scale size permits control of their shape, size and morphology [18]. It has been proven by many authors that nanometric materials show higher catalytic activity than their larger scale correspondents [19]. Figure 4 presents the catalytic systems that have been so far successfully used in heterogeneous catalysis.

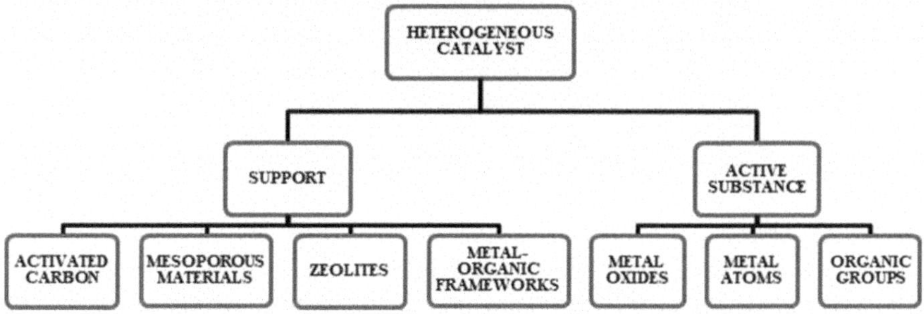

Fig. 4 The catalytic systems most often used in heterogeneous catalysis [20]

It is worth mentioning, that support is typically in the micro or mesometer range, while the active compounds may be nanoparticles. In this way, very often the final nanocatalysts are obtained. Nanocatalysts can be used either in the supported form with the help of solids such as zeolites, carbon, and oxides, or without any support [20]. These materials can be fabricated also by growing a solid material around a molecular template. Nanoscale elements can also be positioned on the catalyst surface using a variety of techniques. It is interesting to note that these environmentally friendly noble metal catalysts are all nanoparticles doped on porous or acid/base solid supports, which can be classified as nanocomposites or nanocatalysts.

3.1 Ordered Mesoporous Silica Materials

Silica-based mesoporous materials reveal a number of attractive properties which have made them very interesting objects of studies. They show for instance [21]: well-developed surface area, the presence of mesopores, high degree structure ordering, high sorption capacity, high hydrothermal and thermal resistance. Thanks to the unique physicochemical properties these materials can be widely used in heterogeneous catalysis [22]. Pure silica precursors are practically useless as catalysts because of their neutral chemical character, however, they can be easily endowed with active centers, e.g. by addition of single metal ions, by modification with metal oxides or with different functional groups. In this way they acquire new properties that can be used in many catalytic processes.

The procedure for the synthesis of mesoporous molecular sieves representing class M41S, published in 1992 by the research team from Mobil Oil Company, has initiated intense studies of ordered mesoporous materials [23]. There is rich literature on the mechanisms of syntheses of different materials and their comprehensive characterization. There are a few groups of mesoporous silicas, e.g. MCM (Mobil Composition of Matter), SBA (Santa Barbara Amorphous) or KIT (Korean Institute of Technology) of which the most popular are MCM-41, MCM-48, SBA-15, SBA-16, KIT-5, KIT-6. These silicas are obtained by different methods which determines their physicochemical properties (specific surface area, pore size and pore volume). The syntheses of such silicas are based on the use of the source of silicon and surfactants which in a properly chosen environment and in appropriate reaction conditions form two- or three-dimensional structures as a result of the processes of polymerization and condensation. Their porous structure is obtained as a result of calcination or extraction [21, 24]. The relatively large pores in mesoporous silica facilitate mass transfer, and the large surface area allows obtaining high concentration of active sites per mass of the material. Mesoporous silica enables catalytic reactions involving bulky substrates and/or products [16].

As mentioned above, one of the trends in the synthesis of mesoporous materials is the introduction of heteroatoms into their structure, which act as catalytically active sites. So far, several procedures have been described for the incorporation of heteroatoms, which also allow the design of catalysts with active sites located on highly accessible surfaces of the internal pore channels. These procedures include impregnation, grafting by liquid-solid reaction, grafting by gas-solid reaction, grafting by solid-solid reaction, isomorphic substitution, molecularly designed dispersion method and template ion exchange method. The aim of such modifications was to synthesize porous materials capable of accommodating large molecules and thus processing heavier petroleum feedstocks. For example, the introduction of aluminum ions into such silica frameworks led to moderately strong acidity compared to that of normal zeolites, which was due to the larger Si-O-Al bond angles. Applications of such materials in petrochemistry were, therefore, limited. Currently, research is focused on the introduction of catalytically active functions within mesopores (e.g. wall functionalization) or metal or oxide nanoparticles. Interesting results have been obtained; however, it must still take some time for this research to result in breakthroughs in catalytic processes. H-Al-MCM-41 materials have been used as conversion catalysts for a wide variety of substrates in reactions as diverse as hydrocarbon cracking, hydrocracking and hydroisomerization, olefin polymerization, oligomerization and isomerization, Friedel–Crafts alkylation and alkylation of phenols and aromatics, acetylation, Beckmann rearrangements, glycosidation, aldol condensation and Prins condensation [15].

3.2 Zeolites

Zeolites are a large group of aluminosilicate minerals with a microporous, crystalline structure. They consist of TO_4 tetrahedral units connected by T-O-T bridges. T-O-T angles are ideally 140–150° and rarely exceed 180° [15] (usually T=Al or Si). Besides the three-dimensional structures whose channels are occupied by water molecules, this group of compounds include aluminosilicates of alkali elements, aluminosilicates of rare-earth elements and many other single- or multivalent metal aluminosilicates [25]. Crystalline zeolites are most often produced by hydrothermal methods at elevated temperatures and in the presence of a mineralizing agent, source of desired atoms T, substance directing structure development (surfactants) [26]. Zeolites show a number of features making them suitable as catalysts supports: well-developed surface area, presence of strong acidic centers, ion-exchange ability, microporous structure and high thermal stability [27].

An interesting feature of zeolites is the versatility of their structure, resulting from a large number of modification procedures that also adjust their catalytic properties. The catalytic activity of zeolites is related to the presence of active centers in their skeleton. Such centers can be acidic and their strength, number and type (Brønsted or Lewis) are controlled either during synthesis by isomorphic replacement of the central Si atom at tetrahedral sites in the zeolite lattice (for example, with trivalent elements, such as Al, B, Fe, Cr, Sb, As or Ga and tetravalent elements such as Ge, Ti, Zr or Hf) and/or post-synthesis (steam treatment, chemical vapor deposition, ion exchange of interstitial cations, reduction of said metal ions to a lower valence state) [15].

To date, the greatest breakthrough in zeolite catalysts has been in the fluid catalytic cracking (FCC) process for the conversion of crude oil into fuels and chemicals. FCC is the main conversion process in a typical petroleum refinery, and as the most critical ingredient of the catalyst, zeolites are responsible for producing majority of the gasoline used around the world as well as taking an important role in the production of other transportation fuels (e.g., diesel, jet fuel, etc.) and building blocks for the petrochemical industry (e.g., propylene, butylenes, aromatics, etc.). Zeolite catalysts have excellent pore structures for diffusivity of feedstocks and intermediate products and strong acid sites for high overall conversion. In addition to cracking feedstock, new generation catalysts have added functionalities such as metal passivation to mitigate the effects of V and Ni from feed. Such catalysts enable the FCC to process heavier and higher metal-content feedstocks without pre-treating for commercially attractive conversions to valuable liquid hydrocarbons [28]. The addition of zeolites (FAU and MFI) to acid clay and silica-alumina cracking catalysts in the mid-1960s resulted in a dramatic increase in gasoline yields and conversions and a reduction in coke and dry gas. The new catalysts gave gasolines with lower octane numbers, mainly due to their lower olefinicity. The first catalysts to increase the octane number of FCC were introduced by catalyst manufacturers in the USA in the mid-1970s. Zeolites played a key role in petrochemicals and organic synthesis in the twentieth century [15] and are also widely used in environmental catalysis [16]. Zeolites not only provided strong solid acidity for high activity and stability, but also introduced a high surface area pore/channel network that brought size and shape selectivity as well as higher density of acid sites (per unit weight of catalyst). What's more, the stability of the zeolite was improved for maximum conversion via the introduction of rare earth oxides. Ion exchange technology in the FCC catalyst manufacturing process allowed for the reduction of sodium, a catalyst poison for the acid sites. Lower sodium enabled higher catalyst activity and stability [28].

Because of their unique properties related to the nanometric size, crystalline zeolites are excellent catalytic substances. The only restriction in their use is the fact that they are not able to cope with diffusion of the molecules whose size is similar to that of zeolite pores [16, 28]. This fact simulated the studies on the synthesis of hierarchical structure zeolites showing secondary porosity mainly in the range of mesopores. Such zeolites facilitate the access of reagent molecules of greater sizes and at the same time preserve the unique properties of this group of materials [27, 29].

It is claimed that combined micro- and mesoporous materials have advantages over purely micro- or mesoporous materials. Reagent molecules are therefore expected to diffuse more easily through the pores to reach the active sites of the zeolite. These materials provide better hydrothermal stability, are multifunctional, allow the processing of a wide variety of substrates, provide a more controlled leaching rate for the steady and gradual release of active ingredients, and are more capable of encapsulating waste in micropores. The principle of creating these types of zeolites is to introduce additional mesoporosity into the structure. This additional mesoporosity in zeolites can be created after synthesis by chemical treatments including hydrothermal, acid, alkaline, desilication/dealumination or recrystallization procedures, templating or carbon-based methods [15, 26].

3.3 Activated Carbons

Another group of very attractive catalyst supports are activated carbons. Although there are many types of carbon materials (graphite, carbon black, activated carbon, carbon fibers, carbon-coated alumina, glassy carbon, pyrolytic carbon, fullerenes, nanotubes, etc.) high surface area activated carbon is the carbon material most commonly chosen as a carrier for heterogeneous systems [30]. These materials are mainly known as very good adsorbents as they show suitable porosity, large surface area, excellent electron conductivity, high stability, moreover they are relatively cheap, non-toxic and are easily utilized. Thanks to all the above characteristics they have become very interesting as catalysts of catalyst supports [17, 31]. Activated carbons are obtained mainly from wood (charcoal) but also from post-agricultural waste such as nutshells, hay, fruit stones or peels [32]. The carbon obtained in this way is then subjected to physical or chemical activation. Activated carbons are versatile materials of many interesting properties and can be subjected to various modifications, which permits designing of a specific material for a particular catalytic process [17].

All activated carbons have a porous structure, usually with relatively few chemically bound heteroatoms (mainly oxygen and hydrogen). In addition, activated carbon can contain up to 15% mineral matter. It is now generally accepted that the average activated carbon structure consists of aromatic sheets and strips, often curved and resembling a mixture of wood shavings and crumpled paper, with gaps of varying sizes. Of course, the highly disorganized structure of the material depends on the precursor and its treatment. The degree of activation will condition the reduction in the number of aromatic sheets in the original carbon, leaving single and generally non-planar layers in some cases. During pyrolysis of the precursor, heteroatoms such as oxygen, hydrogen and nitrogen are removed as volatile gaseous products and the remaining elemental carbon atoms are then grouped into stacks of planar aromatic sheets cross-linked in a random manner. The degree of activation will determine the porosity of the final activated carbon, but it usually has pores belonging to several groups: micropores, mesopores and macropores [30].

The potential of carbon as a catalyst support has not yet been fully exploited, despite the fact that there is a considerable amount of literature dedicated to this field published in the last 20 years. This large amount of research is mainly due to the fact that carbon has certain characteristics that are very valuable and unattainable by any other support, although it is also true that carbon supports cannot be used in hydrogenation reactions >700 K or in the presence of oxygen >500 K, as they may gasify to give methane and carbon dioxide, respectively [30].

Advantages of carbon carriers [30, 33]:

(i) the carbon structure is resistant to acidic and basic media-activity carbon catalysts both in the acidic and basic medium is contributed to the combining effect from the fast electron transfer, synergistic cooperation for the heterostructure, and the abundant effective surface areas of carbon. As shown already, the carbon surface may have different amounts and types of oxygen surface groups and consequently both negatively and positively charged surface sites exist in aqueous solution, depending on the pH,

(ii) its structure is stable at high temperatures (even above 1000 K)—the stability of carbon-based carriers is mainly due to the presence of oxygen surface groups which are not the only centers conditioning the catalytic behavior of carbon-supported catalysts. Thus, when a high surface area carbon black is subjected to heat treatment in an inert atmosphere at temperatures ranging from 1600 K to 2200 k there is not only a decrease in the surface area, but also an increase in crystalline ordering and crystallite size,

(iii) its pore structure can be adjusted to achieve the pore size distribution needed for a given catalytic reaction,

(iv) the porous carbons can be prepared in different physical forms (granules, pellets, extrudates),

(v) the surface chemistry of carbon can be modified to increase hydrophilicity,

(vi) the active phase can be easily recovered from spent catalysts by burning the carbon carrier,

(vii) the cost of carbon supports is generally lower than that of conventional supports.

There are many other structures that have been successfully used in catalytic processes. According to literature the most often used for catalytic purposes are transition metal oxides [34, 35], composites [36] and silicas [37].

3.4 Metal–Organic Frameworks

The metal–organic frameworks (MOFs), also called porous coordination polymers (PCPs), are two- or three-dimensional porous crystalline materials with infinite lattices synthesized from secondary building units, metal cations salts or clusters, and polydentate organic ligands with coordination type connections. The broad range of complex metal cations and connector molecules on one side, and the flexibility of synthesis routes along with the ability for post functionalization of such structures on the other side, give rise to a synthesis of the bewildering arrays of possible MOFs. Materials of this type bridge the gap between micro- and mesoporous materials and have well-developed specific surface areas (up to 10,000 m^2/g) which far exceeds the specific surface area of zeolites and activated carbons. To date, more than 20,000 MOF structures have been described. It is worth mentioning that some of them have been manufactured on a ton scale by BASF and are available from several chemical raw material suppliers. According to the literature, the most commonly used MOFs are those with designations: MIL-53, HKUST-1, Fe-BTC, and ZIF-8 [38].

Materials in this class due to their properties are widely used in gas storage and separations, drug delivery, sensing, energy conversion technologies, water purification and many catalytic processes (organic and molecular reactions, electrocatalysis, and photocatalysis). Major challenges of our society are solved using these well-defined heterogeneous catalysts in the fields of synthesis, energy and environment [38, 39].

MOFs are excellent supports for catalytic systems due to their metal centers, highly ordered porous structure, uniform pore size and surroundings, very large surface area, chemical versatility and high mechanical and thermal stability, which makes them superior to other porous materials [40]. A special feature of MOF nanoparticles is their ability to activate substrates from synergistic interactions with both parts of the MOF backbone (Lewis acidic metal, etc.) and encapsulated nanoparticles, allowing tandem reactions to occur [38]. These heterogeneous catalysts have provided unique opportunities to study fundamentals of reaction mechanisms and substrate binding to well-defined catalytic sites [39].

4 Nanostructured Materials in Catalysis

4.1 Historical Background

As noted earlier, heterogeneous catalysis is essential for most industrial processes and is also vital for energy production processes, irrespective of its source, i.e. fossil fuels such as coal and natural gas or alternative sources such as biomass, solar energy, nuclear energy, wind or hydrothermal energy. Design and development of new technologies with the use of catalysis must be performed taking into account the necessity of protection of the natural environment [18]. The positive impact of heterogeneous catalysis on industrial activities has been observed since the early twentieth century. In 1908 a German physicist Fritz Haber proposed an economic synthesis of ammonia in which nitrogen and hydrogen were pumped into the reactors and under high pressure and temperature, in the presence of an osmium catalyst, were joined forming ammonia. This synthesis is known as the Haber–Bosch method. The studies related to the design of the above-mentioned method were undertaken because of the period of starvation predicted to take place in the 1930s in Europe (nitrogen compounds stimulate the growth of plants). Then Carl Bosch and Alwin Mittasch from BASF, tested over 2500 different materials and disclosed an iron catalyst that showed high catalytic activity and was cheaper than commercial catalysts. In 1913, BASF started production of ammonia from hydrogen and nitrogen from the atmosphere in Ludwigshafen. At that time Fritz Haber was known as the one who "made bread from the air". For the studies in 1918 Haber was given the Nobel Prize in chemistry [41]. Moreover, this process provided substrates for production of explosives which strengthened the position of Germany in the World War I. Nowadays the global production of nitrogen fertilizers by the Haber–Bosch method reaches a few hundred thousand annually and the Mittasch catalyst, with some small modifications, is used on a large scale [42].

The 1930s witnessed the discovery of three important types of catalysts for crude oil refining, for the processes of catalytic cracking, alkylation and dehydrogenation.

Heterogeneous catalysis played also an important role in the World War II. Using the new catalysts of cracking and alkylation, the Allied forces produced high octane airplane fuel which endowed Spitfires with higher efficiency than Messerschmitts in the Battle for England. Moreover, the catalytic dehydrogenation of methylcyclohexane permitted effective production of toluene needed for trinitrotoluene (TNT) production [42].

Another catalytic process of great significance for political reasons was the Fischer–Tropsch synthesis. Both Germany and Japan had access to fossil coal but did not have a reliable access to crude oil. The Fischer–Tropsch process catalyzed by a cobalt-iron system was the conversion of coal into the synthesis gas from which hydrocarbons were obtained in further reactions. The main application of Fischer–Tropsch synthesis was production of liquid fuels. The fuel obtained is free from sulfur and nitrogen compounds so it is more environmentally friendly [43, 44]. Table 1 presents the most important technological processes in which heterogeneous catalysis was applied.

In the following part of the paper the use of heterogeneous catalysis over different catalysts in the syntheses of fine chemicals from biomass and the potential use of nanomaterials are described.

Table 1 Exemplary application of all types of heterogeneous catalysts in chemical processes

Process	Catalysts	Products	Applications	References
Synthesis of ammonia (Haber–Bosch process)	Magnetite (Fe)	NH_3	Gunpowder, fertilizer, explosives	[45]
Synthesis of methanol	$Cu/ZnO/Al_2O_3$	CH_3OH	Fuel, bulk chemicals	[46]
Fischer–Tropsch process	Co, Fe	C_5-C_{11} hydrocarbons	Automotive fuel	[47]
Cracking	Clays	C_7-C_9 alkanes	Fuel, detergents	[48]
Alkylation	Zeolites, clays, silicates	C_7-C_9 isoalkanes	High-octane fuel	[49]
Dehydrogenation	Pt/Al_2O_3	Alkanes	Polymers, bulk chemicals	[50]
Hydrodesulfurization	Co/Mo sulfides	Sulfur-free diesel	Automotive fuel	[51]
Hydrocracking	Pt on zeolites or aluminosilicates	Saturated hydrocarbons	Automotive/ aviation fuel	[52]
Isomerization	H-ZSM-5 zeolites	p-xylene	Polymers, bulk chemicals	[53]
Polymerization	Ti, Ziegler–Natta	Poly(ethylene)	Polymers, bulk chemicals	[54]
Oxidation	Vanadium oxide	Phthalic acids	Polymers	[55]

4.2 *Transformations of Biomass to Fine Chemicals Using Heterogeneous Catalysis*

Biomass is the material of biological origin in a broad sense, including plants, e.g. trees, grass and crops, animal excrements and municipal wastes. As biomass is widely available and renewable it is a universal raw product. It is composed mainly of saccharide polymers (65–85% wt.) and lignin (10–25% wt.) and contains small amounts of triglycerides, sterols, alkaloids, terpenes, terpenoids, waxes, resins and inorganic minerals [56]. According to the US Department of Energy (DOE) 10% of chemical compounds should be produced from biomass till 2020, while till 2050—50% of compounds [57].

Because of their structure, the products of natural origin are interesting substrates to be used in synthesis of fine chemicals [58–61] including terpenes, which are hydrocarbons of plant origin of the general formula $(C_5H_8)_n$, and oligomers of isoprene [62–65]. The most important sources of terpenes are turpentine, oleoresins extracted from coniferous trees and terbinth (called also Chian, Scio or Cyprus turpentine) and essential oils from citruses [66].

As most of essential oils are expensive, they are used directly as e.g. isolated natural chemicals. The cheap natural compounds such as limonene or pinene are the initial substrates for the syntheses of new important chemicals, e.g. aromatic substances, cooking essences, pharmaceuticals and solvents. A large number of catalytic chemical processes for the synthesis of valuable products from terpenes, involving hydrogenation, oxidation, isomerization, hydration, hydroformylation, condensation or cyclization, have been proposed [59–61]. According to literature, the most important terpenes and terpenoids are pinene, limonene, carene, geraniol, nerol, citronellol, citral and citronellal [66, 67]. One of the most important monoterpenes is α-pinene, which is used as a solvent and precursor for camphene—an important ingredient in perfume industry [68]. In view of the above, we will focus on the reactions with pinene.

Examples of heterogeneous catalysts used in biomass transformation processes are given in Fig. 5 and their applications are described below:

(a) **Micro- and mesoporous materials**

Zeolites are an important class of inorganic crystalline materials that have been widely used in petroleum refining, and in the petrochemical and fine chemical industries as catalysts, adsorbents, and ion-exchangers (e.g. TS-1 (Ti-doped MFI-type zeolite), Sn-β (Sn-doped BEA-type zeolite), H-USY (FAU-type zeolite) [69]. Micro- and mesoporous materials are also widely used for the industrial isomerization reaction of glucose to fructose to produce corn syrup. This is an important process because in recent years, glucose isomerization has played a key role in the synthesis of biomass-derived chemical platforms used to produce fuels and chemicals [70].

Among the chemicals from biomass resources, lactic acid is among the products of interest. It is generally obtained by fermentation of sugars and as such is often used in detergents, cosmetics and as a food additive, cosmetics and as a food

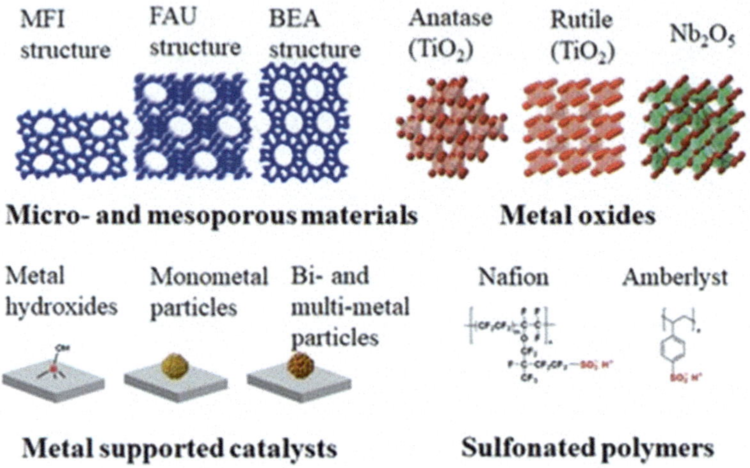

Fig. 5 Examples of heterogeneous catalysts used in biomass transformation processes. Reprinted from Hara et al. [69] under the terms and conditions of the Creative Commons CC-BY license. https://creativecommons.org/licenses/by/4.0/

additive. Recently, interest in lactic acid has increased significantly due to its use in biodegradable plastics. For example, the mesoporous material Sn-MCM-41 was used to synthesize alkyl lactate from triose, where Lewis acidity was obtained by grafting Sn onto the MCM-41 surface [71].

Metal–organic frameworks (MOFs)—organic-inorganic hybrid crystalline porous materials are very interesting catalysts widely used to transform and valorize biomass into valuable chemical compounds, e.g. for example: hydrolyzing cellulose to glucose, fructose or sorbitol; hydrolyzing fructose, glucose or maltose to 5-hydroxymethylfurfural (5-HMF); hydrolyzing sucrose to methyl lactate, furans, levulinic acid, lignin or vanillin and hydrolyzing triglycerides to esters and glycerol—all these aspects are widely described in a review by Herbst and Janik [72].

(b) **Metal oxides**

Metal oxides have both Brønsted and/or Lewis acid sites on the surface that can be catalytically active for various organic reactions. Representative examples are metal oxides consisting of group IV and V elements. Among the numerous metal oxides with proven catalytic activity, one can distinguish:

(i) Nb_2O_5: furfuryl alcohol dehydration [73], HMF formation from glucose [74], production of fuel precursors from biomass-derived chemicals [75],
(ii) TiO_2: hydrogen production from biomass [76], conversion of biomass into formic acid [77], conversion of biomass-derived furfural into cyclopentanone/cyclopentanol [78].

(c) **Metal supported catalysts**

Metal supported catalysts are one of the important catalysts that are widely used in many large-scale processes (including petroleum refining, automotive exhaust gas cleaning, and chemical synthesis such as hydrogenation and oxidation). Not surprisingly, metal nanoparticle catalysis has attracted much attention from researchers around the world. Furthermore, immobilization of metallic nanoparticles on a catalytically inert metal-oxide support often results in a dominant effect on activity due to the strong metal-support interactions that affect the electronic charge of the metallic catalysts. Even more interesting systems are bi- and multimetallic nanoparticles due to their favorable catalytic properties, which usually differ from those of the component metals due to electronic and/or geometrical synergistic effects [69].

There are many literature reports on the application of such systems in the transformation of biomass to useful chemicals, e.g.: formation trioses from glycerol [79], oxidation of HMF (5-hydroxymethylfurfural) to DFF (2,5-diformylfuran) [80], hydrodeoxygenation biomass derived chemicals into biofuels [81, 82], production of lactic acid [83, 84] and many others.

(d) **Sulfonated polymers**

Polymeric ion exchange resins, as reported in the available literature, have been used for a number of industrially important transformations. The two main classes of ion exchange resins are based on styrene-based sulfonic acids (Amberlyst and Dow), which show very high specific activity in esterification and etherification reactions, and perfluorosulfonic acid-based catalysts, including Nafion resin/silica nanocomposites. Resins of this type have proven high activity in the formation of linear alkyl benzene, isomerization and some selected acylation type reactions [69].

Scientists in Zhang's research group [85] synthesized sulfonated polytriphenylamine (SPTPA) and then determined its catalytic activity and reusability in the co-synthesis of 5-HMF and furfural from corn cob in lactone solvents. Furfural is produced from the hemicellulosic fraction of lignocellulosic biomass, whereas 5-HMF is obtained by catalytic conversion of the cellulosic fraction. Despite the fact that furfural production has been industrialized for about 100 years, current commercial furfural production technology is based on highly polluting and energy-intensive methods with relatively low yields. Therefore, laboratory testing is important to overcome the drawbacks of this process.

4.3 Reactions with Pinene as Precursor

In 1995 the global production of terpene oils was close to 330.000 tons. The greatest contribution to this amount was brought by pinene which as a pure substance is obtained by fractional distillation [60]. It can be subjected to a number of reactions leading to fine chemicals (Scheme 1).

Isomerization of α-pinene over acidic catalysts is widely discussed in literature, e.g. [86]. On the industrial scale, pinene isomerization is performed in the presence

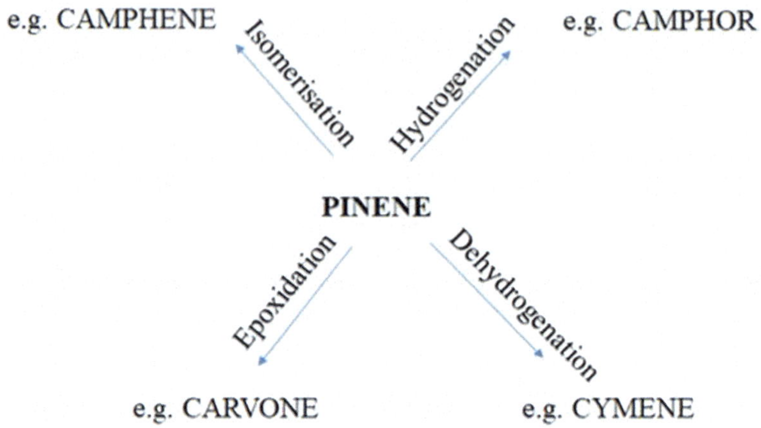

Scheme 1 Exemplary reactions with pinene. Redrawn after Sell [60]

of titanium oxide as a catalyst, in atmospheric pressure, at 373 K. The main products of this reaction are limonene, camphene and tricyclic monoterpenes and in small amounts fenchenes and bornylene (Scheme 2).

As the rate of isomerization is low and the catalyst must be treated with acid in order to generate a layer of titanium acid on the catalyst surface, the search for other materials that would ensure high selectivity to camphene and/or limonene is continued [87, 88]. As catalysts for this reaction also zeolites and modified clays have been tested. These materials have suitable acidic centers and show shape-selectivity [69].

Yilmaz et al. have tested in this reaction Beta zeolite of different SiO_2/Al_2O_3 ratios, modified with B, Ti or V atoms. Isomerization of α-pinene was performed in liquid

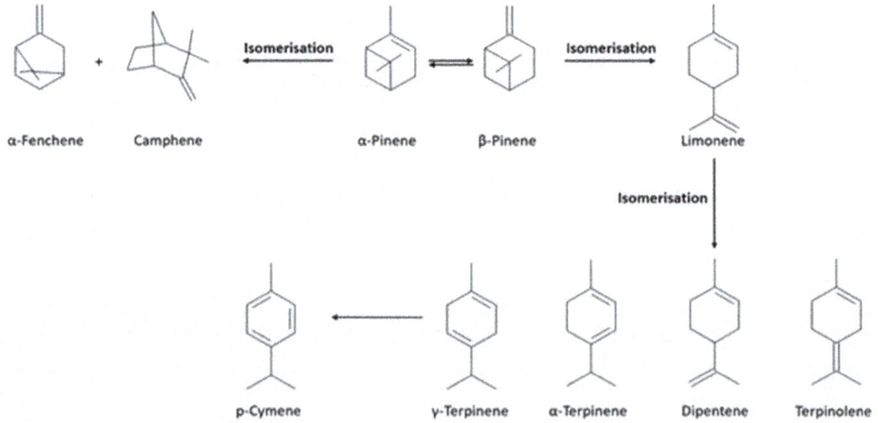

Scheme 2 Possible products of α-pinene isomerization. Redrawn after Jenkin et al. [86]

phase at 373 K [89]. The acidic Beta zeolite (of SiO_2/Al_2O_3 from 55 to 66) showed high catalytic activity and selectivity to desired reaction products. The materials additionally modified with metal atoms showed insignificant catalytic activity. Other catalysts studied in this reaction, along with beta zeolite, are characterized in Table 2.

Pinene can be subjected to hydrogenation over a nickel or palladium catalyst. This reaction leads to self-oxidation of α-pinene to pinene peroxide that undergoes hydrogenation to the alcohol-pinan-2-ol~~u~~. Thermal treatment of this alcohol leads to valuable aromas and linalool which is an intermediate in production of vitamins A and E [67, 102]. Transformation of α-pinene to linalool is shown in Scheme 3.

Epoxidation of pinene leads to the epoxides that are key substrates in organic chemistry for functionalization of substrates and syntheses of alcohols, carbonyl

Table 2 Catalytic systems used in pinene isomerization

Catalyst	Temperature [K]	Conversion of α-pinene [%]	Selectivity to camphene/limonene/terpinene [%]	References
Beta	373	100	27/-/20	[90]
Mordenite	393	100	37/32/-	[91]
Clinoptilolite	323	80–85	35/25/-	[92]
Mesoporous silica FSM-16	353	77	41/41/-	[93]
Alumina	Gas-phase	75	38/-/-	[94]
Acid-treated bentonite	423	99	47/13/11	[95]
Kaolinic clay	373	67–94	65/23/-	[96]
Exchanged clay	353	100	50/20/-	[97]
Amberlyst 15	393	99	31/-/15	[98]
Yb/SiO_2	323	100	62/20/-	[99]
ZSM-5	348	11	33/34/-	[100]
ZSM-12	348	43	43/31/-	
MCM-22	348	77	35/45/-	
Hierarchical MCM-22	363	100	76/18/-	[101]

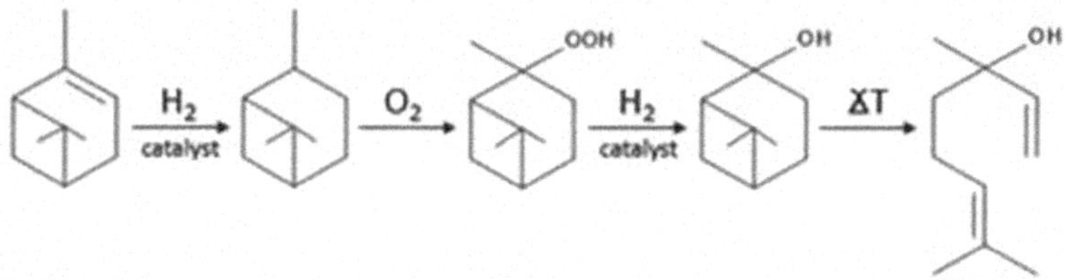

α-pinene linalool

Scheme 3 Transformation of α-pinene to linalool. Redrawn after Sheldon et al. [67]

compounds, ethers etc. Literature provides a number of different heterogeneous catalysts used in pinene epoxidation, including metal complexes supported on or incorporated in micro- or mesoporous materials and titanium silicates [103–109].

4.4 Transformations of Carbohydrates to Fine Chemicals

Carbohydrates are the most abundant class of organic compounds on the Earth. Taking into account that they are cheap, their resources are unlimited, a number of carbohydrates of different chemical structures permit syntheses of new compounds of different properties and wide use and the fact of depletion of crude oil and natural gas resources, carbohydrates have become of increasing importance for chemical synthesis on the industrial scale [67]. The saccharide compounds of industrial importance are sorbitola product of D-glucose hydrogenation and saccharose. Sorbitol is obtained on reduction of glucose with the use of Raney catalyst (Ni) [56, 110]. Although a number of promising catalysts of this reaction have been proposed, the Raney catalyst is predominantly used in industrial processes [111]. Gue et al. [112] have proved that the activity of nickel catalyst is influenced by the addition of such metals as Mo, Fe, Cr and Ti. Although the nickel catalysts are cheap in synthesis, they have serious drawbacks such as fast poisoning and toxic effect on the natural environment because of the necessity of leaching with alkalies. Because of these drawbacks, in particular the latter one, the search for new catalysts showing high selectivity to sorbitol, ensuring high conversion of glucose and being nontoxic, is intensely continued. Table 3 presents characterization of different materials that are used in glucose hydrogenation.

Recently, the interest in application of hierarchical zeolites in catalytic reactions has considerably increased. In comparison to the conventional microporous zeolites the use of these materials leads to increase in the catalytic activity and decrease in

Table 3 Examples of catalytic systems used in glucose hydrogenation

Catalyst	Selectivity to sorbitol [%]	Conversion of glucose [%]	References
Norit rox 0.8—Ru/C	>99.2	100	[113]
Ru/CP97 Engelhard	>98	100	[111]
Nickel supported on aluminosilicate	>98	44	[114]
Ru/Al_2O_3	>99	100	[115]
Ni/SiO_2	92	45	[116]
$Ni–B/SiO_2$	–	~80	[117]
Co-B	–	89	[118]
Ruthenium on mesoporous matrix MCM-41	98	100	[119]

the catalyst susceptibility to deactivation. Such hierarchical zeolites modified with transition metal ions (niobium or tin) have been applied by Feliczak-Guzik et al. in triose (dihydroxyacetone) isomerization to lactic acid and alkyl lactates, with the use of microwave radiation. The use of materials modified with niobium and tin ensured high DHA conversion and high selectivity to lactic acid and alkyl lactates [120, 121]. On the industrial scale lactic acid is obtained by the fermentation of carbohydrates with the use of genetically modified enzymes. However, biotechnological processes are marred with limited yield of the process, need of control of the conditions of fermentation in the reactor (temperature or pH) and high production of waste. Trioses such as glyceride aldehyde (GLA) and dihydroxyacetone (DHA) used in the process are cheap substrates, can be obtained as a result of aerobic oxidation of glycerol, and their molecular structure resembles that of lactic acid [122–124].

5 Heterogeneous Catalysts—Recent Developments

For over a few recent decades, one of the main problems has become the reduction of dependence on nonrenewable energy resources, mainly fossil fuels. The motivation for the effort in this direction is the need of reduction of emission of carbon dioxide coming from combustion of fossil fuels [125] and the serious depletion of fossil fuels resources in global scale [20, 126]. The use of renewable energy resources has no negative effect on climate changes, these resources are commonly available and their loss is replenished as a result of natural processes [126]. Particular attention is paid to utilization of biomass that can be processed to high-quality biofuels. Presently, about 1/5 of energy from fossil fuels is used by transportation [126, 127], so the use of biofuels would substantially reduce the exploitation of fossil fuel resources and the amount of CO_2 that is created, which contributes to global warming (biomass is originally created through photosynthesis, where solar energy converts water and carbon dioxide (found in the atmosphere) into carbohydrates, which are its building blocks—hence the total amount of CO_2 is reduced) [128]. Besides the transport sector, biomass can be used in chemical industry as the demand for chemicals has been gradually increasing. According to the recent trends, fuel and first generation chemicals are obtained from carbohydrates and vegetable oils. The second generation technology has been developed to make use of a cheaper and more abundant raw product that is lignocellulose [129]. The use of lignocellulose containing materials permits production of useful objects from ecologically preferred or acceptable raw materials that have been used for thousands of years [130]. Recently, the most interesting and urgent research problem is conversion of biomass to liquid biofuels. The oil obtained as a result of this process is composed of hundreds of different chemical compounds, mainly organic aromatic ones containing oxygen and belonging to aldehydes, ketones, alcohols, esters and carboxylic acids [131, 132]. The biooil preserves up to 70% of the initial energy of biomass and its chemical composition is more uniform than that of the initial biomaterial [133, 134]. The pyrolytic oil, prior to be used as liquid fuel, must be subjected to modification in order to make its physicochemical properties similar to those of fossil fuels. One of the most promising processes that could improve the biofuel properties is catalytic hydrodeoxygenation (HDO) [20, 126]. This process is aimed at enhancement of the energetic value of the bio-oil through transformation of potential energy carriers into simple hydrocarbons. Hydrodeoxygenation of organic compounds contained in the biomaterial leads to obtaining hydrocarbons, in the presence of gas hydrogen supplied to the system under elevated pressure and temperature. According to literature, the majority of functional groups of the chemical compounds contained in the pyrolytic oil is capable of conversion in their deoxygenated analogues in temperatures ranging from 573 to 723 K and under pressure of 7–20 MPa [135].

The application of elevated pressure and temperature is necessary as the chemical compounds containing high number of oxygen atoms, show low reactivity [126]. Besides, this process needs the use of specific catalysts [136]. Hydrodeoxygenation of compounds occurs with high efficiency provided both high temperatures and high pressures are applied. After all, for the whole process to be sustainable, it is required to achieve the highest possible degree of deoxygenation as well as hydrogenation with reduced energy and hydrogen consumption. Developing a catalytic system that meets the above aspects is quite a challenge. Such a system plays a key role, since it becomes possible to reduce the working temperatures and pressures, and therefore it is possible to reduce costs related to the whole process. Therefore, the selection of a suitable catalyst is essential for the overall system to operate at a satisfactory performance. A number of requirements are placed on potential catalysts for the HDO process: operating efficiency, long lifetime, high degree of deoxygenation, high selectivity towards desired reaction products, regenerability, thermal and chemical resistance, minimization of side reactions, non-toxicity, and low cost (including production cost) [137].

The majority of the processes taking place with the use of gas hydrogen have been performed in the presence of heterogeneous catalysts, whereby all or most of the above expectations are met [138]. The active ingredients usually were transition metal atoms showing high efficiency in the processes involving hydrogen because of their high affinity to molecular hydrogen easily adsorbed on their surface [126, 139]. Most often used supports of active phases were metal oxides, zeolites, activated carbons and mesoporous silicas [20]. The understanding of mechanisms of hydrodeoxygenation with the use of actual raw products is difficult, the reaction pathway is mostly explained by laboratory scale tests using the so-called model chemical compounds that are identical to those present in bio-oil [20]. Most often the studies are carried out for the compounds containing a hydroxyl group (-OH), e.g. phenol, and methoxyl group (-OCH_3) e.g. anisole, as these groups are most often found in lignin-derivatives contained in bio-oil [140, 141]. After optimization of the process of hydrodeoxygenation an attempt can be made to transfer the process parameters from the laboratory to industrial scale [142]. Table 4 presents examples of catalytic systems used in this process.

Transition metal atoms and their nanocomposites have attracted much attention in bio-oil upgrading due to their low cost and high catalytic activity [156, 157]. However, most of the catalysts used in HDO reactions are based on noble metals

Table 4 Examples of catalytic systems used in the process of hydrodeoxygenation

Catalyst	Substrate	Temperature [K]	Pressure [MPa]	References
h-ZSM-5, SBA-15, Al-SBA-15	Anisole	493	5.0	[133]
Activated carbon C, γ-Al_2O_3, SBA-15, Al-SBA-15 TiO_2, CeO_2	Anisole	493–583	0.3	[143]
Ru/meso-TiO_2 with Fe	Anisole	473–623	1.0	[144]
Ni/SiO_2	Anisole	493	0.5–3.0	[145]
Pd/SBA-16, Pt/SBA-16	Anisole	363–403	2.5–6.0	[146]
Ru/MCM-41	Phenol	573	4.5	[147]
Ni/Al_2O_3-HZSM-5	Phenol	473	5.0	[148]
Nb/Ru/Pd or Pt supported on SBA-16	Phenol	363–403	2.5–6.0	[149]
SiO_2/Pd; Nb_2O_5/Pd	Phenol	573	0.1	[150]
Al_2O_3/Pd	Phenol	573	0.1	[151]
MCM-41/Pd	Phenol	553	1.0	[152]
Mo_2C/TiO_2	Phenol	623	2.5	[153]
Pd/SiO_2, Pd/Nb_2O_5, PdNb/SiO_2	Phenol	573	0.1	[154]
Ag/TiO_2	Phenol	573	0.1	[155]

such as Au, Ag, Pd, and Pt, indicating their high production costs and thus hindering their practical applications. Thanks to nanocatalysis, it has become possible to use smaller amounts of these metals, which solves the problem of tremendously high costs and additionally results in better activity of these systems. Indeed, most authors report that nanocatalysts show a significant effect on the activity and selectivity. Unlike conventional catalysts, the highly variable surface area of nanostructured materials promotes catalyst-substrate interactions that effectively increase product yields [158]. Moreover, the development of alternative fuels using nanocatalysts offers some advantages over traditional acid-base catalysts. Nanocatalysts typically improve reaction kinetics by allowing the reaction to occur at a lower temperature (lower process energy intensity), reducing side reactions, and increasing recycling and energy recovery. This is mainly influenced by the highly differentiated surface area and surface energy of catalysts at the nanoscale. This is particularly important from an environmental point of view and is also in line with the principles of Green Chemistry. These characteristics can solve the issues associated with heterogeneous catalysts such as mass transfer resistance, longer reaction time and rapid deactivation [158].

6 Heterogeneous Photocatalysis

Heterogeneous photocatalysis is an interdisciplinary field of science, involving several areas of chemistry, physics, and to some extent photobiology (natural photosynthesis). Historically, heterogeneous photocatalysis is based on four basic pillars: heterogeneous catalysis, photochemistry, molecular spectroscopy of adsorbed molecules and solid state spectroscopy, together with materials science and surface science of semiconductors and insulators [159]. Its development has been significantly influenced by studies of the photostability of dyes and pigments in heterogeneous systems [160].

According to the IUPAC, a photocatalytic reaction involves the absorption of light by either a photocatalyst or a substrate involved in the reaction. The photocatalyst is a substance which allows the reaction to take place in the presence of light. It is not consumable in this process. As heterogeneous photocatalysis can be carried out in various systems using reactants in the gas phase, in the liquid organic phase or in aqueous solution, it covers a wide range of chemical reactions, e.g. oxidation, dehydrogenation, hydrogen transfer, oxygen and deuterium isotopic exchange, metal deposition, water detoxification, removal of gaseous pollutants.

The steps of classical heterogeneous catalysis, comprising 7 points, are summarized in Sect. 2.1. The main difference between it and heterogeneous photocatalysis is the different catalyst activation process. The traditional thermal activation of the catalyst is replaced by activation with photons (radiation quanta) of the appropriate wavelength. The steps of heterogeneous photocatalysis do not include the following steps of classical heterogeneous catalysis: 1, 2, 3, 5, 6, 7; stage 4 includes photoelectron processes: i.e. [161, 162]:

- photon absorption by the photocatalyst and not by the reactants;
- formation of electron-hole pairs (e^- -h^+), which dissociate as photoelectrons and holes;
- recombination reactions between excited centers and adsorbed substances.

A comparison of heterogeneous catalysis with heterogeneous photocatalysis is shown in Fig. 6.

Three major problems are still the challenge in heterogeneous photocatalysis. These include: (i) increasing the spectral sensitivity of photocatalysts to visible light, (ii) photocatalytic environmental purification is only acceptable for low-level pollutants because the amount of ultraviolet photons is limited in both sunlight and indoor lighting, (iii) developing innovative photocatalytic reactors that would significantly reduce the impact of mass transfer [164]. In order to eliminate these problems, it has become the aim of the scientific community to develop modern and efficient photocatalysts capable of absorbing sunlight. The limitations of using a particular semiconductor as a photocatalyst for a particular process can be overcome by modifying the semiconductor surface with metals; coupling semiconductor photocatalysts; doping with transition metal ions, rare earth ions and non-metal ions; sensitizing the semiconductor surface with chemisorbed and physisorbed dyes [165].

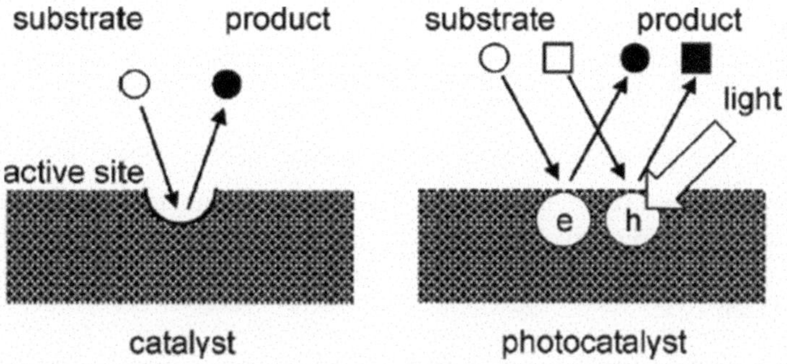

Fig. 6 Comparison of heterogeneous catalysis with heterogeneous photocatalysis (e states for electron, h—hole). Reprinted from B. Ohtani [163], with permission from Elsevier B.V. Copyright 2010

6.1 Photophysical Processes in Solid Photocatalysts

As a result of the action of a radiation beam of a given energy, photoexcitation of the photocatalyst occurs with simultaneous charge transfer towards the reactants in the ground state or excitation of the adsorbate with charge transfer towards the photocatalyst [166]. Thus, the cycle of heterogeneous photocatalytic reactions begins with the absorption of light quanta by the solid photocatalyst and ends with chemical transformations of molecules on its surface. The role of the photocatalyst and the corresponding photophysical processes occurring in solids are often treated in a simplistic manner. The ensemble of photon-absorbing molecules is regarded as a light-harvesting system in which the photocatalyst molecule acts as a sensitizer and as a source of photoelectrons and photofires, i.e. reducing and oxidizing agents. The so-called intrinsic absorption of light by the solid is considered to be of primary importance in heterogeneous photocatalysis. This is a justifiable approach from the point of view of most of the researches focused on the study of chemical reaction mechanisms or on the practical application of heterogeneous photocatalysis [167].

6.2 Advanced Oxidation Processes (AOPs)

Advanced oxidation processes (AOPs) include a number of technologies based on the production of highly reactive entities such as reactive oxygen species (ROS) e.g. hydroxyl radicals, superoxide anion radicals. Some of the most intensively studied advanced oxidation processes are photocatalytic processes, affecting the degradation of environmental pollutants (air, water, wastewater treatment) through oxidative decomposition initiated by ROS using semiconductors [159, 168, 169]. These materials can be nanoporous TiO_2, mesoporous materials, nanorods, dendritic nanostructures, SnO_2, ZnO, ZrO_2, $SrTiO_3$, CdS, MoS_2, Fe_2O_3 and WO_3 [170, 171]. Among these materials, TiO_2 has gained the most recognition because of its properties [172]. It is a biologically and chemically inert, inexpensive, resistant to photocorrosion and chemical corrosion, highly photoactive n-type semiconductor (the number of electrons in the conduction band exceeds the number of holes in the valence band—the so-called electron conduction). It occurs mainly in minerals such as tetragonal rutile and anatase and rhombic brucite, ilmenite, leucoxene, perovskite and sphene (titanite), and is also found in many iron ores [165]. Furthermore, it shows the potential to completely oxidize various organic compounds, including persistent organic pollutants [165, 173].

6.3 *Selected Examples of Applications of Photocatalysts*

Heterogeneous photocatalysts have been applied for the degradation of organic compounds, bacteria and microorganisms and in the reduction of toxic metal ions present in water and wastewater. Selected examples of applications of photocatalysts in catalytic processes are presented below.

6.3.1 Photocatalytic Conversion of Glycerol to Hydrogen

Photocatalytic reactions are one of the ways to extract hydrogen, not only from water (water splitting) but also from organic substances (photocatalytic reforming of organic compounds) using a renewable energy source, the sun [174].

Water splitting is based on the ability of water to be reduced and oxidized by reactions with photogenerated electrons and positively charged "holes", during the irradiation of semiconductors, in the presence of selected co-catalysts. Photocatalytic reforming of organic compounds, on the other hand, exploits the ability of some organic compounds to donate electrons to positive photocatalyst "holes" and get oxidized generating proton ions, while photogenerated electrons reduce the latter to hydrogen in the presence of suitable co-catalysts [174]. Glycerol is the main by-product obtained during biodiesel production. Recently, the demand for biodiesel has grown in parallel with the production of glycerol, whose production reached 2.5 Mt/year in 2020 [175]. In glycerol obtained in this way, there are residues of various impurities (e.g. methanol, water) that must be removed in order for it to be used in the food or pharmaceutical industry. When used to produce hydrogen, no prior purification is necessary. The photocatalytic conversion of glycerin to hydrogen yields, stoichiometrically, 7 molecules of hydrogen and three of carbon dioxide (Eq. 1) [176]. This reaction does not require high energy input, but only a photocatalyst active in the Vis light range.

$$C_3H_8O_3 + 3H_2O \xrightarrow{\text{catalyst},\ h\upsilon} 7H_2 + 3CO_2 \tag{1}$$

Table 5 Photocatalysts used in the photocatalytic conversion of glycerol to hydrogen

Catalyst	Efficiency of hydrogen production	References
CuO_x/TiO_2	From 580 to 900 $\mu mol \cdot g^{-1}$ h	[190]
Pt/TiO_2 Au/TiO_2	15.6 ml after 3 h 8.8 ml after 3 h	[191]
Pt/TiO_2	From 84 to 208 $\mu mol \cdot h^{-1}$	[192]
Pt/TiO_2	Maximum 24.2 $mmol \cdot g^{-1} \cdot h^{-1}$	[186]

Table 5 collects the photocatalysts used in the photocatalytic conversion of glycerol to hydrogen. So far, this reaction has been mainly carried out using titanium(IV) oxide (TiO_2) as a catalyst. This is due to the fact that this material has a high reactivity and chemical stability under ultraviolet light ($\lambda < 387$ nm) [177]. Typically, this carrier, was modified with different metals (co-catalysts): Pt [178, 179], Cu [180], Au [181], Pd [182], Ni [183], Co [184], Ag [185], Mn [186], Cr [186] and W [186]. From among the above-mentioned metals, platinum is used most often. Among others, Fu's research group has shown that doping photocatalysts with platinum affects the efficiency of hydrogen production [187].

Platinum-modified TiO_2 was the most intensively studied catalyst in this reaction, although the yield of the discussed process was low [188, 189]. Among others, researchers from de Oliveira-Melo's [189] research group came to this conclusion when they studied binary (Pt/hex-CdS) and ternary ($Pt/CdS/TiO_2$ and Pt/TiO_2/hex-CdS) hybrid photocatalysts in photo-induced reforming of glycerol under visible light irradiation (>418 nm). Systems with CdS/aqueous solution interfacial contact showed better activities, suggesting that the hydrogen production mechanism can be influenced by hydrolytic surface reactions on CdS. The lower activity of $Pt/CdS/TiO_2$ may have been due to the fact that the potential conduction band (CB) edge of TiO_2 is more positive than that of CdS, creating the potential gradient at the CdS/TiO_2 interface. The CB electrons generated on CdS and trapped by platinum attached to the CdS surface are transferred to TiO_2 CB through the potential gradient. The same potential gradient is created in TiO_2/CdS, but in this case, the photogenerated holes on the CdS surface are less accessible to glycerol, resulting in the photocorrosion of the semiconductor.

Glycerol is also used to produce hydroxyacetaldehyde (HAA). Researchers in Chong's research group [193] have demonstrated that aqueous glycerol can be converted to HAA, H_2, and HCOOH by cleaving C–C bonds on TiO_2-based photocatalysts under anaerobic conditions. Based on the results, they concluded that the selectivity of HAA production is strongly dependent on the dominant TiO_2 facets. Rutile with a high proportion of {110} facets yields more than 90% HAA selectivity, whereas anatase with dominant {001} or {101} facets can yield HAA selectivity of less than 20% or 50%, respectively.

In addition, glycerol can be used as the building block for others fine chemicals such as dihydroxyacetone (DHA). DHA is very commonly used in cosmetics, self-tanning formulations, as well as in the food and pharmaceutical industries. As reported in the literature, DHA is also produced from glycerol using heterogeneous photocatalysis in aqueous media. Researchers in the Augugliaro research group [194] examined the range of UV irradiation, TiO_2 catalyst loading, and glycerol concentration to produce DHA in water and reported a maximum DHA selectivity of 8% with a conversion of 35% after 70 h of irradiation. Similarly, researchers from Zhou's research group [195] using a Pd/TiO_2 photocatalyst to oxidize glycerol in water reported selectivity as high as 9% for DHA production and a glycerol conversion of 21% after 18 h of reaction.

6.3.2 Water, Waste Water and Air Treatment

At present, environmental pollution is becoming an increasingly serious problem [196]. Many literature reports indicate the application of TiO_2 as an effective photocatalyst for the degradation of most common organic compounds, e.g.: aliphatic alcohols, aliphatic alcohols, carboxylic acids, alkenes, phenols [196]. The most important factors affecting the efficiency of the conducted water, wastewater and air treatment processes are: photocatalyst charge, solution pH, temperature, dissolved oxygen, initial pollutant concentration, light wavelength, light intensity [197, 198]. A particularly interesting issue is the effect of photocatalyst charge on process efficiency. Doping is a method used to add another substance that has energy levels almost the same as the valence band or conduction band edge of the main band of a semiconductor, which allows the concentration of charge carriers to be increased either by donating or accepting electrons. However, the opposite effect has been observed, which can promote the recombination of e^- and h^+. Therefore, manipulating the morphology is important in the fabrication of nanosized photocatalysts to reduce the distance from e^- to the surface, thus creating h^+ in the bulk phase. That is the purpose of doping is relatively straightforward: modifying its large bandgap and electronic structure in order to optimize its optical properties for visible light harvest, improving each step in the charge kinetics to reduce the massive recombination of photogenerated carriers, and improving the interface and surface characteristics [199].

Table 6 collects the examples of applications of other photocatalysts used in water, wastewater and air treatment processes.

7 Summary

Production and application of chemicals with the use of heterogeneous catalysis has considerably changed with time. Catalytic processes are used to obtain energy from renewable sources for the production of fibers, medicines or plastics. Development of science including the search for new catalytic systems contributes to progress in chemical technology. It can be concluded that modern and environmentally friendly technology would not be possible without the use of catalysis, which is well documented in literature.

Table 6 Photocatalysts used in water, wastewater and air treatment processes

Water and waste water treatment		
Catalyst	Pollution	References
ZnO	Water-soluble fractions of crude oil	[200–205]
Fe_3O_4 (or based on Fe_3O_4)	Dyes, e.g., methyl red, methyl orange, rhodamine B, methylene blue, methylene violet	[206–208]
Based on MOF: AgBr/ZIF-8 CdS/g-C_3N_4/MOF Ni/MOF	Dyes, e.g., methylene blue, rhodamine B, red-120	[209–211]
Based on zeolites: Zeo-TiO_2, Zeo-ZnO_2 Zeolite/WO_3/Fe_3O_4	Dyes, e.g., rhodamine B	[212, 213]
Air treatment		
TiO_2, in the form of coatings and nanotube	Air pollutants, specially sulfur dioxide and nitrates	[214–216]
ZIF-8	Air pollutant (reduction of E. coli cells)	[217]

With the introduction of the new nanotechnologies briefly discussed in this review, heterogeneous catalysis has a very strong potential to address many modern industrial processes that were not previously available to it for various reasons. In particular, it may be able to offer solutions to problems related to energy extraction, storage and utilization. This could include ways to produce biofuels or hydrogen, which could potentially be produced photocatalytically by harvesting sunlight, as well as improving the efficiency of fuel cells. In terms of environmental issues, further advances are needed on catalysts already developed to clean up automobile exhaust to address the resulting byproducts. Another challenge is to design catalysts capable of treating air, water and also waste water.

As presented in this review, in some respects the field of nanotechnology is ahead of the field of heterogeneous catalysis in this search for sustainable/green chemistry solutions. To date, many synthetic procedures have already been developed to create complex nanostructures with designed properties—some of these approaches have been presented above. Although they still exhibit some limitations, and although it is not yet possible to produce solids with active sites as versatile and selective as those available in homogeneous catalysis, many types of solids can now be produced that were previously unknown. It is an undeniable achievement of the nanotechnology division that it has been able to make so big progress in developing new synthetic methods in such a short period of time.

Acknowledgements This work was supported by the Foundation for Polish Science (TEAM-NET programme; project no: POIR.04.04.00-00-1792/18)

References

1. Y.-C. Lin, G.W. Huber, Energy Environ. Sci. **2**, 68 (2009)
2. V.D. Santo, F. Liguori, C. Pirovano, M. Guidotti, Molecules **15**, 3829 (2010)
3. R.R. Malherbe, Inglomayor Sect. A **16**, 51 (2019)
4. J.C. Védrine, Catalysts **7**, 341 (2017)
5. J.C. Védrine, Appl. Catal. A: Gen. **474**, 40 (2014)
6. S. Wacławek, V.V.T. Padil, M. Černik, Ecol. Chem. Eng. S. **1**, 9 (2018)
7. S.B. Singh, P.K. Tandon, J. Energy Chem. Eng. **2**, 106 (2014)
8. M.S. Spencer, in *Catalyst Handbook II*, ed. M.V. Twigg (Wolfe Publishing Ltd, 1992), p. 17
9. J.A. Dumesic, G.W. Huber, M. Boudart, in *Introduction: Principles of Heterogeneous Catalysis*. ed. by G. Ertl (Wiley-VCH, Weinheim, Germany, 2008), p. 1
10. F. Zaera, Surf. Sci. **500**, 947 (2002)
11. G. Centi, S. Perathoner, Catal. Today **77**, 287 (2003)
12. F. Devred, P. Dulgheru, N. Kruse, in *Comprehensive Inorganic Chemistry II from Elements to Application*, ed. by J. Reedijk, K. Poeppelmeier (Elsevier, 2013), p. 7
13. S.B. Somwanshi, S.B. Somvanshi, P.B. Kharat, J. Phys, Conf. Ser. **1644**, 1 (2020)
14. S.B. Singh, P.K. Tandon, J. Energy Environ. Chem. Eng **2**, 106 (2014)
15. I. Fechete, Y. Wang, J.C. Védrine, Catal. Today **189**, 2 (2012)
16. J. Liang, Z. Liang, R. Zou, Y. Zhao, Adv. Mater. **29**, 1 (2017)
17. I. Matos, M. Bernardo, I. Fonseca, Catal. Today **285**, 194 (2017)
18. M.R. Gogate, Chem. Eng. Commun. **204**, 1 (2017)
19. H. Li, H.-X. Zhang, X.-L. Yan, B.-S. Xu, J.-J. Guo, New Carbon Mater. **33**, 1 (2018)
20. N. Arun, R.V. Sharma, A.K. Dalai, Renew. Sustain. Energy Rev. **48**, 240 (2015)
21. F. Hoffmann, M. Cornelius, J. Morell, M. Fröba, Angew. Chem. Ger. Ed. **45**, 3216 (2006)
22. A. Corma, Chem. Rev. **97**, 2373 (1997)
23. J.S. Beck, J.C. Vartuli, W.J. Roth, M.E. Leonowicz, C.T. Kresge, K.D. Schmitt, C.T.-W. Chu, D.H. Olson, E.W. Sheppard, S.B. McCullen, J.B. Higgins, J.L. Schlenkert, J. Am. Chem. Soc. **114**, 10834 (1992)
24. S. Kumar, M.M. Malik, R. Purohit, Mat. Today: Proc. **4**, 350 (2017)
25. J. Weitkamp, Solid State Ion. **131**, 175 (2000)
26. E. Koohsaryan, M. Anbia, Chin. J. Catal. **37**, 447 (2016)
27. A. Feliczak-Guzik, Micropor. Mesopor. Mater. **259**, 33 (2018)
28. M. Clough, J.C. Pope, L.T.X. Lin, V. Komvokis, S.S. Pan, B. Yilmaz, Micropor. Mesopor. Mater. **254**, 45 (2017)
29. M. Hartmann, A.G. Machoke, W. Schwieger, Chem. Soc. Rev. **45**, 3313 (2016)
30. F. Rodriguez-Reinoso, Carbon **36**, 159 (1998)
31. D.S. Su, G. Wen, S. Wu, F. Peng, R. Schlögl, Angew. Chem. Int. Ed. **56**, 936 (2017)
32. Ö. Sahin, C. Saka, Bioresour. Technol. **136**, 163 (2013)
33. C. Liu, K. Wang, X. Zheng, X. Liu, Q. Lian, Z. Chen, Carbon **139**, 1 (2018)
34. A. Akbari, M. Amini, A. Tarassoli, B. Eftekhari-Sis, N. Ghasemian, E. Jabbari, Nanostructures Nanoobjects **14**, 19 (2018)
35. A. Kusior, J. Banas, A. Trenczek-Zajac, P. Zubrzycka, A. Micek-Ilnicka, M. Radecka, J. Mol. Struct. **1157**, 327 (2018)
36. A. Alam, Y. Zhang, H.-C. Kuan, S.-H. Lee, J. Ma, Prog. Polym. Sci. **77**, 1 (2018)
37. V. Polshettiwar, C. Len, A. Fihri, Coord. Chem. Rev. **253**, 2599 (2009)
38. Q. Wang, D. Astruc, Chem. Rev. **120**, 1438 (2020)
39. T. Goetjen, J. Liu, Y. Wu, J. Sui, X. Zhang, J.T. Hupp, O.K. Farha, Chem. Commun. **56**, 10409 (2020)
40. F. Seidi, M. Jouyandeh, M. Taghizadeh, A. Taghizadeh, H. Vahabi, S. Habibzadeh, K. Formela, M.R. Saeb, Materials **13**, 2881 (2020)
41. B. Friedrich, Angew. Chem. Int. Ed. **44**, 3957 (2005)
42. G. Rothenberg, *Catalysis Concepts and Green Applications* (Wiley-VCH Verlang GmbH & Co. KGaA, Weinheim, 2008)

43. R. Agrawal, N.R. Singh, F.H. Ribeiro, W.N. Delgass, Natl. Acad. Sci. USA **104**, 4828 (2007)
44. O.Y. Poletaeva, D.Z. Latypova, E.M. Movsumzade, Chem. Technol. Fuels Oils **52**, 11 (2016)
45. K.F. Ortega, D. Rein, Ch. L¨uttmann, J. Heese, F. Özcan, M. Heidelmann, J. Folke, K. Kähler, R. Schlögl, M. Behrens, ChemCatChem **9**, 659 (2017)
46. K. Klier, Adv. Catal. **31**, 243 (1982)
47. G.P. Van der Laan, A.A.C.M. Beenackers, Catal. Rev. **41**, 255 (1999)
48. B.W. Wojciechowski, Catal. Rev. **40**, 209 (1998)
49. A. Feller, J.A. Lercher, Adv. Catal. **48**, 229 (2004)
50. J. Yu, R. Wang, S. Ren, X. Sun, Ch. Chen, Q. Ge, W. Fang, J. Zhang, H. Xu, D. Sheng Su, ChemCatChem **4**, 1376 (2012)
51. S. Eijsbouts, S.W. Mayo, K. Fujita, Appl. Catal. A: Gen. **322**, 58 (2007)
52. J.H. Gary, G.E. Handwerk, *Petroleum Refining Technology and Economics* (CRC Press, Boca Raton, 2001)
53. A. Corma, Catal. Lett. **22**, 33 (1993)
54. J. Huang, G.L. Rempel, Prog. Polym. Sci. **20**, 459 (1995)
55. P.J. Gellings, H.J.M. Bouwmeester, Catal. Today **58**, 1 (2000)
56. M. Crocker, *Thermochemical Conversion of Biomass to Liquid Fuels and Chemicals*, (RSC Energy and Environment Series No. 1, RSC Publishing, Royal Society of Chemistry, UK, 2010)
57. DOE (U.S. Department of Energy): Plant/crop-based renewable resources 2020—A vision to enhance U.S. economic security through plant/crop-based resource use. DOE/GO-10098-385, Washington, 1998 i roadmap, DOE (U.S. Department of Energy): The technology
58. A. Corma, M. Renz, M. Susarte, Top. Catal. **52**, 1182 (2009)
59. R.G. Berger, *Flavours and Fragrances* (Springer, 2007)
60. C.S. Sell, *The Chemistry of Fragrances* (RSC Publishing, 2006)
61. E. Breitmaier, *Terpenes, Flavors, Fragrances, Pharmacia, Pheromones* (Wiley-VCH, 2006)
62. A. Mendoza, Y. Ishihara, P.S. Baran, Nat. Chem. **4**, 21 (2012)
63. K. Torssell, *Natural Products Chemistry: A Mechanistic, Biosynthetic and Ecological Approach* (Swedish Pharmaceutical Press, Stockholm, 1997)
64. W.C. Evans, *Trease and Evans' Pharmacognosy* (Londyn: 15th edn. Saunders, 2002)
65. J.J. Mann, R.S. Davidson, J.B. Hobbs, D.V. Banthorpe, J.B. Harborne, *Natural Products: Their Chemistry and Biological Significance* (Longman, Londyn, 1994)
66. A. Corma, S. Iborra, A. Velty, Chem. Rev. **107**, 2411 (2007)
67. R.A. Sheldon, I. Arends, U. Hanefeld, *Green Chemistry and Catalysis* (Wiley-VCH, 2007)
68. F. Tzompantzi, M. Valverde, A. Pérez, J.L. Rico, A. Mantilla, R. Gómez, Top. Catal. **53**, 1176 (2010)
69. M. Hara, K. Nakajima, K. Kamata, Sci. Technol. Adv. Mater. **16**, 1 (2015)
70. M. Moliner, Y. Roman-Leshkov, M.E. Davis, Proc. Natl. Acad. Sci. U.S.A. **107**, 6164 (2010)
71. F. de Clippel, M. Dusselier, R. van Rompaey, P. Vanelderen, J. Dijkmans, E. Makshina, L. Giebeler, S. Oswald, G.V. Baron, J.F.M. Denayer, P.P. Pescarmona, P.A. Jacobs, B.F. Sels, J. Am. Chem. Soc. **134**, 10089 (2012)
72. A. Herbst, C. Janiak, Cryst. Eng. Comm. **19**, 402 (2017)
73. X. Chan, T. Pu, X. Chen, A. James, J. Lee, J.B. Parise, D.H. Kim, T. Kim., Catal. Commun. **97**, 65 (2017)
74. K. Nakajima, Y. Baba, R. Noma, M. Kitano, J.N. Kondo, S. Hayashi, M. Hara, J. Am. Chem. Soc. **133**, 4224 (2011)
75. Y. Jing, Y. Xin, Y. Guo, X. Liu, Y. Wang, Chin. J. Catal. **40**
76. M. Imizcoz, A.V. Puga, Catalysts **9**, 584 (2019)
77. B. Jin, G. Yao, X. Wang, K. Ding, F. Jin, ACS Sustain. Chem. Eng. **5**, 6377 (2017)
78. M.D. Astuti, D. Kristina, R. Rodiansono, D.R. Mujiyanti, Bull. Chem. React. Eng. Catal. **15**, 231 (2020)
79. S. Carrettin, P. McMorn, P. Johnston, K. Giffin, C.J. Kiely, G.J. Hutchings, Phys. Chem. Chem. Phys. **5**, 1329 (2013)
80. G.D. Yadav, R.V. Sharma, Appl. Catal. B Environ. **147**, 293 (2014)

81. S. De, B. Saha, R. Luque, Bioresour. Technol. **178**, 108 (2015)
82. W. Luo, W. Cao, P.C.A. Bruijnincx, L. Lin, A. Wang, T. Zhang, Green Chem. **21**, 3744 (2019)
83. R. Kupila, K. Lappalainen, T. Hu, H. Romar, U. Lassi, Appl. Catal. A: Gen. **612**, 118011 (2021)
84. P. Mäki-Arvela, I.L. Simakova, T. Salmi, D.Y. Murzin, Chem. Rev. **114**, 1909 (2014)
85. L. Zhang, G. Xi, J. Zhang, H. Yu, X. Wang, Bioresour. Technol. **224**, 656 (2017)
86. M.E. Jenkin, D.E. Shallcross, J.N. Harvey, Atmos. Environ. **34**, 2837 (2000)
87. G.A. Rudakov, L.S. Ivanova, T.N. Pisareva, A.G. Borovskaya, Gidroliz. Lesokhim, Prom-st. **4** (1975)
88. A. Severino, J. Vital, L.S. Lobo, Stud. Surf. Sci. Catal. **78**, 685 (1993)
89. S. Yilmaz, S. Ucar, L. Artok, H. Gulec, Appl. Catal. A: Gen. **287**, 261 (2005)
90. G. Gunduz, R. Dimitrova, S. Yilmaz, L. Dimitrov, M. Spassova, J. Mol. Catal.: A **225**, 253 (2005)
91. C.M. Lopez, F.J. Machado, K. Rodriguez, B. Mendez, M. Hasegawa, S. Pekerar, Appl. Catal. A: Gen. **173**, 75 (1998)
92. O. Akpolat, G. Gunduz, F. Ozkan, N. Besun, Appl. Catal. A: Gen. **265**, 11 (2004)
93. T. Yamamoto, T. Tanaka, T. Funabiki, S. Yoshida, J. Phys. Chem. B **102**, 5830 (1998)
94. V. Krishnasamy, J. Indian Chem. Soc. **59**, 1151 (1982)
95. M.K. Yadav, C.D. Chudasama, R.V. Jasra, J. Mol. Catal. A: Chem. **216**, 51 (2004)
96. C. Volzone, O. Masini, N.A. Comelli, L.M. Grzona, E.N. Ponzi, M.I. Ponzi, Mater. Chem. Phys. **93**, 296 (2005)
97. C. Breen, Appl. Clay Sci. **15**, 187 (1999)
98. O. Chimal-Valencia, A. Robau-Sanchez, V. Collins-Martinez, A. Aguilar-Elguezabal, Bioresour. Technol. **93**, 119 (2004)
99. T. Yamamoto, T. Matsuyama, T. Tanaka, T. Funabiki, S. Yoshida, J. Mol. Catal. A: Chem. **155**, 43 (2000)
100. Ł Mokrzycki, B. Sulikowski, Z. Olejniczak, Catal. Lett. **127**, 296 (2009)
101. X. Ma, D. Zhou, X. Chu, D. Li, J. Wang, W. Song, Q. Xia, Micropor. Mesopor. Mater. **237**, 180 (2017)
102. J. Leiner, A. Stolle, B. Ondruschka, T. Netscher, W. Bonrath, Molecules **18**, 8358 (2013)
103. T. Blasco, A. Corma, M.T. Navarro, J.P. Pariente, J. Catal. **156**, 65 (1995)
104. M.P. Kapoor, A. Raj, Appl. Catal. A: Gen. **203**, 311 (2000)
105. D.T. On, M.P. Kapoor, P.N. Joshi, L. Bonneviot, S. Kaliaguine, Catal. Lett. **44**, 171 (1997)
106. R. Raja, G. Sankar, J.M. Thomas, Chem. Commun. **1299**, 829 (1999)
107. T. Sakamoto, C.J. Pac, Tetrahedron Lett. **41**, 10009 (2000)
108. N. Ravasioa, F. Zaccheriab, M. Guidottia, R. Psaro, Top. Catal. **27**, 1 (2004)
109. X.-F. Guo, G.-J. Kim, Top. Catal. **53**, 510 (2010)
110. P. Gallezot, P.J. Cerino, B. Blanc, G. Flèche, P. Fuertes, J. Catal. **146**, 93 (1994)
111. B.W. Hoffer, E. Crezee, F. Devred, P.R.M. Mooijman, W.G. Sloof, P.J. Kooyman, A.D. van Langeveld, F. Kapteijn, J.A. Moulijn, Appl. Catal. A: Gen. **253**, 437 (2003)
112. H. Guo, H. Li, Y. Xu, M. Wang, Mater. Lett. **57**, 392 (2002)
113. P. Gallezot, N. Nicolaus, G. Fleche, P. Fuertes, A. Perrard, J. Catal. **180**, 51 (1998)
114. N. Déchamp, A. Gamez, A. Perrard, P. Gallezot, Catal. Today **24**, 29 (1995)
115. M.C.M. Castoldi, L.D.T. Camara, D.A.G. Aranda, React. Kinet. Catal. Lett. **98**, 83 (2009)
116. S. Schimpf, C. Louis, P. Claus, Appl. Catal. A **318**, 45 (2007)
117. H. Li, H. Li, J.-F. Deng, Catal. Today **74**, 53 (2002)
118. H. Li, H. Li, M. Wang, Appl. Catal. A: Gen. **207**, 129 (2001)
119. J. Zhang, L. Lin, J. Zhang, J. Shi, Carbohydr. Res. **346**, 1327 (2001)
120. A. Feliczak-Guzik, M. Sprynskyy, I. Nowak, B. Buszewski, Catalysts **8**, 31 (2018)
121. A. Feliczak-Guzik, M. Sprynskyy, I. Nowak, M. Jaroniec, B. Buszewski, J. Colloid Interf. Sci. **516**, 379 (2018)
122. P. Maki-Arvela, I.L. Simakova, T. Salmi, D. Murzin, Chem. Rev. **114**, 1909 (2014)
123. R. De Clercq, M. Dusselier, B.F. Sels, Green Chem. **19**, 5012 (2017)

124. X. Yang, J. Bain, J. Huang, W. Xin, T. Lu, Ch. Chen, Y. Su, L. Zhou, F. Wang, J. Xu, Green Chem. **19**, 692 (2017)
125. IEO. International energy outlook 2016 Washington. DC: U.S. Energy Information Administration. https://www.eia.gov/outlooks/ieo/pdf/0484(2016).pdf. Accessed 28 April 2021
126. Z. He, X. Wang, Catal. Sustain. Energy **1**, 28 (2013)
127. P.M. Mortensen, J.-D. Grunwaldt, P.A. Jensen, K.G. Knudsen, A.D. Jensen, Appl. Catal. A: Gen. **407**, 1 (2011)
128. S. Yujie, P. Zhang, Y. Sua, Renew. Sustain. Energy Rev. **50**, 991 (2015)
129. G. Centi, R.A. van Santen, *Catalysis for Renewables: From Feedstock to Energy Production* (WILEY-VCH Verlag GmbH & Co. KGaA, Weinheim, 2007)
130. L.A. Lucia, O.J. Rojas, Nanosci. Technol. Renew. Biomater. (Wiley, UK, 2009)
131. Q. Bu, H. Lei, A.H. Zacher, L. Wang, S. Ren, J. Liang, Y. Wei, Y. Liu, J. Tang, Q. Zhang, R. Ruan, Bioresour. Technol. **124**, 470 (2012)
132. H.Y. Zhao, D. Li, P. Bui, S.T. Oyama, Appl. Catal. A: Gen. **391**, 305 (2011)
133. T. Sankaranarayanan, A. Berenguer, C. Ochoa- Hernández, I. Moreno, P. Jana, J.M. Coronado, D.P. Serrano, P. Pizarro, Catal. Today **243**, 163 (2015)
134. T.-S. Kim, S. Oh, J.-Y. Kim, I.-G. Choi, J.W. Choi, Energy **68**, 437 (2014)
135. A. Sanna, T.P. Vispute, G.W. Huber, Appl. Catal. B Environ. **165**, 446 (2015)
136. Y.-C. Lin, C.-L. Li, H.-P. Wan, H.-T. Lee, C.-F. Liu, Energy & Fuel **25**, 890 (2011)
137. F. Cheng, C.E. Brewer, Renew. Sustain. Energy Rev. **72**, 673 (2017)
138. L. Faba, E. Dia, S. Ordóñez, Renew. Sustain. Energy Rev. **51**, 273 (2015)
139. G. Yao, G. Wu, W. Dai, N. Guan, L. Li, Fuel **150**, 175 (2015)
140. E.H. Lee, R. Park, H. Kim, S.H. Park, S.-C. Jung, J.-K. Jeon, S.C. Kim, Y.-K. Park, J. Ind. Eng. Chem. **37**, 18 (2016)
141. X. Zhu, L.L. Lobban, R.G. Mallinson, D.E. Resasco, J. Catal. **281**, 21 (2011)
142. Y. Han, D.N. McIlroy, A.G. McDonald, J. Anal. Appl. Pyrolysis **117**, 94 (2016)
143. Y. Yang, C. Ochoa-Hernández, V.A. de la Peña O'Shea, P. Pizarro, J.M. Coronado, D.P. Serrano, Appl. Catal. B: Environ. **145**, 91 (2014)
144. T.N. Phan, C.H. Ko, Catal. Today **303**, 219 (2018)
145. S. Jin, Z. Xiao, C. Li, X. Chen, L. Wang, J. Xing, W. Li, C. Liang, Catal. Today **234**, 125 (2014)
146. P. Szczyglewska, A. Feliczak-Guzik, I. Nowak, Micropor. Mesopor. Meter. **312**, 110691 (2021)
147. C. Newman, X. Zhou, B. Goundie, I.T. Ghampson, R.A. Pollock, Z. Ross, M.C. Wheeler, R.W. Meulenberg, R.N. Austin, B.G. Frederick, Appl. Catal. A: Gen. **477**, 64 (2014)
148. C. Zhao, S. Kasakov, J. He, J.A. Lercher, J. Catal. **296**, 12 (2012)
149. A. Feliczak-Guzik, P. Szczyglewska, I. Nowak, Catal. Today **325**, 61 (2018)
150. A.M. Barrios, C.A. Teles, P.M. Souza, R.C. Rabelo-Neto, G. Jacobs, B.H. Davis, L.E.P. Borges, F.B. Noronha, Catal. Today **302**, 115 (2018)
151. P.M. Souza, R.C. Rabelo-Neto, L.E.P. Borges, G. Jacobs, B.H. Davis, U.M. Graham, D.E. Resasco, F.B. Noronha, ACS Catal. **5**, 7385 (2015)
152. Y. Zeng, Z. Wang, W. Lin, W. Song, J.M. Christensen, A.D. Jensen, Catal. Commun. **82**, 46 (2016)
153. S. Boullosa-Eiras, R. Lodeng, H. Bergem, M. Stöcker, L. Hannevold, E.A. Blekkan, Catal. Today **223**, 44 (2014)
154. C.A. Teles, R.C. Rabelo-Neto, N. Duong, J. Quiroz, P.H.C. Camargo, G. Jacobs, D.E. Resasco, F.B. Noronha, Appl. Catal. B Environ. **277**, 119238 (2020)
155. A.N.K. Lup, F. Abnisa, W.M.A.W. Daud, M.K. Aroua, Chin. J. Chem. Eng. **27**, 349 (2019)
156. G. Bharath, K. Rambabu, A. Hai, F. Banat, H. Taher, J.E. Schmidt, P.L. Show, Energy Convers. Manag. **213**, 112859 (2020)
157. P. Yan, E. Kennedy, M. Stockenhuber, Green Chem. **23**, 4673 (2021)
158. M. Mofijur, S.Y.A. Siddiki, M.B.A. Shuvho, F. Djavanroodi, I.M. Fattah, H.C. Ong, M.A. Chowdhury, T.M.I. Mahlia, Chemosphere **270**, 128642 (2021)

159. S.E. Braslavsky, A.M. Braun, A.E. Cassano, A.V. Emeline, M.I. Litter, L. Palmisano, V.N. Parmon, N. Serpone, Pure Appl. Chem. **83**, 931 (2011)
160. J. Harada, R.J. Mell, 21st Aerospace Sciences Meeting. https://doi.org/10.2514/6.1983-74. Accessed 29 April 2021
161. J.M. Herrmann, Catal. Today **53**, 115 (1999)
162. J.C. Colmenares, R. Luque, Chem. Soc. Rev. **43**, 765 (2014)
163. B. Ohtani, J. Photochem. Photobiol. C. **11**, 157 (2010)
164. A. Fujishima, X. Zhang, D.A. Tryk, Int. J. Hydrog. Energy **32**, 2664 (2007)
165. X. Chen, S.S. Mao, Chem. Rev. **107**, 2891 (2007)
166. J.M. Herrmann, Top. Catal. **34**, 49 (2005)
167. A.V. Emeline, V.N. Kuznetsov, V.K. Ryabchuk, N. Serpone, in *Solar Photocatalysis*, Chapter 1—Heterogeneous photocatalysis: Basic approaches and terminology, pp. 1–47 (2013)
168. W. Huang, S. Xiao, H. Zhong, M. Yan, X. Yang, Chem. Eng. J. **418**, 129297 (2021)
169. R. Chauhan, G.K. Dinesh, B. Alawa, S. Chakma, Chemosphere **277**, 130324 (2021)
170. A. Di Paola, E. Garcia-Lopez, G. Marci, L. Palmisano, J. Hazard. Mater. **211–212**, 3 (2012)
171. S.F. Chin, S.C. Pang, F.E.I. Dom, Mater. Lett. **65**, 2673 (2011)
172. G. Guo, B. Yu, P. Yu, X. Chen, Talanta **79**, 570 (2009)
173. M. Sathish, B. Viswanathan, R.P. Viswanathan, Appl. Catal. B Environ. **74**, 307 (2007)
174. K.C. Christoforidis, P. Fornasiero, ChemCatChem **9**, 1523 (2017)
175. D. Cespi, F. Passarini, G. Mastragostino, I. Vassura, S. Larocca, A. Iaconi, A. Chieregato, J.-L. Dubois, F. Cavani, Green Chem. **17**, 343 (2015)
176. K. Mikołajczyk, M. Gmurek, M. Stelmachowski, Inż. Apar. Chem. **54**, 178 (2015)
177. A. Zaleska, M. Nischk, A. Cybula in New and Future Developments in Catalysis, ed. by S.L. Suib (Elsevier, Amsterdam, 2013) p. 63
178. F.J. López-Tenllado, J. Hidalgo-Carrillo, V. Montes, A. Marinas, F.J. Urbano, J.M. Marinas, L. Ilieva, T. Tabakova, F. Reid, Catal. Today **280**, 58 (2017)
179. M.R. Pai, A.M. Banerjee, S.A. Rawool, A. Singhal, C. Nayak, S.H. Ehrman, A.K. Tripathi, S.R. Bharadwaj, Sol. Energy Mater. Sol. Cells **154**, 104 (2016)
180. M. Jung, J.N. Hart, D. Boensch, J. Scott, Y.H. Ng, R. Amal, Appl. Catal. A: Gen. **518**, 221 (2016)
181. W.T. Chen, A. Chan, Z.H.N. Al-Azri, A.G. Dosado, M.A. Nadeem, D. Sun-Waterhouse, H. Idriss, G.I.N. Waterhouse, J. Catal. **329**, 499 (2015)
182. H. Bahruji, M. Bowker, P.R. Davies, L.S. Al-Mazroai, A. Dickinson, J. Greaves, D. James, L. Millard, F. Pedrono, J. Photochem. Photobiol. A: Chem. **216**, 115 (2010)
183. S.I. Fujita, H. Kawamori, D. Honda, H. Yoshida, M. Arai, Appl. Catal. B Environ. **181**, 818 (2016)
184. G. Sadanandam, K. Lalitha, V.D. Kumari, M.V. Shankar, M. Subrahmanyam, Int. J. Hydrog. Energy **38**, 9655 (2013)
185. D.P. Kumar, N.L. Reddy, M. Karthik, B. Neppolian, J. Madhavan, M.V. Shankar, Sol. Energy Mater. Sol. Cells **154**, 78–87 (2016)
186. M. Stelmachowski, M. Marchwicka, E. Grabowska, M. Diak, A. Zaleska, J. Adv. Oxid. Technol. **17**, 179 (2014)
187. X. Fu, J. Long, X. Wang, D.Y.C. Leung, Z. Ding, L. Wu, Z. Zhang, Z. Li, X. Fu, Int. J. Hydrog. Energy **33**, 6484 (2008)
188. R. Liu, H. Yoshida, S.-I. Fujita, M. Arai, Appl. Catal. B Environ. **144**, 41 (2014)
189. M. de Oliveira Melo, L.A. Silva, J. Photochem. Photobiol. A: Chem. **226**, 36 (2011)
190. V. Gombac, L. Sordelli, T. Montini, J.J. Delgado, A. Adamski, G. Adami, M. Cargnello, S. Bernal, P. Fornasiero, J. Phys. Chem. A **114**, 3916 (2009)
191. M. Bowker, P.R. Davies, L.S. Al-Mazroai, Catal. Lett. **128**, 253 (2009)
192. M. Li, Y. Li, S. Peng, G. Lu, S. Li, Front. Chem. China **4**, 32 (2009)
193. R. Chong, J. Liu, X. Zhou, J. Yang, L. Huang, H. Han, F. Zhang, C. Li, Chem. Commun. **50**, 165 (2014)

194. V. Augugliaro, H.H.E. Nazer, V. Loddo, A. Mele, G. Palmisano, L. Palmisano, S. Yurdakal, Catal. Today **151**, 21 (2010)
195. B. Zhou, J. Song, H. Zhou, L. Wu, Z. Liu, B. Han, RSC Adv. **5**, 36347 (2015)
196. J. Spanget-Larsen, M. Gil, A. Gorski, D.M. Blake, J. Raluk, J.G. Radziszewski, JACS **123**, 45 (2001)
197. S. Ahmed, M.G. Rasul, R. Brown, M.A. Hashib, J. Environ. Manag. **92**, 311 (2011)
198. U.G. Akpan, B.H. Hameed, J. Hazard. Mater. **170**, 520 (2009)
199. S.I.S. Mashuri, M.L. Ibrahim, M.F. Kasim, M.S. Mastuli, U. Rashid, A.H. Abdullah, A. Islam, N.A. Mijan, Y.H. Tan, N. Mansir, N.H.M. Kaus, T.-Y.Y. Hin, Catalysts **10**, 1260 (2020)
200. R.L. Ziolli, W.F. Jardim, J. Photochem. Photobiol. A: Chem. **147**, 205 (2001)
201. N. Vela, M. Martínez-Menchón, G. Navarro, G. Pérez-Lucas, S. Navarro, J. Photochem. Photobiol. A: Chem. **232**, 32 (2012)
202. S.M. King, P.A. Leaf, A.C. Olson, P.Z. Ray, M.A. Tarr, Chemosphere **95**, 415 (2014)
203. O.R.S. da Rocha, R.F. Dantas, M.M.M.B. Duarte, M.M.L. Duarte, V.L. da Silva, Chem. Eng. J. **157**, 80 (2010)
204. J. Li, S. Zhang, C. Chen, G. Zhao, X. Yang, J. Li, X. Wang, A.C.S. Appl, Mater. Interf. **4**, 4991 (2012)
205. S. Singh, K.C. Barick, D. Bahadur, J. Mater. Chem. A **10**, 3325 (2013)
206. R.V. Solomon, I.S. Lydia, J.P. Merlin, P. Venuvanalingam, J. Iran. Chem. Soc. **9**, 101 (2012)
207. J. Wang, J. Yang, X. Li, D. Wang, B. Wei, H. Song, X. Li X, S. Fu, Phys. E: Low-dimens. Syst. Nanostruct. **75**, 66 (2016)
208. M. Shekofteh-Gohari, A. Habibi-Yangjeh, Ceram. Int. **43**, 3063 (2017)
209. Y. He, L. Zeng, Z. Feng, Q. Zhang, X. Zhao, S. Ge, X. Hu, H. Lin, Adv. Powder Technol. **31**, 439 (2020)
210. Y. Chen, B. Zhai, Y. Liang, Y. Li, J. Li, J. Solid State Chem. **274**, 32 (2019)
211. R. Ramachandran, T. Sakthivel, M. Li, H. Shan, Z.-X. Xu, F. Wang, Chemosphere **271**, 128509 (2021)
212. F. Alakhras, E. Alhajri, R. Haounati, H. Ouachtak, A.A. Addi, T.A. Saleh, Surf. Interf. **20**, 100611 (2020)
213. M. Rubab, I.A. Bhatti, N. Nadeem, S.A.R. Shah, M. Yaseen, M.Y. Naz, M. Zahid, Nanotechnology **32**, 345705 (2021)
214. N.R. Neti, G.R. Parmar, S.B. Bakardjieva, J. Subrt, Chem. Eng. J. **163**, 219 (2010)
215. W. Chen, J.S. Zhang, Build. Environ. **43**, 246 (2008)
216. F. Shiraishi, T. Ishimatsu, Chem. Eng. Sci. **64**, 2466 (2009)
217. P. Li, J. Li, X. Feng, J. Li, Y. Hao, J. Zhang, H. Wang, A. Yin, J. Zhou, X. Ma, B. Wang, Nat. Commun. **10**, 2177 (2019)

Nanoarchitectured Ferrites, Graphene Oxide, and Their Composites as Enzyme Mimics and Photocatalysts

Vibha Verma, Manpreet Kaur, Sucheta Sharma, and Divya Utreja

Abstract Heterogeneous catalysis, using ferrites nanoparticles (NPs), graphene oxide (GO), and their nanocomposites, has garnered immense interest due to their distinct merits such as non-toxicity, facile synthesis, cost-effectiveness, and ease of recovery. Two important aspects covered under heterogeneous catalysis are enzyme mimic activity and photocatalytic degradation of organic pollutants. Ferrite NPs and GO-based composites have tunable properties as compared to pristine ferrite NPs and GO. The presence of GO in the nanocomposite endows it with numerous surface functional groups for interaction with different species which provide binding sites for metal ions and organic contaminants as well as enhance their enzyme mimic potential. On the other hand, the presence of ferrite NPs in the composites imparts magnetic properties to them. These enzyme mimics can be used in the biomedical application in enzymeless sensors, which are otherwise costly and require sophisticated storage conditions. Graphene-ferrite composites have greater surface area and lower band gap owing to which they have potential application as photocatalysts. The magnetic-GO NC can substitute commercially used TiO_2 which absorbs only ultraviolet radiations due to wide band gap (3.2 eV) for degradation of organic contaminants in the wastewater treatment. This chapter provides an overview of the recent strategies for the synthesis/properties of ferrites NPs/GO and their nanocomposites followed by their applications in the field of photocatalysis and enzyme mimics.

Keywords Enzyme mimics · Ferrites · Graphene · Nanocomposites · Photocatalysis

V. Verma · M. Kaur (✉) · D. Utreja
Department of Chemistry, Punjab Agricultural University, Ludhiana, Punjab 141 004, India
e-mail: manpreetchem@pau.edu

S. Sharma
Department of Biochemistry, Punjab Agricultural University, Ludhiana, Punjab 141 004, India

S. J. Ikhmayies (ed.), *Advances in Catalysts Research*, Advances in Material Research and Technology, https://doi.org/10.1007/978-3-031-49108-5_3

1 Introduction

Ferrites are mixed metal oxides with ferric oxide as their main component. Spinel ferrites (MFe_2O_4) are well-known for their extensive application in the fabrication of information storage devices, batteries, sensors, and high-density magnetic recording due to their magnetic and optical properties [1]. Heterogeneous catalysis, using ferrites nanoparticles (NPs) and their composites, has received immense attention due to their non-toxicity, facile synthesis, cost-effectiveness, and ease of recovery. These are being widely used for catalyzing various reactions [2] due to their selective attachment to the functional molecules. The catalytic potential of ferrite NPs has also been exploited for environment remediation as they catalyze the degradation of organic pollutants [3]. However, their application as photocatalysts is limited due to the recombination of electron-hole pairs. Thus, to make them efficient photocatalysts, the addition of suitable co-catalysts is considered as a preferred approach. Noble metals, e.g. Pt and Ru have been used as dopants to upsurge the electron transport ability and suppress the electron-hole pair recombination [4, 5]. Recently, the use of graphene/graphene oxide (GO) that works as an excellent substitute for noble metals has been explored due to its superior mechanical strength, excellent electronic mobility, and support matrix provided by the two-dimensional structure of graphene/GO for the other components in composite [6].

In graphene, single-layered sp^2 hybridized carbon atoms are packed into hexagonal honeycomb lattice. Its synthesis was firstly reported via "Scotch tape" method in 2004 [7]. The unique combination of various fascinating properties such as large surface area, strong Young's modulus, flexible structure, high thermal conductivity, and fast charged carrier mobility has proved it as a promising matrix in the world of nanotechnology, materials/environmental science and biotechnology [8]. Numerous protocols have been developed for the exfoliation of graphene to get the desired physico-chemical properties. GO, an oxidized form of graphene with oxygen containing functional group, serves as a favorable anchoring center for active species. Owing to elevated electron mobility and extended π-π conjugation system, it can act as a promising electron acceptor, thus suppressing the electron-hole pair recombination [9]. However, the most serious problem confronted is employing exfoliated graphene alone, as a catalyst for treatment of waste–water and in other applications is its recovery and recycling. Nanocomposites of graphene with magnetic NPs are the best way out of this dilemma. Thus, graphene-based nanocomposites work as excellent adsorbents and photocatalysts [10].

Enzymes are the biocatalysts that increase the rate of biochemical reactions. Their applications are not limited to biochemical reactions, but they are extensively used in different industries and are important from medical point of view as they possess therapeutic applications [11]. Synthetic molecules exhibiting properties similar to enzymes are an emerging area of interest, as natural enzymes have some disadvantages. They denature quickly on changing pH, heat, solvent, and other environmental conditions [12–14]. Moreover, their high cost, optimum storage conditions required to retain their activity, make them less viable for practical use. Presently, work is going on to synthesize artificial enzymes by integrating the catalytic center into a diversity of scaffolds for intrinsic enzyme-like activity [15]. Molecules that have rapid electron transfer and substrate adsorbing potential can work well as promising artificial enzymes. GO and ferrite NPs are suitable candidates due to their invincible potential to work as a functional carrier, and they also endow large surface area for the adsorption of substrate [16, 17]. A review on cyclo-dextrins as artificial mimics to glycosidases, epoxidases, oxidases, and esterases has been compiled [18]. In another review, work done on artificial catalysts and bio-macromolecule as hybrid catalysts for the development of artificial enzymes have been reported [19]. A wide range of graphene-based materials exhibiting peroxidase-like activity has also been reviewed [20]. Enzyme mimic activity of polymeric and dendrimeric molecules, nanoparticulate, and supramolecules was discussed [21].

Removal of organic contaminants by conventional water treatment is not safe as they can be transported easily to the aquatic system. Moreover, the degradation products are also found to be toxic and carcinogenic. Thus, they are considered as a major threat to living organisms. Major organic pollutants include dyes, herbicides, insecticides, food processing wastes, pharmaceuticals, detergents, plastics, paints, synthetic fibers, solvents, and volatile organic compounds. Various advanced oxidation processes have been employed for the wastewater treatment [22]. Among these, photocatalytic oxidation is a widely studied method as it is aimed at total oxidation of the substrate into non-toxic end product in the presence of light, and is found to be cost-effective. Till date, a wide range of catalysts particularly TiO_2 as a commercial catalyst, have been used for photocatalytic oxidation of organic contaminants as it absorbs UV radiation to degrade organic contaminants [23–25]. However, the large band gap (3.2 eV) lowers its energy harnessing efficiency for complete degradation of organic contaminants as it can absorb only 5% of solar energy. However, the energy crisis has intensified the need for developing materials that are cost-effective with high-energy harnessing efficiency that are active even in the visible range, and results in complete removal of organic contaminants. Graphene oxide and its magnetic nanocomposites provide enlarged surface area for the adsorption and degradation of organic contaminants and can be reused due to magnetic characteristics. A review by Hashim et al. [26] envisaged a wide range of GO-based nanocomposites exhibiting photocatalytic activity. The use of GO in photocatalysis has also been reported in recent studies [9]. The photocatalysis of organic pollutants using graphene-based composites has also been addressed [27]. The photocatalytic potential of ferrites-based composites as visible light-driven photocatalysts has also been highlighted [28].

The advantages of GO-ferrite nanocomposites (NCs) include the synergistic adsorption and photocatalysis of organic pollutants. One component of NCs enhances the adsorption, while the other improves its catalytic efficiency. Various carbon-based NCs have also been reported that show synergistic effects due to high surface area and charge interactions. Ag-doped CdSe/GO loaded with cellulose acetate NCs have shown potential for photocatalytic degradation for malachite green (MG) dye. The presence of GO enhanced the electron-hole separation efficiency [29]. A similar synergistic process was observed in g-C_3N_4-agar hybrid aerogel due to potential adsorption capacity of 3D network aerogels and desired band gap values against MB dye [30]. The synergistic effect was also observed in graphene aerogels/rGO-Ag@Ag_3PO_4 3D ascribed to its porous network structure which enhanced absorption potential for cationic and anionic dyes. The enhancement attributed to larger surface area and π-π interactions, which promoted the separation of electron-hole pairs due to its conductive nature [31].

In this chapter, recent work done in the field of enzyme mimics and photocatalysis using ferrite NPs, GO and their nanocomposites have been compiled. In addition, an overview of recent strategies for the synthesis of GO/ferrites and magnetic nanocomposites followed by their applications in the field of photocatalysis and enzyme mimics are discussed. The review is divided into three parts viz. synthetic strategies/properties of ferrite NPs, GO and their nanocomposites with different materials followed by their photocatalytic and enzyme mimic activities.

2 Synthesis and Properties for Ferrite NPs, GO, and Their Nanocomposites

2.1 Synthetic Strategies of Ferrite NPs, GO, and Their Nanocomposites

In the last decade, a large number of published articles about magnetic nanoparticles have highlighted efficient protocols to achieve highly stable, shape-controlled, and narrow size distribution of magnetic nanoparticles. Till date, several physical and chemical methods including co-precipitation, microwave, microemulsion, solvothermal, sonochemical, thermal decomposition, chemical vapor deposition, carbon arc, combustion synthesis, and laser pyrolysis synthesis have been reported for the synthesis of ferrite NPs [32, 33]. They are broadly categorized as physical and chemical methods [34]. Physical methods are top-down approach whereas chemical methods are bottom-up approach and are outlined in Scheme 1.

In the physical methods, high-energy ball milling method involves placing mixture of metal salts in the ball mill for high-energy collisions from the balls that result into small size particles into nano range [35]. In laser pyrolysis technique, a compound is excited using laser radiation, and energy is transferred to the reaction medium, which increases its temperature [36]. The reactants thus decompose, and the formation of nanoparticles takes place. Electric arc deposition method involves the use of high vapor pressure by electron bombardment in "high vacuum" to heat the material to be deposited, which is transported by diffusion to be deposited by condensation [37]. In ion implantation method, ions of an element are stepped up into a solid target, thus physical, electrical, or chemical properties of the target are changed [38].

Among the chemical methods, co-precipitation method involves the precipitation of metal cations, as hydroxides, oxalates, carbonates, or citrates. The precipitates are dried and then heated to the required temperature in a suitable atmosphere and

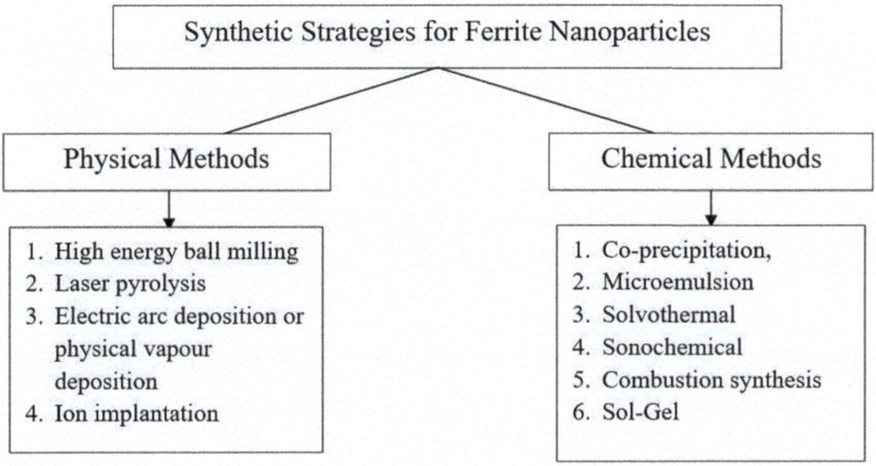

Scheme 1 Different approaches for synthesis of ferrite nanoparticles

the final product is obtained [39]. Microemulsion method is further classified into normal micelle methods [oil-in-water] and reverse micelle methods [water-in-oil]. This method involves the use of surfactants above the critical micelle concentration [40]. In the solvothermal process, a solvent is used under moderate to high pressure (between 1 atm and 10,000 atm), and temperature (between 100 °C to 1000 °C) to facilitate the contact of reactants during synthesis [41]. If solvent used is water, then the process is named as "hydrothermal synthesis." In sonochemical method, sound waves are used [42]. The metal oxide particles come in contact with shock waves produced by cavitation bubbles, and collide at high speed. The energy produced from collisions initiates a reaction that leads to the formation of ferrite NPs. The combustion process involves an exothermic redox reaction between an oxidizer (O) and a fuel (F). Stoichiometry $\phi_e = (O/F)$ is maintained unity by balancing the oxidizing (O) and reducing valency (F) [43]. The sol-gel method involves various processes like hydrolysis, gelation, polymerization, dehydration, drying, and densification [44]. In this method conversion of sol to gel takes place thus so named. The synthesized nanomaterials can be characterized through various techniques such as; Fourier-transform infrared spectroscopy (FT-IR) shown in Fig. 1a (i), X-ray diffraction (XRD) shown in Fig. 1a (ii), scanning electron microscopy—energy dispersive X-ray spectroscopy (SEM-EDX) (Fig. 1b left and right) [43, 45]. The characterization of prepared ferrite NPs has been shown in Fig. 1a, b. The surface of ferrite NPs is observed to be porous as characterized by SEM images. The XRD patterns of ferrite NPs demonstrate a particular plane, corresponding to which the interplanar distance is calculated.

Graphene oxide can be synthesized by chemical oxidation and exfoliation of pristine graphite using various methods, which vary in the reaction conditions and choice of oxidizing agents. For example, Brodie, Staudenmaier, Hummer's methods or some variations of these methods can be employed for the synthesis of GO. Brodie first found that the oxidizing mixture ($KClO_4$ + fuming HNO_3) could form GO only

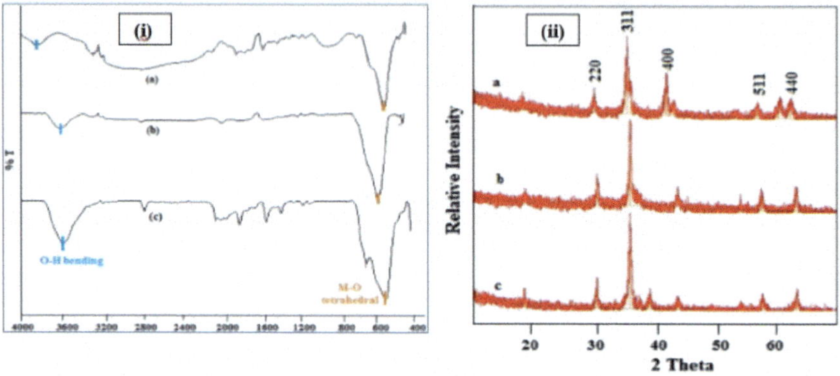

(a)

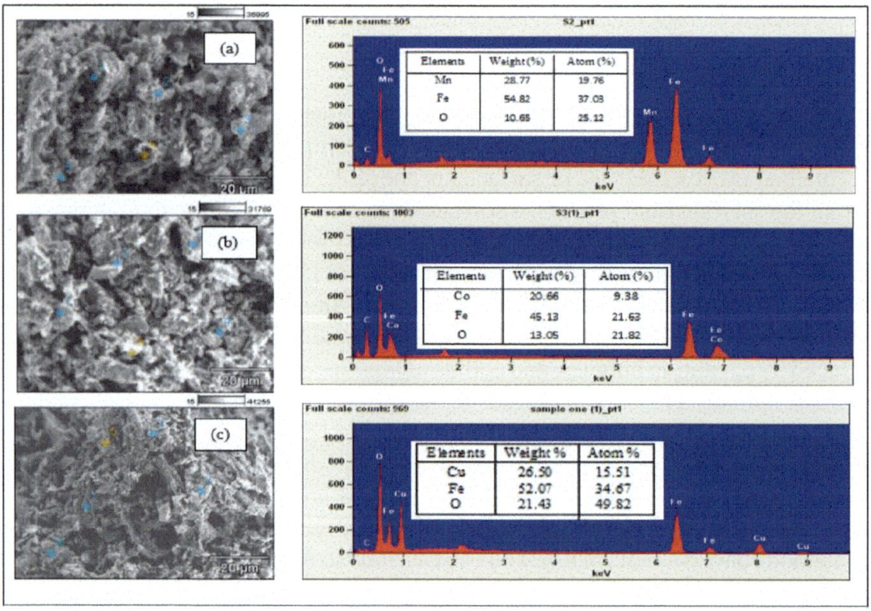

(b)

Fig. 1 **a** FT-IR (i), and XRD images (ii) of **a** $MnFe_2O_4$ **a** $CoFe_2O_4$ **a** $CuFe_2O_4$ NPs. **b** SEM-EDX images of **a** $MnFe_2O_4$ **a** $CoFe_2O_4$ **a** $CuFe_2O_4$ NPs. Reprinted from Verma et al. [43], with permission from Elsevier. Copyright (2019)

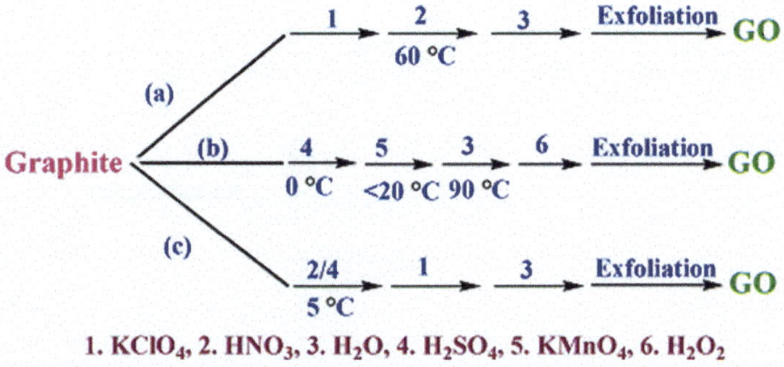

Scheme 2 Different routes for synthesis of graphene oxide

with graphitizable carbons that contain regions of graphitic structure (Scheme 2a) [46]. Staudenmaier then reported the formation of GO (Scheme 2b) when graphite was heated with H_2SO_4, HNO_3, and $KClO_4$ [47]. In Hummers method, a mixture of $KMnO_4$ and H_2SO_4 was used as oxidizing agents (Scheme 2c).

The products of these reactions depend upon the oxidants used, graphite source, and reaction conditions. Graphene oxide produced by this method from flake graphite, had disrupted conjugation in the graphene plane and abundant functional groups, such as epoxide, hydroxyl, carboxyl, and carbonyl on its surfaces. These oxygen containing functional groups could form strong complexes with metal ions and allow GO to act as an adsorbent for heavy metal ions and their removal.

Till date, numerous GO-ferrite nanocomposites have been synthesized using various methods like hydrothermal, solvothermal, sonochemical etc. Cobalt ferrite fabricated on reduced graphene oxide (CoFe@rGO) has been prepared by a three-step chemical method including hydrothermal synthesis, annealing process, and mixing with paraffin [48]. Reduced graphene oxide/strontium ferrite/polyaniline (rGO/SF/PANI) nanocomposites were prepared by in situ polymerization route [49]. Graphene-based cobalt ferrites composites were synthesized by in situ co-precipitation method and sonochemical methods as described in Fig. 2 [50, 51]. Low-temperature reaction process was used for zinc ferrite-GO composites [52]. Sonochemical method was employed for the synthesis of manganese ferrite ($MnFe_2O_4$)/GO nanocomposites [53]. Another solvothermal protocol was reported for the preparation of GO-supported ferrite (MFe_2O_4, M = Fe, Co and Ni) NC [54]. Synthesis of Fe_2O_3@GO core-shell nanosheets using sonication method and reversal of core-shell, i.e. GO@Fe_2O_3 was achieved using N,N'-dicyclohexylcarbodimide as the binding agent [55].

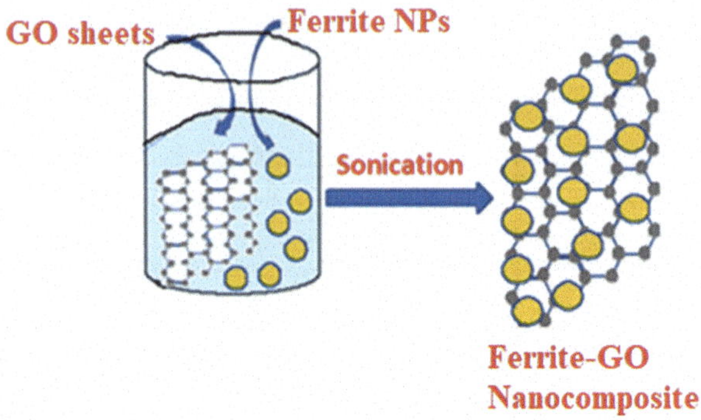

Fig. 2 Synthesis of ferrite-graphene oxide nanocomposites via sonication

2.2 Properties of GO/Ferrite NPs and Their Nanocomposites

Ferrites are mixed metal oxides having iron oxide as their main component. Due to the varying arrangement of oxygen anions around metal cations, crystal structure of ferrites is altered. Based on the position of oxygen anion ferrites have been classified into spinels, garnets, and magnetoplumbites. Spinel ferrites have general formula AB_2O_4 where A and B are metal cations one of which is divalent and the other is trivalent (mainly Fe) [56]. Garnets possess general formula $M_3Fe_5O_{12}$ and are isomorphous with mineral garnet $Ca_3Fe_2(SiO_4)_3$ [57]. They have enhanced magnetization and dielectric properties than spinels. Magnetoplumbites possess general formula $MFe_{12}O_{19}$ (M = Ca, Sr, Pb, Ba) and contain stack of oxygen ions and only oxygen ion coordinated to Fe ion [58]. They are ferrimagnetic materials, *i.e.* they are magnetized and attracted toward the magnet. They retain their catalytic activity after a number of runs, thus are cost-effective. Based on the coercivity value, ferrite materials are categorized as soft ferrites and hard ferrites. Soft ferrites have low coercivity value, *i.e.* magnetization can be easily reversed without dissipating much energy, high electric resistivity, and tremendous magnetic properties in the region of high frequency. On the contrary, hard ferrites have a high value of coercivity and are also called permanent magnets as they retain their magnetism after being magnetized. These are also considered to possess excellent dielectric properties as they are electrical insulators that can be polarized by applying electric field. Further, the dielectric properties of ferrites depend upon their methods of preparation. They possess very low electrical conductivity as compared to metals and thus are classified as semiconductors. Further, electrical conductivity depends on temperature and presence of various metal ions and is found to increase with the rise in sintering temperature.

Crystal structure and properties of GO depend primarily on preparation method and extent of oxidation. Layered structure of parent graphite is preserved in GO, however interlayer spacing is enlarged. Oxygen containing functional groups like epoxide, carbonyl, hydroxyl, and phenol are incorporated using concentrated acids (Fig. 3). The hydrophilic nature can be explained on the basis of the presence

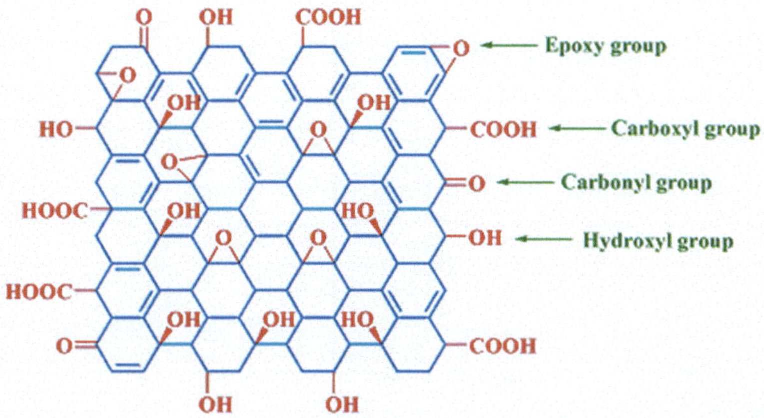

Fig. 3 Structure of graphene oxide showing the presence of oxygen containing functional groups

of oxygen containing functional groups. Carbon electrons in graphene are sp^2 hybridized, thus graphene has planar structure with bond angle 120° [59].

GO possesses high affinity for accepting electrons, thus acts as photocatalyst. It also extends the lifetime of photo-generated charge carriers and subsequently enhances charge extraction and separation. For enhancement of photocatalytic properties, many GO-based composites have also been synthesized, and their photocatalytic activity has been reported. N-doped graphene has less value of work function, and the decrease in work function value depends on N-doping sites. On the contrary, work function can be increased by metal chloride doping [60].

Owing to the disturbed sp^2 bonding network by the incorporation of oxygen containing functional groups, electrical conductivity of GO is reduced as compared to graphite. Restoration of sp^2 bonding network on reducing GO using chemical or thermal methods significantly enhanced its electrical conductivity up to 10,000-fold as compared to GO. Chemical reduction method can partially remove functionalities on GO, but high-density residual defects disrupt its optoelectronic properties. Thus, thermal reduction method is more useful for the enhancement of electrical conductivity. GO can be made n-type or p-type semiconductor using functionalization. GO with additional oxygenated components has small band gap and is a p-type semiconductor. However, the replacement of oxygenated components with nitrogen rich counterparts converted it into n-type semiconductor. Thus, band gap and semiconductor properties of GO derivatives can be varied by tuning its oxygenation. Surface properties of GO can be varied by modification of its surface. GO and its composites are excellent adsorbents and act on adsorbates through π-π stacking interactions for physical adsorption or through surface complexation by chemical adsorption (Fig. 4). Thus, they possess significant applications for the removal or recovery of ionic pollutants. Much work has been reported on the adsorption studies of GO and its composites for waste water treatment [9].

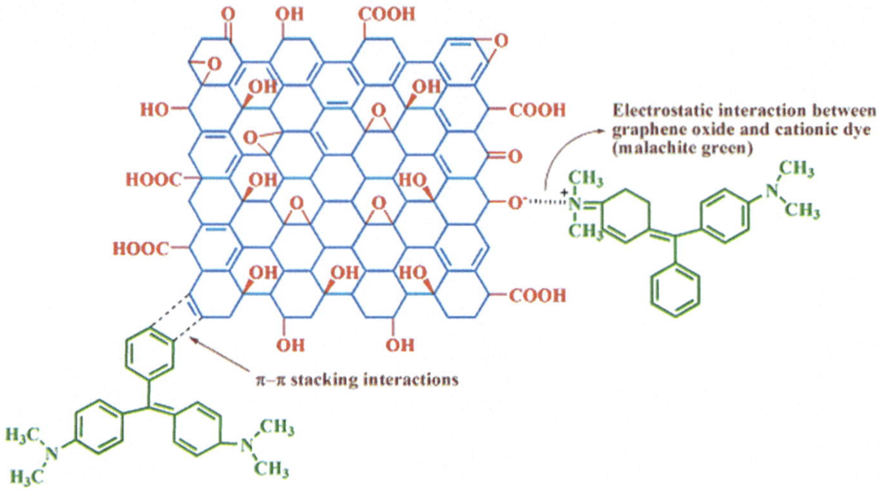

Fig. 4 Different interactions between graphene oxide and dye molecules

Ferrites-GO composites exhibited various improved properties than bare ferrite NPs and GO. The presence of GO in the nanocomposite endow it with numerous surface functional groups for interaction with different species. Functional groups provide binding sites for metal ions and organic contaminants. On the other hand, the presence of ferrite NPs in the composites imparts magnetic properties to them. Zinc ferrite-rGO NC exhibited superior absorption and magnetic properties than parent compounds [61]. Synergistic effect among strontium ferrite (SF), RGO, and polyaniline (PANI) for improved microwave absorption properties of RGO/SF/PANI nanocomposites [49]. Addition of rGO to $CuFe_2O_4$ enhanced strength and toughness of ferrite NPs [50]. Also, addition of $CuFe_2O_4$ to rGO enhanced surface area and electrocatalytic activity. Graphene-ferrite composites have enhanced surface area and lower band gap owing to which, they have potential application in photocatalysis.

3 Possible Catalytic Mechanisms

3.1 Enzyme Mimic Activity of GO/Ferrite NPs and Their Nanocomposite

In this review, catalytic mechanism of peroxidase-like activity has been discussed in brief as the reaction mechanism of other enzyme mimics are substrate-specific and generally include redox reactions. The catalytic pathways for peroxidase-like activity can be categorized into two processes including electron transfer processes and production of reactive oxygen species (ROS) [62]. The induction of Fenton or

Fenton-like reactions in the presence of nanoenzymes is a well-known pathway for peroxidase-like nanoenzymes. Under this mechanism, initially H_2O_2 is adsorbed on the nanoenzyme's surface which leads to the breaking of O-O bond and results in the production of two hydroxyl radicals. The partial electron exchange interactions between generated hydroxyl radicals nanoenzymes stabilizes the produced ROS [63]. In the next step, oxidation of substrate in the presence of hydroxyl radical leads to the production of intermediate species. The generated intermediate species either mineralized to CO_2, H_2O, and inorganic salts or bind to form dimeric products which produce a characteristic color.

Unlike the Fenton reaction mechanism, some nanozymes may exhibit peroxidase-like activity through an electron transfer mechanism. In electron transfer mechanisms, electron transfer between substrate and H_2O_2 occurs via nanozymes which act as a mediator. The whole process occurs without the formation of hydroxyl radicals [64]. The hot electrons generated on the surface of nanozymes are transferred to the molecular orbitals of H_2O_2 through plasmon excitation, which activates the decomposition of H_2O_2 into $OH^{\cdot}$ and OH^- anion. The scavenging of holes by additional electron donors, oxidation of water, or OH^- anions inhibits the recombination of hot electrons and holes, which favors the cleavage of H_2O_2 and improves the reaction rate [65].

3.2 *Photocatalytic Mechanism of GO/Ferrite NPs and Their Nanocomposite*

The photochemical advanced oxidation processes (AOPs) utilize light energy to generate intermediate species. The Fenton type AOPs involve the formation of hydroxyl radicals in an acidic medium, from the decomposition of H_2O_2 in the presence of iron catalyst [66]. In photo-assisted Fenton reaction the catalytic reduction of Fe^{3+} to Fe^{2+} occurs in the presence of UV/Vis light, which boosts the formation of hydroxyl radicals. The light irradiated can also decompose H_2O_2 molecules to generate hydroxyl radicals [67]. The feasibility of Fenton process is due to the efficient mineralization of organic pollutants under normal temperature and pressure conditions. Moreover, these processes are economically cheap, environmental-friendly, and can disinfect water more efficiently than chemical AOPs [68]. The wavelength and intensity of light have a significant impact on the degradation rate of pollutants. The steps given in Eqs. (1–6) are involved during photo-Fenton process [69]:

$$Fe^{2+} + H_2O_2 + hv \rightarrow Fe^{3+} + OH^{\cdot} + OH^- \quad (1)$$

$$OH^{\cdot} + H_2O_2 \rightarrow HO_2 + H_2O \quad (2)$$

$$Fe^{2+} + OH^{\cdot} \rightarrow Fe^{3+} + OH^{-} \tag{3}$$

$$Fe^{3+} + HO_2 \rightarrow Fe^{2+} + O_2 + H^{+} \tag{4}$$

$$OH^{\cdot} + OH^{\cdot} \rightarrow H_2O_2 \tag{5}$$

$$Organic\ Pollutant + OH^{\cdot} \rightarrow Degraded\ Product \tag{6}$$

4 Enzyme Mimic Activity of GO/Ferrite NPs and Their Nanocomposite

Enzyme mimics are the class of catalysts that mimic natural enzymes for catalyzing a wide range of chemical and biological reactions. They can serve as potential and viable alternatives to natural enzymes. They possess similar catalytic activities to their natural counterparts, but have advantages of higher catalytic efficiencies and stability, controlled synthesis at lower cost. Different materials, e.g. metal complexes, polymers, and inorganic NPs can be used as artificial enzymes. Inorganic NPs offer higher stability, large surface areas as enzyme mimics when compared to natural enzymes [70]. Gold NPs have the potential to mimic activity of glucose oxidase, nuclease, peroxidase, catalase, superoxide dismutase, and esterase and can be used in biomedical sciences and biosensing [71]. However, the high cost of gold NPs has led researchers to other mimics such as graphene and ferrite NPs. These mimetic enzymes have been classified into five different categories including oxidoreductases, transferases, hydrolases, isomerases, and induced enzymes, and can be further used in chemical sensors. GO and graphene hemin NCs can be used as modulators to effectively improve the overall catalytic efficiency of enzymes.

Magnetically separable MFe_2O_4 (M = Mg, Ni, Cu) NPs exhibited peroxidase and catalase-like catalytic activity and catalyzed decomposition of H_2O_2 to produce water and oxygen. 2.2-azino-bis(3-ethylbenzothiazoline-6-sulfonic acid) (ABTS) was used as the substrate to study peroxidase-like activity. $MgFe_2O_4$ showed highest affinity toward substrates than other NPs which was attributed to higher value of zeta potential that facilitated stronger electrostatic affinity toward negatively charged ABTS substrate. Due to higher apparent value of Michaelis–Menten constant (K_m) for M = Mg/Ni than horseradish peroxidase (HRP), these NPs showed lower affinity toward H_2O_2, but this value was lower with ABTS, thus exhibiting high affinity for ABTS than HRP. These NPs were stable at a wide range of temperature and pH and, were found to be reusable up to 3 runs with retained catalytic activity. In glucose biosensor, $NiFe_2O_4$ showed linear behavior in the range 9.4×10^{-7} to 2.5×10^{-5} mol/L with detection limit 4.5×10^{-7} mol/L [72]. Au electrode modified with $ZnFe_2O_4$ MNPs immobilized on chitosan-based exhibited improved peroxidase mimic activity than pure Au electrode toward H_2O_2 detection with minimum detection limit of 2.5 nM indicating its potential application in biomedical field [73].

Glassy carbon electrodes were modified using Fe_3O_4 decorated graphite hybrid. They exhibited peroxidase-like activity and were found to be superior to bare carbon electrode toward H_2O_2 reduction. The most excellent results were obtained using graphites, with particle sizes 2 μm and 17 μm, modified with Fe_3O_4 NPs. The amperometric detection of H_2O_2 was observed in the range $(1.1 \pm 0.1) \times 10^5$ μA M^{-1} cm^{-2} mg^{-1} with the detection limit of 0.50 nM at—0.200 V. The obtained values were similar to those of metal or metal oxide-based sensors and carbon nanotubes [74].

Doped Spinel ferrites $Co_xNi_{1-x}Fe_2O_4$ modified carbon paste electrodes (CPE) have been evaluated as artificial hydrogen peroxidase. Tailored electrochemical properties were observed using different Co/Ni molar ratios and $Co_{0.5}Ni_{0.5}Fe_2O_4$/CPE was observed to possess the best electrocatalytic activity. Linear calibration curve was obtained over the wide range of 1.0×10^{-8}–1.0×10^{-3} M with lower detection limit of 3.0×10^{-9} M. Increase in effective electroactive surface area increased electron transfer rate [75]. The $NiCo_2O_4$ nanoparticles fabricated three-dimensional graphene foam was also used as electrochemical electrode for the highly sensitive and selective detections for calcium ion and glucose with a limit up to 4.45 and 0.38 μM, respectively [76].

Use of phosphate containing adenosine analogs with Fe_3O_4 NPs for the oxidation of H_2O_2 and amplex ultrared improved peroxidase-like activity of Fe_3O_4 NPs. Enhanced activity of Fe_3O_4 NPs was attributed to the binding of adenosine analogs to Fe^{2+}/Fe^{3+} sites on the NPs surface and due to activation of H_2O_2 by adenosine 5′-monophosphate (AMP) (Fig. 5). Trend followed by phosphate containing adenosine analogs for improved Fe_3O_4 NP activity: AMP > adenosine 5′-diphosphate > adenosine 5′-triphosphate. The enzyme mimic activity also increased with AMP concentration and polyadenosine length. The K_m value for AMP-attached Fe_3O_4 NPs was 5.3-fold lower and the maximum velocity was 2.7-fold higher than those of pristine Fe_3O_4 NPs [77].

The effect of surfactant coating on enzyme mimic activity of spinel ferrites has also been explored. In this regard, cetyl trimethyl ammonium bromide (CTAB)-coated Fe_2O_3 NPs were synthesized in different w/w ratios to evaluate their peroxidase mimic activity over bare Fe_2O_3 NPs using o-dianisidine dihydrochloride as substrate. The CTAB@Fe_2O_3 NPs with w/w ratio 1:1 showed highest peroxidase mimic activity over other w/w ratio and pristine NPs. The results may be due to increased electrostatic interactions between substrate and peroxidase mimics on the addition of charged surfactant. The enzyme mimic activity also varied with the variation in operational parameters [78].

In another report, bare Cu-$CuFe_2O_4$ NPs exhibited excellent peroxidase mimic activity toward H_2O_2 with lower affinity toward 3, 3′, 5, 5′-Tetramethylbenzidine (TMB) than natural enzyme. However, addition of sodium dodecyl benzene sulfonate (SDBS), an anionic surfactant, to Cu-$CuFe_2O_4$ NPs improved the affinity of NPs toward positive charged TMB along with enhanced peroxidase mimic activity. The modified SDBS-Cu-$CuFe_2O_4$ also possessed more negative zeta potential than bare

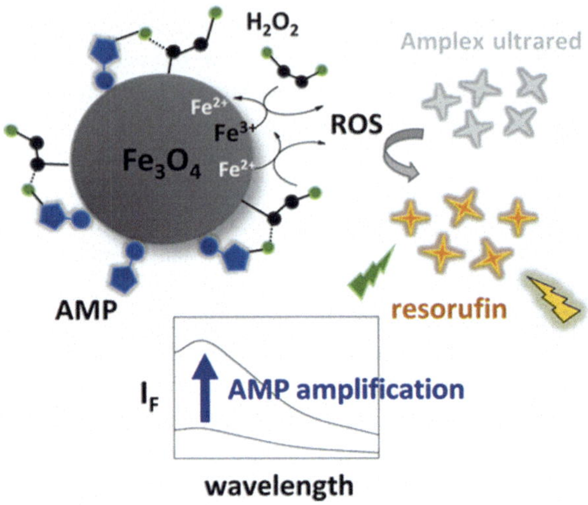

Fig. 5 Peroxidase mimic activity of phosphate containing adenosine analogs with Fe_3O_4 NPs. Reprinted from Yang et al. [77], with permission from American Chemical Society. Copyright (2017)

Cu-$CuFe_2O_4$. The enhanced activity may be due to the increased electrostatic interactions between the negatively charged SDBS-Cu-$CuFe_2O_4$ and positively charged TMB in addition of surfactant. The modified enzyme mimic SDBS-Cu-$CuFe_2O_4$ also showed the rapid detection of H_2O_2 and dopamine with a small limit [79].

Superoxide dismutase (SOD) mimic activity of different transition metal ferrite NPs MFe_2O_4 (M = Mn, Co and Cu) has also been explored. Among synthesized NPs, $CuFe_2O_4$ NPs recorded highest SOD mimic activity followed by $CoFe_2O_4$ and $MnFe_2O_4$ NPs. The trend showed a good correlation with the ease of interconversion of the oxidation state resulting to a stable electronic configuration. The kinetic parameters for $CuFe_2O_4$ NPs calculated from the Michaelis–Menten equation possessed values of maximum rate of reaction V_{max} (0.77 s^{-1}) and K_m (4.20 mM) which indicated its potential use as SOD mimic for numerous applications [80].

The enzymes have also been used as a valuable tool for biomarker detection, and thus NPs possessing enzyme-like activity can also be explored in the same. The doped $Co_{0.25}Zn_{0.75}Fe_2O_4$ NPs exhibited peroxidase-like characteristics in the detection of H_2O_2 using cyclic voltammetry. The K_m value of synthesized NPs was in the suitable range as calculated by Lineweaver–Burk equation. As the doped NPs possessed excellent enzyme mimic activity, they were used in enzymeless disposable microfluidic immune array device (μID) assembled with monoclonal antibody against CYFRA 21-1 covalently immobilized on GO with carbon-based electrodes. Under optimized conditions, the doped NPs showed good response over the bare Fe_3O_4NPs in the range of 3.9–1000 fg mL^{-1} with limit of detection (LOD) of 0.19 fg mL^{-1}. The system also showed satisfactory results in serum samples of healthy and prostate cancer patients with the results obtained with Enzyme-linked Immunosorbent Assay (ELISA). Thus, $Co_{0.25}Zn_{0.75}Fe_2O_4$ NPs associated with μID may prove beneficial for biomarker detection with low-cost and rapid detection for early diagnosis of cancer [81].

Ferrite composites have been known for augmented enzyme mimic activity due to rapid electron transfer owing to the synergistic effect of each component. In this regard, meso-tetrakis(4-chlorophenyl)porphyrin modified $CuFe_2O_4/SiO_2$ (TClPP-$CuFe_2O_4/SiO_2$) NC were synthesized to exploit their enzyme mimic activity by catalyzing the oxidation of TMB. The TClPP-$CuFe_2O_4/SiO_2$NC showed higher peroxidase-like activity as compared to $CuFe_2O_4/SiO_2$. The enhanced activity due to the introduction of porphyrin was attributed to the alteration in crystal structure of $CuFe_2O_4$ from cubic to tetragonal and exposure to high-energy surface. Additionally, TClPP-$CuFe_2O_4/SiO_2$NC showed selective colorimetric detection of glucose over the other glucose analogs such as sucrose, maltose, lactose, and fructose. The NC also showed good correlation for the detection of glucose in human serum with commercial biochemical analyzer [82].

Magnetic composites of graphene GO-Fe_2O_3 with improved peroxidase-like activity (acidic medium) were fabricated. The enzyme mimic activity was observed to be dependent on pH, temperature, and catalyst concentration. The reaction kinetics followed the Michaelis–Menten kinetic model. Electron spin resonance (ESR) technique revealed the presence of free radicals, such as $^{\cdot}OH$ and $^{\cdot}O_2$, as intermediates in the hybrid-H_2O_2 reaction system. It was also observed that the addition of TMB rapidly decreased the concentration of $^{\cdot}OH$ and $^{\cdot}O_2$, and consequently the formation of blue TMB radical was confirmed using absorption spectroscopy and ESR. The nanohybrid also displayed improved catalase-like activity in alkaline and neutral buffers as compared to Fe_2O_3 NPs and GO sheet separately. The H_2O_2 decomposition was demonstrated using ESR and generated $^{\cdot}O_2$ was measured using the dissolved oxygen electrode. Thus, GO-Fe_2O_3 nanohybrids were found to be effective pH-dependent mimetic dual-enzymes [83].

Co_3O_4 nanoparticles-crumpled graphene microspheres (CGM) were evaluated for their potential as enzyme mimics for peroxidase enzyme for the detection of ascorbic acid. The synthesized nanohybrid possessed superior catalytic activity than Co_3O_4/CGM due to the increase in specific surface, which led to more adsorption of TMB and H_2O_2 and rapid electron transfer between the substrate and nanohybrid. The presence of ascorbic acid tempted two electron-reduction of ox-TMB to TMB with fading of blue color. The nanohybrid effective as peroxidase mimic in pH range 1.0–12.0, and temperature 15–80 °C and above 92.5% activity was retained even after one month storage, thus exhibiting long-term storage stability [84].

Immobilization of metalloporohyrin, polypyrrole, and hemin aggregates on GO has led to increased activity of GO. Immobilization was accomplished by π-π stacking interactions. Enzyme mimic activity of prolin tailed metalloporphyrin nickel (II) with rGO was reported in dopamine oxidation. It offered selective interface via hydrogen bonds and biomimetic environment. Oxidation of dopamine followed two electron-two proton mechanisms. Resulted nanohybrid increased the electron mobility due to delocalization of electrons, and rGO increased the effective surface area for redox reactions [85]. Also, hemin aggregates immobilized polyethylene glycol-modified GO (PEG-GO) was evaluated for peroxidase-like activity using TMB and ABTS as substrates in the presence of H_2O_2. The PEG-GO-hemin composite exhibited superior peroxidase-like activity compared to PEG-GO or hemin alone owing to the insertion of hemin enzyme active center with high dispersity. PEG-GO exhibited increased enzyme mimic activity due to tremendous affinity of composite for organic substrate through electrostatic interactions and rapid electron transfer facility. PEG-GO sheets showed superior peroxidase-like activity in combination with hemin containing proteins having hemoglobin and cytochrome c [86].

Peroxidase mimic activity of tetrapyridylporphyrin (TPyP) and its metal complexes of Cu and Mn with GO and thermally reduced GO revealed the noncovalent interaction between graphene and iron-porphyrin involved oxygen functionalities. The two iron-containing composites showed enhanced catalytic activity than each component individually. GO facilitated excellent support for graphene-based NCs with peroxidase mimic activity (K_m = 0.292 mM). The catalytic activity was observed even with very minute amounts of porphyrins (the TPyP:graphene ratio = 1:50). Lipophilic antioxidants, vitamin E, were also detected in a range of 10^{-5}–10^{-4} M, thus has potential application in enzyme free sensors [87].

The Fe_3O_4 NPs fabricated 3D porous graphene (3D GN) exhibited superior peroxidase mimic activity and high affinity toward substrate than 2D graphene-based nanocomposites. The synthesized composite also showed colorimetric detection of glucose with a low detection limit of 0.8 μM. Advanced peroxidase-like activity of graphene supported trimetallic Au-Pd-Fe_3O_4 nanoparticles than its bimetallic counterpart (Pd-Fe_3O_4) was also reported. These NPs oxidized TMB in the presence of H_2O_2 to the final blue product. The enhanced activity of Au-Pd-Fe_3O_4/rGO was attributed to the availability of more active sites and high surface area of graphene so that reactants could approach trimetallic NPs from both sides of the sheet. Thus, obtained composites were found to be efficient and reusable catalyst up to 5 reaction cycles for liquid phase reactions [88].

Nanohybrid Pt-on-Pd/rGO) were evaluated for peroxidase-like activity. Due to the synergistic effect between rGO and Pt-on-Pd, the nanohybrid exhibited better peroxidase-like activity in TMB oxidation and H_2O_2 reduction. The nanofabricated sensor showed a linear range from 0.98 to 130.7 μM of H_2O_2. Electrocatalytic oxidation of methanol was also carried out using Pt-on-Pd/rGO, and the obtained peak current density value, (j_f = 328 mA mg^{-1} Pt) was about 1.85 folds superior to that of commercial Pt black (j_f = 177 mA mg^{-1} Pt). Thus, dual-functional Pt-on-Pd/rGO with enhanced electrocatalytic oxidation potential and peroxidase mimic activity was obtained [89].

Copper oxide, Cu^{2+} ions were used for modifying GO for the enhancement of enzyme mimic activity. For colorimetric detection of cholesterol, CuO:graphene nanosphere (CuO:GNS) hybrid exhibiting peroxidase-like activity for detection of H_2O_2 was reported. The proposed mechanism for the working of sensor involved detection of H_2O_2 produced during free cholesterol oxidation, in the presence of enzyme cholesterol oxidase. The nanohybrid sensor exhibited excellent sensitivity for the detection of cholesterol and displayed a linear calibration curve over the range of 0.1–0.8 mM with LOD up to 78 μM. This nanofabricated sensor also showed sensitivity toward very low concentration of H_2O_2 (0.01–0.1 mM) and LOD of 6.88 μM. The nanohybrid showed superior cytocompatibility than bare CuO

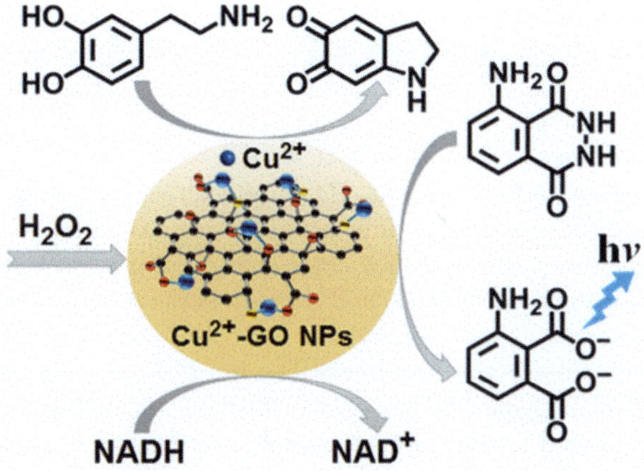

Fig. 6 Peroxidase mimic activity of CuO:graphene nanosphere. Reprinted from Wang et al. [91], with permission from American Chemical Society. Copyright (2017)

[90]. Cu^{2+}-GO NPs were evaluated for their potential as heterogeneous catalyst for horseradish peroxidase and reduced nicotinamide adenine dinucleotide (NADH) peroxidase mimic. These NPs catalyzed the oxidation of dopamine to aminochrome using H_2O_2, and the production of chemiluminescence using luminol and H_2O_2 (Fig. 6) [91].

Cu-hemin metal-organic-frameworks (MOFs)/chitosan (CS)-rGO nanohybrid exhibiting peroxidase-like bioactivity in electrochemical H_2O_2 sensor has been reported. After the Cu-hemin MOFs formation on the CS-rGO surface, Cu-hemin MOFs size decreased and electrical conductivity of the nanohybrid was improved as compared to that of Cu-hemin MOFs. Thus, electrochemical activity toward the reduction of H_2O_2 was found to be superior to some reported enzyme mimics. The nanohybrid exhibited a wide linear range of 0.065–410 μM with an LOD of 0.019 μM [92].

Graphene quantum dots are carbon-based nanomaterials that have attracted researchers for their unique physico-chemical properties. They have size less than 30 nm and high surface area, biocompatibility, improved solubility, and stable fluorescence, due to which they are used as catalysts in various reactions. The nanocomposites of graphene quantum dots-copper oxide (GQDs/CuO) also acted as effective peroxidase mimics by catalyzing the oxidation of TMB in the presence of H_2O_2 to produce blue-colored solution due to the generation of ·OH radical. The NCs demonstrated exact detection of H_2O_2 within a wide range of 0.5–10 μM, with minimum detection limit of 0.17 μM. A colorimetric detection for serum glucose detection was also performed by combining the GQDs/CuO catalytic reaction and oxidation of glucose using glucose oxidase. This method owned advantages due to simplicity, efficient sensitivity, and selectivity for glucose detection within linear range from 2 to 100 μM with LOD of 0.59 μM [93].

Graphene quantum dots of nearly uniform size (~ 5 nm) obtained from wood charcoal were prepared, and their potential as a peroxidase enzyme mimic was inspected for colorimetric detection of glucose and H_2O_2. The GQDs showed sensitive and instant detection of glucose with minimum detection limit of 0.006 mM for wide linear range of 0.01–0.6 mM. The observed lower value of K_m (0.012 mM) and the higher V_{max} (7.2×10^{-7} M s^{-1}) value supported improved peroxidase-like activity of E-GQDs. The significant enzyme mimic activity may be useful in developing colorimetric biosensors [94].

Fe_3C NPs decorated, 3D porous network of N-rich graphene (NGr) was evaluated for peroxidase mimic activity using oxidation of TMB in the presence of H_2O_2. Rapid and distinguished color change with improved stability in exposure to high concentration of H_2O_2 and high temperature was observed. The Michaelis–Menten kinetics was successfully followed over a wide concentration range of 2.0–500.0 μM with a detection limit lower than most of the other similar systems. The reliability of the enzyme free sensor was further evaluated by practical monitoring of glucose in diluted serum samples. Simple preparation, cost-effectiveness, rapid and distinguished color change, high tolerance to high temperature and H_2O_2 concentration, proved Fe_3C/NGr as promising candidates for fast naked eye sensor [95].

The hybridization of GO nanosheets and lysozyme nanofibrils through electrostatic interactions lead to the formation of amyloid-GO composite. Further, amyloid-GO composite was immobilized using high and tunable amount of Au NPs while preserving their high catalytic activity. Enzyme immobilized surfaces of amyloid-GO exhibited high electrochemical behavior, and also served as colorimetric sensor for sensitive and selective detection of glucose. Thus, synthesized nano-catalyst with high aspect ratios (> 10^3) provided an unprecedented possibility for developing biomimetic catalyst for a wide range of applications in biotechnology [96].

Poly(sodium styrene sulfonate)-functionalized graphene nanosheets (PSS-GN) could mimic peroxidase activity, which may be due to the stronger binding between substrate TMB and negatively charged PSS. The PSS-GN NCs could detect glucose and H_2O_2 by colorimetry in the range of 0.005–1 mM and 0.006–0.4 mM, respectively [97]. For enzymeless determination of glucose, graphene oxide-molecular imprinted Polymer (GO-MIP)-based electrochemical sensor has been reported. The proposed electrode exhibited superior electrocatalytic activity for glucose oxidation at optimized conditions with minimum detection limit of 0.1 nM within ~ 2 min. The electrochemical response of GO-MIP-based sensor was linearly proportional to the glucose concentration. The results acquired from GO-MIP-based electrodes in human blood were comparable to commercially available glucose monitors [98].

Selective and ultrasensitive detection of dopamine using noncovalent nanocomposite of poly-copper-2-amino-5-mercapto-1, 3, 4-thiadiazole/rGO (rGO-poly(Cu-AMT)) prepared through π-π stacking interaction has been reported. The rGO-poly(Cu-AMT) nanocomposite showed significant enzyme mimic catalytic activity due to the significant promotion of the electron transfer between the substrate and graphene-based carbon materials. The synergistic electrocatalytic effect between rGO sheet and poly(Cu-AMT) also induced enzyme mimic activity. The biomimic sensor exhibited linear calibration curve over a wide range of 0.01–40 μM and a minimum detection limit of 3.48 nM [99].

Also, rGO/MoS_2 composites and Fe_3O_4NPs nanozymes for synergistic catalysis of electrochemical circulating tumor cells (CTCs) detection has been developed.

The magnetic Fe_3O_4NPs performed the function of both separation and enrichment CTCs, as well as enzyme mimics along with rGO/MoS_2 for synergistic catalysis favoring signal amplification in cytosensors. The developed sensor was selective toward MCF-7 cancer cell detection over the other cancer cells such as other cancer cells: SK-BR-3 cells, MDA-MB-231 cells, H460 cells, and A549 cells. The sensor also showed reproducible potential up to 5 times at cell concentration 20 cell ml^{-1} with accuracy and precision [100].

Another GO composite containing 5, 10, 15, 20-tetrakis (4-carboxylpheyl)-porphyrin modified $ZnFe_2O_4$NPs functionalized on rGO (Por-$ZnFe_2O_4$/rGO) demonstrated peroxidase-like activity through colorimetric detection by catalyzing oxidation of TMB. Comparison indicated that Por-$ZnFe_2O_4$/rGO possessed highest enzyme-like activity over $ZnFe_2O_4$NPs, Por-$ZnFe_2O_4$, and $ZnFe_2O_4$/rGO. Hydroxyl radicals were identified as major reactive species responsible for activity using ESR. The Por-$ZnFe_2O_4$/rGO showed colorimetric detection of H_2O_2 in the range of 0.7–30 μM with the LOD 0.54 μM. Additionally, Por-$ZnFe_2O_4$/rGO-based colorimetric sensor showed linear response to glutathione within the range 2–40 μM with a LOD of 0.76 μM [101].

Recently, GO has been explored as a nuclease mimicking nanozyme for cleavage of DNA in the presence of UV or blue light due to its excellent adsorption potential for DNA. The pH of reaction apparatus affected the adsorption, and thus cleavage process. Both GO and DNA possess negative charge, and thus adsorption is not favorable. However, at lower pH due to protonation of carboxylic group of GO, higher adsorption and thus cleavage was achieved. In case of increasing the carbon length, highest cleavage was recorded for FAM-C_{10} and decreased afterward. Also, well-defined products were recorded even in the absence of metal ion. Thus, GO acted as a potential candidate for nanozyme to be used for substrate adsorption, reaction, and product desorption steps [102].

Recently, Mn_3O_4 NPs, GO, and their NC were exploited for their potential as PPO mimic using catechol as a substrate. Among all, highest PPO mimic activity was recorded for Mn_3O_4 NPs followed by NC. The introduction of GO in NC resulted into less reduction of Mn^{3+} into Mn^{2+} ions which caused lower activity of NC than bare Mn_3O_4 NPs. On the contrary, kinetic studies indicated that NC possessed higher binding affinity in comparison to Mn_3O_4NPs with K_m values of 0.93 mM and 0.7 mM, respectively [103].

Haloperoxidase enzyme is known for the catalytic oxidation of Cl^- by H_2O_2. Double quantum dot composites containing GO and CuO (GOQD-q-CuO) have been reported to possess halo-peroxidase mimic activity using TMB as substrate. The enzymeless selective and sensitive colorimetric sensor based on GOQD-q-CuO for detection of H_2O_2 and glucose possessed the detection limits of 0.5 and 2.5 μM with linear ranges 2.5–200 and 5–400 μM, respectively. In addition, GOQD reversed the zeta potential of q-CuO due to the inclusion of negative charge and thus altered its catalytic potential. Hence, GOQD-q-CuO recorded higher affinity than bare CuO toward positively charged TMB. The presence of Cl^- fastened electron transfer process for reduction of Cu^{2+} to Cu^+ on the GOQD-q-CuO surface, thus ascertains higher detection sensitivity [104].

The enzyme mimic activity of different ferrites and GO-based composites is summarized in Table 1.

5 Photocatalytic Activity of GO/Ferrite NPs and Their Nanocomposites

Photocatalytic properties of semiconductors are influenced by their electronic structure and band gap, i.e. difference between valence band and conduction band. Electron-hole pair formation is influenced by excitation of electrons by incident light. Lowering of band gap is a key factor that prevents electron-hole pair recombination. A semiconductor with band gap E_g (in eV) can utilize incident radiation with wavelength (jn nm) less than $1240/E_g$. Application of photocatalysis in degradation of organic pollutants including dyes, pharmaceuticals, and pesticides, is an emerging area of research. Photocatalyzed reduction and oxidation of variety of organic compounds using ferrite NPs, graphene oxide, and their composites are widely studied. Band gap of ferrite NPs can be tuned either by doping or by making their composites with a variety of materials. Research work in the field of photocatalytic degradation of organic pollutants using these materials is summarized in this section.

Magnetically separable MFe_2O_4 (M = Co, Ni and Fe)-supported TiO_2 nanohybrids were synthesized using co-precipitation method and evaluated as photocatalyst against methylene blue MB) dye. Catalytic efficiency was found in the order $TiO_2/CoFe_2O_4$ (Spinel) > $TiO_2/NiFe_2O_4$ (Inverse Spinel) > TiO_2/Fe_3O_4 (Inverse Spinel). Substitution using TiO_2 created more distortion in case of $CoFe_2O_4$ than other ferrites structures. Thus, either the distribution of cations in ferrites or the formation of hetero-junction at the interface over $TiO_2/CoFe_2O_4$ significantly increases their photocatalytic potential. The reaction followed first-order kinetic equation (Eq. 7). The band gap of TiO_2 was 3.10 eV which was lowered up to 2.72 eV for TiO_2/Fe_3O_4 indicated more degradation efficiency of nanocomposites [105]. However, the addition of $CoFe_2O_4$ and Fe_2O_3 to TiO_2 (TCF) did not significantly increase the degradation of MB. Under UV illumination, 99% removal was achieved with bare TiO_2 in 60 min in comparison to TCF, which showed 98% removal in the same time period. However, under visible-light radiation, 77% and 68% removal were achieved with bare TiO_2 and TCF, respectively. Higher decomposition using bare TiO_2 over TCF may be due to the smaller band gap and large crystal size of TiO_2 as compared to TCF [106].

$$R = k_{obs}[C] \tag{7}$$

R is rate of reaction, k_{obs} is observed rate constant, [C] is concentration of MB molecules in solution

$ZnFe_2O_4$ and its various composites were reported for photocatalysis of methyl orange (MO) dye. Photocatalytic degradation of Orange II was performed using

Table 1 Application of ferrite NPs, GO, and their nanocomposites as enzyme mimics

Serial no.	Nanocomposite	Application as enzyme mimic	References
1	MFe_2O_4 (M = Mg, Ni, Cu)	Peroxidase and catalase	[72]
2	$ZnFe_2O_4$ NPs/chitosan modified Au electrode	Peroxidase	[73]
3	Carbon electrodes modified Fe_3O_4	Peroxidase	[74]
4	$Co_xNi_{1-x}Fe_2O_4$ modified CPE	Peroxidase	[75]
5	$NiCo_2O_4$ NPs fabricated three-dimensional graphene foam	Detections for calcium ion and glucose	[76]
6	Fe_3O_4 using phosphate containing analogs	Peroxidase	[77]
7	CTAB@Fe_2O_3 NPs	Peroxidase	[78]
8	SDBS-Cu-$CuFe_2O_4$ NPs	Peroxidase	[79]
9	MFe_2O_4 (M = Mn, Co and Cu)	Superoxide dismutase	[80]
10	$Co_{0.25}Zn_{0.75}Fe_2O_4$ NPs	Peroxidase	[81]
11	TClPP-$CuFe_2O_4/SiO_2$ NC	Peroxidase	[82]
12	GO-Fe_2O_3	Peroxidase	[83]
13	Co_3O_4-crumpled graphene microspheres (CGM)	Peroxidase	[84]
14	Metalloporohyrin, polypyrrole, and hemin aggregates immobilized GO	Oxidation of dopamine	[85]
15	Hemin aggregates immobilized PEG-GO	Peroxidase	[86]
16	Metal complexes of Tetrapyridylporphyrin-rGO	Peroxidase	[87]
17	Fe_3O_4 Nanoparticle Loaded 3D Porous Graphene	Peroxidase	[88]
18	Pt-on-Pd/rGO	Peroxidase	[89]
19	CuO:Graphene nanosphere	Peroxidase	[90]
20	Cu^{2+}-GO NPs	Peroxidase	[91]
21	Cu-hemin /Chitosan-rGO	Peroxidase	[92]
22	Graphene quantum dots-CuO	Peroxidase	[93]
23	Graphene quantum dots from wood charcoal	Peroxidase	[94]
24	Fe_3C-N-rich graphene	Peroxidase	[95]
25	Au immobilized Amyloid-GO	Glucose oxidase	[96]
26	PSS-GN	Peroxidase	[97]
27	GO-Molecular Imprinted Polymer	Glucose oxidase	[98]
28	Poly-copper-2-amino-5-mercapto-1,3,4-thiadiazole/rGO	Oxidation of dopamine	[99]
29	rGO/MoS_2 composites and Fe_3O_4NPs	Tumor cell detection	[100]
30	Por-$ZnFe_2O_4$/rGO	Peroxidase	[101]
31	GO	Nuclease	[102]
32	Mn_3O_4-GO	PPO	[103]
33	GOQD-q-CuO	Haloperoxidase	[104]

$ZnFe_2O_4$ in the presence of light and persulfate ($S_2O_8^{2-}$). Decolorization efficiency for dye increased to 96.1% in the presence of $S_2O_8^{2-}$ ion. Optimal dose of catalyst was found to be 0.5 g/L and a further increase in dose leads to radical recombination that suppressed the degradation mechanism. Studies were explained on the basis of the presence of $OH^{\cdot}$ and $SO_4^{\cdot -}$ [107]. $ZnFe_2O_4$ without lectoferritin content (ZFO-0LF) was compared with $ZnFe_2O_4$ with lectoferritin content (ZFO-50LF) for their photocatalytic potential. ZFO-0LF was found to possess a larger specific surface area of (127.2 m^2/g) than ZFO-50LF (101.6 m^2/g), and also possessed higher absorption of MO dye, however ZFO-50LF exhibited smaller band gap, larger optical absorption coefficient, and higher photocatalytic degradation efficiency per unit area which led to faster degradation of MO dye as compared to ZFO-0LF [108].

The generation of p-n heterojunction may prove beneficial for charge separation efficiency to enhance photocatalytic potential. As reported hierarchical heterostructures of p-BiOCl/n-$ZnFe_2O_4$ (p-BiOCl/n-$ZnFe_2O_4$ H-Hs) showed superior photocatalytic activity under visible light irradiation for the degradation of Rhodamine B (RhB) over bare $ZnFe_2O_4$ nanofibres as confirmed by higher apparent first-order rate and its normalized constant. The results indicated the improved charge separation efficiency and more active surface sites in hierarchical structure, which led to enhanced photocatalytic performance. Due to the presence of magnetic components, p-BiOCl/n-$ZnFe_2O_4$ H-Hs were easily separated with the use of magnet [109].

Bioinspired synthesis of $ZnFe_2O_4$ NPs using *Monsoniaburkeana* plant extract was done and the resultant NPs possessed spherical as well as rod-like shapes. The synthesized NPs showed the 99.8% removal of MB dye within 45 min (pH = 12, optimum dosage = 25 mg). The NPs possessed reusability character with retainment of catalytic activity up to 5 cycles. Electrons were found to be major reactive species for photocatalytic degradation mechanism. Additionally, the synthesized NPs also showed photocatalytic degradation of sulfisoxazole in water up to 67%. Thus, biosynthesized $ZnFe_2O_4$ NPs may prove beneficial for treatment of water polluted with textile and pharmaceutically contaminants [110]

Photocatalytic potential of 10$ZnFe_2O_4$-90TiO_2 NPs on Rhodamine 6G degradation under visible light was also reported. The photocatalytic activity of NPs exhibited superior activity than TiO_2 and $ZnFe_2O_4$ due to improved values for specific surface area, band gap, and grain sizes necessary for the enhancement of the photocatalytic performances. NPs showed a maximum color removal of 97.87% within 10 min [111].

In another report, porous and hollow ternary Ag/ZnO/$ZnFe_2O_4$ nanocomposites (NCs) with varying Ag content were used for photocatalytic degradation of methylene blue dye. The nanocomposite with Ag content 2.17 wt.% showed preeminent degradation of dye. NC showed efficient capability for charge separation and electron transfer by type-II band alignment from Ag to ZnO Fermi energy level and formed Schottky barrier. $^{\cdot}O_2^-$ radical was found to be active intermediate during photodegradation process. Kinetic studies showed that the reaction followed first-order mechanics [112].

Comparative studies for photocatalyic potential of bare nickel zinc ferrite (NZF) and their ZnO nanorods-based flower-like NC have been reported for degradation of RhB dye. The hierarchical NC showed superior photocatalytic activity for the degradation of RhB under UV and visible light irradiation over NZF. The higher activity of NC due to addition of nanorods was due to the charge separation effect at the interface of two components of NC under Z-scheme system. Bare NZF also possessed photocatalytic potential due to the wide absorption range of NZF. However, in the absence of NZF, ZnO nanorods did not show any photocatalytic activity under light irradiation. Owing to the magnetic nature, NCs were easily separated from the solution, thus were designated as recyclable photocatalysts for removal of organic contaminants [113].

$BiFeO_3$ magnetic nanoparticles were used as visible light-driven photocatalyst coupled with ozonation for the degradation of oxalic acid (OA) and norfloxacin (NFX). The kinetic mechanism followed pseudo-first-order model and rate constants of degrading OA and NFX removal in O_3/Vis/$BiFeO_3$ were found to be 5.48 and 1.65 times as high as the sum of Vis/$BiFeO_3$ and O_3, respectively, (Eq. 8). Experimental results showed synergistic effect between photocatalysis and ozonation. As ozonation coupling suppressed electron-hole pair recombination by electron entrapment, thus increased number of hydroxyl radicals [114].

$$\ln\frac{C}{C_m} = -kt \tag{8}$$

Porous $CuFe_2O_4$ exhibited significant photocatalytic activity to peroxymonosulphate (PMS). More than 98% degradation of atrazine was observed within 15 min with 1 mM PMS and 0.1 g/L $CuFe_2O_4$. However, $CuFe_2O_4$ alone did not show any observable photodegradation to peroxodisulfate or H_2O_2. Experimentally, it was found that the catalytic degradation efficiency of atrazine was enhanced with the increase in PMS and $CuFe_2O_4$ doses, but decreased with the increase in natural organic matters concentration. The extent of degradation was observed maximum at pH = 9.5, thus surface hydroxyl groups in neutral state were accounted for PMS degradation Bicarbonate ions facilitated atrazine degradation by PMS/$CuFe_2O_4$ at low concentrations, but suppressed the degradation at high concentrations. $OH^{\cdot}$ and $SO_4^{\cdot-}$ were found to be reactive species for atrazine degradation in PMS/$CuFe_2O_4$ system [115].

Also, Cu and Zn doped $CoFe_2O_4$ NPs have been reported to exhibit photocatalytic potential against the degradation of MB dye. Bare $CoFe_2O_4$ NPs only showed the adsorption of MB and not the degradation, which was confirmed by variation of concentration by the adsorption/desorption over time. On the contrary, doped NPs showed up to 99% photocatalytic degradation of MB under visible light irradiation. The alteration in pH induced the variation in adsorption and photocatalytic activity, thus demonstrated the electrostatic attractions between adsorbate and adsorbent as a key factor for photocatalytic mechanism. In situ trapping experiments verified the holes and hydroxyl radicals as major species responsible for the dye's degradation. Besides, the magnetic properties of doped NPs resulted into the efficient separation of the photocatalyst with retainment of activity up to 4 cycles with low iron leaching [116].

Doping of ferrites with f-block elements, e.g. yttrium, cerium, lanthanum, and samarium offer high surface area, rapid electron transfer, and suppress electron-hole pair recombination, which enhanced the rate of photocatalytic degradation of various pollutants. Effect of doping of f-block elements on photocatalytic activity of ferrites has been discussed.

Effect of yttrium doping in $CoFe_2O_4$ on the photocatalytic activity for the degradation of H_2O_2 and potassium peroxymonosulphate ($KHSO_5$) has been observed. Surface area was maximum for the composite in the ratio $Y_{0.2}Co_{0.8}Fe_2O_4$. Various factors such as catalyst dose, pH of solution, H_2O_2 concentration, and light were found to influence Fenton's process. Degradation extent was highest at ferrite concentration 0.50 g/L, at neutral pH, H_2O_2 concentration of 8.8 mM, and the catalyst was found to be effective even after 4 runs. Rapid electron transfer between oxidant and catalyst was encouraged in the presence of light that led to generation of more active species. Main reactive species in the case of H_2O_2 was $OH^{\cdot}$, and in the case of $KHSO_5$ was $OH^{\cdot}$ and $SO_4^{\cdot -}$. Band gap of the sample was also found in the range of visible region [117].

Perovskite type Ce doped $SrFeO_3$ mixed oxides have been explored for their photocatalytic activity against Orange II and Rhodamine B (RhB) dyes. Dyes degraded after the adsorption of nanoparticles through electrostatic interactions. Their experimental studies favored the formation of $^{\cdot}O_2^-$ for RhB and $OH^{\cdot}$ for Orange II during the degradation mechanism. Orange II was degraded to higher extent than RhB. The reaction kinetics followed first-order mechanics [118].

Photocatalytic potential of graphitic carbon nitride (g-C_3N_4)-$CaFe_2O_4$ magnetic nanohybrid was compared with pure g-C_3N_4 and $CaFe_2O_4$ against MB dye. Nanohybrid exhibited improved photocatalytic activity than pure g-C_3N_4 and $CaFe_2O_4$ due to better separation efficiency of electron-hole pair, and rapid interface charge transfer, because of the resulted effective heteroconjugation. In the case of nanocomposite, electrons accumulated in the conduction band of $CaFe_2O_4$, and holes accumulated in valence band of g-C_3N_4 lead to the effective separation of electron–hole pairs, and enhanced photocatalytic activity. Positively charged holes governed photocatalytic mechanism [119].

Photocatalytic degradation of 2-propanol in gas–solid regime using $LaFe_{1-x}Cu_xO_{3-\delta}$ as catalyst has been reported. $LaFeO_3$ acted only as adsorbent, but was ineffective for the degradation, however $LaFe_{1-x}Cu_xO_{3-\delta}$ oxidized 2-propanol to propanone and finally to CO_2. The highest degradation was shown by metal oxide $LaFe_{.90}Cu_{0.10}O_3$. The photocatalytic activity attributed to electron entrapment by Cu (II) and oxygen vacancies suppressed recombination of electron-hole pair. Higher oxygen vacancies encouraged more adsorption of molecular O_2, thus more number of $OH^{\cdot}$ produced, which increased the degradation mechanism [120]. $Sr_{0.85}Ce_{0.15}FeO_3$ mixed oxides were evaluated for photocatalytic potential toward Orange II and RhB dye. Nanohybrid showed excellent degradation in the absence of light in the 55–80 °C temperature range. SF tempted the formation of hydroxyl radicals, both by irradiation and thermal activation. Degradation of Orange II was found to be dependent on $OH^{\cdot}$ and followed first-order kinetic equation and in case of RhB, its degradation was affected by the presence of a small amount of superoxide radicals and only adsorption phenomenon was observed. Adsorption was affected by electrostatic interaction between charged surface of SF and dyes. Thus, RhB (positively charged) adsorbed to more extent than Orange II (negatively charged) on SF (negatively charged) at the experimental pH [121].

Fe_3O_4 fabricated $BiVO_4$ nanocomposites were evaluated for their potential as visible light-driven photocatalyst against Acid Red B dye. The nanocomposite $BiVO_4$/30% Fe_3O_4 showed excellent photocatalytic activity due to effective separation of electron-hole pairs. Also, the nanocomposite was found to be effective for 5 runs and thus found to be reusable. The adsorption capacity of nanocomposites was also increased due to the increase in specific surface area. It was also reported that degradation percentage of dye was decreased with the increase in initial concentration of dye. The reaction followed first-order reaction kinetic equation. $^{\cdot}O_2^-$ radical was main radical participated in photocatalysis mechanism [122].

$BiOBr/CoFe_2O_4$ microspheres with varying ratio were synthesized and evaluated for their potential as novel photocatalyst against Congo Red dye by Jiang (2016). Experimental coercivity value of 232 Oe and saturation magnetization value of 33.79 emu g^{-1} indicated their easy separation and recovery from the reaction solution. The synthesized microsphere possessed 2.45 times higher value of initial rate constant that demonstrates excellent adsorption capacity and showed 90.78% degradation of dye. At lower reactant concentration, reaction followed the Langmuir–Hinshelwood kinetic model (Eq. 8). Using various radical scavengers H^+ and $^{\cdot}O_2^-$ were found to be the main active species [123].

Magnetic near IR photocatalyst of $Er^{3+}/Tm^{3+}/Yb^{3+}$-($CaF_2/ZnFe_2O_4$/ZnO) (ETY-FCZ) found to exhibit excellent photocatalytic performance for the degradation of MO and salicylic acid. $ZnFe_2O_4$ exhibited the heterojunction with ZnO, thus ETY-FCZ owned high electron-hole pair separation efficiency, thus resulted in higher removal rates of above-mentioned organic pollutants. Also, due to magnetic properties, ETY-FCZ can be separated efficiently using external magnetic field and reused after recycling, thus it has promising applications in the wastewater treatment [124].

Nanofabricated $Er^{3+}/Tm^{3+}/Yb^{3+}$-M-$Fe_3O_4@SiO_2/CaF_2/TiO_2/Ln_2Ti_2O_7$ (Ln = Er, Tm, or Yb) (ETY-FCTL) composites have been evaluated for their catalytic potential for the degradation of MO as compared to M-$Fe_3O_4@SiO_2/CaTiO_3$, Er^{3+}/Tm^{3+}/Yb^{3+}-M-$Fe_3O_4@SiO_2/CaTiO_3/TiO_2$ and $Er^{3+}/Tm^{3+}/Yb^{3+}$-M-$Fe_3O_4@SiO_2$/$CaF_2/Ln_2Ti_2O_7$. ETY-FCTL nanocomposite exhibited highest photocatalytic activity with the degradation of 99.0% as compared to others. The degradation mechanism followed pseudo-first-order kinetics with the maximum rate constant value 0.01931 min^{-1}. CaF_2 and lanthanides produced oxygen vacancies for electron trapping thus lead to efficient electron-hole pair separation [125].

Literature studies also reported the photodegradation of Acid Orange 95 (AO95), Acid Red 18 (AR18), and Direct Red 81 (DR81) dyes using polyoxometalate ($H_3PW_{12}O_{40}$) immobilized modified cobalt ferrite (MCF) [126]. Cobalt ferrite (CF) nanoparticles were silanized using (3-aminopropyl)trimethoxy silane to link methoxy group with hydroxyl group of CF. Experimental studies revealed that degradation does not depend on the catalyst dose and thus follows zero-order kinetic equation (Eq. 9). The catalyst was found to be reusable as it retained catalytic activity after

4 runs. With the increase in initial dye concentration, the extent of degradation decreased due to competitive adsorption. The maximum values of degradation were 90% (for AO95), 99% (for AR18), and 96% (for DR81).

$$C - C_0 = k_0 t \tag{9}$$

Reduced graphene oxide (rGO) is another candidate explored for photocatalytic degradation of organic pollutants. rGO was synthesized from GO by different reducing agents viz. hydrazine hydrate, sodium hydroxide, potassium hydroxide, etc. The presence of GO or its derivatives has been evidenced to alter surface properties as well as band gap of photocatalyst. In another study, the photocatalytic degradation of Remazol Black B using magnetic GO-MFe_2O_4 (M = Ni, Co, Fe) nanocomposites has been reported [54]. Degradation was found to be dependent on pH, amount of nanocomposite used, and contact time. Maximum removal was obtained at pH = 3, initial dye concentration = 10 ppm, and amount of nanocomposites 100 mg, due to more availability of surface area and active sites for adsorption. Pseudo-first-order kinetic model was found to be well agreement with observed values. The correlation coefficient of pseudo-first-order (Eq. 8) was maximum for Co (0.993) as compared to other metal ions [54].

Cu_2O fabricated rGO nanostructures were found to exhibit photocatalytic activity for the degradation of MO dye [127]. The degradation extent of dye in case of synthesized heterostructure was found to be superior to bare Cu_2O and rGO. Photoexcited electrons shifted from carboxylate π system to Cu^{2+} through ligand to metal charge transfer under UV irradiation which led to efficient charge separation. In this study also $^{\cdot}O_2^-$ radical was found to be involved. rGO/$MgFe_2O_4$ nanocomposite was evaluated as visible light-driven photocatalyst against methylene blue. Electrochemical impedance studies showed a significant decrease in charge transfer resistance in nanocomposite due to the presence of rGO which inhibited the recombination of electron-hole pair. $MgFe_2O_4$ worked as visible light absorber, and rGO facilitated rapid charge transfer. Adsorption of dye on the active centers of nanocomposites occurred due to electrostatic attraction. Owing to the magnetic property, the above-mentioned composite can be separated using external magnetic field and thus can be reused repeatedly [128].

The photocatalytic mechanism also varies with the content of graphene present in the composite. The effect of varying GO content in $Zn_{0.5}Cd_{0.5}S$-rGO NPs on photocatalytic performance has been studied. The NPs with rGO content of 0.5 wt.% exhibited highest degradation of MB, MO, and RhB dyes due to the uniform dispersion of NPs on rGO sheets and effective separation of photoinduced electrons and holes. $Zn_{0.5}Cd_{0.5}S$ due to aggregation showed reduced surface area, however rGO worked as transport channel between metal sulfide semiconductor, and prevented aggregation of $Zn_{0.5}Cd_{0.5}S$. Also, higher surface area of rGO facilitated higher adsorption capacity and more active sites for degradation mechanisms. Using $Zn_{0.5}Cd_{0.5}S$-rGO the observed removal of RhB was 80%, and that of MB was 98% [129].

Also, $CaFe_2O_4$ (CF)-fabricated graphene nanoplatelets $(CF)_{1-x}(GNPs)_x$ NC showed superior photocatalytic activity over bare $CaFe_2O_4$ and GNPs due to synergistic effects of both components under visible light illumination. Among different composition ratios, $(CF)_{0.75}(GNPs)_{0.25}$ possessed highest photocatalytic activity, but decreased on further increasing the GNPs content. Initially, on introducing the GNPs content the catalytic activity enhanced due to the presence of highly conducting e^- acceptor, which led to the efficient separation of e^--h^+ pair. On further increasing the content, the GNPs hindered the interaction of light with active sites of photocatalyst, thus suppresses the photocatalytic degradation of MB dye [130].

Three-dimensional graphene (3DG)-TiO_2 nanocomposites were evaluated for their photocatalytic potential in CO_2 reduction. 3DG-TiO_2 nanocomposites exhibited higher photocatalytic activity than bare P25 TiO_2 due to elevated dispersion of P25 TiO_2 nanoparticles on 3DG [131].

Photocatalytic efficiency of Ag-TiO_2/rGO nanohybrid for the removal of RhB dye, and reduction of CO_2 was evaluated. The photocatalytic and adsorbent capability of Ag-TiO_2/rGO nanohybrid was better than hollow TiO_2. The improved adsorbent efficiency was due to the large Brunauer–Emmett–Teller surface area. and increased photocatalytic activity attributed to the major properties of graphene and localized surface Plasmon (LSPR) effect of Ag nanoparticles. The photo-generated electrons could be transferred by graphene and accepted by Ag thus suppressing electron-hole pair recombination. Reaction kinetics followed first-order model (Eq. 7) for photocatalytic mechanism and the reported result was the removal of 99.5% of dyes [132]. N-doped $Bi_2O_2CO_3$/graphene quantum dots composite (N-BOC/GQDs) were also evaluated as visible light-driven photocatalysts for the removal of NO as compared to the bare N-BOC [133]. Charge separation efficiency of N-BOC/GQDs was improved for photocatalytic process, and superoxide radicals were found to be the active species during photocatalysis (Fig. 7). Also, during the process, the specific changes of NO_x-related species were noticed.

The decrease in band gap of ferrite $Ni_{0.65}Zn_{0.35}Fe_2O_4$ NPs has been observed upon inclusion of rGO from 1.91 eV to 1.84 eV, which resulted in superior photocatalytic

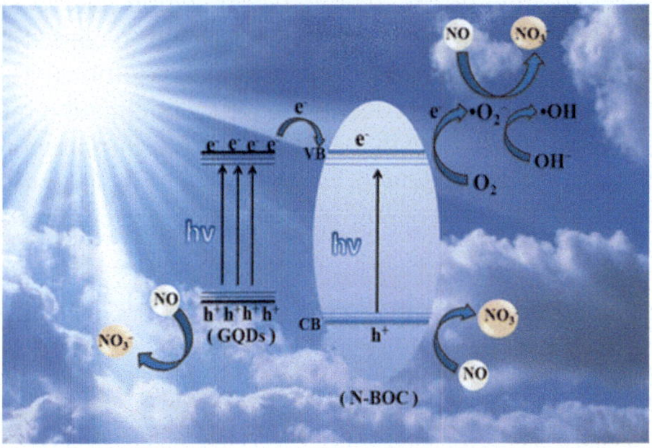

Fig. 7 Photocatalytic activity of N-doped $Bi_2O_2CO_3$/graphene quantum dots composite. Reprinted from Liu et al. [133], with permission from American Chemical Society. Copyright (2017)

efficiency of $Ni_{0.65}Zn_{0.35}Fe_2O_4$@rGO nanohybrids over pristine $Ni_{0.65}Zn_{0.35}Fe_2O_4$ NPs for degradation of MB. Using pristine $Ni_{0.65}Zn_{0.35}Fe_2O_4$ NPs, only 55% degradation was achieved, which increased by more than 90% using NC. The efficient separation of photo-generated electron-hole pair may be the possible reason behind these results. Additionally, magnetic nature of ferrite NPs favored their ease of recovery from aqueous dye solution by conventional magnets with retainment of photocatalytic activity up to 5 cycles [134].

Comparative studies on the photocatalytic potential of cobalt ferrite (CF), CF-rGO, and CF-carbon nanotube (CF-CNT) have been reported [135]. The inclusion of rGO and CNT to CF reduced the band gap from 1.55 to 1.50 and 1.45 eV, respectively. The recombination rate of photo-generated electron-hole pairs was more suppressed in the case of CNT as compared to rGO. The results of band gap and recombination rate were fairly related to photocatalytic degradation of MB dye. With bare CF, 38% degradation of MB was achieved under optimized conditions, which increased up to 58% and 97% with CF-rGO and CF-CNT, respectively. All the synthesized materials were reusable up to 3 runs and stable with Co leaching. Thus, carbon-based materials possessing different structures may also prove beneficial in altering the catalytic activity.

Another work, based on the photocatalytic performance of AgBr-CF/ZF-NG and their individual components against photocatalytic degradation of MG and MO dye have also been reported. Overall, the photocatalytic activity against both dyes was maximum in the case of AgBr/CF/NG, which was comparable to AgBr/ZF/NG. The degradation using bare components as catalyst was lower than these hybrids. The formation of p-n junction between p-type ferrite semiconductors and n-type AgBr contributed to the higher activity of hybrid over the sole components. However, during recycle experiments, higher activity was retained in AgBr/ZF/NG followed by AgBr/CF/ after ten runs. In situ trapping experiments verified the superoxide radical and hydroxyl radical as major reactive species responsible for MO and MG degradation [136].

The $Ni_{0.96}Cd_{0.04}Gd_{0.04}Fe_{1.96}O_4$ and their rGO NC (NCGF/rGO) have been compared for photocatalytic potential. The heterojunction-based NCGF/rGO photocatalyst achieved 92.27% MB degradation and 53.18% Rh–B degradation under visible light. On the contrary, bare NCGF recorded 20.25% MB and 11.93% Rh–B dye degradation under the identical conditions. The superior activity of NCGF/rGO was due to the reduction in the recombination rate of electron–hole pair due to the inclusion of rGO in the NCGF matrix. Furthermore, the NCGF/rGO hybrid was found to be reusable with retained 92% of its catalytic after 5 runs [137].

The applications of ferrites and GO-based composites are summarized in Table 2.

Table 2 Application of ferrite NPs, GO, and their nanocomposites as photocatalyst

Serial no.	Nano-composites	Organic contaminant	References
1	TiO_2-MFe_2O_4 (M = Co, Ni and Fe)	MB	[105]
2	TCF	MB	[106]
3	$ZnFe_2O_4$	Orange II	[107]
4	Lacto-ferritin assisted ZFO	MO	[108]
5	p-BiOCl/n-$ZnFe_2O_4$	RhB	[109]
6	$ZnFe_2O_4$ NPs	MB, sulfisoxazole	[110]
7	$10ZnFe_2O_4$-$90TiO_2$	Rh 6G	[111]
8	Ag/ZnO/$ZnFe_2O_4$	MB	[112]
9	NZF-ZnO nanorods	RhB	[113]
10	O_3/Vis/$BiFeO_3$	Oxalic acid and norfloxacin	[114]
11	Porous $CuFe_2O_4$	Atrazine	[115]
12	Cu and Zn doped $CoFe_2O_4$	MB	[116]
13	$Y_{0.2}Co_{0.8}Fe_2O_4$	H_2O_2 and $KHSO_5$	[117]
14	Ce doped $SrFeO_3$	Orange II and RhB	[118]
15	(g-C_3N_4)-$CaFe_2O_4$	MB	[119]
16	$LaFe_{1-x}Cu_xO_{3-\delta}$	2-propanol	[120]
17	$Sr_{0.85}Ce_{0.15}FeO_3$	Orange II and RhB	[121]
18	Fe_3O_4 fabricated $BiVO_4$	Acid Red B	[122]
19	BiOBr/$CoFe_2O_4$ microspheres	CR	[123]
20	Er^{3+}/Tm^{3+}/Yb^{3+}-(CaF_2/$ZnFe_2O_4$/ZnO)	MB and salicylic acid	[124]
21	Er^{3+}/Tm^{3+}/Yb^{3+}-M-Fe_3O_4@SiO_2/CaF_2/TiO_2/$Ln_2Ti_2O_7$	MO	[125]
22	$H_3PW_{12}O_{40}$ immobalized $CoFe_2O_4$	Acid Orange 95, Acid Red 18, and Direct Red 81	[126]
23	GO-MFe_2O_4 (M = Ni, Co, Fe)	Remazol Black B	[127]
24	Cu_2O fabricated rGO	MO	[55]
25	rGO/$MgFe_2O_4$	MB	[128]
26	ZnCdS-rGO	MB, Mo and RhB	[129]
27	$(CF)_{1-x}(GNPs)_x$	MB	[130]
28	3DG–TiO_2 nanocomposite	CO_2	[131]
29	Ag-TiO_2/rGO	RhB and CO_2	[132]
30	Nitrogen-doped graphene	Phenol	[133]
31	$Ni_{0.65}Zn_{0.35}Fe_2O_4$@rGO	MB	[134]
32	CF, CF-rGO, CF-CNT	MB	[135]
33	AgBr-CF/ZF-NG	MG and MO	[136]
34	NCGF/rGO	MB and RhB	[137]

6 Conclusion and Future Perspectives

Ferrites, graphene oxide, and their nanocomposites have been extensively used in the field of photocatalysis and as enzyme mimics. This chapter highlights an overview on the synthesis of ferrite NPs, GO, and their nanocomposites followed by their applications as enzyme mimics and photocatalysts for the degradation of organic contaminants. Literature review revealed that the photocatalytic activity is influenced by the surface area and band gap of the composite, whereas, their potential as enzyme mimic mainly depends on the surface interactions and surface area of contact. For the future perspectives, the following points should be addressed for enhancing the efficiency as enzyme mimics and photocatalysts:

- For better enzyme mimic activity, composites with high surface area and superior electron transfer should be synthesized and tested on real life samples.
- Literature is available on ferrites and graphene oxide separately as enzyme mimics. However, scarce work is available on enzyme mimic activity of graphene-ferrite composites.
- Photocatalysis studies using pure/binary doped ferrite NPs and their GO composites have been studied in detail, however, scanty information is available on the composites of ternary doped ferrite NPs and heteroatom-doped GO.
- The degradation products of organic contaminants should be studied for their non-toxicity for future applications in treatment plants.

References

1. S.P. Dalawai, S. Kumar, M.A.S. Aly, Md.Z.H. Khan, R. Xing, P.N. Vasambekar, S. Liu, A review of spinel-type of ferrite thick film technology: fabrication and application. J. Mater. Sci. Mater. Electron. **30**, 7752–7779 (2019). https://doi.org/10.1007/s10854-019-01092-8
2. B.I. Kharisov, H.V.R. Dias, O.V. Kharissova, Mini-review: ferrite nanoparticles in the catalysis. Arab. J. Chem. **12**(7), 1234–1246 (2019). https://doi.org/10.1016/j.arabjc.2014.10.049
3. M. Kumar, H.S. Dosanjh, Sonika, J. Singh, K. Monir, H. Singh, Review on magnetic nanoferrites and their composites as alternatives in waste water treatment: Synthesis, modifications and applications. Environ. Sci. Water Res. Technol. **6**, 491–514 (2020). https://doi.org/10.1039/C9EW00858F
4. L. Mao, H. Liu, S. Liu, Q. Ba, H. Wang, L. Gao, X. Lia, C. Huang, W. Chen, Pt-Ru bi-metal co-catalyst: Preparation, characterization and its effect on CdS's activity for water splitting under visible light. Mater. Res. Bull. **93**, 9–15 (2017). https://doi.org/10.1016/j.materresbull.2017.03.035
5. T.M. Suzuki, A. Iwase, H. Tanaka, S. Sato, A. Kudo, T. Morikawa, Z-scheme water splitting under visible light irradiation over powdered metal-complex/semiconductor hybrid photocatalysts mediated by reduced graphene oxide. J. Mater. Chem. A. **3**, 13283–13290 (2015). https://doi.org/10.1039/C5TA02045J
6. J. Yu, J. Jin, B. Chenga, M. Jaroniec, A noble metal-free reduced graphene oxide–CdS nanorod composite for the enhanced visible-light photocatalytic reduction of CO_2 to solar fuel. J. Mater. Chem. A. **2**, 3407–3416 (2014). https://doi.org/10.1039/C3TA14493C

7. K.S. Novoselov, A.K. Geim, S.V. Morozov, D. Jiang, Y. Zhang, S.V. Dubonos, I.V. Grigorieval, A.A. Firsov, Electric field effect in atomically thin carbon films. Science **306**, 666–669 (2004). https://doi.org/10.1126/science.1102896
8. T. Sattar, Current review on synthesis, composites and multifunctional properties of graphene. Top. Curr. Chem. (Z). **377**, 10 (2019). https://doi.org/10.1007/s41061-019-0235-6
9. R.G. Shrestha, L.K. Shrestha, K. Ariga, Carbon nanoarchitectonics for energy and related applications. C. **7**(4), 73 (2021). https://doi.org/10.3390/c7040073
10. P. Singh, P. Shandilya, P. Raizada, A. Sudhaik, A. Rahmani-Sani, A. Hosseini-Bandegharaei, Review on various strategies for enhancing photocatalytic activity of graphene-based nanocomposites for water purification. Arab. J. Chem. **13**(1), 3498–3520 (2020). https://doi.org/10.1016/j.arabjc.2018.12.001
11. R. Singh, M. Kumar, A. Mittal, P.K. Mehta microbial enzymes: industrial progress in 21st century. 3 Biotech **6**(2), 1–15 (2016). https://doi.org/10.1007/s13205-016-0485-8
12. R.M. Daniel, M. Dines, H.H. Petach, The denaturation and degradation of stable enzymes at high temperatures. Biochem. J. **317**(1), 1–11 (1996). https://doi.org/10.1042/bj3170001
13. M.E. Peterson, R.M. Daniel, M.J. Danson, R. Eisenthal, The dependence of enzyme activity on temperature: determination and validation of parameters. Biochem. J. **402**(2), 331–337 (2007). https://doi.org/10.1042/BJ20061143
14. T.P. Silverstein, Falling enzyme activity as temperature rises: negative activation energy or denaturation. J. Chem. Educ. **89**(9), 1097–1099 (2012). https://doi.org/10.1021/ed200497r
15. V. Sharma, M. Bachwani, Artificial enzymes: a review. Curr. Enzym. Inhib. **7**(3), 178–189 (2011). https://doi.org/10.2174/157340811798807623
16. H.Y. Shin, T.J. Park, M.I. Kim, Recent research trends and future prospects in nanozymes. J. Nanomater. 1–11 (2015). https://doi.org/10.1155/2015/756278
17. L. Gao, K. Fan, X. Yan, Iron oxide nanozyme: a multifunctional enzyme mimetic for biomedical applications. Theranostics **7**(13), 3207–3322 (2017). https://doi.org/10.7150/thno.19738
18. J. Bjerre, C. Rousseau, L. Marinescu, M. Bols, Artificial enzymes, "Chemzymes": current state and perspectives. Appl. Microbiol. Biotechnol. **81**, 1–11 (2008). https://doi.org/10.1007/s00253-008-1653-5
19. M. Raynal, P. Ballester, A. Vidal-Ferranab, P.W.N.M. Leeuwen, Supramolecular catalysis. Part 2: Artificial enzyme mimics. Chem. Soc. Rev. **43**, 1734–1787 (2014). https://doi.org/10.1039/C3CS60037H
20. B. Garg, T. Bisht, Y.C. Ling, Graphene-based nanomaterials as efficient peroxidase mimetic catalysts for biosensing applications: an overview. Molecules **20**, 14155–14190 (2015). https://doi.org/10.3390/molecules200814155
21. E. Kuah, S. Toh, J. Yee, Q. Ma, Z. Gao, Enzyme mimics: advances and applications. Chemistry **22**(25), 8404–8430 (2016). https://doi.org/10.1002/chem.201504394
22. S.C. Ameta, in *Advanced oxidation processes for waste water treatment*, ed. by S.C. Ameta, R. Ameta. (Academic Press, Elsevier Inc, 2018), pp.1–12
23. M.F. Sanad, A.E. Shalan, S.M. Bazid, S.M. Abdelbasir, Pollutant degradation of different organic dyes using the photocatalytic activity of ZnO@ZnS nanocomposite materials. J. Environ. Chem. Eng. **6**, 3981–3990 (2018). https://doi.org/10.1016/j.jece.2018.05.035
24. J. Gomes, J. Lincho, E. Domingues, R.M. Quinta-Ferreira, R.C. Martins, N–TiO_2 Photocatalysts: a review of their characteristics and capacity for emerging contaminants removal. Water **11**(2), 373 (2019). https://doi.org/10.3390/w11020373
25. B. Bakbolat, C. Daulbayev, F. Sultanov, R. Beissenov, A. Umirzakov, A. Mereke, A. Bekbaev, I. Chuprakov, Recent developments of TiO_2-based photocatalysis in the hydrogen evolution and photodegradation: a review. Nanomaterials. **10**(9), 1790 (2020). https://doi.org/10.3390/nano10091790
26. N. Hashim, Z. Muda, M.Z. Hussein, I.M. Isa, A. Mohamed, A. Kamari, S.A. Bakar, M. Mamat, A.M. Jaafar, A brief review on recent graphene oxide-based material nanocomposites: synthesis and applications. J. Mater. Environ. Sci. **7**(9), 3225–3243 (2016)

27. K. Thakur, B. Kandasubramanian, Graphene and graphene oxide-based composites for removal of organic pollutants: a review. J. Chem. Eng. Data **64**(3), 833–867 (2019). https://doi.org/10.1021/acs.jced.8b01057
28. M.J. Jacinto, L.F. Ferreira, V.C. Silva, Magnetic materials for photocatalytic applications—a review. J. Sol-Gel Sci. Technol. **96**, 1–14 (2020). https://doi.org/10.1007/s10971-020-05333-9
29. M.K. Ahmed, A.E. Shalan, M. Afifi, M.M. El-Desoky, S. Lanceros-Méndez, Silver-doped cadmium selenide/graphene oxide-filled cellulose acetate nanocomposites for photocatalytic degradation of malachite green toward wastewater treatment. ACS Omega **6**(36), 23129–23138 (2021). https://doi.org/10.1021/acsomega.1c02667
30. Y.F. Tan, P. Chandrasekharan, D. Maity, C.X. Yong, K.-H. Chuang, Y. Zhao, S. Wang, J. Ding, S.S. Feng, Multimodal tumor imaging by iron oxides and quantum dots formulated in poly (lactic acid)-d-alpha-tocopheryl polyethylene glycol 1000 succinate NPs. Biomater. **32**, 2969–2978 (2011). https://doi.org/10.1016/j.biomaterials.2010.12.055
31. F. Chen, S. Li, Q. Chen, X. Zheng, P. Liu, S. Fang, 3D graphene aerogels-supported Ag and Ag@Ag3PO4 heterostructure for the efficient adsorption-photocatalysis capture of different dye pollutants in water. Mat. Res. Bull. **105**, 334–341 (2018). https://doi.org/10.1016/j.materresbull.2018.05.013
32. A. Akbarzadeh, M. Samiei, S. Davaran, Magnetic nanoparticles: preparation, physical properties and applications in biomedicine. Nanoscale Res. Lett. **7**(1), 1–13 (2012). https://doi.org/10.1186/1556-276X-7-144
33. S. Majidi, F.Z. Sehrig, S.M. Farkhani, M.S. Goloujeh, A. Akbarzadeh, Current methods for synthesis of magnetic nanoparticles. Artif. Cells. Nanomed. Biotechnol. **44**(2), 722–734 (2016). https://doi.org/10.3109/21691401.2014.982802
34. M. Kaur, N. Kaur, Vibha, in *Ferrites: Synthesis and Applications for Environmental Remediation.* ed. by V.K. Sharma, R. Doong, H. Kim, R.S. Varma, D.D. Dionysion. Ferrites and Ferrates: Chemistry and Applications in Sustainable Energy and Environmental Remediation. ACS Symposium Series (American Chemical Society, Washington DC), pp. 113–136. https://doi.org/10.1021/bk-2016-1238.ch004
35. N. Salah, S.S. Habib, Z.H. Khan, A. Memic, A. Azam, E. Alarfaj, N. Zahed, S. Al-Hamedi, High-energy ball milling technique for ZnO nanoparticles as antibacterial material. Int. J. Nanomed. **6**, 863–869 (2011). https://doi.org/10.2147/IJN.S18267
36. R. D'Amato, M. Falconieri, S. Gagliardi, E. Popovici, E. Serra, G. Terranova, E. Borsella, Synthesis of ceramic nanoparticles by laser pyrolysis: from research to applications. J. Anal. Appl. Pyrolysis **104**, 461–469 (2013). https://doi.org/10.1016/j.jaap.2013.05.026
37. M.V. Antisari, R. Marazzi, R. Krsmanovic, Synthesis of multiwall carbon nanotubes by electric arc discharge in liquid environments. Carbon **41**(12), 2393–2401 (2003). https://doi.org/10.1016/S0008-6223(03)00297-5
38. S. Sawai, Y. Nakahara, N. Matsumoto, J. Choi, T. Kato, M. Kawaguchi, Synthesis and characterization of carbon nanoparticle films prepared by plasma-based ion implantation. Surf. Interfac. Anal. **46**(10–11), 961–965 (2014). https://doi.org/10.1002/sia.5457
39. J. Kaur, M. Kaur, M.K. Ubhi, N. Kaur, J.-M. Greneche, Composition optimization of activated carbon-iron oxide nanocomposite for effective removal of Cr(VI)ions. Mater. Chem. Phys. **258**, 124002 (2021). https://doi.org/10.1016/j.matchemphys.2020.124002
40. J. Cao, J. Chen, L. Yi, P. Li, L.W. Qi, Comparison of oil-in-water and water-in-oil microemulsion electrokinetic chromatography as methods for the analysis of eight phenolic acids and five diterpenoids. Electrophoresis **29**(11), 2310–2320 (2008). https://doi.org/10.1002/elps.200700749
41. S. Komarneni, Y.D. Noh, J.Y. Kim, S.H. Kim, H. Katsuki, Solvothermal/hydrothermal synthesis of metal oxides and metal powders with and without microwaves. Z. Naturforsch. **65**, 1033–1037 (2010). https://doi.org/10.1002/chin.201043010
42. P. Nag, S. Banerjee, Y. Lee, A. Bumajdad, Y. Lee, P.S. Devi, Sonochemical synthesis and properties of nanoparticles of $FeSbO_4$. Inorg. Chem. **51**(2), 844–850 (2012). https://doi.org/10.1021/ic201353u

43. V. Verma, M. Kaur, J.M. Greneche, Tailored structural, optical and magnetic properties of ternary nanohybrid $Mn_{0.4}Co_{0.6-x}Cu_xFe_2O_4$ (x= 0, 0.2, 0.4, 0.6) spinel ferrites. Ceram. Int. **45**(8), 10865–10875 (2019). https://doi.org/10.1016/j.ceramint.2019.02.164
44. B.C. Brodie, On the atomic weight of graphite. Philos. Trans. R. Soc. Lond. **149**, 249–259 (1859). https://doi.org/10.1098/rstl.1859.0013
45. R. Eivazzadeh-Keihan, S. Asgharnasl, M.S. Bani, F. Radinekiyan, A. Maleki, M. Mahdavi, P. Babaniamansour, H. Bahreinizad, A.E. Shalan, S. Lanceros-Méndez, Magnetic copper ferrite nanoparticles functionalized by aromatic polyamide chains for hyperthermia applications. Langmuir, **37**(29), 8847–8854 (2021). https://doi.org/10.1021/acs.langmuir.1c01251
46. L. Staudenmaier, Verfahrenzurdarstellung der graphitsäure. Ber Deutsche Chem Ges **31**, 1481–1487 (1898). https://doi.org/10.1002/cber.18980310237
47. W.S. Hummers, R.E. Offeman, Preparation of graphitic oxide. J. Am. Chem. Soc. **80**, 1339–1339 (1958). https://doi.org/10.1021/ja01539a017
48. Y. Ding, Q. Liao, S. Liu, H. Guo, Y. Sun, G. Zhang, Y. Zhang, Reduced graphene oxide functionalized with cobalt ferrite nanocomposites for enhanced efficient and lightweight electromagnetic wave absorption. Sci. Rep. **6**, 1–9 (2016). https://doi.org/10.1038/srep32381
49. J. Luo, P. Shen, W. Yao, C. Jiang, J. Xu, Synthesis, characterization, and microwave absorption properties of reduced graphene oxide/strontium ferrite/polyaniline nanocomposites. Nanoscale Res. Lett. **11**(144), 1–14 (2016). https://doi.org/10.1186/s11671-016-1340-x
50. M. Siyar, N. Khan, A. Maqsood, M. Younas, M. Daud, Structural properties and mechanical characterizations of graphene based cobalt ferrites nanocomposites for load baring applications. Int. J. Eng. Res. Technol. **5**(7), 376–380 (2014)
51. G. Wang, Y. Ma, Z. Wei, M. Qi, Development of multifunctional cobalt ferrite/graphene oxide nanocomposites for magnetic resonance imaging and controlled drug delivery. Chem. Eng. J. **289**, 150–160 (2016). https://doi.org/10.1016/j.cej.2015.12.072
52. J. Sannigrahi, D. Bhadra, M.G. Masud, B.K. Chaudhuri, Anomalous transport properties of graphene oxide magnetic ferrite nanocomposites. AIP Conf. Proc. **1447**(953) (2012). https://doi.org/10.1063/1.4710319
53. G. Wang, Y. Ma, L. Zhang, J. Mu, Z. Zhang, X. Zhang, H. Che, Y. Bai, J. Hou, Facile synthesis of manganese ferrite/graphene oxide nanocomposites for controlled targeted drug delivery. J. Magn. Magn. Mater. **401**, 647–650 (2016). https://doi.org/10.1016/j.jmmm.2015.10.096
54. S. Sheshmani, B. Falahat, F.R. Nikmaram, Preparation of magnetic graphene oxide-ferrite nanocomposites for oxidative decomposition of Remazol Black B. Int. J. Biol. Macromolec. **97**, 671–678 (2017). https://doi.org/10.1016/j.ijbiomac.2017.01.041
55. J. Kaur, M. Kaur, Effect of core-shell reversal on the structural, magnetic and adsorptive properties of Fe_2O_3-GO nanocomposites. Ceram. Int. **43**(18), 16611–16621 (2017). https://doi.org/10.1016/j.ceramint.2017.09.051
56. S.S. Shinde, Crystal structure and magnetic interactions of ferrites. Int. J. Sci. Res. **5**(11), 1034–1036 (2016). https://www.ijsr.net/archive/v5i11/ART20163030.pdf
57. F. Kitaeva, E.V. Zharikov, I.L. Chistyi, The properties of crystals with garnet structure. Phys. Status Solidi **92**(2), 475–488 (1985). https://doi.org/10.1002/pssa.2210920217
58. R. Gerber, Z. Simsa, L. Jensovsky, A note on the magnetoplumbite crystal structure. Czechoslovak J. Phys. **44**(10), 937–940 (1994). https://doi.org/10.1007/BF01715487
59. X. Ji, Y. Xu, W. Zhang, L. Cui, J. Liu, Review of functionalization, structure and properties of graphene/polymer composite fibers. Compos. Part A Appl. Sci. Manuf. **87**, 29–45 (2016). https://doi.org/10.1016/j.compositesa.2016.04.011
60. H. Wang, T. Maiyalagan, X. Wang, Review on recent progress in nitrogen-doped graphene: synthesis, characterization, and its potential applications. ACS Catal. **2**(5), 781–794 (2012). https://doi.org/10.1021/cs200652y
61. P. Fei, Q. Wang, Zhong, B. Su, Preparation and adsorption properties of enhanced magnetic zinc ferrite-reduced graphene oxide nanocomposites via a facile one-pot solvothermal method. J. Alloys. Compd. **685**, 411–417 (2016). https://doi.org/10.1016/j.jallcom.2016.05.279
62. T.M. Chen, X.J. Wu, J.X. Wang, G.W. Yang, WSe_2 few layers with enzyme mimic activity for high-sensitive and high-selective visual detection of glucose. Nanoscale **9**, 11806 (2017). https://doi.org/10.1039/C7NR03179C

63. K. Herget, P. Hubach, S. Pusch, P. Deglmann, H. Gotz, T.E. Gorelik, I.A. Gural'skiy, F. Pfitzner, T. Link, S. Schenk, M. Panthofer, V. Ksenofontov, U. Kolb, T. Opatz, R. Andre, W. Tremel, Haloperoxidase mimicry by CeO_{2-x} nanorods combats biofouling. Adv. Mater. **29**, 1603823 (2016). https://doi.org/10.1002/adma.201603823
64. H. Jia, D. Yang, X. Han, J. Cai, H. Liu, W. He, Peroxidase-like activity of the Co_3O_4 nanoparticles used for biodetection and evaluation of antioxidant behavior. Nanoscale **8**, 5938 (2016). https://doi.org/10.1039/C6NR00860G
65. C. Wang, Y. Shi, Y.Y. Dan, X.G. Nie, J. Li, X.H. Xia, Enhanced peroxidase-like performance of gold nanoparticles by hot electrons. Chem. Eur. J. **23**, 6717 (2017). https://doi.org/10.1002/chem.201605380
66. E. Domingues, E. Fernandes, J. Gomes, R.C. Martins, Advanced oxidation processes perspective regarding swine wastewater treatment. Sci. Total Environ. **776**, 145958 (2021). https://doi.org/10.1016/j.scitotenv.2021.145958
67. Y. Deng, R. Zhao, Advanced oxidation processes (AOPs) in wastewater treatment. Curr. Pollut. Rep. **1**, 167–176 (2015). https://doi.org/10.1007/s40726-015-0015-z
68. M. Xu, C. Wu, Y. Zhou, Advancements in the Fenton process for wastewater treatment. Advanced Oxidation Processes—Applications, Trends, and Prospects, Ciro Bustillo-Lecompte, IntechOpen, London, UK (2020). https://doi.org/10.5772/intechopen.90256.
69. R. Ameta, A.K. Chohadia, A. Jain, P.B. Punjabi, Fenton and photo-Fenton processes, in *Advanced oxidation processes for wastewater treatment: Emerging green chemical technology* (Elsevier, Amsterdam, The Netherlands, 2018), pp. 49–87. https://doi.org/10.1016/B978-0-12-810499-6.00003-6
70. R. Ragg, M.N. Tahir, W. Tremel, Solids go Bio: inorganic nanoparticles as enzyme mimics. Eur. J. Inorg. Chem. **13–14**, 1906–1915 (2016). https://doi.org/10.1002/ejic.201501237
71. Y. Lin, J. Ren, X. Qu, Nano-gold as artificial enzymes: hidden talents. Adv. Mater. **26**(25), 4200–4217 (2014). https://doi.org/10.1002/adma.201400238
72. L. Su, W. Qin, H. Zhang, Z.U. Rahman, C. Ren, S. Ma, X. Chen, The peroxidase/catalase-like activities of MFe_2O_4 (M=Mg, Ni, Cu) MNPs and their application in colorimetric biosensing of glucose. Biosens. Bioelectron. **63**, 384–391 (2015). https://doi.org/10.1016/j.bios.2014.07.048
73. M.B. Gholivand, A. Akbari, M. Faizi, F. Jafari, Introduction of a simple sensing device for monitoring of hydrogen peroxide based on $ZnFe_2O_4$ nanoparticles/chitosan modified gold electrode. J. Electroanal. Chem. **796**, 17–23 (2017). https://doi.org/10.1016/j.jelechem.2017.05.004
74. L. Venosta, M.V. Bracamonte, M.C. Rodriguez, S.E. Jacobo, P.G. Bercoff, Comparative studies of hybrid functional materials based on different carbon structures decorated with nano-magnetite. Suitable application as platforms for enzyme-free electrochemical sensing of hydrogen peroxide. Sens. Actuators. B. Chem. **248**, 460–469 (2017). https://doi.org/10.1016/j.snb.2017.03.159
75. L. Luo, Y. Zhang, F. Li, X. Si, Y. Ding, D. Deng, T. Wang, Enzyme mimics of spinel-type $Co_xNi_{1-x}Fe_2O_4$ magnetic nanomaterial for eletroctrocatalytic oxidation of hydrogen peroxide. Anal. Chim. Acta **788**, 46–51 (2013). https://doi.org/10.1016/j.aca.2013.06.028
76. M. Wu, S. Meng, Q. Wang, W. Si, W. Huang, X. Dong, Nickel–cobalt oxide decorated three-dimensional graphene as an enzyme mimic for glucose and calcium detection. ACS Appl. Mater. Interfaces **7**(38), 21089–21094 (2015). https://doi.org/10.1021/acsami.5b06299
77. Y.C. Yang, Y.T. Wang, W.L. Tseng, Amplified peroxidase-like activity in iron oxide nanoparticles using adenosine monophosphate: Application to urinary protein sensing. ACS Appl. Mater. Interfaces **9**(11), 10069–10077 (2017). https://doi.org/10.1021/acsami.6b15654
78. D. Garg, M. Kaur, S. Sharma, V. Verma, Effect of CTAB coating on structural, magnetic and peroxidase mimic activity of ferric oxide nanoparticles. Bull. Mater. Sci. **41**, 134 (2018). https://doi.org/10.1007/s12034-018-1650-y
79. F. Xia, Q. Shi, Z. Nan, Improvement of peroxidase-like activity and application for detection of H_2O_2 and dopamine for SDBS-Cu-$CuFe_2O_4$. Surfaces Interfaces **24**, 101109 (2021). https://doi.org/10.1016/j.surfin.2021.101109

80. V. Verma, M. Kaur, S. Sharma, Superoxide dismutase mimic activity of spinel ferrite MFe_2O_4 (M= Mn, Co and Cu) nanoparticles. Bull. Mater. Sci. **42**, 120 (2019). https://doi.org/10.1007/s12034-019-1783-7
81. C.A. Proença, T.A. Baldo, T.A. Freitas, E.M. Materón, A. Wong, A.A. Durán, M.E. Melendez, G. Zambrano, R.C. Faria, Novel enzyme-free immunomagnetic microfluidic device based on $Co_{0.25}Zn_{0.75}Fe_2O_4$ for cancer biomarker detection. Anal. Chim. Acta. **1071**, 59–69 (2019). https://doi.org/10.1016/j.aca.2019.04.047
82. Z. Xu, X. Lyu, B. Yang, W. Cao, R. Li, X. Zhang, X. Zhang, G. Fan, X. Kong, Q. Liu, Meso-tetrakis(4-chlorophenyl)porphyrin functionalized $CuFe_2O_4/SiO_2$ nanocomposites with enhanced peroxidase-like activity conveniently using for visual biosensing at room temperature. Colloid. Surf. A: Physicochem. Eng. Aspects. **569**, 28–34 (2019). https://doi.org/10.1016/j.colsurfa.2019.02.052
83. L. Song, C. Huang, W. Zhang, M. Ma, Z. Chen, N. Gua, Y. Zhang, Graphene oxide-based Fe_2O_3 hybrid enzyme mimetic with enhanced peroxidase and catalase-like activities. Colloids Surf A Physicochem Eng Asp **506**, 747–755 (2016)
84. S. Fan, M. Zhao, L. Ding, H. Li, S. Chen, Preparation of Co_3O_4/crumpled graphene microsphere as peroxidase mimetic for colorimetric assay of ascorbic acid. Biosens. Bioelectron. **89**(2), 846–852 (2017). https://doi.org/10.1016/j.bios.2016.09.108
85. X. Yan, Y. Gu, C. Li, L. Tang, B. Zheng, Y. Li, Z. Zhang, M. Yang, Synergetic catalysis based on the proline tailed metalloporphyrin with graphene sheet as efficient mimetic enzyme for ultrasensitive electrochemical detection of dopamine. Biosens. Bioelectron. **77**, 1032–1038 (2016). https://doi.org/10.1016/j.bios.2015.10.085
86. W. Zhang, Y. Sun, Z. Lou, L. Song, Y. Wu, N. Gu, Y. Zhang, In vitro cytotoxicity evaluation of graphene oxide from the peroxidase-like activity perspective. Colloids Surf. B Biointerfaces **151**, 215–223 (2017). https://doi.org/10.1016/j.colsurfb.2016.12.025
87. C. Socaci, F. Pogacean, A.R. Biris, M. Coros, M.C. Rosu, L. Magerusan, G. Katona, S. Pruneanu, Graphene oxide vs. reduced graphene oxide as carbon support in porphyrin peroxidase biomimetic nanomaterials. Talanta. **148**, 511–517 (2016). https://doi.org/10.1016/j.talanta.2015.11.023
88. J. Zhang, J. Ma, X. Fan, W. Peng, G. Zhang, F. Zhang, Y. Li, Graphene supported Au-Pd-Fe_3O_4 alloy trimetallic nanoparticles with peroxidase-like activities as mimic enzyme. Catal. Commun. **89**, 148–151 (2017). https://doi.org/10.1016/j.catcom.2016.08.027
89. X. Zhang, G. Wu, Z. Cai, X. Chen, Dual-functional Pt-on-Pd supported on reduced graphene oxide hybrids: peroxidase-mimic activity and an enhanced electrocatalytic oxidation characteristic. Talanta **134**, 132–135 (2015). https://doi.org/10.1016/j.talanta.2014.11.002
90. V. Sharma, S.M. Mobin, Cytocompatible peroxidase mimic CuO: graphene nanosphere composite as colorimetric dual sensor for hydrogen peroxide and cholesterol with its logic gate implementation. Sens. Actuators. B Chem. **240**, 338–348 (2017). https://doi.org/10.1016/j.snb.2016.08.169
91. S. Wang, R. Cazelles, W.C. Liao, M. Vázquez-González, A. Zoabi, R. Abu-Reziq, I. Willner, Mimicking horseradish peroxidase and NADH peroxidase by heterogeneous Cu^{2+}-modified graphene oxide nanoparticles. Nano Lett. **17**(3), 2043–2048 (2017). https://doi.org/10.1021/acs.nanolett.7b00093
92. L. Wang, H. Yang, J. He, Y. Zhang, J. Yu, Y. Song, Cu-hemin metal-organic-frameworks/chitosan-reduced graphene oxide nanocomposites with peroxidase-like bioactivity for electrochemical sensing. Electrochim. Acta **213**, 691–697 (2016). https://doi.org/10.1016/j.electacta.2016.07.162
93. L. Zhang, X. Hai, C. Xia, X.W. Chen, J.H. Wang, Growth of CuO nanoneedles on graphene quantum dots as peroxidase mimics for sensitive colorimetric detection of hydrogen peroxide and glucose. Sens. Actuators. B. Chem. **248**, 374–384 (2017). https://doi.org/10.1016/j.snb.2017.04.011
94. N.R. Nirala, G. Khandelwal, B. Kumar, Vinita, R. Prakash, V. Kumar, One step electro-oxidative preparation of graphene quantum dots from wood charcoal as a peroxidase mimetic. Talanta. **173**, 36–43 (2017). https://doi.org/10.1016/j.talanta.2017.05.061

95. S. Wu, H. Huang, X. Feng, C. Du, W. Song, Facile visual colorimetric sensor based on iron carbide nanoparticles encapsulated in porous nitrogen-rich graphene. Talanta **167**, 385–391 (2017). https://doi.org/10.1016/j.talanta.2017.02.003
96. X. Wu, M. Li, Z. Li, L. Lv, Y. Zhang, C. Li, Amyloid-graphene oxide as immobilization platform of Au nanocatalysts and enzymes for improved glucose-sensing activity. J. Colloid. Interfac. Sci. **490**, 336–342 (2017). https://doi.org/10.1016/j.jcis.2016.11.058
97. J. Chen, J. Ge, L. Zhang, Z. Li, S. Zhou, L. Qu, PSS-GN nanocomposites as highly-efficient peroxidase mimics and their applications in colorimetric detection of glucose in serum. RSC Adv. **5**, 90400–90407 (2015). https://doi.org/10.1039/C5RA15837K
98. S. Alexander, P. Baraneedharan, S. Balasubrahmanyan, S. Ramaprabhu, Highly sensitive and selective non enzymatic electrochemical glucose sensors based on graphene oxide-molecular imprinted polymer. Mater. Sci. Eng. C **78**, 124–129 (2017). https://doi.org/10.1016/j.msec.2017.04.045
99. Y. Li, Y. Gu, B. Zheng, L. Luo, C. Li, X. Yan, T. Zhang, N. Lu, Z. Zhang, A novel electrochemical biomimetic sensor based on poly(Cu-AMT) with reduced graphene oxide for ultrasensitive detection of dopamine. Talanta **162**, 80–89 (2017). https://doi.org/10.1016/j.talanta.2016.10.016
100. L. Tian, J. Qi, K. Qian, O. Oderinde, Y. Cai, C. Yao, W. Song, Y. Wang, An ultrasensitive electrochemical cytosensor based on the magnetic field assisted binanozymes synergistic catalysis of Fe_3O_4 nanozyme and reduced graphene oxide/molybdenum disulfide nanozyme. Sensors Actuators B Chem. **260**, 676–684 (2018). https://doi.org/10.1016/j.snb.2018.01.092
101. B. Bian, Q. Liu, S. Yu, Peroxidase mimetic activity of porphyrin modified $ZnFe_2O_4$/reduced graphene oxide and its application for colorimetric detection of H_2O_2 and glutathione. Colloid. Surf. B: Biointerface **181**, 567–575 (2019). https://doi.org/10.1016/j.colsurfb.2019.06.008
102. J. Zhang, S. Wu, L. Ma, P. Wu, J. Liu, Graphene oxide as a photocatalytic nuclease mimicking nanozyme for DNA cleavage. Nano Res. **13**, 455–460 (2020)
103. M. Singh, M. Kaur, M.K. Sangha, M.K. Ubhi, Comparative evaluation of manganese oxide and its graphene oxide nanocomposite as polyphenol oxidase mimics. Mater. Today. Commun. **27**, 102237 (2021)
104. T. Guo, T. Xu, W. Xia, A.J. Carrier, L. Wang, X. Zhang, Graphene oxide and CuO double quantum dot composites (GOQD-q-CuO) with enhanced haloperoxidase-like activity and its application in colorimetric detection of H_2O_2 and glucose. **260**, 124126 (2021)
105. Y.J. Shih, C.C. Su, C.W. Chen, C.D. Dong, Synthesis of magnetically recoverable ferrite (MFe_2O_4, M = Co, Ni and Fe)-supported TiO_2 photocatalysts for decolorization of methylene blue. Catal. Commun. **72**, 127–132 (2015). https://doi.org/10.1016/j.catcom.2015.09.017
106. Z. Nasrollahi, A. Ebrahimian Pirbazari, A. Hasan-Zadeh, A. Salehi, One-pot hydrothermal synthesis and characterization of magnetic nanocomposite of titania-deposited copper ferrite/ferrite oxide for photocatalytic decomposition of methylene blue dye. Int. Nano. Lett. **9**, 327–338 (2019). https://doi.org/10.1007/s40089-019-00287-5
107. C. Cai, Z. Zhang, H. Zhang, Electro-assisted heterogeneous activation of persulfate by Fe/SBA-15 for the degradation of Orange II. J. Hazard. Mater. **313**, 209–218 (2016). https://doi.org/10.1016/j.jhazmat.2016.04.007
108. X. Wang, L. Chen, Q. Fan, J. Fan, G. Xu, M. Yan, M.J. Henderson, J. Courtois, K. Xiong, Lactoferrin-assisted synthesis of zinc ferrite nanocrystal: its magnetic performance and photocatalytic activity. J. Alloys Compd. **652**, 132–138 (2015). https://doi.org/10.1016/j.jallcom.2015.08.228
109. Y. Sun, C. Shao, X. Li, X. Guo, X. Zhou, X. Li, Y. Liu, Hierarchical heterostructures of p-type bismuth oxychloride nanosheets on n-type zinc ferrite electrospun nanofibers with enhanced visible-light photocatalytic activities and magnetic separation properties. J. Colloid. Interfac. Sci. **516**, 110–120 (2018). https://doi.org/10.1016/j.jcis.2018.01.033
110. A. Makofane, D.E. Motaung, N.C. Hintsho-Mbita, Photocatalytic degradation of methylene blue and sulfisoxazole from water using biosynthesized zinc ferrite nanoparticles. Ceram. Int. (2021). https://doi.org/10.1016/j.ceramint.2021.04.274

111. M. Ignat, R. Rotaru, P. Samoila, L. Sacarescu, D. Timpu, V. Harabagiu, Relationship between the component synthesis order of zinc ferrite–titania nanocomposites and their performances as visible light-driven photocatalysts for relevant organic pollutant degradation. C. R. Chim. **21**, 263–269 (2018). https://doi.org/10.1016/j.crci.2016.11.004
112. S. Wu, X. Shen, G. Zhu, H. Zhou, Z. Ji, K. Chen, A. Yuan, Synthesis of ternary Ag/ZnO/ $ZnFe_2O_4$ porous and hollow nanostructures with enhanced photocatalytic activity. Appl. Catal. B Environ. **184**, 328–336 (2016). https://doi.org/10.1016/j.apcatb.2015.11.035
113. W. Wang, N. Li, K. Hong, H. Guo, R. Ding, Z. Xia, Z-scheme recyclable photocatalysts based on flower-like nickel zinc ferrite nanoparticles/ZnO nanorods: enhanced activity under UV and visible irradiation. J. Alloy. Comp. **777**, 1108–1114 (2019). https://doi.org/10.1016/j.jallcom.2018.11.075
114. J. Yin, G. Liao, J. Zhou, C. Huang, Y. Ling, P. Lu, L. Li, High performance of magnetic $BiFeO_3$ nanoparticle-mediated photocatalytic ozonation for wastewater decontamination. Sep. Purif. Technol. **168**, 134–140 (2016). https://doi.org/10.1016/j.seppur.2016.05.049
115. Y.H. Guan, J. Ma, Y.M. Ren, Y.L. Liu, J.Y. Xiao, L.Q. Lin, C. Zhang, Efficient degradation of atrazine by magnetic porous copper ferrite catalyzed peroxymonosulfate oxidation via the formation of hydroxyl and sulfate radicals. Water Res. **47**(14), 5431–5438 (2013). https://doi.org/10.1016/j.watres.2013.06.023
116. F. Almeida, E.C. Grzebielucka, S.R.M. Antunes, C.P.F. Borges, A.V.C. Andrade, E.C.F. Souza, Visible light activated magnetic photocatalysts for water treatment. J. Environ. Manage. **273**, 111143 (2020). https://doi.org/10.1016/j.jenvman.2020.111143
117. R. Sharma, S. Bansal, S. Singhal, Augmenting the catalytic activity of $CoFe_2O_4$ by substituting rare earth cations into the spinel structure. RSC Adv. **6**, 71676–71691 (2017). https://doi.org/10.1039/C6RA14325C
118. F. Deganello, L.F. Liotta, S.G. Leonardi, G. Neri, Electrochemical properties of Ce-doped $SrFeO_3$ perovskites-modified electrodes towards hydrogen peroxide oxidation. Electrochim. Acta **190**, 939–947 (2016). https://doi.org/10.1016/j.electacta.2015.12.101
119. D.M. Vadivel, A. Habibi-Yangjeh, B. Paul, S.S. Dhar, K. Selvam, Facile synthesis of novel $CaFe_2O_4$/g-C_3N_4 nanocomposites for degradation of methylene blue under visible-light irradiation. J. Colloid. Interface Sci. **480**, 126–136 (2016). https://doi.org/10.1016/j.jcis.2016.07.012
120. F. Parrino, E. García-López, G. Marcì, L. Palmisano, V. Felice, I.N. Sora, L. Armelao, Cu-substituted lanthanum ferrite perovskites: preparation, characterization and photocatalytic activity in gas-solid regime under simulated solar light irradiation. J. Alloys. Comp. **682**, 686–694 (2016)
121. M.L. Tummino, E. Laurenti, F. Deganello, A.B. Prevot, G. Magnacca, Revisiting the catalytic activity of a doped $SrFeO_3$ for water pollutants removal: effect of light and temperature. Appl. Catal. B Environ. **207**, 174–181 (2017). https://doi.org/10.1016/j.apcatb.2017.02.007
122. Y. Zhai, Y. Yin, X. Liu, Y. Li, J. Wang, C. Liu, G. Bian, Novel magnetically separable $BiVO_4$/ Fe_3O_4 photocatalyst: synthesis and photocatalytic performance under visible-light irradiation. Mater. Res. Bull. **89**, 297–306 (2017). https://doi.org/10.1016/j.materresbull.2017.01.011
123. R. Jiang, H.Y. Zhu, J.B. Li, F.Q. Fu, Yao, S.T. Jiang, G.M. Zeng, Fabrication of novel magnetically separable BiOBr/$CoFe_2O_4$ microspheres and its application in the efficient removal of dye from aqueous phase by an environment-friendly and economical approach. Appl. Surf. Sci. **364**, 604–612 (2016). https://doi.org/10.1016/j.apsusc.2015.12.200
124. S. Huang, H. Wang, N. Zhu, Z. Lou, L. Li, A. Shan, H. Yuan, Metal recovery based magnetite near-infrared photocatalyst with broadband spectrum utilization property. App. Catal. B. Environ. **181**, 456–464 (2016). https://doi.org/10.1016/j.apcatb.2015.08.015
125. S. Huang, Z. Lou, N. Zhu, A. Shan, L. Li, Preparation of $CaF_2/TiO_2/Ln_2Ti_2O_7$ (Ln = Er, Tm, Yb) based magnetite near-infrared photo-catalyst supported on waste ferrite. Mater. Res. Bull. **86**, 107–112 (2017). https://doi.org/10.1016/j.materresbull.2016.10.008
126. N.M. Mahmoodi, M.A. Rezvani, M. Oveisi, A. Valipour, M.A. Asli, Immobilized polyoxometalate onto the modified magnetic nanoparticle as a photocatalyst for dye degradation. Mater. Res. Bull. **84**, 422–428 (2016). https://doi.org/10.1016/j.materresbull.2016.08.042

127. Y.C. Pu, H.Y. Chou, W.S. Kuo, K.H. Wei, Y.J. Hsu, Interfacial charge carrier dynamics of cuprous oxide-reduced graphene oxide (Cu_2O-rGO) nanoheterostructures and their related visible-light-driven photocatalysis. Appl. Catal. B Environ. **204**, 21–32 (2017). https://doi.org/10.1016/j.apcatb.2016.11.012
128. N. Ain, W. Shaheen, B. Bashir, N.M. Abdelsalam, M.F. Warsi, M.A. Khan, M. Shahid, Electrical, magnetic and photoelectrochemical activity of rGO/$MgFe_2O_4$ nanocomposites under visible light irradiation. Ceram. Intern. **42**(10), 12401–12408 (2016). https://doi.org/10.1016/j.ceramint.2016.04.179
129. X.F. Zhang, Y. Chen, Y. Feng, X. Zhang, J. Qiu, M. Jia, J. Yao, Facile preparation of $Zn_{0.5}Cd_{0.5}S$@rGO nanocomposites as efficient visible light driven photocatalysts. J. Alloys. Compd. **705**, 392–398 (2017). https://doi.org/10.1016/j.jallcom.2017.02.207
130. M. Israr, J. Iqbal, A. Arshad, P. Gómez-Romero, Sheet-on-sheet like calcium ferrite and graphene nanoplatelets nanocomposite: a multifunctional nanocomposite for high-performance supercapacitor and visible light driven photocatalysis. J. Solid State Chem. **293**, 121646 (2021). https://doi.org/10.1016/j.jssc.2020.121646
131. J. Park, T. Jin, C. Liu, G. Li, M. Yan, Three-dimensional graphene–TiO_2 nanocomposite photocatalyst synthesized by covalent attachment. ACS Omega **1**(3), 351–356 (2016). https://doi.org/10.1021/acsomega.6b00113
132. J. Zhang, F.X. Xiao, G. Xiao, B. Liu, Self-assembly of Ag nanoparticle-modified and graphene-wrapped TiO_2 nanobelt ternary heterostructure: surface charge tuning toward efficient photocatalysis. Nanoscale **6**(19), 11293–11302 (2014). https://doi.org/10.1039/c4nr03115f
133. Y. Liu, S. Yu, Z. Zhao, F. Dong, X.A. Dong, Y. Zhou, N-Doped $Bi_2O_2CO_3$/graphene quantum dot composite photocatalyst: enhanced visible-light photocatalytic NO oxidation and in situ DRIFTS studies. J. Phys. Chem. C **121**(22), 12168–12177 (2017). https://doi.org/10.1021/acs.jpcc.7b02285
134. H. Javed, A. Rehman, S. Mussadiq, M. Shahid, M.A. Khan, I. Shakir, P.O. Agboola, M.F.A. Aboud, M.F. Warsi, Reduced graphene oxide-spinel ferrite nano-hybrids as magnetically separable and recyclable visible light driven photocatalyst. Synth. Met. **254**, 1–9 (2019). https://doi.org/10.1016/j.synthmet.2019.05.013
135. F. Jelokhani, S. Sheibani, A. Ataie, Adsorption and photocatalytic characteristics of cobalt ferrite-reduced graphene oxide and cobalt ferrite-carbon nanotube nanocomposites. J Photochem. Photobiol. A: Chem. **403**, 112867 (2020). https://doi.org/10.1016/j.jphotochem.2020.112867
136. N. Chnadel, V. Dutta, S. Sharma, P. Raizada, Sonu, A. Hosseini-Bandegharaei, R. Kumar, P. Singh, V.K. Thakur, Z-scheme photocatalytic dye degradation on AgBr/Zn(Co)Fe_2O_4 photocatalysts supported on nitrogen-doped graphene. Mater. Today Sustain. **9**, 100043 (2020). https://doi.org/10.1016/j.mtsust.2020.100043
137. A. Rahman, S. Zulfiqar, A.U. Haq, I.A. Alsafari, U.Y. Qazi, M.F. Warsi, M. Shahid, Cd-Gd-doped nickel spinel ferrite nanoparticles and their nanocomposites with reduced graphene oxide for catalysis and antibacterial activity studies. Ceram. Int. **47**, 9513–9521 (2021). https://doi.org/10.1016/j.ceramint.2020.12.085

Advances in Polyphenol Oxidase Mimic as Catalyst

Harmilan Kaur, Vibha Verma, Manpreet Kaur, and Sucheta Sharma

Abstract Artificial enzymes have received immense interest due to their exceptional properties such as high stability and low cost. Polyphenol oxidase enzyme is well known for its contribution in the field of food industry, biosensors, medicine, and water remediation. Thus, polyphenol oxidase mimics are now envisioned as a relevant alternative with a wide range of implementations in the field of biosensing. Till date, various nanomaterials along with metal complexes and metal–organic frameworks have been explored to mimic polyphenol oxidase activity. This chapter summarizes the latest progress in the field of polyphenol oxidase mimics, and highlights the factors affecting the polyphenol oxidase mimic activity. The kinetic studies and mechanism of polyphenol oxidase as well as its mimic has been compared. The biosensing techniques to evaluate the polyphenol oxidase and mimic activity of synthesized materials have also been reported. The future research challenges and opportunities to enhance the polyphenol oxidase mimic activity are summarized.

Keywords Nanocomposite · Nanozymes · Polyphenol oxidase mimics · Kinetic analysis · Mode of action

1 Introduction

The term 'Enzyme mimics' refers to the synthetic molecules, which show the similar functions of the corresponding natural enzymes, though having different structures. Natural enzymes have been widely used in agro industries, chemical industries, and water treatment owing to their high efficiency and high substrate specificity [1]. These days, artificial enzymes are envisioned as promising alternatives to natural ones due to their facile synthesis, cost-effectiveness, structure robustness, and unusual high

H. Kaur · V. Verma · M. Kaur (✉)
Department of Chemistry, Punjab Agricultural University, Ludhiana 141004, India
e-mail: manpreetchem@pau.edu

S. Sharma
Department of Biochemistry, Punjab Agricultural University, Ludhiana 141004, India

S. J. Ikhmayies (ed.), *Advances in Catalysts Research*, Advances in Material Research and Technology, https://doi.org/10.1007/978-3-031-49108-5_4

catalytic performance. The enzyme polyphenol oxidase (PPO) is of prime interest due to its wide range of implementations such as dephenolization of industrial wastewaters, deodorization, decolourization, and removal of various textile and nontextile dyes [2–7]. Polyphenol oxidase catalyses includes two reactions: the hydroxylation of monophenolic compounds to o-diphenolic compounds such as catechol, and catechol is converted to O-quinones through oxidation. During the previous ten years, most of the research work was focussed on peroxidase, catalase, and superoxide dismutase-like activity of NPs, however NPs having PPO-like activity have received relatively less attention [8].

Till date, a variety of nanomaterials have been explored as PPO mimics. Nanoparticles (NPs) have size in the range of 1–100 nm. Enzyme mimics NPs have lower molecular weight and possess less complex structure as compared to the natural enzymes. Typically, on reduction to nanoscale, the properties of these particles get altered as compared to bulk material. They own exceptionally small size and enormous area per unit volume and have fascinating physio-chemical and biological properties [9]. Particles, rods, fibres, sheets, and tubes are the various forms of nanomaterials. NPs are extremely reactive because of their inherent ability to scavenge reactive oxygen species (ROS) and used to imitate the enzyme activity. Moreover, the artificial enzymes may possess high substrate adsorbing ability. Enzyme mimic activity of nanomaterials has also led to their application in immunoassay techniques and cell/tissue growth [10, 11]. Apart from that, metal–organic framework (MOF) has been known to exhibit enzyme mimic activity. These MOFs are composed of metal and organic bridging ligands, which exhibit enzyme like properties due to the presence of metal ions. Due to their structural and functional diversity, their properties can be tailored [12–14].

Different environmental parameters such as pH and temperature are known to affect the PPO/PPO-mimic activity apart from enzyme mimic and substrate dose. The conditions at which enzyme shows maximum activity are known as optimum conditions, and shows negligible activity if the parameters are altered [15, 16]. As at elevated temperatures above 50 °C, the enzymes loose their catalytic efficiency due to fragility in structure. Likewise, enzymes show their maximum activity at an optimum level of substrate and pH. Thus, the knowledge about the steps involved in the mechanism of PPO action is crucial in order to compare PPO activity with the PPO-mimic activity. The PPOs are the enzymes with a variable number of histidine residues as ligands and contain copper in various PPO types. Thus, Cu plays a vital role in every mechanism of action. The mechanism proposed for the action of N. crassa polyphenol oxidase appears to fit the data for most of the polyphenol oxidases [17, 57]. Polyphenol oxidase enzyme is widely utilized for bioremediation of phenolic pollutants in industrial effluent. The PPO enzyme exists in two forms, tyrosinase and laccase, and is found in a wide range of microbes, plants, and animals. In this chapter, we report the PPO mimic activity of various metal-based materials and the role of PPO mimics in environmental degradation. On the basis of structure, we have discussed PPO mimics in two categories, namely metal nanomaterials and metal–organic framework. Michaelis–Menton equation is well known for the determination of parameters related to kinetic analysis. The kinetics of the reaction for catalytic oxidation of catechol using PPO mimics has been determined through Lineweaver–Burk plot employing Michaelis–Menton equation [27]. Various detection techniques such as amperometric, fluorimetric, and colourimetric detections for detecting PPO/PPO mimics activity are also discussed. Thus, the chapter provides in-depth knowledge regarding the structure activity relation of reported PPO mimics and future perspectives to develop molecules with higher PPO mimic activity for multidisciplinary usage.

2 PPO-Mimic Activity of Different Materials

2.1 *Nanomaterials as PPO-Mimic*

The biological enzymes are the macromolecular catalysts that help in catalysing the biochemical reactions of natural enzymes. All the natural enzymes exhibit catalytic activity in a narrow range of physiological conditions, thus the development of suitable alternatives to enzymes are needed [18, 19]. The PPOs present in plants have catechol oxidase activity mainly localized to plastids and play a major role in post-harvest browning. The PPOs often get induced by wounding or any pathogenic attack so they are believed to play a vital role in plant defence mechanisms. The PPO mimic nanomaterials possess a wide range of applications in cell or tissue growth, for specific disease diagnostics, biosensing, therapeutics, and removal of pollutants from industries [20–24].

Metal nanoparticles (NPs) are known for their excellent properties and wide range of applications in catalytic reactions, thus they have received considerable attention in the past recent years. The Pt NPs showed PPO-like activity by oxidizing the polyphenolic compounds into corresponding o-quinones. The study utilized four major approaches to evaluate the PPO-like activity namely ultraviolet–visible (UV–vis) spectroscopy, ultra-high performance liquid chromatography (HPLC), Ultra Performance Liquid Chromatography (UPLC), and oximetry techniques. For the catalytic oxidation of polyphenols, such as quercetin, L-tyrosine, L-dopa (-), and epicatechin, the potential of Pt NPs as PPO mimics was compared to mushroom tyrosinase. The oxidation of polyphenols catalysed by Pt NPs was monitored using UV–vis spectroscopy, and the oxidized products of polyphenols were identified using UHPLC separation, and high-resolution mass spectrometry (HRMS). The electron spin resonance (ESR) oximetry approach confirmed the intermediate products of semiquinone radicals produced during the oxidation of polyphenols and confirmed the O_2 consumption throughout the oxidation reaction. When compared to mushroom tyrosinase, Pt NPs had a greater catalytic performance [25]

Au@Pt NPs have also been tested for their PPO mimic action. Using catechol as reagent and monitoring the absorbance rise at 420 nm, the PPO mimic performance of Au@Pt NPs was evaluated. The oxidation of catechol was catalysed by the Au@Pt NPs, which resulted in a change in the colour of the solution from colourless to grey.

The data was quantitatively expressed as mean ± standard deviation (SD) of triplicate measurements. The results evaluated in one-way analysis of variance (ANOVA) with Duncan's multiple-test were used to compare the set of experimental data using SAS version 8.0 [26]. These NPs had PPO mimic activity more than 110 units of PPO at concentrations greater than 125 mg L^{-1}. The Au@Pt NPs displayed PPO activity across a wide pH and temperature range when compared to PPO. The NPs also displayed a greater affinity for substrates than the PPO. As a result, these stable NPs showed potential as powerful PPO-mimics that can be employed in a variety of nanotechnology applications.

The biomimetic polyphenolic activities of egg white inorganic hybrid nanoflowers (EW-hNFs) were evaluated colourimetrically by utilizing catechol and tyrosine as substrates. When EW-hNFs were added as an enzyme mimic, the absorbance at 265 nm decreased compared to the control in the absence of ascorbic acid and EW-hNFs. The researchers discovered that EW-hNFs, which are formed entirely of egg white proteins, contain polyphenol oxidase activity. EW-hNFs may have advantages in a variety of biological applications, as well as in the discolouration of direct textile dyes, due to their low cost and accessibility [27].

Mn_3O_4 has the ability to transfer electrons, so, it has been employed in a variety of catalytic reactions. Mn_3O_4 NPs, graphene oxide (GO), and their nanocomposites (NCs) were made utilizing sol–gel method, Hummer's technique, and ultrasonication processes to assess their potential as PPO mimics using catechol as a substrate. The synthesized materials catalysed the conversion of catechol to ortho-quinones due to oxidation of catechol. At 390 nm, the increase in absorbance was measured to monitor the process, which was accompanied by a change in colour of the reaction solution from colourless to yellow. Pristine Mn_3O_4 NPs exhibited the highest PPO mimic activity followed by NC. The results attributed to the tendency of materials to participate in electron transfer processes. Due to the availability of d orbitals, pristine Mn_3O_4 NPs take electrons by direct coordination of catecholate, however, the inclusion of GO hindered the activities, since it can donate electrons, but not participate in electron transfer processes. The result of the students' *t*-test and probability was used to compare each set of experimental data using Microcal origin software. Thus, the results demonstrated the use of Mn_3O_4 NPs as PPO mimic [28].

2.2 Other Materials as PPO-Mimics

The catalytic oxidation of catechols was used to assess the catechol oxidase biomimetic catalytic activity of the reported cobalt(II) and copper(II) complexes (containing the N_2O_2 donor sets of imine ligands) as functional models. Among the several catechols utilized in catechol oxidase model studies, 3,5-di-tertbutylcatechol is the most commonly employed catechol for catecholase activity of tyrosinase.

The quinone-catechol pair is easily oxidized to the corresponding quinone 3,5-di-tert-butyl-o-quinone (DTBQ) due to its low redox potential, and other oxidation steps such as ring-opening are difficult due to its bulky substituents. 3,5-di-tert-butyl-o-quinone (3,5-DTBQ) is a somewhat stable chemical compound with a high absorption maximum wavelength $\lambda_{max} = 400$ nm. As a result, electronic spectroscopy can be used to determine activities and reaction rates by tracking the appearance of 3,5-ditert-butyl-o- quinone (3,5-DTBQ) absorption. As these complexes and substrates are soluble, the reactivity studies were carried out in dimethylformamide (DMF) solution. This study synthesized and characterized a series of cobalt(II) and copper(II) complexes with tetradentate imine ligands. The biomimetic catalytic activity of the synthesized complexes was evaluated. According to the results, all complexes catalyse the aerobic oxidation of catechol to the light-absorbing o-quinone. As a result, the study emphasizes the significance of metal ions in metal–organic framework PPO mimic action [29].

Binuclear complexes having formula $[Cu_2(AnthenMe_2)_2(\mu_{1,3}\text{-}SCN)_2(NCS)_2]$ (1) and $[Cu_2(AnthenMe_2)_2(\mu_{1,1}\text{-}N_3)_2(N_3)_2]$ (2), where $AnthenMe_2$ is the bidentate N,N-donor Schiff base formed by anthracene-9-aldehyde and N,N-dimethylethylenediamine were explored as PPO mimics. Despite the fact that these 2 complexes have quite distinct Cu-Cu separations, their catalytic activities for the oxidation of 3,5-ditertiarybutyl catechol ($DTBCH_2$) to the quinone (DTBQ) are rather similar. The compound catecholase activity was measured using 3,5-di-tert-butylcatechol (3,5-$DTBCH_2$) as substrate. The initial rate technique was used to determine the kinetics of the oxidation of the substrate 3,5-$DTBCH_2$. The concentration of the substrate 3,5-$DTBCH_2$ was kept at least 10 times higher than the concentration of the complex, and the rise in the concentration of the quinone was measured with a spectrophotometer of absorbance maximum at wavelength $\lambda_{max} = 400$ nm. As a result, anthracene-containing pseudo-halide bridged copper (II) complexes are investigated as PPO mimics [30].

Another anti-inflammatory binary copper (II) complex of the non-steroidal drug naproxen (Nap) having formula $[Cu_2(Nap)_4]_n$ and its ternary complex with 3-pyridylmethanol (3-pym) of formula $[Cu(Nap)_2(3\text{-}pym)_2]_n$ have been evaluated as PPO mimic towards the aerobic oxidation of 3,5-di-tert-butylcatechol (DTBC) to 3,5-di-tert-butylquinone (DTBQ). The results showed that both the complexes possessed moderate catalytic activities. The catecholase activity of the complexes for the air oxidation of DTBC to the corresponding DTBQ was monitored spectrophotometrically at 25 °C by following the increase of the DTBQ absorption band at wavelength $\lambda = 400$ nm. The binuclear complex exhibited higher enzyme mimic activity over mononuclear complex as suggested by higher oxidation rate of substrate [31].

As a PPO mimic, a series of copper(II) complexes with large cyclic ligands containing N_3S_2-donating atoms in the large macrocyclic ring were utilized. The substrates 3,5-di-tert-buty1catehol and o-aminophenol were utilized to test the PPO mimic activity. The increase in the usual quinone absorption band at 400 nm and for ascorbate peroxidase (APX) at 433 nm was used to monitor the production of 3,5-di-tert-butyl-quinone (3,5-DTBQ). The substrate concentration dependency of both examined substrates showed the saturation type behaviour. In another study, kinetic measurements of catechol oxidase mimic activity of N_3S_2 donors macrocyclic Cu(II) complexes revealed that increasing the acidic character (Lewis acidity) of the Cu(II) centre enhances electron transport between catalyst and catecholate ring. The higher activity of five-coordinate complexes over the octahedral complexes was due to the coordinated saturation of octahedral complexes, which inhibits the binding of reacting species during the catalytic mechanism. The complex first dissociates to form low coordinate complex, which requires energy in contrast to the five-coordinate complex having already a vacant coordination site. Addition of triethylamine to the oxidation reaction medium accelerated the oxidation rate [32].

The catalytic activities of Cu(II) the formation of ferrocenecarboxylate complex with nitrogen-based ligands for oxidative coupling of 2,6-dimethylphenol are explored as PPO mimetics. Spectral studies suggested that, Cu(II) ion was coordinated in the plane with four nitrogen atoms of pyrazoles, and the axial sites are occupied by oxygen atoms from two ferrocenecarboxylate groups to yield chromophore [33].

Another MOF having the formula $Zn(HL)_2Cl_2$, where the Schiff base (HL) = 2-(2-methoxybenzylideneamino)phenol has been evaluated for its potential as PPO mimic. In this molecule, three donor centres and Zn centre exhibited a distorted tetrahedral geometry. The molecule adopted zwitter ionic form through protonation, and exhibited monodentate coordination. For PPO mimic activity, the oxidation of 4-methylcatechol (4-MC) in methanol catalysed in the presence of Zn-Schiff base was monitored. Various studies including electro-chemical study, electron paramagnetic resonance (EPR) analysis, and electrospray ionization (ESI) mass spectrometry confirmed the catalytic oxidation of substrate used through enzyme-substrate binding along with the generation of radicals [34].

The sensing of dopamine (DA) fluorescence was done using a fluorescent nanocomposite based on polymer dots of adenosine monophosphate (AMP) and Cu^{2+} (Pdots@AMP-Cu) having PPO mimic activity. The oxidation of the substrate 2,4-dichlorophenol and 4-aminoantipyrine catalysed by the nanocomposite showed fluorescence emission at wavelength $\lambda = 668$ nm resulted in a change in the colour of the solution from colourless to red. Due to electron transfer, dopamine was oxidized to eumelanin, which then quenched fluorescence Pdots@AMP-Cu. The nanocomposites showed strong selectivity for organic molecules, uric acid, and other chemicals, suggesting that they could be useful for detecting dopamine in human serum samples [35].

Mononuclear Cu (I) and Cu (II) complexes having N_4 donors in structural moiety possessed catechol oxidase mimic activity by converting 3,5-DTBC to 3,5-DTBQ as confirmed due to the increase of quinone absorption at wavelength 400 nm. The rate of oxidation of the reaction was linearly proportional to the rate of OH^- or Cl^- dissociation. Apart from that geometry, complex structure also influenced catalytic activity. Planar mononuclear complexes did not show catalytic activity possibly due to the non-effective steric match of the substrate with catalyst, whereas, complex with square-planar geometry were effective catalysts. The coordination of the fifth ligand to Cu(II) in trigonal bipyramidal complexes reduces the strength of the equatorial field and ability to alter structural geometry as observed in copper superoxide dismutase enzyme [36].

Copper (II) complexes of the pentadentate pyridine-based ligand, containing two thiolate S, pyridine N, and two azomethines N, as well as the series of tetradentate ligands containing two thiolate S, and two azomethine N, showed PPO mimic catalytic activity by converting 3,5-di-tert-buty1catehol (3,5-$DTBCH_2$) to 3,5-di-tert-butyl-quinone (3,5-DTBQ) due to the increase in quinone absorption band at 400 nm. On one hand, the kinetic studies showed first-order dependence of catalytic activity of complexes on substrate concentration, on the other hand, rate constant of the oxidation reaction increased linearly with increasing catalyst concentration. In a preequilibrium process, the production of an intermediate complex-substrate adduct was also observed. The binuclear Cu(II) complexes showed prominent results as compared to the mononuclear complexes, which were in accordance with the fact that the mechanism of reaction requires two metal ions in close proximity. The presence of phenyl group improved the catalytic activity due to the conjugation effect in oxidation reaction. The catalytic efficiency of complexes was also altered by Lewis acidity of Cu(II) centre surrounded by ligand substituents [37].

Diclofenac-based complexes of several metal ions $[Co(diCl)_2(2\text{-pyet})_2]$, $[Ni(diCl)_2(2\text{-pyet})_2]$, $[Cu_2(diCl)_2(2\text{-pyet})_2]$, and $[Cu_2(diCl)_2(2\text{-pypr})_2]$ exhibited catechol oxidase like activity towards the aerobic oxidation of $DTBCH_2$ (di-tert-butycatechol) under catalytic conditions. With the substrate to complex ratio, the rate-determining step altered. The results of the kinetic study were similar to those of previously reported catechol oxidase mimics in the literature, but they were lower than those of the natural catechol oxidase enzyme from *ipomoea batatus* (sweet potatoes) having a rate constant $k_{cat} = 8.26 \times 10^{-6}\ h^{-1}$. The study paved way to design new analogues having the enhanced activity [38]. The activities of catechol oxidase were investigated using 3,5-DTBC as a substrate. The 3,5-DTBC has a low quinone-catechol reduction potential and may be quickly oxidized to the equivalent o-quinone, 3,5-DTBQ, which is stable, and has a maximum absorption at wavelength $\lambda = 410$ nm, allowing for easy spectrophotometric monitoring of the reaction. The reaction rate was found to have a first-order dependence on substrate concentration when the substrate concentration was low. At high substrate concentrations, all complexes showed saturation kinetics. The results show that, iron complex $[Fe_2(\mu\text{-}Cl)_2(hbpg)_2]\ Cl_2(H_2O)_2$ exhibits higher polyphenol oxidase activity than copper complexes $Cu_2(\mu\text{-}Cl)_2(hbpg)_2$ and $[Cu_2(\mu\text{-}OH_2)_2(hbpg)_2](NO_3)_2(H_2O)_2$, and manganese complex $[Mn_2(\mu\text{-}Cl)_2(hbpg)_2](H_2O)_2$, because the reduction potential of the iron centre is higher. The findings imply that the metal centres redox potential is an important element in regulating catechol oxidase activity in metal complexes [39].

3 Factors Affecting PPO Mimic Activity

3.1 *Effect of pH*

The pH scale determines whether a sample is acidic or alkaline. Due to pH-induced structural changes, enzymes and enzyme mimics' activity is affected by pH. With changes in pH, bell-shaped activity curves are obtained for enzymatic reactions. The pH value at which maximum enzyme activity is attained is referred to as optimum pH. The activity of enzymes is reduced to a larger extent when the pH value deviates from optimum values [40–43]. Likewise, artificial enzymes too possess optimum pH value due to pH-induced alteration in surface properties of enzyme mimics. PPO's optimal pH ranges from about 4.0 to 7.0, and the value of the pH depends upon the material's origin, extraction process, and substrate. Furthermore, the optimal pH of plant polyphenol oxidases is between 6 and 7. Reduction of Mn_3O_4 involved a one-electron transfer process due to the presence of semiquinone radicals. When it comes to nanocomposite, there was not much reduction of Mn_3O_4 to Mn^{2+} ions, which reduced nanocomposite performance as compared to the pristine Mn_3O_4 NPs, as GO restricted electron transfer [28]. The stability of the egg white hybrid nanoflower (EW-hNFS) with biomimetic PPO-like activity was tested throughout a pH range of 5.0–9.0, with a loss of activity of 6% to 7% [27]. The ideal pH for Au@Pt NPs and PPO was found to be 6.0 and 5.0, respectively [26].

3.2 *Effect of Temperature*

Temperature affects enzyme as well as enzyme mimic activity. The thermal sensitivity of enzyme causes denaturation of enzyme structure due to bond breakings, and thus lower enzyme activity has been observed at extreme high temperature. Rise in temperature of about 10 °C will increase the activity of most enzymes by 50–100%. Lower temperatures lead to slow chemical reactions and enzymes become inactive at freezing temperatures. Thus, the temperature corresponding to which maximum activity is obtained, known as optimum temperature for particular enzyme [44]. Likewise, for enzyme mimics, variation in temperature results into bond formation and bond breaking processes with the alteration in kinetics of the reaction. For bare Mn_3O_4 NPs and Mn_3O_4-GO NC, optimum temperature was found to be 20 °C. The production of surface complexes between catechol and oxides diminished when the temperature was raised, because only a little quantity of catecholate was oxidized to semiquinone free radicals [28]. The stability of EW-hNFs bimomimetic polyphenol oxidase was tested across a wide temperature range, and it was determined to be best up to 40 °C. As a result, these EW-hNFs are more versatile than natural enzymes and can be used in a variety of situations [27]. The ability of Au@Pt NPs to mimic PPO was investigated at temperatures ranging from 0 °C to 60 °C. At the same temperature range, the PPO mimetic activity of Au@Pt NPs was matched to 15 units of polyphenol oxidase. The temperatures 20 °C and 40 °C were reported to be optimum temperatures for Au@Pt NPs and PPO, respectively [26].

3.3 Effect of Substrate Concentration and Enzyme Mimic Concentration

The effect of substrate concentration on enzyme activity is complex and depends on various factors. In case of enzyme mimics at constant catalyst concentration, the rate of reaction generally increases as the substrate concentration increases until equilibrium is attained. After further increase in the substrate concentration, invariable rate of reaction is attained. After saturation point, the interaction between excess substrate and catalyst molecules inhibited [45]. In this context, for pristine Mn_3O_4 NPs and Mn_3O_4-GO NC as PPO mimics, the concentration of catechol dose was varied from 1 to 5 mM, and optimized catechol dose was found to be 3 mM [28]. In addition, 60 μL of Au@Pt NPs, 180 μL of 1/5 mM buffer solution (pH = 7), and 60 μL catechol make up the reaction volume 300 μL for Au@ Pt NPs to mimic the PPO by observing the lower absorbance at 420 nm [26]. The biomimetic tyrosinase or catechol oxidase efficiency of EW-hNFs was colourimetrically tested using catechol and tyrosine as substrates. The effectiveness of phenol oxidase/peroxidase in EW-hNFs was also evaluated colourimetrically with pyrogallol and o-dianisidine at 0.2 mM in the phosphate buffer solution (pH = 6.5, 50 mM) at 37 °C. At 450 nm, variations in absorbance over 3 min were observed for pyrogallol and o-dianisidine [27].

The enzyme activity is known to increase with the increase in enzyme concentration until saturation is achieved between enzyme and substrate molecules. Afterwards, there is negligible effect on raising the enzyme concentration. Enzyme mimic activity follows the same trend with the increase in concentration. Effect of catalyst dose on PPO activity in case of both Mn_3O_4 NPs and Mn_3O_4-GO NC as PPO mimics was observed by varying their concentration from 1 to 7 mg. The maximum activity was observed at 7 mg. When concentration of the catalyst varies in the amount, the initial rate is found to be linearly dependent on the catalyst concentration. The rate of reaction with respect to the catalyst concentration indicates the first order dependence. Also, without the catalyst, there is no quantifiable rate of oxidation [28].

4 Detection Systems for Evaluation of PPO and PPO-Mimic Activity

The nanozymes (nanomaterials having enzyme activity) have received a lot of attention for replicating traditional enzymes owing to their inexpensive cost and stability in catalytic activities. Due to its porosity, it has significant advantages over other nanomaterials in terms of tunability. The electrical and optical characteristics of Pdots are different. Bioluminescence imaging, biosensors, medication administration, and a variety of other electroluminescent applications all utilize Pdots. The Pdots are a novel type of fluorescent nanoparticle with a variety of properties, including exceptional photostability, high luminosity, and minimal toxicity. It was observed that AMP-Cu may oxidize DA to eumelanin in a fluorescent nanocomposite (Pdots@AMP-Cu) due to its PPO-mimicking effect. Enzymatic reactions and amperometric detection of the resulting product are often used in this type of analysis. Alkaline phosphatse (ALP) is a key enzyme in the hydrolysis and transphosphorylation of proteins. A high level of ALP in a biological system is linked to a number of human disorders, such as diabetes, liver disease, bone disease, and so on. The ALP level is commonly employed as a diagnostic sign. The synthesis of nanozymes using Cu^{2+} and guanosine monophosphate was reported by researchers [46, 47], indicating that the analytical method is both sensitive and selective for ALP, and it is crucial to investigate its uses in clinical diagnosis and therapy good manufacturing practice (GMP). To replicate PPO, crude egg white was used as the organic component and Cu^{2+} as the inorganic component in the creation of organic–inorganic hybrid nanoflowers (EW-hNF). ATP, ADP, and AMP are coupled to Cu^{2+} to produce fluorescent nanozymes with PPO mimetic activity. As a result, the activity of PPO and PPO mimics in a variety of nanomaterials may now be assessed using a number of detection modalities [46–49].

4.1 Amperometric Detection

Amperometric sensors are stand-alone analytical devices that measure current response resulting from the oxidation or reduction of an electroactive element and provide specific quantitative analytical information. To date, amperometric biosensors have been widely used for evaluation of enzyme/enzyme mimic activity. Glucose oxidase-based sensor is the most used biosensor system. It produces a current proportional to the concentration of the substance to be detected. Polyphenol oxidase mimic activity of various synthesized materials has been evaluated employing amperometric detection. The designed sensor measures the produced current corresponding to the concentration of reactant or product of the reaction catalysed by PPO mimic.

The creation of an amperometric biosensor for the detection of tyramine (Tyr) in food and various different types of beverages is based on the immobilization of tyrosinase (Polyphenol oxidase) on glassy carbon electrode altered by nanocomposite consisting of gold nanoparticles (AuNPs) synthesized using a green method and poly(8-anilino-1-naphthalene sulphonic acid) altered glassy carbon electrode. The biosensor had a linear response to tyramine in the range of 10–120 M under optimum experimental circumstances for fixed potential amperometric detection, and the limit of detection was determined to be 0.71 M. The new platform demonstrated high selectivity, long-term stability, and repeatability. The higher value of the Michaeli–Menten constant comes out to be 79.3 M, which demonstrated the robust interaction between tyrosinase and the nanocomposite. The biosensor was effectively used to determine Tyr in dairy products and fermented beverages with high recoveries, making it a viable biosensor for tyramine measurement [50].

The detection of phenolic chemicals is important due to their toxicity, even at low concentrations. The determination of phenols by amperometry is a straightforward procedure. Direct oxidation of phenols is one option, but polyphenol oxidase (tyrosinase) enzyme biosensors that oxidize phenolic compounds to their adjacent quinones in another way. The amplification of the amperometric signal is achieved by reducing the resulting quinones, as long as the reduced product is catechol, which can then be oxidized by the polyphenol oxidase immobilized on the biosensor's surface. The simultaneous determination of multiple phenols was accomplished by merging biosensor signals. Each molecule was extracted and quantified using Artificial Neural Networks (ANN). The three analyses of the three-analyte research case solved in this chapter were phenol, catechol, and m-cresol. Because of good prediction ability, separate quantification of these three phenols was possible [51].

4.2 Fluorimetric Detection

Fluorescence-based instruments have attracted researchers' attention in sensing field due to their higher sensitivity. A fluorescence microscope can distinguish two different particles spaced by less than 10 nm distance, due to super-resolution technology. The most essential component in a fluorometer is the presence of fluorophore molecule and is based on the measurement of fluorescence intensity at fixed wavelength values of excitation and emission. Fluorescence lifetime is determined through the decay of emission intensity, which is a unique property of different fluorophores. The fluorescence detection of dopamine was performed employing Pdots@AMP-Cu as PPO-mimic. A combination of fluorescent polymer dots, adenosine monophosphate (AMP), along with the Cu^{2+} coordination nanostructures were used to make Pdots@AMP-Cu. The Pdots@ AMP-Cu showed fluorescence emission at 668 nm which is red in colour, as well as efficient PPO-like activity by catalysing the substrates (2,4-dichlorophenol) 2,4-DP and (4-aminoantipyrine) 4-AP, which leads to colour change from colourless to reddish colour. A novel detection method for dopamine (DA) was then developed based on Pdots@AMP-Cu. According to the findings, Pdots@AMP-Cu can oxidize DA to form eumelanin, and eumelanin can effectively quench the fluorescence of Pdots@AMP-Cu due to the occurrence of electron transfer process. The fluorescent intensity was found to have a linear relationship with the DA concentration, which ranged from 10 to 400 M, and the detection threshold for DA was 4 M. The suggested approach demonstrated great selectivity for amino acids, ascorbic acid, and a variety of other acids, and it could be utilized to detect DA in human serum samples with a higher recovery rate [35].

High-sensitivity monitoring for organophosphorus pesticides (OPs) as well as on-site screening are deemed necessary to preserve the ecosystem and prevent illnesses. The enzyme tyrosinase (Polyphenol oxidase) was utilized to regulate the quenching of gold nanoclusters (AuNCs) to create a novel fluorimetric sensing platform for detection and quantification of Ops (Fig. 1). A one-step green synthetic method for the synthesis of gold clusters (AuNCs) was devised using chicken egg white (CEW) as a template and stabilizing agent. TYR catalyses the conversion of dopamine to dopaminechrome, which uses a dynamic quenching technique to effectively quench the fluorescence intensity of AuNCs at 630 nm. The activity of TYR was reduced in the presence of OPs, resulting in AuNC fluorescence regeneration. This proposed fluorescent framework was proven to be capable of rapid OP detection with a detection limit of 0.1 mg mL^{-1} (using paraoxon as a model). The fluorescent probe was also utilized to produce paper-based OP test strips, indicating its vast potential for real-time and on-site applications [52].

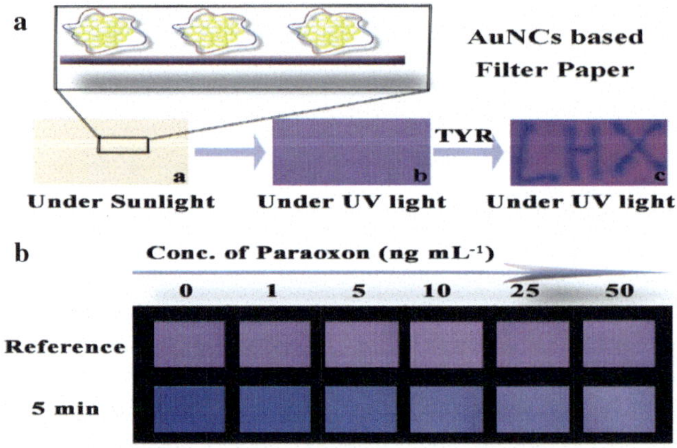

Fig. 1 **a** The feasibility of using TYR/DA reaction solution as ink for test strips. The images of AuNCs-based test strips were taken for test strips (a) exposed to sun light, (b) exposed to the ultraviolet light of wavelength 365 nm, and (c) in the presence of TYR/DA solution under ultraviolet light. **b** Visual detection of paraoxon using AuNCs-based test strips. Paraoxon concentration values are 0, 1.0, 5.0, 10.0, 25.0, and 50.0 ng mL^{-1}, respectively. Reprinted from Yan et al. [52], with permission from Elsevier. Copyright (2017)

4.3 *Colourimetric Detection*

Colourimetric detection utilizes chromogenic substrates as these compounds exhibit a characteristic colour change when subjected to chemical reaction in the presence of an enzyme/enzyme mimic. The detection is based on Lambert beers law as the concentration of a particular molecule is measured corresponding to the absorbance value. A colourimetric analysis of alkaline phosphatase activity was carried out using a nucleotide coordinated copper ion that mimicked PPO-like activity. Copper-coordinated nucleotides such as adenosine triphosphate (ATP), adenosine diphosphate (ADP), and adenosine monophosphate (AMP) were used to make the nanozymes. At 3 mM nucleotide levels, the catalytic activity of ADP-Cu and AMP-Cu was greater than that of ATP-Cu. More nanozymes were created when ALP hydrolysed ATP into ADP and AMP, and the system's catalytic activity increased, resulting in an increase in colourimetric signal. The colourimetric detection limit for ALP was determined to be 0.3 U L^{-1}, and the signal intensity was proportional to the ALP concentration in the range of 0–30 U L^{-1}. In this manner, the colourimetric sensing ALP method was established. This approach was also used to detect ALP activity in human serum samples, implying that it has a wide range of diagnostic and practical uses [53].

Enzyme-simulating nanomaterials can replace complicated natural enzymes and catalyse enzyme-like activities in a number of severe applications due to their obvious benefits of high reactivity, hardness, and cost-effectiveness. To increase the performance of catechol oxidase mimic enzyme activity, pyrazolate-based metal–organic frameworks (MOFs), and the coordination regulating method were used to produce a novel colourimetric dual-readout sensing platform for dopamine detection. Auxiliary ligands such as benzoic acid or 4-Formyl-1(H)-pyrazole were used to change the growth mode of pyrazolate-based porphyrinic MOF materials. The Cu-TPP(PA) MOF was tested to see if it could detect dopamine (DA) in human serum samples, and if it could be used as paper-based analytical tool. This innovation will broaden the scope of MOF applications in analytical chemistry and pave the path for the production of highly active and stable nano-enzymes [54].

5 Kinetic Analysis of PPO-Mimic Activity

The necessity of kinetic analysis of the enzyme/enzyme mimic catalysed reaction lies in their ability to determine the equilibrium state of chemical reaction. Kinetic studies provide evidences regarding the mechanisms of various chemical reactions. Also the knowledge of reaction mechanisms is of practical use in order to determine the most efficient way for a reaction to occur. Michaelis–Menton equation is well known for the determination of parameters related to kinetic analysis. The kinetics of the reaction for catalytic oxidation of catechol using PPO mimics have been determined through Lineweaver–Burk plot (Fig. 2) employing Michaelis–Menton

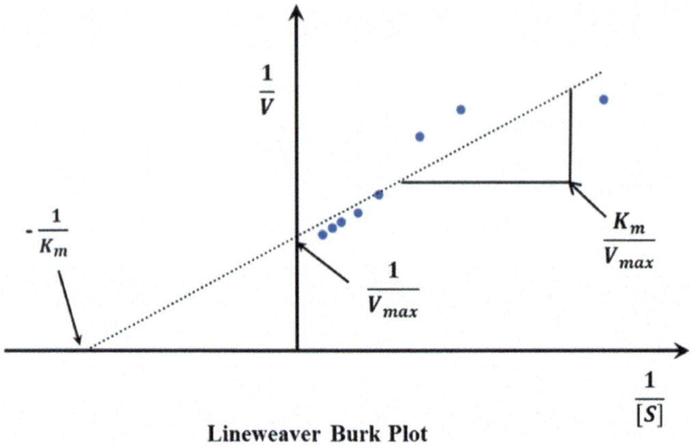

Fig. 2 Lineweaver–Burk plot

equation as follows:

$$\frac{1}{V} = \frac{K_m}{V_{max}[S]} + \frac{1}{V_{Max}} \tag{1}$$

where:

V is the Initial Velocity

$V_{max.}$ is the maximum velocity of the reaction.

[S] is the substrate concentration and

K_m is called as Michaelis–Menten constant, which indicates the enzyme affinity for its particular substrate. The smaller the value of K_m, the greater is the affinity between enzyme and substrate. The kinetic parameters of some PPO mimics have been reported in the literature (Table 1)

The reaction kinetics for the catalytic oxidation of substrates by Au@Pt NPs or PPO were assessed by measuring the absorbance fluctuations over a 1-min timeframe. The Michaelis–Menten constant was used to calculate the kinetics.When catechol and Levodopa (L-DOPA) were employed as substrates, PPO's K_m values were considerably lower than when pyrogallol was used. Due to its lower K_m values, L-DOPA was chosen as the preferable substrate. In the oxidation of all substrates, Au@Pt NPs showed lower catalytic activity, but greater affinity for the substrates, particularly catechol and pyrogallol, as compared to PPO. The values of V_{max} and K_m using catechol as substrates were 36.05×10^0 and 1.29×10^2 respectively [26]. The kinetic characteristics of egg white hybrid nanoflowers (EW-hNFs) showed phenol oxidase-like activity, which was calculated using Catechol (0.01–0.8 mM) as a substrate. The ideal temperature of biomimetic EW-hNFs was determined using catechol at temperatures ranging from +4 °C to 60 °C. The kinetic data of the phenol-like EW-hNFs were determined by changing the concentration of the substrate catechol. The values for V_{max} and K_m were 0.118 U/mg protein and 0.069 mM, respectively [27].

Table 1 Kinetic parameters of some PPO mimics reported in literature

Serial. No	Substrate	Complex	K_m	V_m	K_{cat}	Ref
1	3,5-DTBC	$Cu_2(\mu\text{-}Cl)_2(hbpg)_2$	31.59 mmol L^{-1}	1.08×10^{-2} mmol L^{-1} s^{-1}	390.2 h^{-1}	[39]
		$[Fe_2(\mu\text{-}Cl)_2(hbpg)_2]Cl_2(H_2O)_2$	34.47 mmol L^{-1}	1.30×10^{-2} mmol L^{-1} s^{-1}	468.7 h^{-1}	
		$[Mn_2(\mu\text{-}Cl)_2(hbpg)_2](H_2O)_2$	24.67 mmol L^{-1}	1.06×10^{-2} mmol L^{-1} s^{-1}	381.6 h^{-1}	
2	3,5-DTBC	$[CuL^1\ H_2O]H_2O$	14.3×10^{-3} M	29.5×10^{-3} M s^{-1}	11.80 s^{-1}	[37]
		$[CuL^2\ H_2O]$	24.0×10^{-3} M	35.0×10^{-3} M s^{-1}	14.00 s^{-1}	
		$[CuL^3\ H_2O]H_2O$	25.0×10^{-3} M	22.0×10^{-3} M s^{-1}	8.80 s^{-1}	
		$[CuL^4\ H_2O]H_2O$	330.0×10^{-3}	125.0×10^{-3} M s^{-1}	500.0 s^{-1}	
3	3,5-DTBC	$[LCu(OH)](ClO_4)$	10.0×10^{-3} mol L^{-1}	3.5×10^{-4} mol L^{-1} s^{-1}	21.0 h^{-1}	[36]
		[LCuCl]Cl	9.5×10^{-3} mol L^{-1}	14.7×10^{-4} mol L^{-1} s^{-1}	88.2 h^{-1}	
		$[LCu](ClO_4)_2H_2O$	10.0×10^{-3} mol L^{-1}	2.50×10^{-4} mol L^{-1} s^{-1}	15.0 h^{-1}	
		$[LCu]_2(ClO_4)_2$	9.3×10^{-3} mol L^{-1}	28.7×10^{-4} mol L^{-1} s^{-1}	172 h^{-1}	
4	Phenol	CuII–MTS	–	–	1.58×10^{-2} s^{-1}	[56]
	Phenol	CuII–CYS	–	–	1.27×10^{-2} s^{-1}	
	Catechol	CuII–MTS	–	–	1.58×10^{-2} s^{-1}	
	Catechol	CuII–CYS	–	–	1.43×10^{-2} s^{-1}	
5	o-aminophenol	$[CuL^1Cl]Cl$	19.5×10^3 M	19.2×10^3 M s^{-1}	384 s^{-1}	[32]
		$[CuL^1\ Br]Br.H_2O$	10.1×10^{-3} M	17.7×10^{-3} M s^{-1}	354 s^{-1}	
		$[CuL^1\ NO_3]NO_3.H_2O$	15.8×10^{-3} M	55×10^{-3} M s^{-1}	1100 s^{-1}	
		$[CuL^1\ ](AcO)_2.H_2O$	7.4×10^{-3} M	12×10^{-3} M s^{-1}	240 s^{-1}	

(continued)

Table 1 (continued)

Serial. No	Substrate	Complex	K_m	V_m	K_{cat}	Ref
6	DMF	$[Cu_2(dicl)_2(2\text{-}pyet)_2]$	7×10^{-4} M	2.06×10^{-6} M s^{-1}	148.1 h^{-1}	[38]
		$[Cu_2(dicl)_2(2\text{-}pypr)_2]$	2.32×10^{-4} M	1.87×10^{-6} M s^{-1}	135.24 h^{-1}	
7	Quercetin	Pt NPs	54.37 μM	5.79 μM s^{-1}	244.82 s^{-1}	[25]
		Tyrosinase	26.23 μM	0.54 μM s^{-1}	12.13 s^{-1}	
8	Catechol	Au@Pt NPs	1.29×10^{-2} mM	13.97 mol L^{-1} min^{-1}	–	[26]
	L-DOPA	Au@Pt NPs	6.70×10^{0} mM	9.91×10^{0} mol L^{-1} min^{-1}	–	
9	Catechol	Mn_3O_4	1.14 mM	0.52 mol L^{-1} min^{-1}		[28]
		NC	0.84 mM	0.41 mol L^{-1} min^{-1}		

Abbrevations 3,5-DTBC (3,5-di-tert-butylcatechol)
L^1 (Pentadentate pyridine-based ligand)
L^2 (Ligand containing two thiolate sulfur)
L^3 (Ligand containing two pyridine nitrogen)
L^4 (Ligand containing two azomethines nitrogen)
L (Ligand)
MTS (DL-methioninoylsulfadiazine)
CYS (L-cystinoylsulfadiazine)
DMF (Dimethylformamide)
dicl (Diclofenac)

The kinetic parameters for the PPO mimic activity of Cu(ll), Mn(ll), Fe(lll) complexes with N_2O_2 were also evaluated. The final concentration of the metal complex was 0.1 mmol L^{-1}, 3,5-DTBC was 10 mmol L^{-1}, and 2,6-DMP was 100 mmol L^{-1}. At low substrate concentration, the reaction displayed first order kinetics, however, saturation kinetics was observed at high substrate concentration for all complexes. The complexes showed comparable values for K_m and V_{max}, while K_{cat} for Fe containing complex was higher than the other metal complexes, thus suggesting the higher activity [39].

The kinetic measurements of Cu (II) complexes containing sulfur/nitrogen donor sets mimicking the function of catechol oxidase were made. The studies showed that by varying the catalyst concentration and keeping the 3,5-$DTBCH_2$ (3,5-di-tert-butycatechol) concentration constant, the reaction followed a first–order kinetics. The catalytic activities found to be strongly dependent on the nature of both the ligand substituents within the carbonyl moiety and the junction between the central N donors of the azomethine linkages of the in situ formed diimine ligands. The outcomes from Linewevear–Burk plots are $V_{max} = (22.0–125.0)10^{-3}$ M s^{-1}, $K_M = (14.3–30.0) \times 10^{-3}$ M, $k_{cat} = 8.8–500.0$ s^{-1}, and $k_{cat}/K_M = (3.52–15.16)10^2$. By comparing these results with other catechol oxidase model systems of mononuclear Cu(II) complexes and the present Cu(II) complexes, they showed the modest catechol oxidase biomimetic catalytic activity, but much lower than those reported for the binuclear Cu(II) complexes ($k_{cat} = 200–6000$), and at least many orders of magnitude less than the natural enzyme. ($k_{cat} = 8250$) [37].

The initial rate of oxidation was used to determine the catalytic activity of Cu(II)–MTS (methioninoylsulfadiazine) and Cu(II)–CYS (cystinoylsulfadiazine) in the homogeneous oxidation of phenol and catechol in ethanolic solution at 25 °C. MBTH was added to copper(II) complex together with very dilute solution of phenol or catechol dissolved in ethanol, then H_2O_2 was added to the solution in a UV-cell to perform the oxidation. due to the formation of an adduct between 3-methyl-2-benzothiazolinone hydrazone (MBTH) with the oxidized form of phenol or catechol with time was obtained on Varian Cary 3 E spectrophotometer. The initial rate was determined from a linear increase in concentration with time. The rate law for this reaction can be obtained with steady-state approximation similar to the Michaelis-Menton kinetics and K_{cat} comes out to be 1.58×10^{-2} s^{-1} for Cu(II)-MTS, and K_{cat} for Cu(II)-CYS comes out to be 1.27×10^{-2} s^{-1}. The rate constants are comparable or much higher than those published [56].

6 Mechanism of Action of PPO and PPO Mimics

6.1 Mechanism of Tyrosinase Enzyme

Tyrosinase (EC 1.14.18.1) is a cuproprotein that, with the assistance of molecular oxygen, catalyses the hydroxylation of monophenols to o-diphenols (monophenolase activity), and the oxidation of o-diphenols to o-quinones (diphenolase activity). This enzyme may be found all around the evolutionary tree. In the catalytic cycle, the enzyme tyrosinase exists in three forms: meta, deoxy, and oxy, which are distinguished by the degree of copper oxidation: Eox: oxytyrosinase ($Cu^{2+}Cu^{2+}O_2^{-2}$); Em: metatyrosinase ($Cu^{2+}Cu^{2+}O_2^{-2}$); Ed: deoxytyrosinase (Cu^{1+} Cu^{1+}); Em: deoxytyrosinase (Cu^{1+} Cu^{1+}); Em: deoxytyrosinase (Cu^{1+} Cu^{1+}); Em: deoxytyrosinase. In the case of mushroom tyrosinase, the enzyme is saturated at the oxygen concentrations seen in the solutions. Em and Eox, two enzyme forms that oxidize o-diphenols to oquinones, are active on o-diphenols, Em without oxygen, and Eox with oxygen. The Em form, on the other hand, is inactive when it comes to monophenols, generating an Em M dead-path complex. The Eox form is active on monophenols, hydroxylating them into o-diphenols by an electrophilic aromatic substitution. The kinetic complexity of the tyrosinase monophenolase activity is due to the difference in activities of the Em and Eox forms. The Em M complex is inactive, thus in order for the enzyme to work on monophenols, the Em form must be reduced to produce the Ed form, which then binds to oxygen to form Eox, completing the catalytic cycle as shown in Scheme 1.

Diphenolase activity is unaffected since the two forms Em and Eox are both active on o-diphenol (D) as shown in Scheme 2. This problem (between the Em and Eox forms) has been solved in nature, as the product of the enzyme when it acts on its physiological substrates (L-tyrosine (M), Ldopa (D) is o-quinones, which evolves to

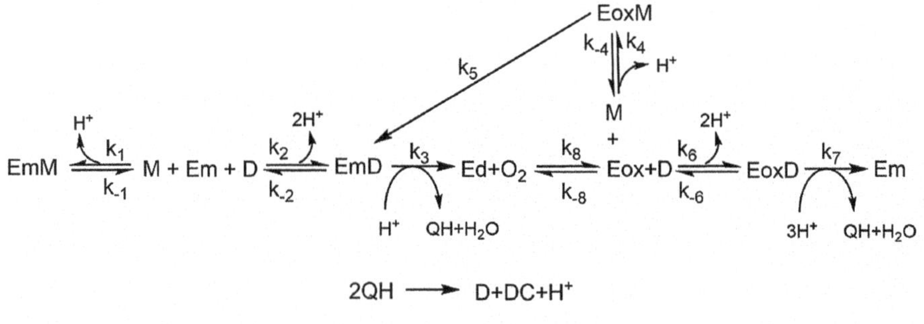

Scheme 1. Monophenolase and diphenolase activities of tyrosinase. M is monophenol, D is o-diphenol, QH is o-dopaquinone protonated, and DC is dopahrome. The enzymatic forms are: Em (metatyrosinase), Ed (deoxytyrosinase), and Eox (oxytyrosinase). Reproduced from García-Molina et al. [57], under the terms and conditions of the Creative Commons Attribution (CC BY) license. https://creativecommons.org/licenses/by/4.0/)

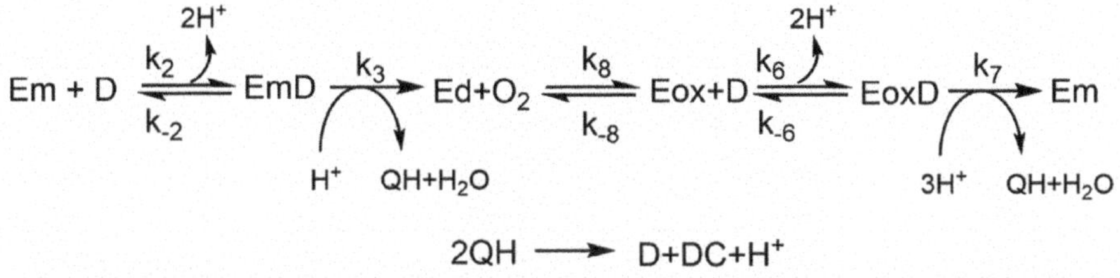

Scheme 2. Diphenolase activity of tyrosinase. D is o-diphenol, QH is o-dopaquinone protonated and DC is dopachrome. Reproduced from García-Molina et al. [57], under the terms and conditions of the Creative Commons Attribution (CC BY) license. https://creativecommons.org/licenses/by/4.0/)

go through chemical reactions in the medium, generating D, which reduces Em to Ed and thus Eox is formed.

When tyrosinase reacts with M, it produces D, which then becomes o-Q, which then produces dopachrome (DC) and accumulates D in the medium (Scheme 1). That is, when tyrosinase works on M (monophenolase activity), enzymatic processes involving diphenolase activity, as well as chemical reactions involving evolution from o-Q, will occur. Because of the overlap in activities (monophenolase and diphenolase), several researchers have attempted to segregate the monophenolase activity from the others in order to gain direct kinetic information on this enzyme activity. The initial attempt at isolating monophenolase activity employed hydroxylamine to convert Em to Ed and borate at pH = 8 to inhibit the D produced by the enzyme from M. These researchers quantify oxygen consumption (O_2) and conduct steady-state rate assessments in the lab [57].

6.2 Mechanism of PPO-Like Activity of Synthesized NPs

The interacting nature between the catechol and the synthesized NCs was expected by electron transfer process. Figure 3 shows the probable mechanism for PPO activity of synthesized NCs. In the presence of NCs catechol donates its electrons to NCs by electron transfer process to become catecholate. The electron transfer process caused this catecholate to produce semiquinone free radicals along with the o-quinones in the presence of dissolved O_2.

In the case of Mn_3O_4 NPs, electron gain by Mn^{3+} is accomplished by the coordination of catecholate with metal directly into the sigma orbital which is vacant ($3d^4$ $3d^5$) without changing the spin state. The Mn^{3+} ion is reduced to Mn^{+2} ion by accepting an electron from catechol.

Iron (Fe) has electronic configuration of [Ar] $3d^6 4s^2$, but in the case of Fe_2O_3 NPs, Fe is present as Fe^{3+}, which already has a stable electronic configuration ($3d^5$), which inhibits its participation in electron transfer reactions. In the case of GO, it acts as Lewis base and donates electrons. In the case of nanocomposite, however,

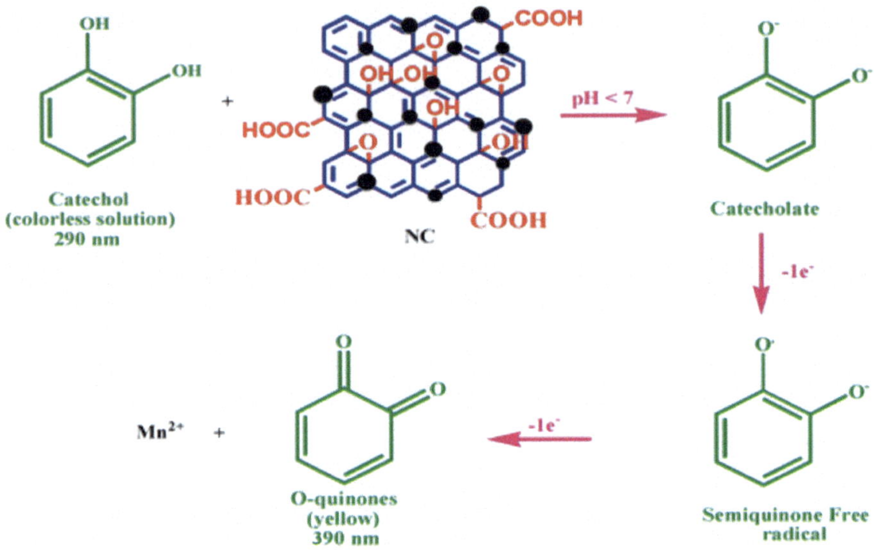

Fig. 3 The process of electron transfer between catechol and the NC. Reprinted with permission from Singh et al. [28] Copyright (2021), Elsevier

the activity is reduced because GO obstructs the electron transfer pathway. The only activity shown by NC is due to the presence of Mn_3O_4 NPs. The activity of PPO depends mainly on the process of electron transfer and surface area, but the process of electron transfer was dominated. Without the electron transfer process, catechol cannot be reduced to more catecholate in the semiquinone free radicals and o-quinone [28].

Catechol solution was characterized by an absorption peak at wavelength = 284 nm. The oxidation of catechol solution in the presence of Mn_3O_4 and NC was determined spectrophotometrically by the formation of o-quinones at wavelength = 390 nm. The higher the absorbance, the more is the PPO-like activity [28].

6.3 Another Mechanism

Several studies on catechol oxidation have shown that the degree of liability of the fifth ligand has an effect on the rate of catalysis for five-coordinate Cu (II) complexes with tetradentate ligands. Furthermore, only once catechol and copper(II) produce a copper(II) catecholate intermediate, can electron transport from catechol to copper(II) begin. The dissociation capability of coordinated hydroxo or chloro of [LCu(OH)](ClO_4) **(1)** and [LCuCl]Cl **(2)** has a significant impact on the capacity of the complexes to catalyse this oxidation (Fig. 4). Although the five-coordinate [LCu(OH)](ClO_4) **(1)** and [LCuCl]Cl **(2)** (containing tetradentate ligand) have a vacant coordination site, the possibility of binding the highly basic catecholate anion to copper(II) in [CuLX] appears remote. This is because the resulting intermediate [catecholate–$(CuLX)_2$] is highly unstable due to excessive electron density on copper(II) and the Jahn–Teller effect, the intermediate [catecholate-$(CuLX)_2$] is highly unstable. Oxidation begins with the dissociation of OH^- or Cl^-. As a result, the bigger the OH^- or Cl^- binding constant, the slower the reaction will be. The susceptibility of chloro is greater than that of hydroxo according to a kinetic investigation of ligand substitution **1** and **2**; the same order holds in their interaction with 3,5-DTBC. These findings show that catecholase mimetic activity is influenced by the rate of hydroxo or chloro dissociation. Cu(II) Complexes' catalytic characteristics are heavily influenced by their reduction potentials [36].

7 Reactions Catalysed by PPO

Polyphenol oxidase catalyses the various types of oxidation reduction reactions (Fig. 5). The well-known substrates and their products are monophenols (p-cresol) to ortho quinone (**2**), 4-methyl catechol into 4-methyl-1, 2-benzoquinone (**3**). It catalyses diphenols (e.g. Catechol) into 1, 2-benzoquinone (**4**) and also catalyses triphenols (e.g. Pyragallol) into hydroxy-o-benzoquinone (**5**). The hydroxylation of monophenolic compounds occurs at a slower rate than oxidation of o-diphenolic compounds. Some polyphenol oxidases that hydroxylate monohydroxy phenols (such as those found in mushrooms and human skin) can produce BH_2 over time, resulting in enzyme activation. The initial rate of product generation for these enzymes shows hysteresis, which accelerates with time. However, 3-tritiated p-cresol without or with the catalytic quantities of catechol undergo redox reaction. Chemical moieties such as H, longer alkyl chains, COOH, F, and NO_2 can be substituted for the 4-methyl group in p-cresol and 4-methyl catechol. The 1,4-dihydroxyphenols and 1,3,5-trihydroxyphenols are inhibitors of polyphenol oxidases or create michael addition compounds with the production of enzymatically generated o-benzoquinones. The two-electron oxidation of 3,5-di-tert-butylcatechol by catechol oxidase biomimetic catalytic activity was also investigated (6). The 3,5-di-tert-butyl-o-quinone (3,5-DTBQ) is a stable oxidation product with a strong absorption at $\lambda_{max} = 400$ nm. The extraordinarily high room-temperature stability of 3,5 DTBQ shows that a single-oxidation reaction route is followed, and that the resulting o-quinone does not undergo additional oxidative cleavage. As a result, the polyphenol oxidase enzyme may catalyse a wide range of reactions when given the right substrates [58, 59].

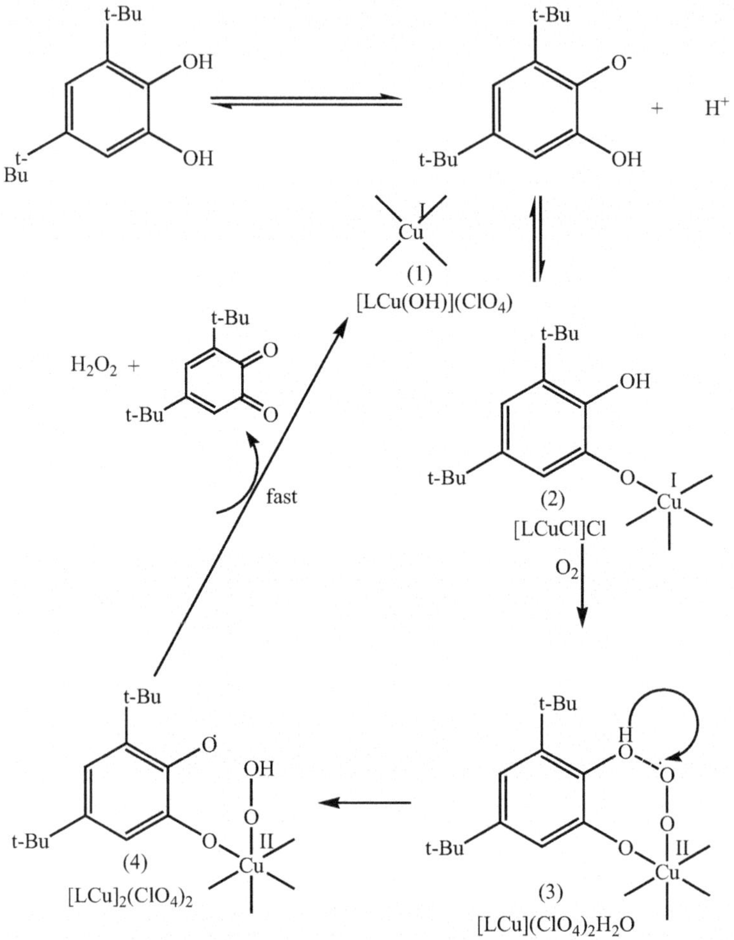

Fig. 4 Probable mechanism for aerobic catalytic oxidation of 3,5-DTBC in the presence of 4. Reprinted with permission from Ramadan et al. [36], with permission from Taylor & Francis. Copyright (2012)

8 Role of PPO Mimics in Environmental Decontamination

Phenolic compounds and related phenolics are among the most common contaminants found in industrial wastewater from industries such as steel, metals, petroleum refining, resin, plastic, various types of the chemical products, medications, and dyes, and many others. The concentration of these contaminants is usually somewhere between 100 and 1,000 mg. These molecules are linked to a variety of health problems, including heart arrhythmias, kidney illness, skin cancer, and even death in extreme circumstances. When they enter a live system, they quickly spread

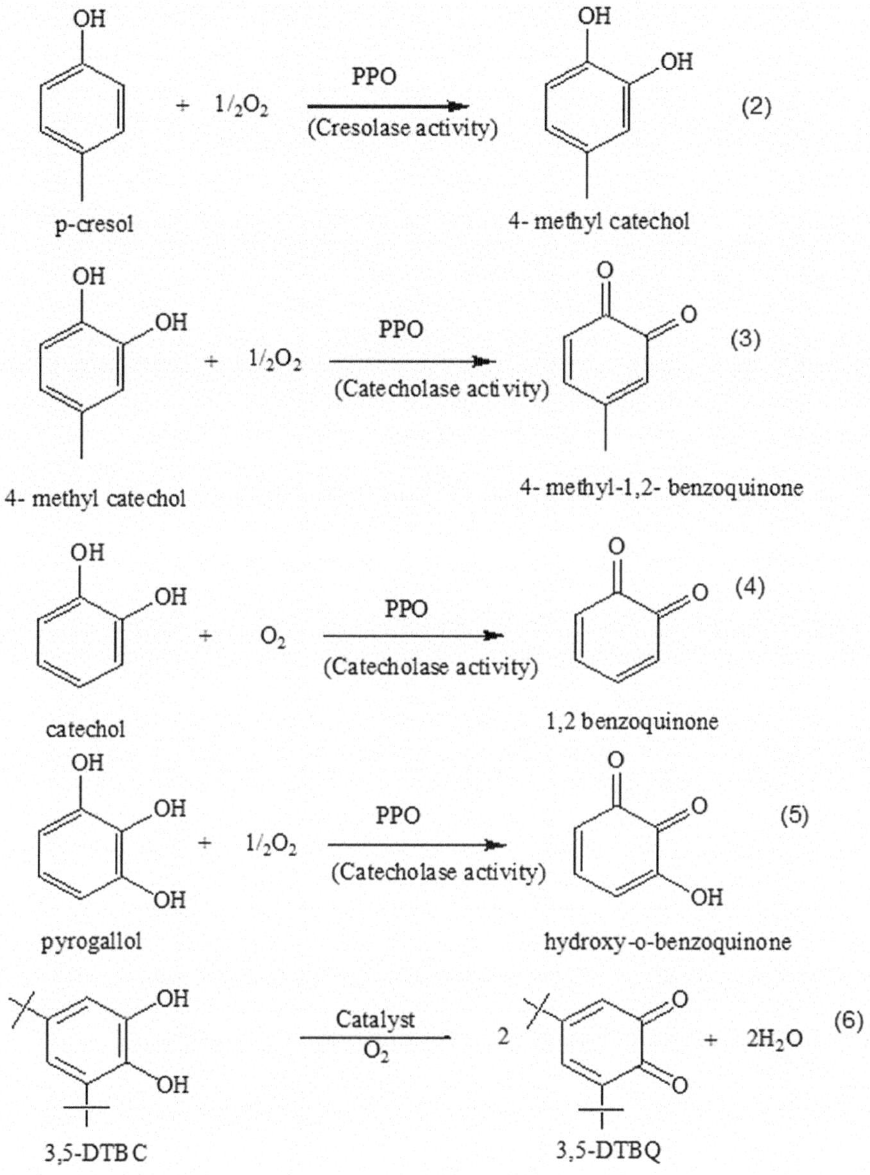

Fig. 5 Reactions catalysed by PPO/PPO mimic [59]

throughout the body and begin to harm it. As a result, the elimination of such pollutants is critical. Toxic pollutants can be removed from industrial wastewater using a variety of physical, chemical, and biological processes. Quick sorption by activated sludge, adsorption by low-cost adsorbents, powdered activated carbon, and pyrolysed rusk of rice with subsequent biodegradation are some of the more modern physical methods. Pulsed high-voltage discharge is an effective physical approach for removing a variety of organic pollutants. The photo–fenton reaction is used in the chemical approach to remove various synthetic, aromatic, or natural chemicals. The solvent-impregnated resin system is used to remove different phenolic and thio phenolic compounds from water by creating complexes and hydrophobic interactions of phosphine oxides and phosphates. It is feasible to purify wastewater using ion exchange membranes as solid polymer electrolytes. The biodegradation of different phenolic compounds is mostly accomplished through phytoremediation and microbiological elimination methods [60–64].

As a result, the need of the hour is to investigate innovative strategies with high effectiveness, efficiency, and application. Enzymatic treatment is a better biological method for removing different phenolic chemicals from industrial effluent. The employment of oxidoreductive enzymes can effectively treat phenolic chemicals, making it a cost-effective and efficient approach. Polypehnol oxidase is far more effective at removing phenolic pollutants from waste water. Polyphenol oxidase enzyme is widely utilized for bioremediation of phenolic pollutants in industrial effluent. This PPO enzyme is found in 2 types: tyrosinase and laccase, and it is found in a wide range of microbes, plants, and animals. These enzymes' oxidoreductive nature permits them to function in a broader pH and temperature range. They also destroy a wide range of mono and diphenolic compounds when immobilized on various carriers or matrices. The high cost of manufacture prevents these enzymes from being widely used for decontamination on a large scale. Meanwhile, several bench and field investigations have indicated that enzymatic wastewater treatment is the best option for phenolic compound biodegradation via the biological route. PPO conjugates from various nanomaterials have also been used to remove some phenolic chemicals, reducing the downsides of various enzymatic water treatments [65–68]. Other than polyphenol oxidase, several other enzymes have been found, including phenol oxidase, phenol hydroxylase, and cytochrome P450. These enzymes are extremely effective in biodegrading phenolic pollutants from industrial water treatment plants [69, 70]. There is also a plethora of information about the usage of both free and immobilized PPO in various applications. However, there is still a need for a comprehensive analysis of the elimination of diverse phenolic pollutants from industrial wastewater employing PPOs such as laccase and tyrosinase.

9 Tyrosinase-Based Biosensors

9.1 Tyrosinase-Based Amperometric biosensors

Various amperometric biosensors based on tyrosinase have been described in various studies. For the detection of phenolics and herbicides, tyrosinase was immobilized using an entrapment approach in a polymer called poly 3,4-ethylenedioxythiophene (PEDT). For various parameters, a glassy carbon electrode was tuned. Amperometric tests were carried out. The biosensor was sensitive to phenolic chemicals at values of 5–500 nM. Herbicides inhibit tyrosinase, hence their effectiveness was determined by this. The herbicides atrazine and diuron had detection limits of 1 and 0.5 mg L^{-1}, respectively. The cross-linking method of immobilization was used to create a tyrosinase biosensor. To detect phenolic chemicals during flow injection, the enzyme was cross-linked at a 3-mercaptopropionic acid (MPA) self-assembled monolayer (SAM) on a gold disc electrode. The detection potential (100 mV against Ag|AgCl|KCl 3 M), flow rate (1.02 ml min^{-1}), injection volume (350 μl), and carrier solution pH (0.05 M phosphate buffer of pH = 7.0) were all optimized in the experiments. The biosensor was very reproducible and did not require any pre-treatment processes. The biosensor had a five-day lifespan. The tyrosinase biosensor demonstrated high consistency of flow-injection (FI) readings in these settings, without the requirement for pretreatment operations. For five days, the biosensor remained steady. The biosensor was beneficial for measuring phenolic chemicals in real time [71].

9.2 Tyrosinase-Based Voltametric Biosensors

Using a boron-doped diamond (BDD) electrode modified by tyrosinase, a tyrosinase-based biosensor was produced (Ty). The immobilization was carried out using polyaniline (PANI) doped with polyvinyl sulfonate (PVS) composite sheets, with voltametric techniques used for detection. The biosensor was sensitive to L-phenylalanine and L-tyrosine as a substrate. For both substances, the detection limit was 1.0×10^{-2} M. After 24 h, up to 85% of the activity was preserved. This biosensor eliminates the sample preparation and separation steps necessary for other detection techniques, and it may be used to determine phenol in red wine. Adsorbing and immobilizing the enzyme on the surface of high isoelectric point ZnO nanoparticles was used to create a tyrosinase biosensor (nano-ZnO). Electrostatic interactions and the development of a layer by chitosan on a glassy carbon electrode assisted this adsorption. The biosensor does not require any additional electron mediator. There were optimization studies carried out. The biosensor's detection limit was 5.0×10^{-8} mol L^{-1} [71].

9.3 Tyrosinase-Based Optical Biosensors

A tyrosinase-based optical biosensor was created employing a hybrid nafion/sol–gel immobilization system for 3-methyl-2-benzothiazolinone hydrazone (MBTH) and chitosan for tyrosinase immobilization. The quinones formed as a result of the oxidation process were stabilized by MBTH by producing a quionone MBTH adduct. As the concentration of substrate rises, the colour intensity increases. When the concentrations of phenol, catechol, and m-cresol were 0.5–7.0, 0.5–10.0, and 1.0–13.0 mg/L, the biosensor was responsive, with limits of detection of 0.18, 0.23, and

0.43 mg/L, respectively. For three months, the biosensor remained steady. Fiorentino et al. created a phenolic chemical detection optical biosensor based on tyrosinase. Layer by layer assembly of polycation polymer (dimethyldiallylammonium chloride). Absorption and fluorescence spectroscopy were used to investigate different aspects of the biosensor. The biosensor was utilized to detect L-DOPA and showed excellent reusability and stability. The studies using absorbance yielded a detection limit of 23 μM and a linear response up to 350 μM. Fluorescence measurements yielded a limit of detection of 3 μM and a linear response in the tens of μM range [71, 72].

9.4 Tyrosinase-Based Thermal Biosensors

The heat created or emitted because of a biological reaction is monitored in thermometric biosensors. The course of chemical processes and the structural dynamics of molecules were studied using the thermodynamic characteristics of molecules. The utilization of these features in biosensors, on the other hand, established the foundation for the creation of thermometric devices. Xie et al. [73] created a biosensor that combines electrochemical and calorimetric techniques to detect both electric and thermal signals concurrently. The enzyme was immobilized on a reticulated vitreous carbon (RVC) matrix coated with poly (pyrrole). Platinum was used to create the enzyme column. Quinones are formed when catechol is oxidized. Variations in temperature are measured calorimetrically, whereas current changes are detected electrochemically. The biosensor's sensitivity is improved by catechol cycling. This hybrid biosensor may be utilized to conduct comparative research. The fact that so few studies have been conducted on using tyrosinase as a thermal biosensor shows that this is an area in need of further research [71].

10 Conclusion and Future Perspectives

Natural enzymes have inherent constraints, such as their high cost, limited stability, and difficulty in storage, which sparked the creation and development of PPO analogues. Different metal oxides and other nanomaterials as PPO-mimics, as well as parameters impacting PPO-mimetic activity are discussed in detail. The mode of action and kinetic analysis of PPO mimic activity and their practical applications in different matrices is an important domain for future research to develop highly active PPO mimics. Metal-based organic complexes and nanozymes have emerged as PPO-mimics. The most noteworthy aspect is that the detection techniques such as amperometric, fluorometric, and colourimeteric have proved to be a very essential to measure the PPO-mimic efficiency of various nanomaterials. Research on practical applications of PPO-mimics need to be strengthened to replace with natural PPO in different laboratory assays and other potential applications. The environmental applications of PPO mimics are another emerging research area, which has potential applications in water remediation.

References

1. X. Liu, W. Qi, H. Zhao, L. Zhang, Y. Su, Y. Lu, BSA-templated MnO_2 nanoparticles as both peroxidase and oxidase mimics. Analyst **137**(19), 4552–4558 (2012). https://doi.org/10.1039/C2AN35700C
2. H. Tan, Q. Li, Z. Zhou, C. Ma, Y. Song, F. Xu, L. Wang, A sensitive fluorescent assay for thiamine based on metal-organic frameworks with intrinsic peroxidase-like activity. Anal. Chim. Acta **856**, 90–95 (2015). https://doi.org/10.1016/j.aca.2014.11.026
3. H. Wei, E. Wang, Nanomaterials with enzyme-like characteristics (nanozymes): Next-generation artificial enzymes-II. Chem. Soc. Rev. **48**(4), 1004–1076 (2019). https://doi.org/10.1039/C8CS00457A
4. R.M. Daniels, M. Dines, H.H. Petach, The denaturation and dehydration of stable enzyme at high temperatures. Biochem. J. **317**, 1–11 (1996). https://doi.org/10.1042/Fbj3170001
5. J.X. Xie, X.D. Zhang, H. Wang, H. Zheng, Y. Huang, J. Xie, Analytical and environmental applications of nanoparticles as enzyme mimetics, TrAC Trend. Anal. Chem. **39**, 114–129 (2012). https://doi.org/10.1016/j.trac.2012.03.021
6. Y. Lin, J. Ren, X. Qu, Catalytically active nanomaterials: A promising candidate for artificial enzymes. Acc. Chem. Res. **47**(4), 1097–1105 (2014). https://doi.org/10.1021/ar400250z
7. Z. Chen, J.J. Yin, Y.T. Zhou, Y. Zhang, L. Song, M. Song, S. Hu, N. Gu, Dual enzyme-like activities of iron oxide nanoparticles and their implication for diminishing cytotoxicity. ACS Nano **6**, 4001–4012 (2012). https://doi.org/10.1021/nn300291r
8. W. Chen, J. Chen, Y. B. Feng, L. Hong, Q. Y. Chen, L.F. Wu, X.H Lin, X.H Xia Peroxidase-like activity of water-soluble cupric oxide nanoparticles and its analytical application for detection of hydrogen peroxide and glucose. Analyst. **137**(7)**,** 1706–1712 (2012). doi: https://doi.org/10.1039/c2an35072f
9. A. Dancs, K. Selmeczi, D. Arus, D. Szunyogh, T. Gajda, Catechol Oxidase and SOD mimicking by Copper (II) complexes of multihistidine peptides. Int. J. Pept. Res. Ther. **24**(4), 571–575 (2017). https://doi.org/10.1007/s10989-017-9645-x
10. F. Manea, F.B. Houillon, L. Pasquato, P. Scrimin, Angew, Nanozymes: Gold nanoparticle-based transphosphorylation. Angew. Chem. Int. Ed. Engl. **43**(45), 6165–6169 (2004). https://doi.org/10.1002/anie.200460649
11. K. Xiong, L. Zhou, J. Wang, A. Ma, D. Fang, L. Xiong, Q. Sun, Construction of food-grade pH-sensitive nanoparticles for delivering functional food ingredients. Trends Food Sci. Technol. **96**, 102–113 (2020). https://doi.org/10.1016/j.tifs.2019.12.019
12. L. Su, W. Dong, C. Wu, Y. Gong, Y. Zhang, L. Li, G. Mao, S. Feng, The peroxidase and oxidase-like activity of $NiCO_2O_4$ mesoporous spheres: Mechanistic understanding and colorimetric biosensing. Anal. Chim. Acta **951**, 124–132 (2017). https://doi.org/10.1016/j.aca.2016.11.035
13. J. Golchin, K. Golchin, N Alidadian, S. Ghaderi, S. Eslamk hah, M. Islamkhah, A. Akbarzadeh, Nanozyme applications in biology and medicine: An overview. Artif. Cells Nanomed. Biotechnol. **45**(6)**,** 1069–1076 (2017). doi:https://doi.org/10.1080/21691401.2017.1313268
14. A. Hayat, J. Cunningham, G. Bulbul, S. Andreescu, Evaluation of the oxidase-like activity of nanoceria and its application in colorimetric assays. Anal. Chim. Acta **885**, 140–147 (2015). https://doi.org/10.1016/j.aca.2015.04.052
15. M.P. Fraignier, L. Marques, A. Fleuriet, J.J. Machiex, Biochemical and immunochemical characteristics of polyphenol oxidases from the different fruits of *Prunus*. J. Agric. Food Chem. **43**, 2375–2380 (1995). https://doi.org/10.1021/jf00057a011
16. Md. N. Karim, S.R. Anders on, S. Singh, R. Ramanathan, V. Bansal, Nanostructured silver fabric as a free-standing Nanozyme for colorimetric detection of glucose in urine. Biosens. Bioelectron. **110,** 8–15(2018). doi:https://doi.org/10.1016/j.bios.2018.03.025

17. M. Comotti, C. Della Pina, E. Falletta, M. Rossi, Aerobic oxidation of glucose with gold catalyst hydrogen peroxide as intermediate and reagent. Adv. Synth. Catal. **348**(3), 313–316 (2006). doi:https://doi.org/10.1002/adsc.200505389
18. C. Xu, C. Zhao, M. Li, L. Wu, J. Ren, X. Qu, Artificial evolution of graphene oxide chemzyme with enantioselectivity and near-infrared photothermal effect for cascade biocatalysis reactions. **10**(9), 1841–1847(2014). doi:https://doi.org/10.1002/smll.201302750
19. S. Liu, J. Tian, L. Wang, Y. Luo, X. Sun, A general strategy for the production of photoluminescent carbon nitride dots from organic amines and their application as novel peroxidase-like catalysts for colorimetric detection of H_2O_2 and glucose. RSC Adv. **2**(2), 411–413 (2012). https://doi.org/10.1039/C1RA00709B
20. Y. Song, X. Wang, C. Zhao, K. Qu, J. Ren, X. Qu, Label-free colorimetric detection of single nucleotide polymorphism by using single-walled carbon nanotube intrinsic peroxidase-like activity. Chem. Eur. J. **16**(12), 3617–3621 (2010). https://doi.org/10.1002/chem.200902643
21. Y. Li, Z. Kang, L. Kong, H. Shi, Y. Zhang, M. Cui, D.P. Yang, MXene-Ti_3C_2/ CuS nanocomposites: Enhanced peroxidase-like activity and sensitive colorimetric cholesterol detection. Mater. Sci. Eng. C **104**, 110000 (2019). https://doi.org/10.1016/j.msec.2019.110000
22. Q. Liu, Y. Ding, Y. Yang, L. Zhang, L. Sun, P. Chen, C. Gao, Enhanced peroxidase-like activity of porphyrin functionalized ceria nanorods for sensitive and selective colorimetric detection of glucose. Mater. Sci. Eng. C Mater Biol Appl. **59**, 445–453 (2015). doi:https://doi.org/10.1039/C7AY02459B
23. R. Xiangling, W. Mingqian, H. Xing, Li Zheng, Z. Jing, Z. Wei, C. Xudong, R. Hong, M. Xianwei, Superoxide dismutase mimetic ability of Mn-doped ZnS QDs. Chin. Chem. Lett. **29**(12),1865–1868(2018).doi:https://doi.org/10.1016/j.cclet.2018.12.007
24. D. Song, T. Li, Y.Y. Wei, Z.R. Xu, Controlled formation of porous $CuCo_2O_4$ nanorods with enhanced oxidase and catalase catalytic activities using bimetal-organic frameworks as templates. Colloids Surf. B Biointerfaces **188**, 110764 (2020). https://doi.org/10.1016/j.colsurfb.2019.110764
25. Y. Liu, H. Wu, Y. Chong, W. G. Wamer, Q. Xia, L. Cai, Z. Nie, P.P. Fu, J.J Yin, Platinum nanoparticles: Efficient and stable catechol oxidase mimetics. ACS Appl. Mater. Interfaces **7**, 19709–19717 (2015) doi:https://doi.org/10.1021/acsami.5b05180.
26. J.W. Lee, S. Yoon, Y.M. Lo, H. Wu, S.Y. Lee, B. Moon, Intrinsic polyphenol oxidase-like activity of gold@platinum nanoparticles. RSC Adv. **5**(78), 63757–63764 (2015). https://doi.org/10.1039/C5RA07636F
27. C. Altinkayank, E. Kocazorbaz, N. Ozedemir, F. Zihnioglu, Egg white nanoflower(EW-hNFs) with biomimetic polyphenol oxidase activity, Synthetic characterization and potential use in decolorization of synthetic dyes. Int. J. Biol. Macromol. **109**, 205–211 (2018). https://doi.org/10.1016/j.ijbiomac.2017.12.072
28. M. Singh, M Kaur, M. kaur Sangha, MK. Ubhi, Comparitive evaluation of Manganese oxide and its graphene oxide nanocomposite as polyphenol oxidase mimics. Materials Today Commun. **27**, 102237 (2021). doi :https://doi.org/10.1016/j.mtcomm.2021.102237
29. M.I. Ayad, Synthesis, Characterization and catechol oxidase biomimetic catalytic activity of cobalt(II) and copper(II) complexes containing N_2O_2 donor sets of imine ligands. Arabian J. Chem. **9**, S1297–S1306 (2012). https://doi.org/10.1016/j.arabjc.2012.02.007
30. P. Adak, C. Das, B. Ghosh, S. Mondal, B. Pakhira, E. Sinn, A. J. Blake, AE. O'Connor, SK. Chattopadhya, Two pseudohalide-bridged Cu(II) complexes bearing the anthracene moiety: Synthesis, crystal structures and catecholase-like activity. Polyhedron **119**, 39–48 (2016) doi :https://doi.org/10.1016/j.poly.2016.08.015
31. A.L. Abuhijleh, J. Khalaf, Copper (II) complexes of the anti-inflammatory drug naproxen and 3-pyridylmethanol as auxiliary ligand. Characterization, superoxide dismutase and catecholase mimetic activities. Euro. J. Med. Chem. **45**(9), 3811–3817 (2010). doi:https://doi.org/10.1016/j.ejmech.2010.05.031
32. A.E.M. Ramadan, S.Y. Shaban, S.M.E. Khalil, M. Shebl, R.A.S. El-Naem, Synthesis and characterization of N_3S_2 donors macrocyclic copper(II) complexes. Catechol oxidase and phenoxazinone synthase biomimetic catalytic activity. J Chinese Adv. Mater. Soc. **5**(4), 215–240 (2017). doi :https://doi.org/10.1080/22243682.2017.1331318

33. A.L. Abuhijleh, Biomimetic catalytic activities of copper (II) ferrocenecarboxylate complexes with nitrogen based ligands as catechol oxidase and phenoxazinone synthase and for oxidative coupling of 2,6-dimethylphenol. (2015) http://hdl.handle.net/20.500.11889/4157
34. S. Mahato, N. Meheta, M. Kotakonda, M. Joshi, M. Shit, A.R Choudhury, B. Biswas, Synthesis, structure, polyphenol oxidase mimicking and bactericidal activity of a zinc-schiff base complex. (2020) doi: https://doi.org/10.1016/j.poly.2020.114933
35. H. Huang, J. Bai1, J. Li, L. Lei, W. Zhang, S. Yan, Y. Li, Fluorescence detection of dopamine based on the polyphenol oxidase–mimicking enzyme. Anal. Bioanal. Chem. **412**(22), 5291–5297 (2020). doi:https://doi.org/10.1007/s00216-020-02742-1
36. A.E.M. Ramadan, M.M. Ibrahim, I.M. El-Mehasse, New mononuclear copper(I) and copper(II) complexes containing N_4 donors; crystal structure and catechol oxidase biomimetic catalytic activity. J. Coordination Chem. **65**(13), 2256–2279 (2012). https://doi.org/10.1080/00958972.2012.690513
37. A.E. Motaleb, Y.L. Aly, S.M.E. Khalil, M. Shebl, R.A.S. El Naem, Synthesis and characterization of copper (II) complexes containing the sulfur/nitrogen donor sets. Mimicking the function of catechol oxidase. Int. J. Adv. Res. **3**(7), 10–32 (2015)
38. S. Caglar, E. Dilek, B. Caglar, E. Adiguzel, E. Temel, O. Buyukgungor, A. Tabak, New metal complexes with diclofenac containing 2-pyridineethanol or 2-pyridinepropanol: Synthesis, structural, spectroscopic, thermal properties, catechol oxidase and carbonic anhydrase activities. J. Coordination Chem. **69**(22), 3321–3335 (2016). https://doi.org/10.1080/00958972.2016.1227802
39. L. Lu, Y. Song, Hui Liu, J. Zhang, Synthesis, characterization, and polyphenol oxidase activity of Cu^{II} , Mn^{II}, and Fe^{III} complexes with a N_2O_2 ligand. J. Coordination Chem. **65**(7), 1278–1288. doi: https://doi.org/10.1080/00958972.2012.671480
40. J.O. Carvalho, J.F.F. Orlanda, Heat stability and effect of pH on enzyme activity of polyphenol oxidase in buriti (Mauritia flexuosa Linnaeus f.) fruit extract. Food Chem. **233**, 159–163 (2017). doi:https://doi.org/10.1016/j.foodchem.2017.04.101
41. J.F. Dai, G.J. Wang, C.K. Wu, Investigation of the surface properties of graphene oxide and graphene by inverse gas chromatography. Chromatographia **77**(3–4), 299–307 (2014). https://doi.org/10.1007/s10337-013-2597-1
42. E. Breslmayr, M. Hanzek, A. Hanrahan, C. Lietner, R. Kittl, B. Santek, C. Oostenbrink, R. Ludwig, A fast and sensitive activity assay for lytic polysaccharide monooxygenase. Biotechnol. Biofuels **11**, 79 (2018). https://doi.org/10.1186/s13068-018-1063-6
43. L. Duan, Z. Wang, Y. Hou, Z. Wang, Zepeng, G. Gao, W. Chen, P.J. Alvarez, The oxidation capacity of Mn_3O_4 nanoparticles is significantly enhanced by anchoring them onto reduced graphene oxide to facilitate regeneration of surface-associated Mn(III). Water Res. **103**, 101–108 (2016). doi:https://doi.org/10.1016/j.watres.2016.07.023
44. A.T. Stone, Reductive dissolution of manganese(III/IV) oxides by substituted phenols. Environ. Sci. Technol. **21**, 979–988 (1987). https://doi.org/10.1021/es50001a011
45. Y. Ono, T. Matsumura, S. Fukuzumi T. Keil, Electron spin resonance studies on the mechanism of the formation of p-benzosemiquinone anion over manganese dioxide. J. Chem. Soc., Perkin Trans. **2**(11), 1421 doi:https://doi.org/10.1039/p29770001421
46. K.H. Kung, M.B. Mcbride, Electron transfer process between hydroquinone and hausmannite (Mn_3O_4). Clay's Clay Miner. **36**, 297–302 (1998)
47. X. Zhou, P. Ma, A. Wang, C. Yu, T. Qian, S. Wu, J. Shen, Dopaminen fluorescent sensors based on polypyrrole/graphene quantum dots core/shell hybrids. Biosens. Bioelectron. **64**, 404–410 (2016). https://doi.org/10.1016/j.bios.2014.09.038
48. B. Wang, Y. Chen, Y. Wu, B. Weng, Y. Liu, C. Li, Synthesis of nitrogen- and iron containing carbon dots, and their application to colorimetric and fluorometric determination of dopamine. Microchim. Acta **183**, 2491–2500 (2016). https://doi.org/10.1007/s00604-016-1885-5
49. K. Syslova, L. Rambousek, M. Kuzma, V. Najmanova, V. Bubenikova-Valesova, R. Slamberova, P. Kaser, Monitoring of dopamine and its metabolites in brain microdialysates: Method combining freezedrying with liquid chromatography-tandem mass spectrometry. J. Chromatogr. A **1218**(21), 3382–3391 (2011). https://doi.org/10.1016/j.chroma.2011.02.006

50. W. Silva , M. Emilia Ghica, R.F. Ajayi, E. I. Iwuohab,C.M.A. Brett, Tyrosinase based amperometric biosensor for determination of tyramine in fermented food and beverages with gold nanoparticle doped poly(8-anilino-1-naphthalene sulphonic acid) modified electrode. Food Chem. **282**, 18–26 (2019) doi:https://doi.org/10.1016/j.foodchem.2018.12.104
51. A. Gutes, F. Cespedes, S. Alegret, M. del Valle, Determination of phenolic compounds by a polyphenol oxidase amperometric biosensor and artificial neural network analysis. Biosen. Bioelectron. **20**(8), 1668–1673 (2005). https://doi.org/10.1016/j.bios.2004.07.026
52. X. Yan, H. Li, T. Hu, X. Su, A novel fluorimetric sensing platform for highly sensitive detection of organophosphorus pesticides by using egg white-encapsulated gold nanoclusters. Biosen. Bioelectron. **91**, 232–237 (2016). doi: https://doi.org/10.1016/j.bios.2016.11.058
53. H. Huang, J. Bai, J. Li, L. Lei, W. Zhang, S. Yana, Y. Li, Fluorometric and colorimetric analysis of alkaline phosphatase activity based on a nucleotide coordinated copper ion mimicking polyphenol oxidase. J. Mater. Chem. **7**(42), 6508–6514 (2019). https://doi.org/10.1039/C9TB01390C
54. D. Zhang, P. Du, J. Chen, H. Guo, X. Lu, Pyrazolate-based porphyrinic metal-organic frameworks as catechol oxidase mimic enzyme for fluorescent and calorimetric dual mode detection of dopamine with high sensitivity and specificity. Sens. Actuators B Chem. **341**, 130000 (2021). https://doi.org/10.1016/.snb.2021.130000
55. J.M.G. Rodriguez, N.P. Hux, S.J. Philips, M.H. Towns, Michaelis−Menten Graphs, Lineweaver−Burk plots, and reaction Schemes: Investigating introductory biochemistry students' conceptions of representations in enzyme kinetics. J. Chem. Educ. **96**(9), 1833–1845 (2019). https://doi.org/10.1021/acs.jchemed.9b00396
56. A.I. Hanafy, Z.M. El-Bahy, I.O. Ali, Synthesis and characterization of Copper (II) complexes of sulfadiazine with amino acids: Catalytic activity towards phenol and catechol. J. Coordination Chem. **65**(8), 1459–1474 (2012). https://doi.org/10.1080/00958972.2012.671936
57. P.G.-Molina, J.L.M.-Munoz, J.A. Ortuno, J.N.R-Lopez, P.A.G-Ruiz, F.G-Canovas, F.G-Molina, Considerations about the continuous assay methods, spectrophotometric and spectrofluorometric, of the menophenolase activity of tyrosinase. Biomol. **11**, 1269 (2021) doi: https://doi.org/10.3390/biom11091269
58. V.A. Edalli, S.I. Mulla, R. sharma, Y. Shouche, Evaluation of p-cresol degradation with polyphenol oxidase (PPO) immobolized in various matrices. 3 Biotech. **6**, 229 (2016). doi: https://doi.org/10.1007/s13205-016-0547-y
59. S. Saiedian, E. Keyhani, J. Keyhani, Polyphenol oxidase activity in dormant saffron (*Crocus sativus L.*) corm. Acta. Physiol. Plant. **29**(5), 463–471 (2007). doi:https://doi.org/10.1007/s11738-007-0056-z
60. A.A. Amer, S.M. Reda, M.A. Mousa, M.M. Mohamed, Mn_3O_4/graphene nanocomposites: Outstanding performances as highly efficient photocatalysts and microwave absorbers. RSC Adv. **7**(2), 826–839 (2017). https://doi.org/10.1039/C6RA24815B
61. G. Akay, E. Erhan, B. Keshinkler, O.F. Algur, Removal of the phenol from the wastewater using the membrane immobilized enzymes: Part II cross-flow filteration. J. Membrane. Sci. **206**(1–2), 61–68 (2002). https://doi.org/10.1016/S0376-7388(01)00626-3
62. S. Mukherjee, B. Basak, B. Bhunia, A. Dey, B. Mondal, Potential use of polyphenol oxidases (PPO) in the bioremediation of phenolic contaminants containing industrial wastewater. Rev. Environ. Sci. Bio-Technol. **12**, 61–73 (2013). https://doi.org/10.1007/s11157-012-9302-y
63. N. Duran, E. Esposito, Potential applications of oxidative enzymes and phenoloxidase-like compounds in wastewater and soil treatment. Appl. Catal. B Environ. **28**(2), 83–99 (2000). https://doi.org/10.1016/S0926-3373(00)00168-5
64. N. Duran, M.A. Rosa, A. D'Annibale, L. Gianfreda, Applications of laccases and tyrosinases (phenoloxidases) immobilized on different supports. Enzyme Microb. Technol. **31**(7), 907–931 (2002). https://doi.org/10.1016/S0141-0229(02)00214-4
65. W.S. El-Sayed, M. Ismaeil, F. El-Beih, Isolation of 4-chlorophenol-degrading bacteria, Bacillus subtilis OS1 and Alcaligenes sp. OS_2 from petroleum oil-contaminated soil and characterization of its catabolic pathway. Aus. J. Basic. Appl. Sci. **3**, 776–783 (2009)

66. L. Ensuncho, M. Alvarez-Cuenca, R.L. Legge, Removal of aqueous phenol using immobilized enzymes in a bench scale and pilot scale three-phase fluidized bed reactor. Bioprocess Biosyst. Eng. **27**(3), 185–191 (2005). https://doi.org/10.1007/s00449-005-0400-x
67. S. Jin, C. Wu, Z. Ye, Ying, Designed inorganic nanomaterials for intrinsic peroxidase mimics: A review. Sens. Actuators B Chem. **283**, (2018) doi:https://doi.org/10.1016/j.snb.2018.10.040
68. H. Xia, N. Li, X. Zhong, Y. Jiang, Metal-Organic frameworks: A potential platform for enzyme immobilization and related applications. Front. Bioeng. Biotechnol. **8**, 695 (2020). https://doi.org/10.3389/fbioe.2020.00695
69. E. Erhan, B. Keskinler, G. Akay, O.F. Algur, Removal of phenol from water by membrane-immobilized enzymes: Part I dead-end filtration. J. Membrane Sci. **206**(1–2), 361–373 (2002). https://doi.org/10.1016/S0376-7388(01)00779-7
70. P. Galliker, G. Hommes, D. Schlosser, P.F. Corvini, P. Shahgaldian, Laccase-modified silica nanoparticles efficiently catalyze the transformation of phenolic compounds. J. Colloid Interface Sci. **349**(1), 98–105 (2010). https://doi.org/10.1016/j.jcis.2010.05.031
71. I. Gul, M.S. Ahmad, S.M.S. Naqvi, A. Hussain, R. Wali, A.A. Farooqi, I. Ahmed, Polyphenol oxidase (PPO) based biosensors for detection of phenolic compounds: A review. J. App Biol. & Biotechnol. **5**(03), 072–085 (2017). https://doi.org/10.7324/JABB.2017.50313
72. D. Fiorentino, A. Gallone, D. Fiocco, G. Palazzo, A. Mallardi, Mushroom tyrosinase in polyelectrolyte multilayers as an optical biosensor for *o*-diphenols. Biosens. Bioelectron. **25**(09), 2033–2037 (2010). https://doi.org/10.1016/j.bios.2010.01.033
73. B. Xie, X. Tang, U. Wollenberger, G. Johansson, L. Gorton, F. Scheller, B. Danielsson, Hybrid biosensor for simultaneous electrochemical and thermometric detection. Anal. Lett. **30**(12), 2141–2158 (1997). https://doi.org/10.1080/00032719708001729

Ru-Doped ZnS as an Enhanced Visible Light-Driven Photocatalyst

Sujoy Kumar Mandal, Supriya Ghosal, Devdas Karmakar, and Debnarayan Jana

Abstract Recently, visible light-activated semiconductor photocatalyts have attracted the scientific community due to their immense success in wastewater treatment as well as decomposition of hazardous organic pollutants. In this work, sphalerite ZnS nanoparticles (NPs) with various ruthenium (Ru) concentrations have been synthesized via solvothermal method. The feasible doping of Ru into the ZnS matrix has been confirmed by structural characterizations. The electronic and optical properties have been investigated both experimentally as well as theoretically. Ru plays a vital role in the reduction of the optical band gap and simultaneous enhancement of the absorptivity towards the visible region of the solar spectrum. The photocatalytic degradation efficiency of Ru-doped ZnS has been tested for the degradation of methylene blue (MB) under visible irradiation of a light-emitting diode (LED). Photoluminescence quenching as well as Bader charge analysis have shown Ru acting as a charge centre, and plays a key role in the transfer mechanism. Moreover, due to incorporation of Ru, nonmagnetic ZnS becomes magnetic. The optical properties of the system significantly enhanced due to Ru. Thus proposed Ru-doped ZnS system can be a cost-effective superior photocatalyst as well as a good potential candidate for optoelectronic applications.

Keywords Ru-doped ZnS · Methelene blue · Photodegradation · Density functional theory · Band structure

1 Introduction

During the last century, a huge amount of industrial wastewater has been discharged into the rivers, lakes, and coastal areas. In today's world with the growing industrialization, the treatment of water pollution remains a critical issue. The released excess dyes and harmful chemicals into the aquatic ecosystem not only affects human life, but also the entire biosphere as plants and organisms living in these water bodies.

S. K. Mandal · S. Ghosal · D. Karmakar · D. Jana (✉)
Department of Physics, University of Calcutta, 92, A.P.C. Road, Kolkata 700 092, India
e-mail: djphy@caluniv.ac.in

S. J. Ikhmayies (ed.), *Advances in Catalysts Research*, Advances in Material Research and Technology, https://doi.org/10.1007/978-3-031-49108-5_5

A large scale of wastewater is generated in utilizing dyes in colouring. Among various branches of dyes, the synthetic dyes containing a long chain carbon group are extremely harmful and also extensively used in textile, food and paper industries. It is estimated that 10–25% of the textile dyes are released during the dying process and 2–20% are directly discharged into the natural environment. Over the past few decades, many traditional techniques and methods such as adsorption [1, 2], chemical oxidation [3, 4], and biodegradation [5, 6] have been developed by the researchers over the past few decades to degrade organic pollutants from wastewater. However, these methods have limitations in feasible industrial application due to slow degradation rate, incomplete degradation and secondary pollution. In this context, photocatalytic activity is the most efficient method for alleviating the negative environmental effect by decomposing these toxic pollutants into non-hazardous substances [7]. A breakthrough in photocatalysis research occurred in 1972, when Akira Fujishima and Kenichi Honda discovered the electrochemical photolysis of water [8, 9]. Since then many semiconductor oxides such as titanium dioxide (TiO_2), zinc oxide (ZnO), tungsten oxide (WO_3), and perovskite oxides like strontium titanate ($SrTiO_3$) have been utilized as efficient photocatalysts.

1.1 Photocatalysis and Its Mechanism

In photocatalysis, catalysts are activated by the light of sufficient energy and stimulated the chemical reactions on the active surface of the photocatalyst. The overall process in photocatalysis can be described by five independent steps: (i) Mass transfer of the reaction from the bulk phase (either aqueous or gaseous), (ii) Adsorption of the reactants, (iii) photocatalytic reaction, (iv) product desorption and finally (v) mass transfer of the products from the interface to the bulk phase. Among advanced oxidation processes (AOPs) semiconductor (SC) photocatalysts have been extensively used for the decontamination of water. When a semiconductor is excited with energy greater or equals to its band gap energy electron–hole ($e^- - h^+$) pairs are generated. Where, e^-_{CB} is the electron in the conduction band (CB) and h^+_{VB} is the hole in the valence band (VB). The photogenerated electron-hole pairs can migrate to the active surface of the semiconductor and then be adsorbed by it or can recombine producing heat without further reaction. The positive hole can oxidize the organic and inorganic pollutants directly, but in most of the cases, this happens by the production of hydroxyl ($OH^{\cdot}$) radicals. The $OH^{\cdot}$ radical is formed by the direct oxidation of adsorbed water (i.e. hydroxide ion, OH^-) on the surface of the semiconductor catalyst by the valence band hole. Also, the photogenerated conduction band electron indirectly produces $OH^{\cdot}$ radicals. The redox potential of the conduction band electron is −0.52 V, which is negative enough to reduce dissolved oxygen to superoxide (O_2^-) radical [10]. The superoxide radical reacts further with water to form hydroxyl radical. Moreover, produced superoxide radical is highly reactive and it can directly oxidize dye or any hazadous molecule and at the end decomposes into CO_2 and H_2O. Finally, these highly reactive radicals degrade the dyes or organic pollutants into environmentally friendly products. The overall reaction can be summarized as follows [11]:

$$SC \xrightarrow{h\nu} SC\left(e^-_{CB} + h^+_{VB}\right) \tag{1}$$

Hole from valence band

$$h^+ + H_2O/OH^- \rightarrow OH^{\cdot} + H^+ \tag{2}$$

Eletron from conduction band

$$e^-_{CB} + O_2 \rightarrow O_2^{\cdot -} \tag{3}$$

$$O_2^{\cdot -} + H_2O \rightarrow HO_2^{\cdot} + OH^- \tag{4}$$

$$2e^- + H^+ + HO_2^{\cdot} \rightarrow OH^{\cdot} + OH^- \tag{5}$$

$$h^+ \text{ or } O_2^{\cdot} \text{ or } OH^{\cdot} + \text{Dye} \rightarrow CO_2 + H_2O \tag{6}$$

The oxidation–reduction process has been illustrated schematically in Fig. 1.

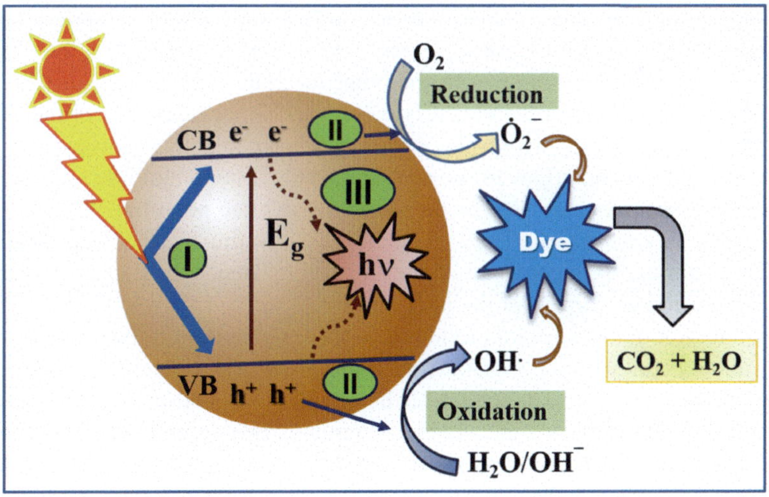

Fig. 1 Mechanism of photocatalysis: (I) Formation of electron (e^-)—hole (h^+) pairs by photoexcitation. (II) Oxidation caused by hole, and reduction caused by electron. (III) Photoinduced charge recombination

1.2 Essential Factors Involved in a Semiconductor Photocatalyst

Fundamentally, the efficiency of a semiconductor as a photocatalyst depends mainly on three key factors (i) band gap, (ii) conduction band and valence band positions, and (iii) charge carrier dynamics, such as diffusion length, charge carrier mobility, and lifetime as well as the rates of surface charge recombination and interfacial charge transfer process. Besides this, the photocatalytic degradation efficiency also depends on the surface area, as photocatalysis is a complete surface phenomenon. Band gap determines whether a photocatalyst will be activated with ultraviolet (UV) or visible irradiation. However, the band edge position of the semiconductor is the most important for using as a photocatalyst. The energies of the conduction and valence bands govern the electron injection ability of the material at the surface. Also, the ability of a semiconductor to undergo photoinduced charge carrier transfer to adsorbed species on its surface is governed by the band energy positions of the semiconductor and the redox potentials of the adsorbed species. Therefore, the knowledge on the absolute positions of E_{CB} and E_{VB} band edges is essential to explore the potential applications of the concerned semiconductor as an efficient photocatalyst. The energy level at the bottom of the conduction band actually determines the reduction potential of photoelectrons and the energy level at the top of the valence band determines the oxidizing ability of photoholes, each value reflecting the ability of the system to promote reductions and oxidations, respectively [12].

When a semiconductor is in contact with the ionic interactions at the interface of the two phases this leads to electrostatic equilibrium within the material. At the semiconductor/electrolyte interface, in particular, electrons flow from the phase of more negative Fermi energy (E_F) to the other to attain equilibrium, in which the semiconductor E_F matches the electrolyte $E_{F,\mathrm{redox}}$. This causes the formation of a space charge layer (SCL) within the semiconductor phase that is associated with the bending of the band edges in the semiconductor. The magnitude and direction of band bending can be adjusted by an externally applied potential. The externally applied potential known as flatband potential (E_{FB}), diminishes the band bending in a semiconductor that is in contact with the electrolyte. Hence the E_{FB} determines the energy of two charge carriers at the interface.

From a thermodynamic point of view, adsorbed molecules can be reduced photocatalytically by conduction band electrons if they have redox potentials lower (more positive) than the flat band potential of the conduction band and can be oxidized by valence band holes if they have redox potentials higher (more negative) than the flatband potential of the valence band [13–15]. However, the primary criteria to get a good semiconductor photocatalyst for organic

Compound degradation is that the redox potentials of the $H_2O/OH^{\cdot}$ ($HO^- = OH^{\cdot} + e^-$, $E_0 = -2.8V$) [18] couple lies within the band gap domain of the material, and also they are stable over prolonged periods of time [19]. The band gap and the corresponding band edge positions of some frequently used semiconductors are presented in Fig. 2. An ideal photocatalyst should possess the following

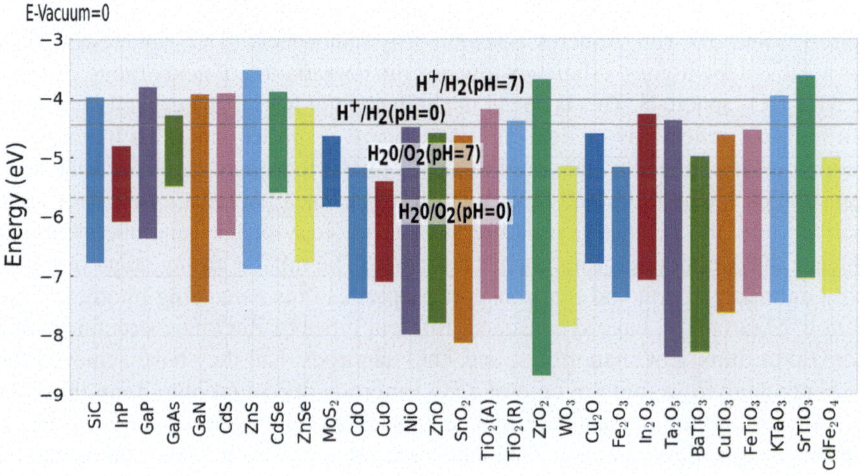

Fig. 2 Diagram for band gap and band edge positions of various semiconductors with respect to vacuum level. We have reproduced the plot using the data from Refs. [16, 17]

characteristics: (i) photoactive, (ii) has the ability to utilize visible and/or near-UV light (UV-B), (iii) biologically and chemically inert, (iv) photostable (not prone to photoanodic corrosion), (v) inexpensive, and (vi) non-toxic. To improve the photocatalytic performance, we have to consider the basic principles of photocatalysis. Also, the temperature, and pH of the solution affect the photocatalytic behaviour. Over the decades, various approaches have been taken to form semiconductor heterostructures which will be discussed in the following section.

1.3 Classification of Semiconductor Photocatalysts

Over the past decades, many semiconducting materials such as TiO_2, ZnO, WO_3, Fe_2O_3, CdS, CdSe, etc., have been used as photocatalysts. Among various semiconductors, TiO_2 is the most studied material as an efficient photocatalyst in many applications such as environmental salvation, energy production, and self-cleaning [20]. The reason is that the nanostructured TiO_2 with high surface area is easily synthesizable, nontoxic, low cost, and highly stable. After TiO_2, ZnO is the second most studied semiconducting material used as a photocatalyst. The unique characteristics of ZnO, such as direct and wide band gap, strong oxidation ability, large free exciton binding energy make ZnO as a potential photocatalyst in environment management. In the early stage of photocatalysis, researchers focussed to synthesize various nanostructures with high surface area to improve the photocatalytic activity, as it is a surface phenomenon. For example, Gao et al. [21] have synthesized various nanostructures of TiO_2 including one-dimensional (1D) TiO_2 nanotubes, 1D TiO_2

nanowires, three-dimensional (3D) TiO_2 spheres assembled by nanoparticles (TiO_2 sphere-P), and 3D TiO_2 spheres assembled by nanosheets (TiO_2 sphere-S). They have achieved the highest efficiency towards the degradation of acid orange 7 (AO_7) dye with TiO_2 sphere-S. This is due to the highest light harvesting capacity resulting from multiple reflections of light on TiO_2 sphere-S surface, and also hierarchical mesoporous structure. Paramasivam et al. [22] have thoroughly reviewed the self-organized TiO_2 nanotubes and other nanostructures as a photocatalysts. They have described how the geometry, structure and morphology of the nanotube assembly influence the overall photocatalytic performance. The tubular arrangement provides highly defined electron and ion transport properties thus enhancing photocatalytic reaction. Mauro et al. have synthesized different types of ZnO nanostructures such as ZnO thin films, ZnO nanofibres, and ZnO nanorods, and they have achieved the best photocatalytic activity in case of ZnO nanorods deposited onto 3 nm thin film [23]. Xia et al. in their research article had discussed the recent development in the synthesis routes of hierarchical ZnO nanostructures, as well as their photocatalytic ability [24].

But, the band gap of the most semiconducting photocatalysts lies in the UV region. So, these materials are activated only in the presence of UV light. But, our solar spectrum contains only 4–5% UV and almost 44% of the visible region. So, keeping in mind the energy issue, the band gap tuning of semiconducting materials have been done by doping with various metals and non-metals [25–31]. By doping the efficiency of a photocatalyst can be improved up to a certain limit. Excess doping tends to deform crystal structure and acts as a recombination centre, which lowers the photocatalytic activity. Moreover, high electron-hole pair recombination rate, insufficient absorption of solar light limits their photocatalytic activity [32]. After that researchers have focussed to synthesize various types of semiconductors heterostructures to further improve the photocatalyst. Typically, the heterostructures are characterized in four types, (i) semiconductor–metal, (ii) semiconductor–semiconductor, (iii) semiconductor carbon heterostructures, and (iv) multicomponent (comprising of metals and semiconductors) heterostructure.

1.3.1 Semiconductor-Metal Heterostructures

Semiconductor metal heterostructures have been formed by deposition of novel metal nanoparticles on semiconducting surfaces. Normally, n-type semiconductor is used to form these types of heterostructures. Work function of metal is greater than that of the semiconductor. Usually, metals have lower Fermi energies (E_F) compared to those of semiconductors. As a result, electrons will flow from the conduction band of semiconductor to the metal until the E_F of the semiconductor reaches the thermal equilibrium with that of the metal, which leads to a constant value of E_F for both devices. Thus an excess negative charge density occurs in the metal and positive charge density in the semiconductor. This form creates a potential difference at the interface and this is called Schottky barrier. This Schottky barrier acts as electron

Fig. 3 Schematic diagram for Schottky barrier. Redrawn after Wang et al. [33]

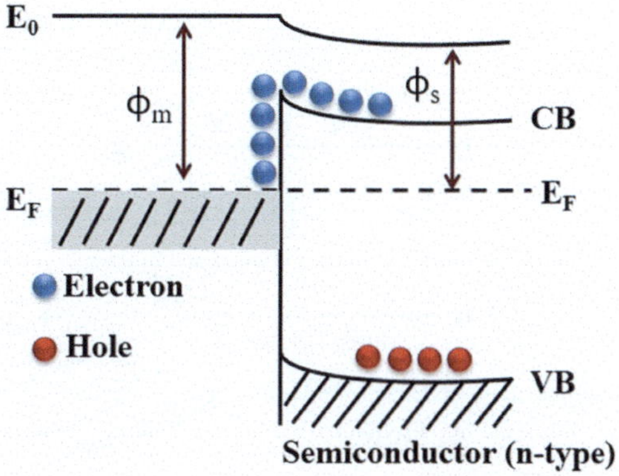

trapping centre and reduces electron-hole recombination. This phenomenon helps in enhancing photocatalytic behaviour. The overall process has been shown in Fig. 3.

For example, Georgekutty et al. have synthesized silver modified ZnO and concluded that the presence of silver facilitates the interfacial charge transfer processes in such a way to utmost utilize CB electrons in enhancing the photocatalytic activity [34].

1.3.2 Semiconductor–Semiconductor Heterostructures

Apart from oxide semiconductors, sulphide materials (such as CdS, MoS_2, Bi_2S_3, $CuInS_2$, SnS2, SnS, etc.) are also very useful owing to low band gap and quite high light absorption coefficient. So many efforts have been made to design the different heterostructures such as ZnO/CdS, CuS/ZnS, CuS/ZnO, ZnO/ZnS, or CdS/TiO_2 [35–39]. Thus, coupling of the wide band gap oxide materials with these types of low band gap sulphides, we can get energetically favourable band structure, and the photocatalytic efficiency of the materials can be enhanced. Karmakar et al. have synthesized flower-like ZnO/MoS_2 heterostructure [40]. Their result says that under sunlight irradiation the methylene blue degradation rate is 2.0 and 1.83 times higher than that of bare ZnO and MoS_2, respectively. Though cadmium-based semiconductor photocatalyst shows promising photocatalytic activity in visible irradiation, it's use is limited in waste water management system due to inherent toxicity.

Semiconductor heterostructures can be done in two types; (i) *p–n* type and (ii) non-*p–n* type.

The *p–n* type heterostructure is very much effective for photocatalysis because at the interface they form a *p–n* junction with a space-charge region due to the diffusion of electrons and holes, and thus create a built-in electric potential. This field drives the electrons and holes in the opposite direction (shown in Fig. 4). The electrons are transferred to the conduction band of the n-type semiconductor while holes are transferred to the valence band (VB) of the p-type semiconductor. As a result

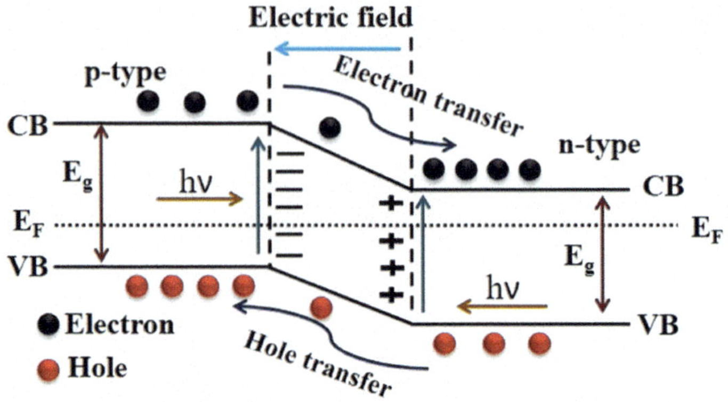

Fig. 4 Schematic diagram for *p–n* heterostructure showing energy band structure and electron pair separation. Redrawn after Wang et al. [33]

electron-hole pair recombination is reduced, occurrence of rapid charge transfer to the catalyst, lifetime of the catalyst is increased, and consequently, photocatalytic activity is enhanced. In recent times, visible light responsive *p–n* junctions such as Ag_2O/TiO_2 [41], $CulnS_2/TiO_2$ [42], CuO/ZnO [43], $CuInS_2/ZnO$ [44], and BiOI/ZnO [13] have been prepared for various photocatalytic applications from hydrogen production to dye degradation.

In case of non-*p–n* type semiconductor (SC) heterostructures, both the band gap and band alignments of the two semiconductors play a crucial role in effective charge transfer as well as reduction of electron-hole pair recombination. Non-*p–n* type's heterostructures can be formed in three different ways depending upon their band gap and electron affinity as illustrated schematically in Fig. 5. In the case of type-I heterojunction, it consists of two semiconductors whereby the CB of semiconductor 2 (SC2) is higher than that of semiconductor 1 (SC1) and the VB of 2 is lower than that of 1, therefore holes and electrons will transfer and accumulate on component 1. In a type-II junction, photoexcited electrons can travel from 2 to 1 due to the more negative CB position of 2. Holes can travel in the opposite direction from the more positive VB of 1–2, leading to a complete electron-hole separation in two different semiconductors enhancing the photocatalytic activity. The third type, type-III, is identical to type-II except for the much more pronounced difference in VB and CB positions, which needs a higher driving force for charge transfer. Most of the heterostructures regarding photocatalytic study have been synthesized in type-II form. Various non-*p–n* type heterostructures such as Bi_2WO_2–TiO_2 [45], SnO_2–TiO_2 [46] TiO_2–ZnO [47] have been used as an efficient photocatalyst.

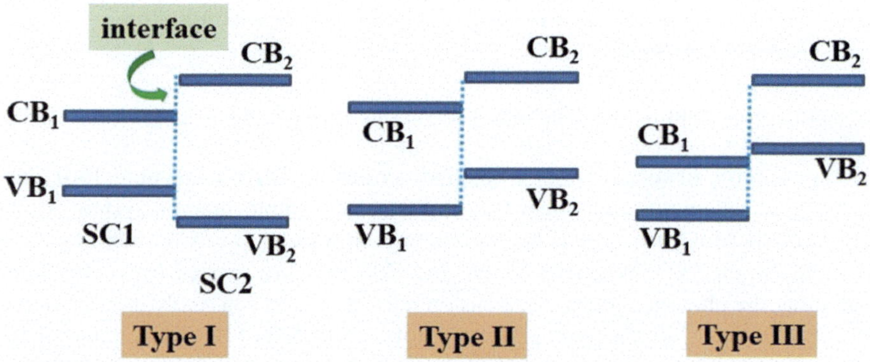

Fig. 5 Schematic energy band diagram of three types of semiconductor heterostructures

1.3.3 Semiconductor-Carbon (S-C) Heterostructures

In an S-C heterostructure, different forms of carbon like activated carbon, carbon nanotubes (CNTs), and graphene have been used. All these carbon forms have a very large specific surface area. So, it can be expected that combining semiconductors with the carbon-based materials yields an increase in adsorbed amounts of pollutants, and thus enhances their photocatalytic activity [33]. Semiconductor-CNTs can form a Schottky barrier junction, which is an effective method of increasing recombination time. Furthermore, CNTs have a large electron-storage capacity and therefore, can accept photon-excited electrons in heterostructure form, thus retarding the recombination. Hoffmann et al. [18] have proposed that in TiO_2-CNTs heterostructure photogenerated electrons in interface region are transferred into the CNTs until the Fermi levels of the TiO_2 and that of the CNTs are on the same level, and holes remain on the TiO_2 to take part in redox reactions. Graphene, being a two-dimensional (2D) structure of sp^2 hybridized carbon atoms, has unique properties such as high surface area, adsorption, and good electrical conductivity. Even reduced graphene oxide (rGO) in composite form, also possesses the ability to accept the photogenerated electrons to prevent the recombination, and provides a favourable adsorption of dye through π–π conjugation between the dye and rGO surface [7]. Graphene shows almost zero conduction resistance for storing and transporting electrons within its π-rich structure due to its potentially ballistic electron transport capability [48]. As a result, its composites naturally reduce the electron-hole pair recombination time and consequently enhance its photocatalytic property towards the degradation of toxic dyes [49].

1.3.4 Multicomponent (Comprising of Metals and Semiconductors) Heterostructure

In a multicomponent heterostructure, a novel nanoparticle metal (Ag, Au, or Pt) is sandwiched between two materials. This combination further reduces the recombination possibility of semiconductor heterostructures. In 2016, Wang et al. [50] have fabricated Z-scheme tricomponent CdS/Ag/Bi_2MoO_2 photocatalytic system. Basically, CdS-Bi_2MoO_6 is a type II heterostructure, in which CB of CdS lies above the CB of Bi_2MoO_6 (S1). When Bi_2MoO_6 and CdS (S2) are excited by visible light irradiation, the photogenerated electrons on the CB of CdS can easily migrate to the CB of Bi_2MoO_6, while the photogenerated holes in the VB of Bi_2MoO_6 can transfer to the VB of CdS. As a result, the photogenerated electrons and holes are spatially separated and undesirable recombination is greatly weakened. But, as the top of the valence band potential of CdS is less positive than that of Bi_2MoO_6 and the bottom of the conduction band potential of Bi_2MoO_6 is less negative than that of CdS, it is difficult to simultaneously possess high charge-separation efficiency and strong redox ability, resulting in a weak redox ability, especially after charge transfer. In the sandwiched structure, photogenerated electrons from CB of S1 can easily flow into metal through the Schottky barrier because CB of S1 is higher than that of Ag. Thus, plenty of electrons in the CB of S1 can be stored in the metal component. As a result, more holes with a strong oxidation power in the VB of S1 escape from the pair recombination and are available to oxidize the pollutants. On the other hand, since the energy level of metal is above the VB of S2, holes in the VB of S2 also easily flow into the metal. Thus, more electrons with a strong reduction power in the CB of S2 can escape from the pair recombination and are available to reduce some absorbed compounds (such as O_2, H^+, etc.). Therefore, simultaneous electron transfers can occur as a result of UV/visible-light excitation of both S1 and S2. This further improves the efficient separation of electrons from the CB of CdS and holes from the VB of Bi_2MoO_6. This is why Z-scheme catalyst S1/Ag/S2 exhibits the highest photocatalytic performance. The detailed charge transfer process is shown schematically in Fig. 6.

1.4 Chalcogenide Material as a Photocatalyst

Having a lot of advantages of oxide materials as a photocatalyst at the same time, chalcogenide materials are also advantageous since they exhibit low photocorrosion effect under band gap excitation unlike ZnO, and thus become suitable for water splitting. Besides this, metal chalcogenide can offer band gap tunability towards visible light activity, rapid electron-hole pair generation, and good photosensivity. In order to enhance photocatalytic properties of metal chalcogenides, there are exist several ways in which metal and non-metal ion doping, dye sensitization, and surface defects are feasible. Among these, sulphur-based metal chalcogenides (such as ZnS,

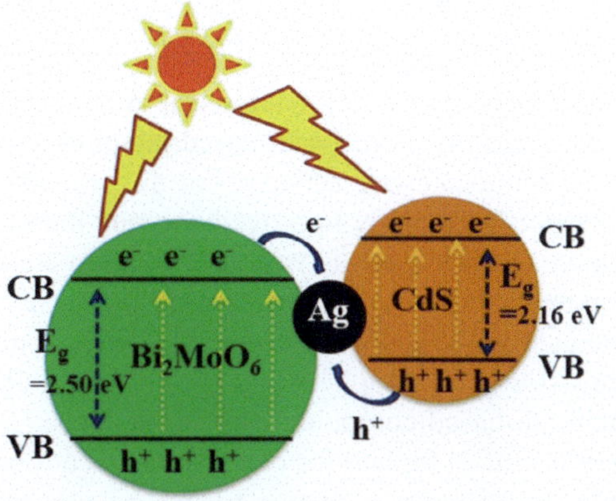

Fig. 6 Role of Ag in the Bi_2MoO_6–Ag–CdS sandwich structure

MoS_2) are important due to several applications such as rapid generation of electron-hole pairs, high carrier mobility, and better photovoltaic applications [50]. Recently, zinc sulphide (ZnS) has received considerable concern as a photocatalyst due to its excellent properties such as wide direct band gap (3.6 eV), relatively negative redox potential of its conduction band, rapid electron-hole pair generation, high stability and less photocorrosion, and lower secondary pollution [51–58]. In general, ZnS shows efficient catalytic properties in the presence of ultraviolet (UV) light. External atom substitution or heterostructure formation is an effective way to resolve band gap-related difficulty of ZnS. In order to choose suitable substitutional atom for ZnS structure, various approaches have been found. As an example, Cu-doped ZnS exhibits band gap tunability from UV to visible light region and enhances hydrogen evolution rate. Sun et al. have verified enhanced visible light activity in (N, C) codoped ZnS system [59]. In search of better potential candidates along with these theoretical works, a few experimental investigations have also been done such as transition metals doped (Mn, Co, Ni, Cu, Ag, and Cd) ZnS nanoparticles are successfully synthesized by Ramasamy et al. [60] Enhanced catalytic performance in Cr-doped ZnS with a variation in Cr concentrations has been reported by Zhao et al. [61]. Besides this, various approaches such as nanostructure formation, metal and non-metal doping, and semiconductor formation have been taken to make ZnS as a visible light-activated photocatalyst. Recently, Kaur et al. [62] and Lee et al. [54, 63] in their review article have discussed the various synthetic methods of ZnS nanocrystalline semiconductors and their composites and the role of these materials in photocatalytic applications from degradation of organic components to hydrogen evolution by water splitting.

1.5 Importance of Ru in Photocatalysis

Doping with rare earth metal ions is another effective way to enhance photocatalytic behaviour [64]. Rare Earth-based metal ions doped nanomaterials not only exhibit better photocatalytic activity, but also the optoelectronic properties of nanomaterials are significantly improved. Among these, transition metal ions are feasible for doping as they possess various oxidation states. Ruthenium (Ru) as a rare transition metal element has a lot of applications in many intriguing fields such as optical attenuation [65], electrochemical supercapacitors [66, 67], ion-sensitive field effect transistor (FET) applications [68, 69], etc. Ru can be a better choice to replace platinum-based compounds in future cancer treatments [70]. Besides, Ru can also be a useful candidate for catalytic applications. Ru-based catalysts are effective in light-driven water oxidation [71]. Ru nanoparticles are efficient for photo-induced water-splitting process [72]. Ru metal particles based on various carbons and oxides are the most effective catalysts to achieve a rapid conversion of carbonyl linkage into the corresponding alcohols [73]. Ru-based catalysts showed effectively better performance than Pt/C catalysts in hydrogen evaluation reactions [74]. Besides these applications, Ru can be used to improve optoelectronic properties. Ru has great importance in Ru-grafted rGO nanocomposites for useful application in constructing solar cells [75]. Ru-incorporated CdO thin films are effective for optoelectronic and gas-sensing applications [76]. Ru doping can be used to improve non-linear optical behaviour of C20 fullerene [77]. Nanocrystalline Ru is used in various types of sensing applications [76, 78].

Motivated from these versatile applications, we have preferred Ru doping as an interesting way to improve photocatalytic as well as optoelectronic characteristics of ZnS materials. Only few research works exist with Ru-doped ZnS systems. Ru-doped ZnS quantum dots have been synthesized for the first time by Sahraei et al. [79]. They have confirmed successful incorporation of Ru in ZnS NPs by X-ray diffraction (XRD) analysis. They have only identified optical properties of Ru-incorporated ZnS NPs. In this work, using both the experimental analysis and first-principle calculations, we have explored photocatalytic as well as optoelectronic behaviour of Ru-doped ZnS matrix. Incorporation of Ru in ZnS matrix creates a localized charge centre. As a result, its photogenerated electron-hole pair recombination time decreases. This pheneomenon helps to enhance the photocatalytic behaviour [80]. Moreover, as the concentration of Ru increases, the system undergoes semiconductor to metallic transition. Ru induces magnetic properties in the nonmagnetic ZnS system [81]. Due to Ru, low energy optical behaviour of the doped system significantly enhanced. Therefore, sufficient amount of Ru can enhance the optoelectronic properties of the system.

2 Experimental Procedures

2.1 *Synthesis of ZnS Nanoparticles*

Nanoparticles (NPs) of ZnS have been prepared via solvothermal method [82]. Freshly prepared aqueous solutions of chemicals have been used for the synthesis of nanoparticles. At first, 22 ml ethylenediamine (EDA) is mixed with 22 ml water under vigorous stirring conditions for 30 min. Then, 33.0 mmol of $ZnNO_36H_2O$ (0.980 g) have been added to the above solution. This solution has been stirred for 15 min. After that, equivalent molar amount of thiourea (0.251 g) has been added to the previous mixture solution and the resulting mixture solution is allowed to stand another 30 min for stirring. Then, the solution was been autoclaved in a Teflon chamber of 55 ml capacity and placed in the preheated oven at 200 °C for 16 h. Finally, the white precipitate has been collected by centrifugation followed by washing with ethanol and distilled water. The wet precipitate has been dried in an oven overnight at 70 °C and collected by mortaring.

2.2 *Synthesis of Ru-Doped ZnS Nanoparticles*

Various amounts of Ru-doped ZnS nanoparticles have been synthesized followed by the synthesis procedure of ZnS nanoparticles. Typically, 33.0 × mmol of ruthenium salt (where $x = 0.005, 0.010$, and 0.015) has been added just after the addition of $33.0(1 - x)$ mmol of zinc nitrate to the abovementioned EDA-water mixture. After 30 min stirring of Zn–Ru salt mixture, 33 mmol of thiourea is added. The next steps are similar to the synthesis of pure ZnS NPs. Synthesized Ru-doped ZnS NPs are collected by mortaring followed by drying overnight at 70 °C and named $Zn_{1-x}Ru_x$ S (where $x = 0.005, 0.010$, and 0.015).

2.3 *Characterization of the Materials*

Structural characterization of pristine and Ru-doped samples have been carried out using (XRD) measurements collected by an X-ray diffractometer [Model: X'Pert Pro (PAN Alytical)] with Cu-K_α radiation of 1.54 Å. The Raman spectra have been taken using RENISHAW inVia Raman Microscope with an Argon laser at an excitation wavelength of 532 nm. The size, morphology, doping percentage, and elemental mapping of ZnS and Ru-doped ZnS NPs have been characterized by field emission scanning electron microscopy (FESEM) equipped with Energy Dispersive Analysis of X-Ray (EDAX) measurement [Model: JEOL JSM-7600F]. Perkin Elmer Lambda 25 UV/VIS Spectrometer has been utilized to accumulate the UV-Vis absorption spectra of the synthesized samples as well as photocatalytic dye degradation. Photoluminescence (PL) spectra of all the samples have been carried out at room temperature using 300 nm excitation source from Horiba FL 1000 fluorescence spectrometer.

2.4 Photocatalytic Framework

The photocatalytic activity of the synthesized NPs has been evaluated by measuring the degradation of methylene blue (MB) dye in aquatic medium under visible light exposure. Here, 30 W LED spotlight of colour temperature 6000 K has been utilized as a visible light source. The emission profile of LED light taken by the excitation profile measurement from fluorescence spectroscopy has been shown in Fig. 7. The reaction system containing aqueous solution of (MB) dye (1×10^{-5} M, 25 mL) and 10 mg of $Zn_{1-x}Ru_x$ S NPs have been magnetically stirred in the dark for 30 min to reach the adsorption equilibrium of dye by the catalyst. Initially, the LED has been kept 10 cm apart from the upper surface of the reaction solution to reduce the thermal effect. Approximately 1.5 ml mixture of catalyst and dye solution has been collected at different intervals from the photoreactor and then centrifuged at a speed of 8000 rpm for 4 min to remove the photocatalyst from the solution. UV-Vis absorption spectra have been recorded each 20 min interval to monitor the variation of concentration of dye.

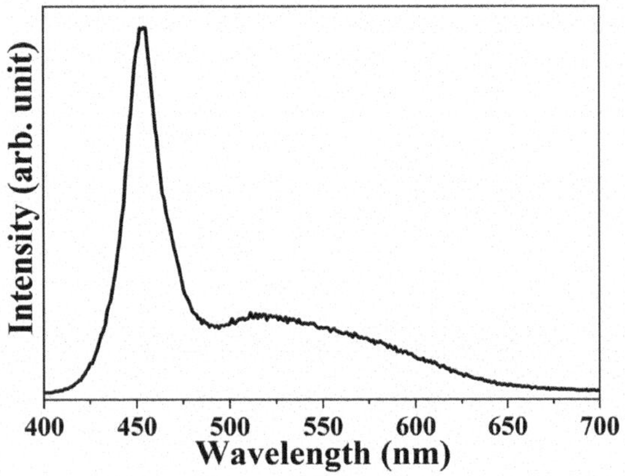

Fig. 7 The emission profile of LED light using fluorescence spectroscopy

3 Experimental Results

3.1 Structural Properties

3.1.1 XRD Patterns

X-ray diffraction patterns for the synthesized ZnS and Ru-doped ZnS nanoparticles ($Zn_{1-x}Ru_xS$, where $x = 0$, 0.005, 0.010, and 0.015) have been recorded and shown in Fig. 8. The observed diffraction peaks located at $2\theta = 28.59°$, 33.11°, 47.50°, 56.37°, 69.38°, and 76.72° correspond to the (111), (200), (220), (311), (400), and (331) planes of cubic ZnS that have been fitted to JCPDS card No. 65-9585. No other extra peaks in the patterns of the Rudoped samples indicate the true doping of ruthenium in the ZnS lattice structure rather confirming the nonexistence of any other phase formation [83]. The average crystalline size of the samples has been calculated using the Scherrer's formula,

$$D = \frac{0.9\lambda}{\beta \cos\theta} \tag{7}$$

where D is the average crystalline size, λ is X-Ray wavelength (Kα line) and β is the full width at half maximum (FWHM) of the (111) plane, the most intense peak and θ is the diffraction angle [84].

The estimated average crystalline sizes of the $Zn_{1-x}Ru_xS$ samples are approximately 16, 15, 19, and 18 nm, respectively. The slight decrease in crystalline size

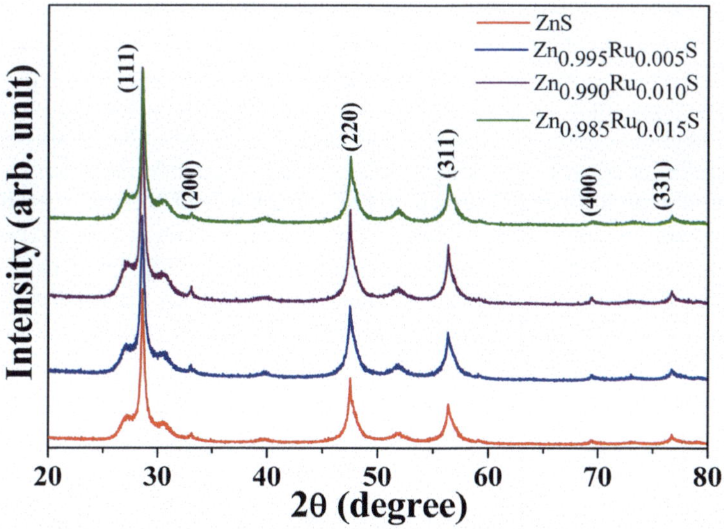

Fig. 8 XRD patterns of pure ZnS and different Ru-doped ZnS NPs

in case of 0.5 % Ru-doped ZnS is due to the lattice rearrangement by incorporation of low doping. No significant change in size distribution in other doped materials is due to the low doping of Ru, and comparatively matching ionic radii of Ru (82 pm) with respect to Zn (74 pm). The detailed information about the size and morphology has been discussed later from the FESEM analysis.

3.1.2 Raman Spectra

Raman scattering is used to study mechanical properties of material in solid, liquid or gaseous form. Here, the Raman spectra of Ru-doped ZnS nanoparticles in the wavenumber range 200–700 cm^{-1} have been shown in Fig. 9. All the samples have been excited with 532 nm laser source. In cubic ZnS, the optical branches are degenerate at $k = 0$ due to the presence of two particles per unit cell. However, immediately away from the centre of the Brillouin zone, the degeneracy breaks down due to the polarization field in ionic or partially ionic crystals [85]. Therefore, it is expected that one longitudinal optical (LO) mode and one transverse optical (TO) mode as Raman active at $\overline{\overline{k}} = 0$. In this experiment, the two prominent peaks of the undoped and different Ru-doped ZnS nanoparticles are at 252 cm^{-1} and 348 cm^{-1} related to LO and TO modes respectively. In bulk material as reported by Brafman et al. [85], Dimitrievska et al. [86], and Cheng et al. [87], the prominent peaks related to the fundamental transverse optical (TO) mode of A_1 and E_1 are at 273, 277 and 278 cm^{-1}, respectively, while the longitudinal optical (LO) mode of A_1 and E_1, is attributed to the vibration of only S anions in the lattice at 351, 348, and 351 cm^{-1}, respectively [85–87]. The red shifting of TO mode from bulk to nanomaterial of ZnS may be due to the phonon confinement effect due to the very small size of the nanoparticles. The asymmetry of the peak broadening to the right side of the peak for LO at 348 cm^{-1} may be due to the loss of translational symmetry in nanomaterial [88]. The peak positioned at 216 cm^{-1} is associated with two phonon processes and this is previously reported by Brafman et al. [85], and Radhu et al. [89]. Moreover, the intensity of this mode is gradually increased with the doping percentage of Ru. A second-order Raman scattering consists of a longitudinal optical and transverse acoustic (LO + TA) phonon mode has been observed at 418, 420, 414, and 405 cm^{-1} for the pure and Ru-doped ZnS nanoparticles respectively. A similar result had been reported by Kim et al. which indicates a Raman peak near 418 cm^{-1} [90]. The redshift of the second order Raman scattering may be due to the strain developed by Ru doping in the nanoparticles. The broad band centred at 618 cm^{-1} can be attributed to the higher order Raman scattering and the surface active phonon [86].

3.1.3 Morphology and Composition

The size and morphological features of the synthesized ZnS and various Ru-doped ZnS systems have been analysed by Field Emission Scanning Electron Microscopy

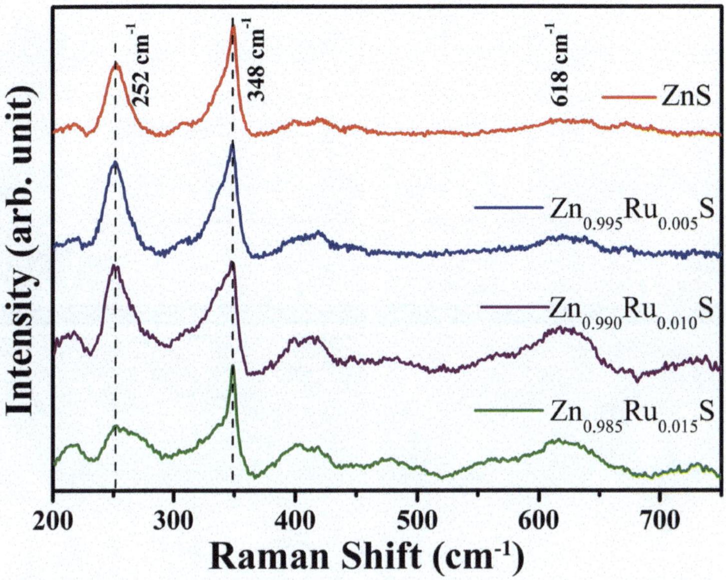

Fig. 9 Raman spectra of pure and different Ru-doped ZnS NPs

(FESEM) images and displaced in Fig. 10a–d. All the FESEM images have been represented in the same scale.

The particles are nearly spherical in nature as depicted from the FESEM images shown in Fig. 10. There was no significant variation in particle size due to low doping of Ru, and in case of all samples, the particle size is within 20 nm, which is clear from FESEM images. This observation is quite consistent with crystalline size prediction from XRD measurement. We have also studied EDAX measurements of the mentioned samples. The corresponding Ru doping percentages are summarized in Table 1.

The measured doping concentration was consistent with the adopted doping concentration at the time of synthesis. Elemental mapping in case of optimized photocatalyst ($Zn_{0.990}Ru_{0.010}S$) has been shown in Fig. 11a–d, which again confirms the presence of Ru in the sample as well as the uniform distribution of the elements over the sample. The EDAX elemental mapping has confirmed the successful incorporation of Ru into the ZnS matrix.

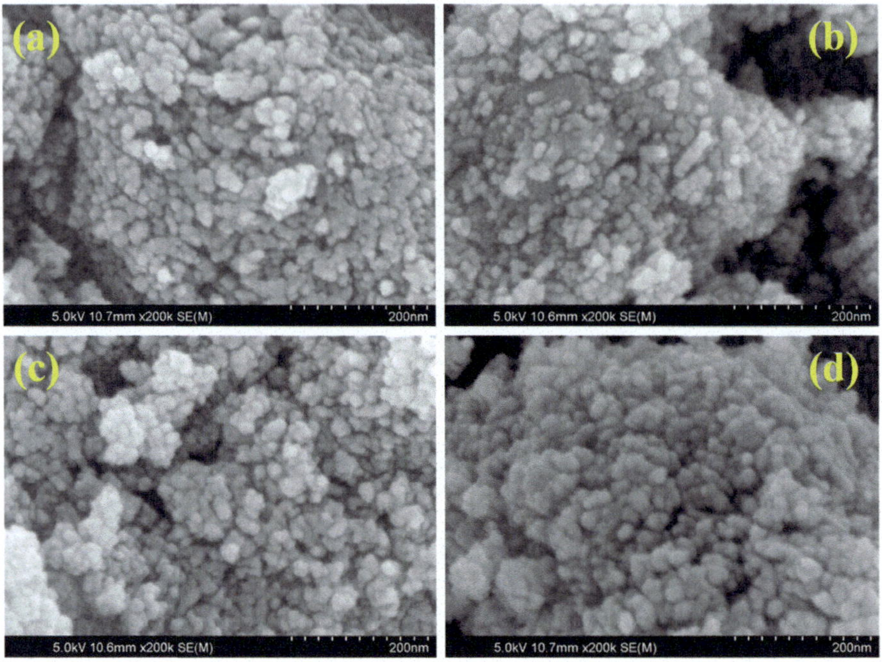

Fig. 10 FESEM images of NPs of **a** ZnS. **b** $Zn_{0.995}Ru_{0.005}S$. **c** $Zn_{0.990}Ru_{0.010}S$. **d** $Zn_{0.985}Ru_{0.015}S$

Table 1 Summary of Ruthenium doping percentage in ZnS crystal structure

Sample	Zn (at. %) (Zn_K)	S (at. %) (S_K)	Ru (at. %) (Ru_L)
ZnS	50.01	49.99	–
$Ru_{0.995}Zn_{0.005}S$	50.86	48.84	0.3
$Ru_{0.990}Zn_{0.010}S$	47.86	51.34	0.8
$Ru_{0.985}Zn_{0.015}S$	47.53	51.27	1.2

3.2 *Optical Properties*

3.2.1 UV-Vis Absorption Spectra

Figure 12a shows the UV-Vis absorption spectra of pure and Ru-doped ZnS NPs. The differential absorbance spectra (first derivative of the absorbance with respect to wavelength of light) of these materials have been plotted in Fig. 12b. The gradual increase in absorption towards higher wavelength with Ru concentration indicates that the absorption of more visible light with the increase in Ru doping concentration. The enhancement of absorbance in the visible region increases the number of visible irradiation assisted photogenerated electrons and holes, which plays a vital role in the enhancement of the photocatalytic activity.

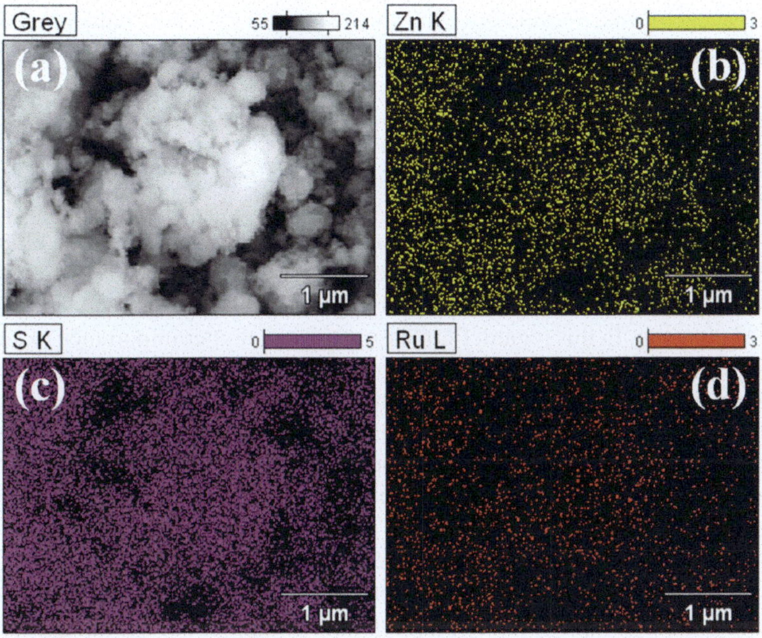

Fig. 11 FESEM image of $Zn_{0.990}Ru_{0.010}S$ NPs. Elemental mapping from FESEM image shown in (**a**) of; **b** Zn (K), **c** S (K), and **d** Ru (L)

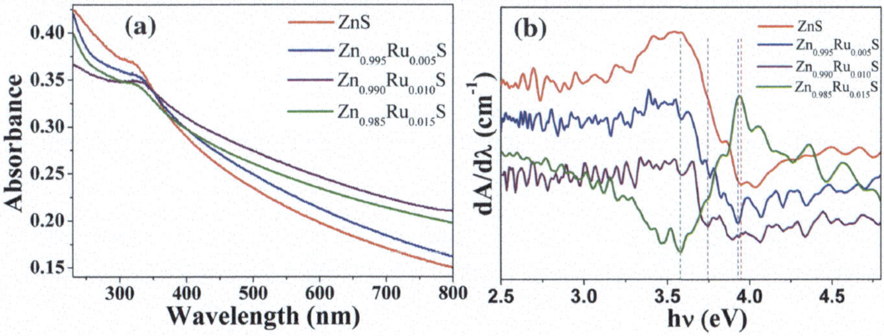

Fig. 12 **a** UV-Vis absorption spectra of ZnS and Ru-doped ZnS NPs. **b** Differential absorption spectra of ZnS and Ru-doped ZnS NPs

The band gap of the $Zn_{1-x}Ru_xS$ material (where $x = 0, 0.005, 0.010$, and 0.015) has been estimated from the differential plot [7], and the corresponding values are 3.94, 3.93, 3.74, and 3.58 eV, respectively. The estimated band gap is slightly higher than the bulk counterpart (3.65 eV for cubic structure) due to the quantum confinement effect at nano-scale [91]. In fact, Ru dopant behaves as an acceptor and creates localized acceptor like a level above the valence band (VB), thus reducing the band gap energy.

3.2.2 Photoluminescence (PL) Spectra

The photocatalytic activity of any sample depends on the effectiveness of charge separation produced by photons. The photoluminescence (PL) shows the efficiency of charge recombination of the sample. Actually, suppressing the recombination of photogenerated charges is one of the most important routes to get enhanced photocatalytic behaviour. In this regard, PL measurement is of utter importance to monitor the charge carrier dynamics as well as understanding the underlying mechanism and designing efficient photocatalysts. Here, from the PL spectra (Fig. 13), we have studied the charge carrier dynamics in detail, including charge carrier generation, trapping and recombination inside the photocatalyst, as well as interfacial charge transfer. The emission arises due to the recombination of electrons and holes giving rise to photons.

The defects and vacancies behave as a reservoir of electrons and holes. Thus photoluminescence takes place for the recombination of electrons and holes between conduction and valance bands and also different energy levels formed by the defect states [92], impurity states [93], surface defects [54, 79], etc. The anion (here S^{2-}) vacancy (V_S) acts as a trap for electrons, and cation (Zn^{2+}) vacancies (V_{Zn}) acts as a trap for holes [92]. In this sample, the PL spectra of pure and Ru-doped ZnS have been shown in the Fig. 13. The asymmetric nature of wavelength versus intensity curve shows that the intensity profile is made up of a number of Gaussian or Lorentzian curves [94]. The emission spectra mainly dominated by a broad peak with a maximum at wavelength $\lambda = 368$ nm and a sharp peak at $\lambda = 467$ nm along with comparatively less intense peaks centred around $\lambda = 396, 419, 450, 481,$ and 492 nm. The emission arising at $\lambda = 368$ nm is related to near band emission i.e. due to the

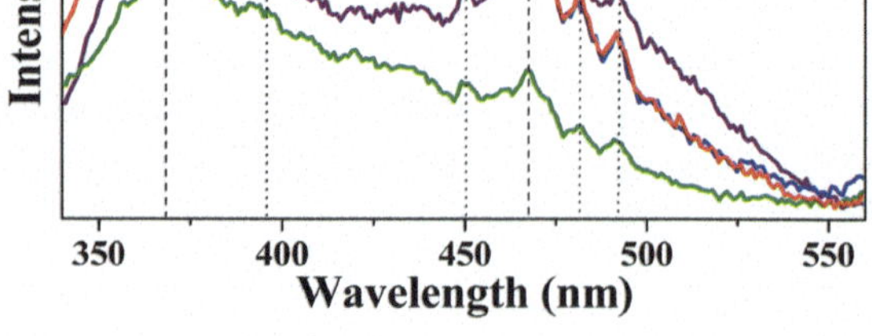

Fig. 13 Photoluminescence spectra of ZnS and Ru-doped ZnS NPs

recombination of trapped electrons in sulphur vacancy (V_s) to the photogenerated holes in the valence band [79, 95]. Sahraei et al. [79] have shown a similar peak at $\lambda = 350$ nm arises due to the same reason in case of ZnS quantum dot. The violet emission at $\lambda = 396$ nm has been attributed to the recombination of conduction band electrons to the hole trapped at zinc vacancy (V_{Zn}) [92], whereas, the 419 nm peak corresponds to sulphur vacancy (V_S). The peak around $\lambda = 450$ nm may be due to native acceptor levels in ZnS NPs [94, 96]. The peak around 467 nm corresponds to the stoichiometric vacancies or interstitial vacancies [94, 96]. The emission peak at $\lambda = 481$ and 492 nm may be due to the recombination of shallow delocalized donor levels to V_{Zn} [97]. The similar nature of the emission profile for different Ru-doped ZnS NPs compared to pure ZnS indicates that the Ru doping does not give rise to any extra defect state. It only alters the population in the defect states already existing in the host material. The slight increase of emission intensity around $\lambda = 368$ nm in case of $Ru_{0.995}Zn_{0.005}S$ is due to the creation of sulphur vacancy in ZnS crystal structure by low doping of ruthenium. This type of observation has been found in the case of average particle size distribution from XRD and FESEM measurements. Formation of sulphur vacancy in $Ru_{0.995}Zn_{0.005}S$ has been also confirmed from EDAX measurement. Further, increase in Ru doping reduces this emission intensity indicating the transfer of photoexcited electrons to Ru ioninduced trap centres [34, 98]. As a result, the quenching of luminescence has been observed, and consequently, enhanced photocatalytic behaviour has been achieved. This type of trapping has been further checked from the theoretical point of view which has been discussed later. Moreover, doping with impurity atoms can change the stoichiometric ratio of Zn and S, this may influence the intensity of the PL emission [92, 95]. The intensity ratio of the peaks at $\lambda = 368$ nm (3.37 eV) and $\lambda = 467$ nm (2.67 eV) are 1.33, 1.40, 1.06, and 1.32 for pure, 0.5%, 1.0%, and 1.5% samples, respectively. The ratio is lowest for 1.0% Ru doping. This observation indicates the immense recombination from V_S to V_{Zn} in the case of 1% Ru doped sample. This is due to the presence of large number of ion pairs near 467 nm. Again, the emission profile of LED light used for irradiation has a maximum of around $\lambda = 470$ nm. As a result, the efficient number of electrons and holes take part in oxidation–redox reaction showing better photocatalytic behaviour under visible irradiation.

3.3 *Photocatalytic Properties*

The photocatalytic activity towards the degradation of MB dye using $Zn_{1-x}Ru_xS$ (where, $x = 0$, 0.005, 0.010, and 0.015) as a catalyst has been performed during 3 h of visible light irradiation. In all the cases, the amount of catalyst used is 10 mg along with other previously mentioned photocatalytic conditions remaining unchanged. Figure 14a represents the C/C_0 versus irradiation time (t) plot; where, C_0 is the concentration of MB at adsorption equilibrium and C is the concentration of dye remaining unchanged after certain time of irradiation [83]. The degradation efficiency has been calculated using the formula,

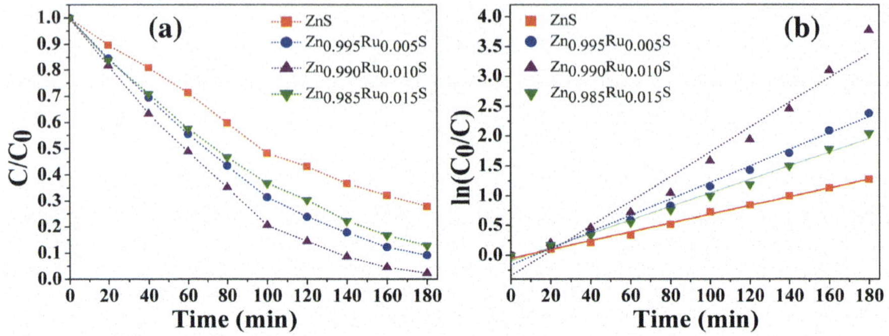

Fig. 14 **a** Photocatalytic degradation curve of (C/C_0 versus time) of MB using 10 mg of Zn_{1-x} Ru_xS under visible LED irradiation. **b** ln (C_0/C) versus time plot for MB degradation

$$\text{Degradation (\%)} = \left(1 - \frac{C}{C_0}\right) \times 100\% \quad (8)$$

The highest degradation efficiency has been achieved for $x = 0.010$ and almost 90% dye has been degraded within 180 min of visible irradiation. Also pure ZnS has shown moderate degradation efficiency due to the presence of various types of defects in the crystal structure, as shown by PL measurements. Figure 14b represents ln(C_0/C) versus time (t) plot for these materials. The rate constant has been determined using the equation ln(C_0/C) = kt, where k is the degradation rate constant [99]. The rate constant calculated from the linear fitting of the logarithmic plot for $Zn_{1-x}Ru_xS$ (where, $x = 0, 0.005, 0.010$, and 0.015) are $(7.4 + 0.2) \times 10^{-3}$, $(13.5 \pm 0.5) \times 10^{-3}$, $(20.7 \pm 1.3) \times 10^{-3}$ and $(11.4 \pm 0.4) \times 10^{-3}$ min^{-1}, respectively, and it agrees with the pseudo-first-order kinetics [99, 100]. $Zn_{1-x}Ru_xS$ (for $x = 0.010$) shows the best photocatalytic activity among all the composites.

In search of enhanced efficiency of dye degradation keeping the concentration of dye unchanged the optimized catalyst ($Zn_{1-x}Ru_xS$, where $x = 0.01$) has been chosen for further experiments. In this work, the rate of the dye degradation efficiency has been studied at different acidic and basic environment in the pH range 4–12. Figure 15 shows the pH-dependent photocatalytic degradation using the optimized Ru-doped ZnS nanoparticle. Inset shows the logarithmic plot [ln(C_0/C)] w.r.t. irradiation time.

The reaction rate has been studied at pH = 4.7, 6.7, 8.9, 10.2, and 12.0, and the corresponding values, calculated from linear fitting of logarithmic plot are $(12.0 \pm 0.2) \times 10^{-3}$, $(20.7 \pm 1.3) \times 10^{-3}$, $(21.6 \pm 1.1) \times 10^{-3}$, $(28.1 \pm 1.2) \times 10^{-3}$, and $(24.6 \pm 1.1) \times 10^{-3}$ min^{-1}, respectively. The calculated rate constant indicates that efficiency increases with pH up to an optimum limit and then decreases. Previously, these types of results have been found by Li et al. [101], Malash et al. [102], and Nguyen et al. [103]. The MB dye, usually available in synthetic form is basic in nature [104]. In aqueous solution the cationic part of MB has been attracted towards the catalyst surface by electrostatic force, which facilitates charge transfer mechanism [105]. During this course, the MB has to compete with the H^+ ions present in ample

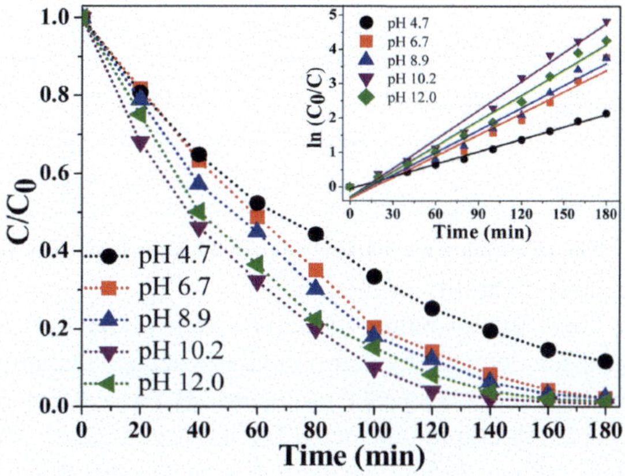

Fig. 15 The influence of pH of the reaction mixture on photocatalytic degradation of MB dye using 10 mg of $Zn_{0.990}Ru_{0.010}S$ NPs as a catalyst amount

amounts in case of acidic medium. MB particle is bulky with respect to the H^+ atom, thus has less mobility. The electron of the generated electron-hole pair produced combines with H^+ ion producing hydrogen atom. As a result, the photocatalytic efficiency of MB at pH = 4.7 decreases. On the contrary, in case of basic medium the cations of MB rush towards the OH^- surrounded catalyst surface. As a consequence, the best degradation efficiency has been obtained at pH = 10.2. At this optimum pH, almost 98% degradation has been achieved within 140 min of irradiation. Further increase in pH (pH = 12.0), dissociation possibility of the MB decreases. In the same time, the efficiency decreases due to the excess amount of OH^- radicals present in the solution which neutralizes the positive charge centre of active MB. As a result, the possibility of charge flow decreases, reducing the MB degradation.

In photocatalysis, hydroxyl radicals ($OH^\cdot$) generated by the illumination of photocatalyst has been widely accepted as the main oxidation agents for the photodegradation of organic pollutants. The mechanism of the formation of $OH^\cdot$ on the surface of the photocatalysts and subsequent degradation of MB can be expressed mathematically as follows:

$$Zn_{1-x}Ru_xS + h\nu \rightarrow Zn_{1-x}Ru_xS\left(e^-_{CB} + h^+_{VB}\right) \tag{9}$$

Superoxide radical $\left(O_2^{\cdot -}\right)$ *formation*

$$Zn_{1-x}R_xS\left(e^-_{CB}\right) + O_2 \rightarrow O_2^{\cdot -} \tag{10}$$

$OH^\cdot$ *radical formation*

$$O_2^\cdot + H_2O \rightarrow HO_2^\cdot + OH^- \tag{11}$$

$$HO_2^\cdot + HO_2^\cdot \rightarrow 2OH^\cdot + O_2 \tag{12}$$

$$Zn_{1-x}Ru_xS(h^+_{VB}) + OH^- \rightarrow OH^{\cdot} + ZnS \quad (13)$$

Dye degradations

$$OH^{\cdot}/O_2^{\cdot-} + MB \rightarrow CO_2 + H_2O \quad (14)$$

So, to ensure the efficiency of the catalyst, it is necessary to know about the $OH^{\cdot}$ generation ability of the material.

Here, the formation of active species ($OH^{\cdot}$ radical) during photocatalytic process has been detected by using the fluorescence technique, where terephthalic acid (TA) has been used as a probe molecule. By the reaction of TA with hydroxyl $OH^{\cdot}$ radical, 2-hydroxyterephthalic acid (TAOH) is generated. This TAOH acid has a fluorescence peak at about 425 nm on excitation at 315 nm [106]. Fluorescence intensity of TAOH is directly proportional to the amount of $OH^{\cdot}$ radical produced on the surface of the photocatalysts during the light illumination. Figure 16 shows the fluorescence intensity ratio (I_t/I_0) of TaOH generation with respect to irradiation time monitored at 425 nm. Where I_0 is the intensity of TaOH without irradiation, I_t is the intensity of TaOH after certain time of irradiation. Inset figure shows the fluorescence profile of TaOH after 180 min irradiation. The intensity ratio shows that in the presence of the photocatalyst fluorescence intensity of TAOH reaches at the highest intensity, which confirms the generation of $OH^{\cdot}$ radical during photocatalytic experiment. This experiment reveals that $Zn_{0.990}Ru_{0.010}S$ system generates sufficient amount of $OH^{\cdot}$ radical and shows effective photocatalytic activity.

In view of the technological feasibility, the cycling performance of the photocatalyst is of utter importance. In this regard, the reusability of the $Zn_{0.990}Ru_{0.0010}S$ photocatalyst has been investigated in a neutral aqueous medium. In Fig. 17, $Zn_{0.990}Ru_{0.0010}S$ photocatalyst inevitably showed a higher degree of stability and photocatalytic activity towards the degradation of MB dye for five consecutive runs.

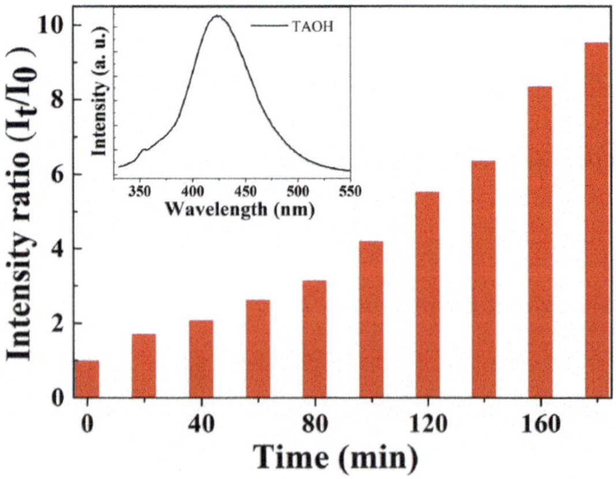

Fig. 16 The photoluminescence intensity ratio of 2-hydroxyterephthalic acid with irradiation time. Inset shows the fluorescence profile of TaOH after 180 min irradiation

Almost 6% loss in photodegradation efficiency was observed after the five cycles, making this photocatalyst as a feasible industrial material. Moreover, the structural stability of the optimized photocatalyst was also tested by XRD patterns which are presented in Fig. 18a. The XRD pattern of Ru-doped ZnS after fourth recycle test (after irradiation) is similar to that of the unirradiated sample. A slight decrease in intensity is due to the surface adsorption of residual MB dye which weakens the diffraction peaks than the unirradiated [80]. The FESEM image of $Zn_{0.990}Ru_{0.0010}S$ catalyst after 4th recycle test is shown in Fig. 18b. The FESEM image clearly indicates that the structure is almost comparable to the synthesized unirradiated material. Slight agglomeration of NPs is observed due to the interaction of residual MB with NPs. This result is in accordance with the XRD pattern of the not recycled one. So, these measurements indicate that Ru-doped ZnS NPs are quite stable against photocorrosion. Also, the stability of the structure is verified using a theoretical approach, which has been discussed in the next section.

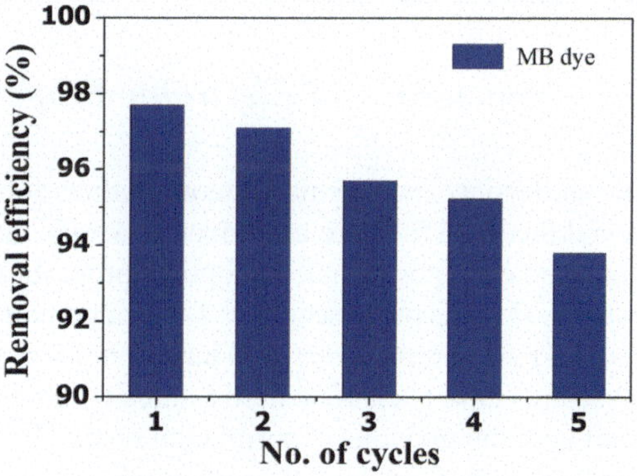

Fig. 17 Reusability of $Zn_{0.990}Ru_{0.0010}S$ photocatalyst for five successive cycles for the degradation of MB

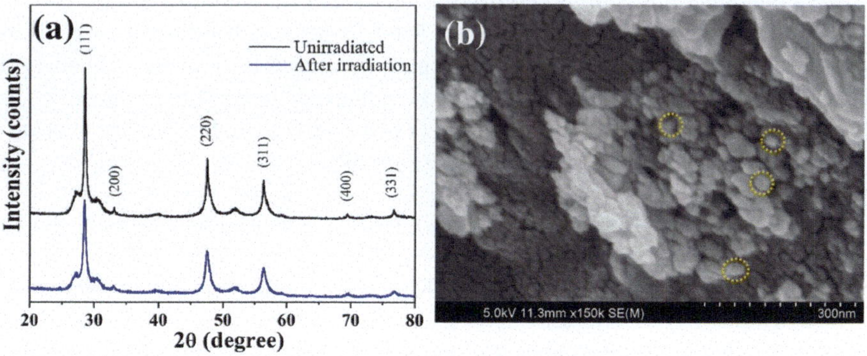

Fig. 18 **a** XRD patterns of $Zn_{0.990}Ru_{0.0010}S$ NPs as synthesized before (Unirradiated), and after fourth recycle test after irradiation), indicated by black and blue colours, respectively. **b** FESEM image of $Zn_{0.990}Ru_{0.0010}S$ NPs after fourth recycle test

In brief, Ru doping in ZnS crystalline structure creates trapping charge centres. With the irradiation by visible light, electron-hole pairs are generated. Due to transfer of charge through Ru ions, the photogenerated electrons and holes in case of $Zn_{0.990}Ru_{0.0010}S$ NPs can effectively react with hydroxyl ions (OH^-), dissolved oxygen, and H^+ ions present in the solution, and hence produce hydroxyl radicals ($OH^{\cdot}$), superoxide radicals ($O_2^{\cdot-}$) and holes (h^{++}), respectively. These active radicals react with surface adsorbed MB dye, and decompose into carbon dioxide and water. In summing up, among these Ru-doped ZnS systems, $Zn_{0.990}Ru_{0.0010}S$ NPs show excellent photodegradation of MB at pH = 10.2, and catalyst loading of 10 mg under visible LED irradiation. From the point of feasible application, Ru-doped ZnS NPs can be projected as a metal-free, cost-effective and visible light activated photocatalyst.

4 Theoretical Results

4.1 Structural and Electronic Properties

Inspired from experimental outcomes, theoretical calculations based on density functional theory (DFT) are employed to verify the electronic properties (i.e. band gap and band edge positions), and optical properties of pristine and Ru-incorporated ZnS system. For this purpose, a $2 \times 2 \times 2$ supercell of Zincblende ZnS unit cell ($a = b = c = 5.41$ Å, $\alpha = \beta = \gamma = 90°$) has been taken. Geometrical relaxation of each of the structures have been done first. Lattice constant of the relaxed unit cell of ZnS agrees well with the previous experimental result [107], indicating that our methodology is quite efficient for calculating electronic structures as well as optical calculations. To observe the effect of Ru doping on the ZnS structure, the concentration of Ru was varied in the range 3.125–12.500% (optimized structures are shown in Fig. 19).

For 3.125% Ru doping configuration, Ru–S bond length is found to be 1.2% suppressed as compared with Zn–S bond length. This decrement in the local network of Ru is mainly occurred due to slightly greater ionic radii of Ru with respect to Zn. In order to examine structural stability of Ru-doped ZnS system, the formation energy was evaluated using the following relationship [108, 109].

$$E_{df}^{n} = E(\mathrm{Ru}_n\mathrm{Zn}_{32-n}\mathrm{S}_{32}) - E(\mathrm{Zn}_{32}\mathrm{S}_{32}) + n\mu_{\mathrm{Zn}} - n\mu_{\mathrm{Ru}} + q\left[\mu_e + E_{\upsilon}^{*}\right] \quad (15)$$

where total energies of pristine ZnS, and Ru incorporated ZnS structure are represented by $E(\mathrm{Zn}_{32}\mathrm{S}_{32})$, $E(\mathrm{Ru}_n\mathrm{Zn}_{32-n}\mathrm{S}_{32})$, and (the number of Zn atoms to be replaced by Ru is denoted by n). Similarly, μ_{Zn} and μ_{Ru} represent the chemical potential of isolated Zn and Ru atoms, respectively. Electron chemical potential of each structure μ_e represents the Fermi level. E_{υ}^{*} is the correction to the maximum value of the valence band in pristine ZnS structure. A correction term δq is included in E_{υ}^{*} to align the potential in doped structure. Defect formation energy was calculated for various

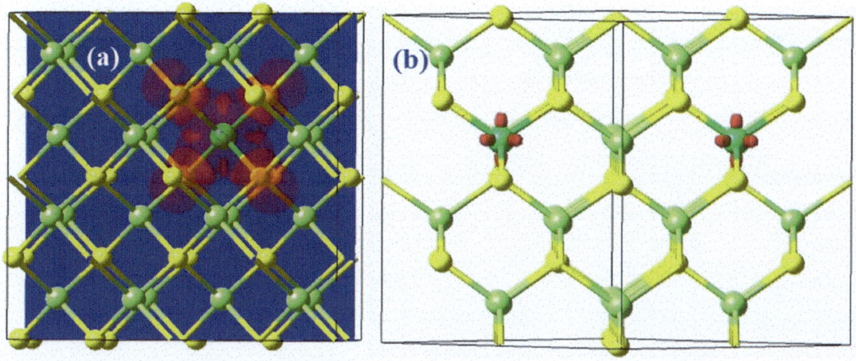

Fig. 19 Geometrically optimized structures of; **a** pristine, and **b** Ru-doped ZnS structures. Zn, Ru, and S atoms are represented by blue, cyan, and yellow-coloured spheres, respectively. The numbers 1–4 represent the substitutional position for Ru atoms

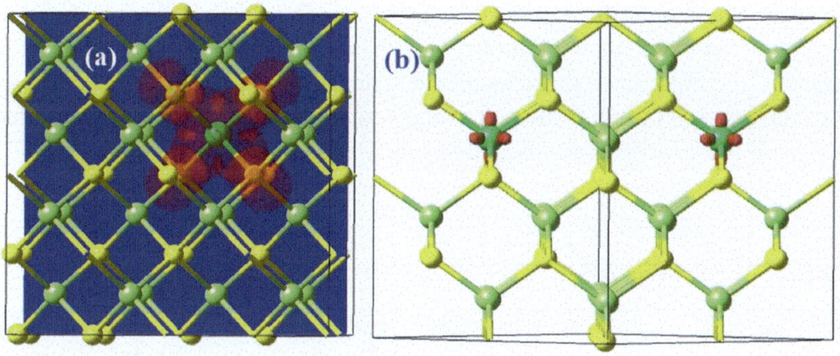

Fig. 20 **a** Local magnetization density for 3.125% Ru-doped ZnS structure, and **b** difference charge density for 6.25% Ru-doped ZnS structure

Ru-doped structures. The results are quite interesting. As the doping concentration increases in the supercell, formation energy exhibits a transition from positive to negative values (shown in Table 2). Hence it can be concluded that Ru incorporation leads the structure to be more stable.

When a semiconducting material is doped with transition metal atom, p orbital of the host atom interacts with d orbital of the dopant atom and causes p-d hybridization in the system. In Ru-doped ZnS system, p-d hybridization induces ferromagnetic coupling in the system [110, 111]. Hybridization between Ru-d and S-p states is verified from local magnetization density (LMD) plot (Fig. 20a). Bader charge analysis suggests that the Ru acted as a charge centre, and participated in the charge transfer mechanism. Figure 20b shows some nonzero charge around Ru atoms relative to Zn and S atoms and indicates charge trapping in the system.

Furthermore, to examine the effect of Ru doping in ZnS matrix, band structure and density of states (DOS) were computed for pristine ZnS as well as Ru-doped ZnS

Table 2 Comparison of structural and electronic properties of various Ru incorporated and pristine ZnS structures

Structure	Doping concentration (%)	Bond length(Å)	Formation energy(eV)	Fermi energy(eV)
ZnS	0.00	2.342	–	2.127
$Ru_1Zn_{31}S_{32}$	3.125	2.312	3.870	3.780
$Ru_2Zn_{30}S_{32}$	6.250	2.297	−0.447	3.861
$Ru_3Zn_{29}S_{32}$	9.375	2.296	−5.732	3.870
$Ru_4Zn_{28}S_{32}$	12.500	2.298	−8.921	4.026

structures using DFT. Both band structures are plotted along the high symmetry path L-Γ-F-W-K-Γ. It can be easily understood from Fig. 21 that pristine ZnS exhibits semiconducting nature with a direct band gap of 2.093 eV at Γ point. But in the band structure of Ru-doped ZnS, some localized states have been occurred in the vicinity of Fermi level. Due to these localized states around Fermi level, Ru-doped ZnS structures possess a metallic nature [81]. These localized states are originated due to the interaction between Ru and S atoms in RuS_4 tetrahedron. As the concentration of Ru increases, the number of these localized states around Fermi level also increases. Thus it can be concluded from electronic structure and charge analysis that Ru interacts with the nearby S atoms and strongly affects the band spectra near Fermi level.

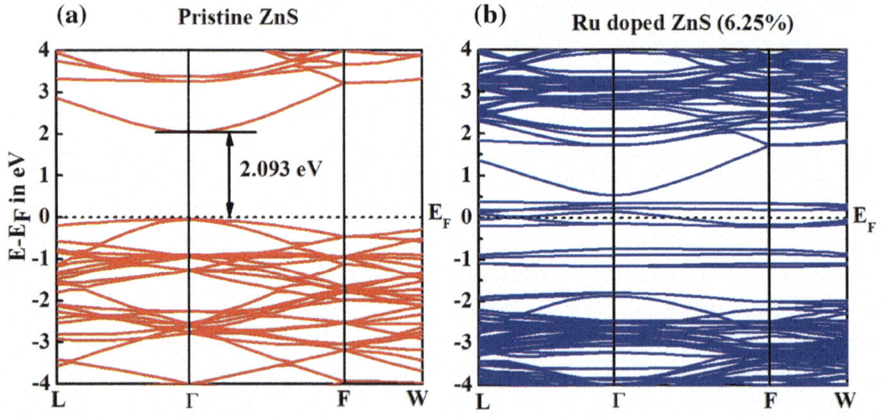

Fig. 21 Electronic band structure of **a** pristine, and **b** 6.25 % Ru-doped ZnS structure

4.2 Optical Properties

Significant enhancement of electronic band structure strongly motivates to study the optical responses associated with pristine and Ru-doped ZnS system. As well known, DFT is a ground state theory, optical calculations on the basis of DFT do not incorporate any theory for excited states of the system. DFT only gives first-hand background before starting optical calculations. Structural isotropy of ZnS restricts calculating optical properties in perpendicular polarization of the electric field only. In order to get better accuracy in optical results, a large number of empty bands have been taken in the calculations.

4.2.1 Imaginary Part of Dielectric Function

Complex dielectric function ($\varepsilon(\omega) = \varepsilon_1(\omega) + i\varepsilon_2(\omega)$) is regarded as the basics of the optical calculations, where $\varepsilon_1(\omega)$ and $\varepsilon_2(\omega)$ are the real and imaginary part of the complex dielectric function, and ω is the angular frequency of incident light wave. Imaginary part of the dielectric function $\varepsilon_2(\omega)$ in the long wavelength limit ($q \rightarrow 0$) can be represented by Eq. (10) [112].

$$\varepsilon_2(q \rightarrow 0, \omega) = \frac{2e^2\pi}{\Omega\varepsilon_0} \sum_{K,VB,CB} \left|\langle\psi_K^{VB}| \vec{u} \,.\, \vec{r}\, |\psi_K^{CB}\rangle\right|^2 \delta\left(E_K^{CB} - E_K^{VB} - \omega\right) \quad (16)$$

where the wavefunctions for valence band and conduction band at K points are represented by ψ_K^{VB} and ψ_K^{CB}, and Ω, e, ε_0, $\vec{u}$, $\vec{r}$ indicate the volume of the supercell, the electronic charge, free space dielectric constant, polarization vector of electric field, and position vector, respectively. The imaginary part of the dielectric function ($\varepsilon_2(\omega)$) was obtained using Eq. (10) and plotted against energy as shown in Fig. 22. This quantity signifies the behaviour of energy dissipated in the system. In pure ZnS structure, main intense peak is occurrs around an energy value of 5.89 eV. As the concentration of Ru increases, the height of this peak decreases. Simultaneously, a low energy peak appears around energy value of 0.60 eV.

4.2.2 Real Part of Dielectric Function and EEL Spectra

Real part of dielectric function $\varepsilon_1(\omega)$ plays a very important role in determining collective behaviour of charge carriers in the system. Once the imaginary part of the dielectric function was obtained, we evaluated the real part of $\varepsilon(\omega)$ using Kramers–Kronig (KK) relation [113].

$$\varepsilon_1(\omega) = 1 + \frac{2}{\pi}\int_0^\infty \frac{\acute{\omega}\varepsilon_2(\acute{\omega})}{\acute{\omega}^2 - \omega^2} d\acute{\omega} \quad (17)$$

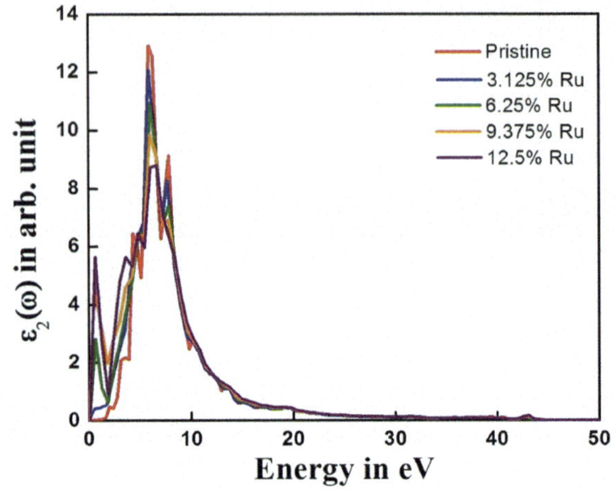

Fig. 22 Variation of the imaginary part of dielectric function ($\varepsilon_2(\omega)$) as a function of incident light wave energy ($\hbar\omega$) in eV, where $\hbar = h/2\pi$, for pristine as well as various Ru-doped ZnS system

Figure 23 depicts $\varepsilon_1(\omega)$ against energy of incident light for pristine and various Ru-doped ZnS systems. Static dielectric constant $\varepsilon_1(0)$ for pristine ZnS has been found to be 6.28. As Ru concentration in the system increases, $\varepsilon_1(0)$ value is significantly enhanced due to excess charge carriers. It can be observed from Fig. 23a that $\varepsilon_1(\omega)$ value reaches maximum (nearly six times the value for pristine ZnS) for 6.25% Ru-doped ZnS system. Moreover, plasma frequencies (ω_p) have been detected from $\varepsilon_1(\omega)$ plot for pristine and Ru doped system. The optical frequencies for which $\varepsilon_1(\omega)$ possesses a transition from negative to positive values with $\varepsilon_2(\omega) > 0$ are termed as plasma frequencies (ω_p). Plasma frequencies for pristine and doped system have been reported in Table 3.

Simultaneously electron energy loss (EEL) spectra were calculated using Eq. (12) and presented in (Fig. 22b).

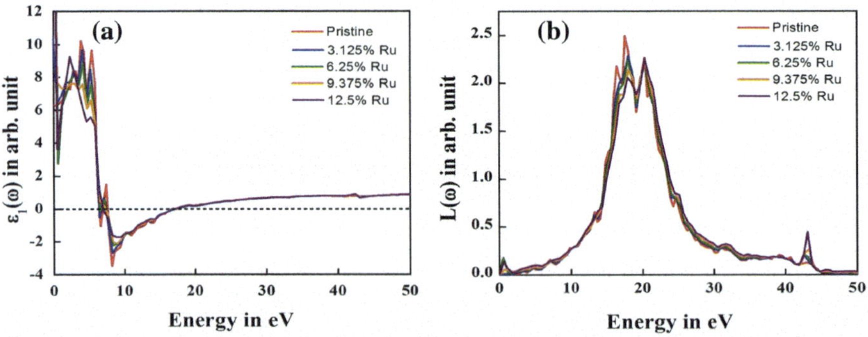

Fig. 23 Variation of: **a** Real part of dielectric function ($\varepsilon_1(\omega)$). **b** Electron energy loss EEL function L(ω) as a function of energy (eV) for pristine as well as various Ru-doped ZnS system

Table 3 Plasma frequencies for pristine and Ru-doped system

Structure (concentration %)	Plasma frequencies (eV)
ZnS (pristine)	6.98, 17.59
Ru_1Zn31S_{32} (3.125%)	6.88, 16.81
$Ru_2Zn_{30}S_{32}$ (6.250%)	6.72, 16.86
$Ru_3Zn_{29}S_{32}$ (9.375%)	17.57
$Ru_4Zn_{28}S_{32}$ (12.500%)	17.50

$$\text{EEL function: } L(\omega) = \text{Im}\left\{-\frac{1}{\varepsilon(\omega)}\right\} = \frac{\varepsilon_2(\omega)}{\varepsilon_1^2(\omega) + \omega_2^2(\omega)} \tag{18}$$

Plasmon peaks are verified with the detected plasma frequencies from $\varepsilon_1(\omega)$ plot. Main EELS peak has been located at 17.59 eV optical energy. Along with this, in Ru-doped ZnS a low energy peak has occurred. Height of this low energy peak reaches a maximum for 6.25% Ru-doped ZnS system. Each structure of the Ru-doped ZnS system has carried this low-energy behaviour. Hence, this low-energy peak symbolizes the presence of Ru and energy loss of fast-moving charge carriers in the system.

4.2.3 Optical Absorption and Reflectivity

In this section, another two optical properties of pristine and Ru-incorporated ZnS system will be addressed. Absorption coefficient of any system signifies the penetration depth of incident photon (carrying a specific wavelength) in the system before getting absorbed. Absorption coefficient $\alpha(\omega)$ has been evaluated using the following equation.

$$\text{Absorption coefficient: } \alpha(\omega) = \frac{2\omega}{\hbar c}\left(\frac{\sqrt{\varepsilon_1^2 + \varepsilon_2^2} - \varepsilon_1}{2}\right)^{1/2} \tag{19}$$

$\alpha(\omega)$ was calculated for all the pristine, and Ru-doped ZnS system and presented in Fig. 14a. Similar to $\varepsilon_2(\omega)$ and EELS plots, the $\alpha(\omega)$ plot for Ru-doped system also contains low energy peak. In this peak, peak height varies with doping concentration of Ru.

Also, optical band gap E_g of the system has been obtained using Tauc plot [114] shown in the inset of Fig. 24a.

$$(\alpha h\nu)^{1/2} = \beta(h\nu - E_g) \tag{20}$$

where β is the band tailing parameter and E_g is the optical band gap energy. Optical band gap of pristine structure has been found to be 1.51 eV. But with increasing

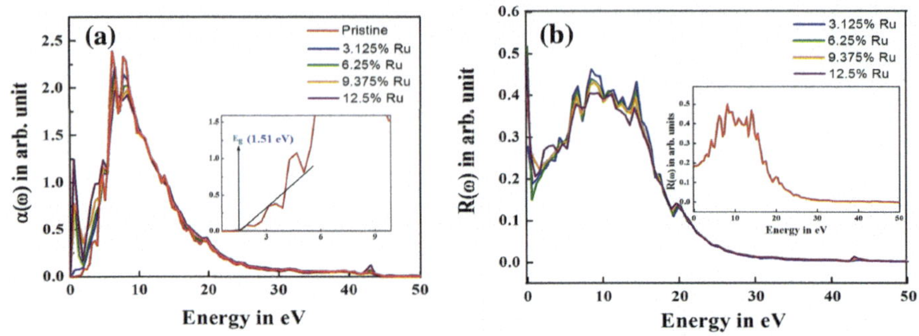

Fig. 24 Variation of **a** absorption coefficient ($\alpha(\omega)$). Inset represents the Tauc plot to find the optical band gap of pristine ZnS. **b** reflectivity $R(\omega)$ with energy (eV) for pristine (inset), as well as various Ru-doped ZnS system

Ru doping concentration optical band gap vanishes. Simultaneously, the reflectivity $R(\omega)$ for pristine and doped structures was calculated using Eq. (15) and shown in Fig. 24b.

$$\text{Reflectivity:} R(\omega) = \left| \frac{\sqrt{\varepsilon_1 + i\varepsilon_2} - 1}{\sqrt{\varepsilon_1 + i\varepsilon_2} + 1} \right|^2 \quad (21)$$

Zero energy reflectance in pristine ZnS has been observed to be 0.18, whereas for 6.25% Ru-doped ZnS this quantity reaches maximum. High value of this quantity measures the metallic character of the doped system. Therefore, not only electronic structures, but also optical responses verify the semiconductor to metallic transition with the variation in Ru concentration.

5 Correlation Between Theoretical and Experimental Outcomes

It is not so easy to comment on the theoretical analogy of some experimental observations. But, in this study, we have tried to explore some correlating points between theoretical and experimental investigations.

(i) From the experimental analysis, some quenching effect has been observed in PL spectra of Ru-doped ZnS NPs as compared to pristine samples. This PL quenching indicates some kind of charge transfers in the Ru-doped ZnS system. Theoretically using Bader charge analysis (shown in Fig. 21b), we have verified the fact and concluded that as a charge centre, Ru participated in the charge transfer mechanism. Beside this, the band structure of Ru incorporated ZnS computed according to Generalized Gradient Approximation-Perdew-Burke-Ernzerhof (GGA-PBE) for exchange-correlation potential suggests that some localized states around Fermi level have occurred due to the presence of Ru. These states also play a vital role to support Ru-induced PL quenching and bringing metallic character in the system.

(ii) We have investigated the electronic configuration for Ru ions. In general, Ru ions exist in

Two oxidation states, Ru^{2+} and Ru^{3+}. Electronic configurations of Ru, Ru^{2+}, and Ru^{3+} are shown.

$$\mathrm{Ru}: [Kr]4d^7 5s^1$$

$$\mathrm{Ru}^{2+}: [Kr]4d^6 5s^0$$

$$\mathrm{Ru}^{3+}: [Kr]4d^5 5s^0$$

Ru^{3+} ion has 1 unpaired electron such that it exhibits non-zero magnetic moment whereas a study suggests that due to the presence of d^6 electrons, Ru^{2+} possesses zero magnetic moment [115]. Bader charge analysis also suggests that charge transfers in Ru atom are not comparable to the Zn atom [80]. Thus Ru and Zn ions cannot possess same oxidation state. It is worthy to mention that Zn ions exist in Zn^{2+} oxidation state. Thus magnetic and charge analysis reveal that Ru atom in Ru-doped ZnS sample exists in Ru^{3+} ionic form [81]. Experimentally Ru doping is possible by utilizing $RuCl_3$ salt (i.e. Ru is in Ru^{3+} ionic state). Therefore, existence of Ru^{3+} ion is verified by theoretical as well as experimental findings.

6 Comparison of Photocatalytic Activity of Ru-Doped ZnS Relative to Various Recent ZnS-Based Photocatalyst

Prior to this work, we had made an extensive literature survey and compared our synthesized Ru-doped ZnS NPs with some recent reports about photocatalytic degradation of MB dye using doped ZnS and ZnS-related heterostructures. The notable results are tabulated in Table 4 for comparison. The table ensures that our reported material shows the comparatively better efficiency in photocatalytic degradation of MB dye under low power visible irradiation. Moreover, this result suggests that this structure of 1% Ru-doped ZnS nanoparticles can be further used for various types of heterostructure formation in order to enhance its photocatalytic activity.

Table 4 Comparison of photocatalytic performance of MB dye with previously reported literature about ZnS-based photocatalyst

Catalyst nature (Ref.)	Concentration and volume of MB dye	Irradiation source	Degradation time (min)	Degradation (%)
5% Cu-doped ZnS (2.5 mg) [84]	3.2 mg/L, 100 ml	20 W white light	60	73.5
1.5% Ni-doped ZnS (30 mg) [116]	10 mg/L, 100 ml	Solar light	180	87.4
N-doped ZnS (150 mg) [99]	25 mg/L, 500 ml	Solar irradiation	180	58.0
Co-doped ZnS (10 mg) [117]	20 mg/L, 40 ml	Stimulated solar light(λ = 320–780 nm)	80	88.2
Zr-doped ZnS (20 mg/L) [118]	1 mg/L, 100 ml	UV-visible (λ = 300–400 nm)	240	76.1
Gd-doped ZnS (100 mg) [119]	15 mg/L, 100 ml	8 W UV-vis lamp	300	87.4
1% Ru-doped ZnS (10 mg) [80]	1×10^{-5} M, 25 ml	30 W visible LED	140	98.0

7 Role of Metal Doping Versus Metal Loading in Semiconductor Photocatalyst

The efficiency of a semiconductor photocatalyst can be increased by the modification of the surface of the semiconductor. Primarily, metal-semiconductor modifications are used to inhibit the charge recombination and increase the selectivity of a particular product. The metal doping can lead to significant variations of the structural, morphological, spectroscopic, and surface properties with respect to the pristine semiconductor. Doping also reduced the bandgap of the semiconductor with the creation of new energy levels (also called the impurity state), between the valence band and the conduction band. This increases the photo-responsiveness of the photocatalyst to the visible region. Sometimes, the foreign metal atom dopant generates new energy levels inside the forbidden region, such as acceptor or donor level, depending upon the dopant ion, and acts as a charge centre. For example, in this work, Ru acts as a positive charge centre in the ZnS matrix. These new energy levels play an important role in increasing the lifetime of the charge carriers. As a result, the photocatalytic efficiency of the semiconductor increases.

On the other hand, surface modification can be done by deposition of metal nanoparticles on the semiconductor surface. Generally, novel metals play a crucial role in forming these types of heterostructures. Because other metals can be easily

Table 5 Comparison of photocatalytic activity of some novel metal loaded ZnS-related heterostructures

Type of sample (Ref.)	Photocatalytic application	Catalyst used	Photocatalytic performance
Gold nanoparticles on ZnS nanobelt[120]	4-nitrophenol degradation(300 W xenon lamp 10 mg catalyst 80 ml dye solution)	ZnS ZnS@Au (0.1%) ZnS@Au (0.2%) ZnS@Au (1%)	**Degradation in 40 min** ~25% ~40% ~55% ~35%
Silver decorated ZnS nanoballs [121]	Methylene blue degradation (15 ppm dye solution in 100 ml, 100 mg catalyst, 300 W xenon lamp with 400 nm cut-off filter)	ZnS ZnS@Ag	**Degradation in 30 min** 20% 60–70%
Ru-loaded ZnS [122]	Hydrogen production	ZnS ZnS@Ru (0.5) ZnS@Ru (0.7) ZnS@Ru (1.5)	**Production rate** 0.81 μmol h^{-1} 0.51 μmol h^{-1} 0.77 μmol h^{-1} 0.26 μmol h^{-1}
Pt-loaded ZnO–ZnS heterostructure[123]	Water splitting	ZnO–ZnS ZnO–ZnS@Ru (1%)	**Production rate** 295 μmol h^{-1} 535 μmol h^{-1}
Pd-modified ZnS [124]	CO_2 photoreduction	ZnS ZnS–Pd (1.5%)	**Amount of CH_4 generated (10^{-6} mol)** 0.640 1.305

oxidized in ambient conditions, in this case, a Schottky barrier is formed between the semiconductor-metal interfaces. Photogenerated electrons flow from the semiconductor to the metal surface at the interface. Thus, the lifetime of the photogenerated electrons increases by trapping the excited electron in the metal. As a result, the recombination rate is reduced, thus enhancing photocatalytic behaviour.

Here, photocatalytic activity of some novel metal-loaded ZnS-related heterostructures are shown in tabular form in Table 5.

8 Concluding Remarks

Photocatalysis is a promising and ecofriendly method for the decomposition of various organic pollutants and toxic dyes present in waste water as it mineralizes these harmful contaminants into non-toxic products. In this book chapter, a quick introduction has been given about photocatalysis and its importance in our daily lives. The basic mechanism of photocatalysis and the essential factors, which affect the efficiency of a semiconductor photocatalyst have been explained thoroughly. To give knowledge to the beginner in this research area, a brief review about the development of semiconductor photocatalyst (mostly oxide and sulphide materials) over the decades including all types of heterostructure formation has been cautiously done. Along with this, the importance of Ru as a dopant in constructing heterogeneous photocatalyst has been carefully demonstrated.

In this work, the electronic and optical properties of Ru-doped ZnS NPs have been investigated via experimental as well as theoretical approach. The Ru-doped ZnS NPs with various Ru-doping concentration have been synthesized via one-step solvothermal method. Optimization of visible LED-assisted photodegradation has been achieved for 1% Ru-incorporated ZnS samples. Elemental mapping and XRD peaks have verified the successful incorporation of Ru in the ZnS matrix. In the presence of Ru, quenching effect is observed in the PL spectra compared to pristine samples. Besides these observations, some theoretical investigations suggest significant modification in electronic and low energy optical behaviour of ZnS due to Ru. Bader charge analysis verifies participation of Ru in the charge transfer mechanism. Electronic band structure of Ru-doped ZnS dictate presence of localized states around Fermi level. Futhermore, some correlation points relating experimental and theoretical aspects of Ru-doped ZnS samples are demonstrated. Quenching effect in PL spectra successfully verified with the help of charge analysis and theoretical band structure. Magnetic measurements are utilized to confirm the oxidation state of Ru in Ru-doped ZnS matrix. Therefore, Ru-doped ZnS has been projected as a metal-free visible light-driven photocatalyst for further various types of heterostructure formation as well as optoelectronic and spintronic materials for future industrial applications.

Acknowledgements Authors are grateful to the Council of Scientific & Industrial Research (CSIR), the Government of India, and the University of Calcutta for financial support. One of the authors (SG) is thankful to Dr. Dirtha Sanyal of VECC, INDIA for computational support. Authors are thankful to Mr. Sujoy Datta of the University of Calcutta for the fruitful suggestions to improve the quality of the book chapter.

References

1. B. Auguié, B.L. Darby, E.C. Le Ru, Nanoscale **11**(25), 12177 (2019)
2. S. Elhani, H. Ishitobi, Y. Inouye, A. Ono, S. Hayashi, Z. Sekkat, Sci. Rep. **10**(1), 1 (2020)
3. P. Devi, U. Das, A.K. Dalai, Sci. Total Environ. **571**, 643 (2016)
4. I. Michael-Kordatou, P. Karaolia, D. Fatta-Kassinos, Water Res. **129**, 208 (2018)
5. L. Goswami, R.V. Kumar, K. Pakshirajan, G. Pugazhenthi, J. Hazard. Mater. **365**, 707 (2019)
6. S. Kahl, S. Kleinsteuber, J. Nivala, M. van Afferden, T. Reemtsma, Environ. Sci. Technol. **52**(5), 2717 (2018)
7. S.K. Mandal, K. Dutta, S. Pal, S. Mandal, A. Naskar, P.K. Pal, T. Bhattacharya, A. Singha, R. Saikh, S. De et al., Mater. Chem. Phys. **223**, 456 (2019)
8. A. Fujishima, K. Honda, Nature **238**, 37 (1972)
9. L. Yu, Y. Ding, M. Zheng, Appl. Catal. B **209**, 45 (2017)

10. J. Remillard, J. McBride, K. Nietering, A. Drews, X. Zhang, J. Phys. Chem. B **104**(18), 4440 (2000)
11. R. Nakamura, A. Imanishi, K. Murakoshi, Y. Nakato, J. Am. Chem. Soc. **125**(24), 7443 (2003)
12. N. Serpone, J. Photochem. Photobiol. A **104**(1–3), 1 (1997)
13. J. Jiang, X. Zhang, P. Sun, L. Zhang, J. Phys. Chem. C **115**(42), 20555 (2011)
14. S.S. Lo, T. Mirkovic, C.H. Chuang, C. Burda, G.D. Scholes, Adv. Mater. **23**(2), 180 (2011)
15. Y. Tak, H. Kim, D. Lee, K. Yong, Chem. Commun. **38**, 4585 (2008)
16. V. Stevanović, S. Lany, D.S. Ginley, W. Tumas, A. Zunger, Phys. Chem. Chem. Phys. **16**(8), 3706 (2014)
17. V.J. Babu, S. Vempati, T. Uyar, S. Ramakrishna, Phys. Chem. Chem. Phys. **17**(5), 2960 (2015)
18. M.R. Hoffmann, S.T. Martin, W. Choi, D.W. Bahnemann, Chem. Rev. **95**(1), 69 (1995)
19. S. Kim, B. Fisher, H.J. Eisler, M. Bawendi, J. Am. Chem. Soc. **125**(38), 11466 (2003)
20. N. Negishi, T. Iyoda, K. Hashimoto, A. Fujishima, Chem. Lett. **24**(9), 841 (1995)
21. P. Gao, A. Li, D.D. Sun, W.J. Ng, J. Hazard. Mater. **279**, 96 (2014)
22. I. Paramasivam, H. Jha, N. Liu, P. Schmuki, Small **8**(20), 3073 (2012)
23. A. Di Mauro, M.E. Fragala, V. Privitera, G. Impellizzeri, Mater. Sci. Semicond. Process. **69**, 44 (2017)
24. Y. Xia, J. Wang, R. Chen, D. Zhou, L. Xiang, Crystals **6**(11), 148 (2016)
25. D.H. Kim, D.K. Choi, S.J. Kim, K.S. Lee, Catal. Commun. **9**(5), 654 (2008)
26. K. Siddhapara, D. Shah, J. Cryst. Growth **452**, 158 (2016)
27. B. Donkova, D. Dimitrov, M. Kostadinov, E. Mitkova, D. Mehandjiev, Mater. Chem. Phys. **123**(2–3), 563 (2010)
28. F. Achouri, S. Corbel, L. Balan, K. Mozet, E. Girot, G. Medjahdi, M.B. Said, A. Ghrabi, R. Schneider, Mater. Des. **101**, 309 (2016)
29. H. Qin, W. Li, Y. Xia, T. He, ACS Appl. Mater. Interfaces **3**(8), 3152 (2011)
30. X. Zong, C. Sun, H. Yu, Z.G. Chen, Z. Xing, D. Ye, G.Q. Lu, X. Li, L. Wang, J. Phys. Chem. C **117**(10), 4937 (2013)
31. Q. Xiang, J. Yu, M. Jaroniec, Phys. Chem. Chem. Phys. **13**(11), 4853 (2011)
32. D.C. Hurum, A.G. Agrios, K.A. Gray, T. Rajh, M.C. Thurnauer, J. Phys. Chem. B **107**(19), 4545 (2003)
33. H. Wang, L. Zhang, Z. Chen, J. Hu, S. Li, Z. Wang, J. Liu, X. Wang, Chem. Soc. Rev. **43**(15), 5234 (2014)
34. R. Georgekutty, M.K. Seery, S.C. Pillai, J. Phys. Chem. C **112**(35), 13563 (2008)
35. B. Li, Y. Wang, J. Phys. Chem. Solids **72**(10), 1165 (2011)
36. J. Zhang, J. Yu, Y. Zhang, Q. Li, J.R. Gong, Nano Lett. **11**(11), 4774 (2011)
37. M. Lee, K. Yong, Nanotechnology **23**(19), 194014 (2012)
38. X. Shuai, W. Shen, J. Phys. Chem. C **115**(14), 6415 (2011)
39. V.M. Daskalaki, M. Antoniadou, G. Li Puma, D.I. Kondarides, P. Lianos, Environ. Sci. Technol. **44**(19), 7200 (2010)
40. S. Karmakar, S. Ghosh, P. Kumbhakar, J. Nanopart. Res. **22**(1), 1 (2020)
41. D. Sarkar, C.K. Ghosh, S. Mukherjee, K.K. Chattopadhyay, ACS Appl. Mater. Interfaces **5**(2), 331 (2013)
42. F. Shen, W. Que, Y. Liao, X. Yin, Ind. Eng. Chem. Res. **50**(15), 9131 (2011)
43. S. Jung, K. Yong, Chem. Commun. **47**(9), 2643 (2011)
44. F. Shen, W. Que, Y. He, Y. Yuan, X. Yin, G. Wang, ACS Appl. Mater. Interfaces **4**(8), 4087 (2012)
45. M. Shang, W. Wang, L. Zhang, S. Sun, L. Wang, L. Zhou, J. Phys. Chem. C **113**(33), 14727 (2009)
46. C. Wang, C. Shao, X. Zhang, Y. Liu, Inorg. Chem. **48**(15), 7261 (2009)
47. L. Wei, C. Yu, Q. Zhang, H. Liu, Y. Wang, J. Mater. Chem. A **6**(45), 22411 (2018)
48. K.S. Kim, Y. Zhao, H. Jang, S.Y. Lee, J.M. Kim, K.S. Kim, J.H. Ahn, P. Kim, J.Y. Choi, B.H. Hong, Nature **457**(7230), 706 (2009)
49. B. Wang, W. Feng, L. Zhang, Y. Zhang, X. Huang, Z. Fang, P. Liu, Appl. Catal. B **206**, 510 (2017)

50. D. Wang, H. Shen, L. Guo, F. Fu, Y. Liang, New J. Chem. **40**, 8614 (2016)
51. D.H. Cho, W.J. Lee, S.W. Park, J.H. Wi, W.S. Han, J. Kim, M.H. Cho, D. Kim, Y.D. Chung, J. Mater. Chem. A **2**(35), 14593 (2014)
52. Q. Zhao, Y. Xie, Z. Zhang, X. Bai, Cryst. Growth Des. **7**(1), 153 (2007)
53. J. Tanne, D. Schafer, W. Khalid, W. Parak, F. Lisdat, Anal. Chem. **83**(20), 7778 (2011)
54. G.J. Lee, J.J. Wu, Powder Technol. **318**, 8 (2017)
55. X. Fang, T. Zhai, U.K. Gautam, L. Li, L. Wu, Y. Bando, D. Golberg, Prog. Mater Sci. **56**(2), 175 (2011)
56. F. Zhang, X. Wang, H. Liu, C. Liu, Y. Wan, Y. Long, Z. Cai, Appl. Sci. **9**(12), 2489 (2019)
57. X. Wang, H. Huang, B. Liang, Z. Liu, D. Chen, G. Shen, Crit. Rev. Solid State Mater. Sci. **38**(1), 57 (2013)
58. S. Hamad, S.M. Woodley, C.R.A. Catlow, Mol. Simul. **35**(12–13), 1015 (2009)
59. H. Sun, X. Zhao, L. Zhang, W. Fan, Journal of Physical Chemistry C **115**(5), 2218 (2011)
60. V. Ramasamy, K. Praba, G. Murugadoss, Spectrochim. Acta Part A Mol. Biomol. Spectrosc. **96**, 963 (2012)
61. L. Zhao, Y. Wang, A. Wang, X. Li, C. Song, Y. Hu, Catal. Today **337**, 83 (2019)
62. S. Kaur, P. Banur, Int. J. Eng. Sci. Technol. **2**(4), 75 (2017)
63. G.J. Lee, Y.H. Hou, H.T. Huang, W. Wang, C. Lyu, J.J. Wu, Catalysts **10**(7), 789 (2020)
64. N. Muhd Julkapli, S. Bagheri, S. Bee Abd Hamid, Sci. World J. **2014**, 692307 (2014)
65. Z.J. Li, J.Y. Shao, Y.W. Zhong, Inorg. Chem. **56**(14), 8538 (2017)
66. K. Brousse, S. Nguyen, A. Gillet, S. Pinaud, R. Tan, A. Meffre, K. Soulantica, B. Chaudret, P.L. Taberna, M. Respaud et al., Electrochim. Acta **281**, 816 (2018)
67. Q. Li, S. Zheng, Y. Xu, H. Xue, H. Pang, Chem. Eng. J. **333**, 505 (2018)
68. J.C. Chou, C.W. Chen, IEEE Sens. J. **9**(3), 277 (2009)
69. L. Manjakkal, D. Szwagierczak, R. Dahiya, Prog. Mater Sci. **109**, 100635 (2020)
70. P. Thangavel, B. Viswanath, S. Kim, Int. J. Nanomed. **12**, 2749 (2017)
71. M. Burian, Z. Syrgiannis, G. La Ganga, F. Puntoriero, M. Natali, F. Scandola, S. Campagna, M. Prato, M. Bonchio, H. Amenitsch et al., Inorg. Chim. Acta **454**, 171 (2017)
72. J. Creus, J. De Tovar, N. Romero, J. García-Antón, K. Philippot, R. Bofill, X. Sala, Chemsuschem **12**(12), 2493 (2019)
73. C. Michel, P. Gallezot, ACS Catal. **5**(7), 4130 (2015)
74. J. Mahmood, F. Li, S.M. Jung, M.S. Okyay, I. Ahmad, S.J. Kim, N. Park, H.Y. Jeong, J.B. Baek, Nat. Nanotechnol. **12**(5), 441 (2017)
75. R. Vinoth, S.G. Babu, V. Bharti, V. Gupta, M. Navaneethan, S.V. Bhat, C. Muthamizhchelvan, P.C. Ramamurthy, C. Sharma, D.K. Aswal et al., Sci. Rep. **7**(1), 1 (2017)
76. P. Velusamy, R.R. Babu, K. Ramamurthi, E. Elangovan, J. Viegas, M. Sridharan, J. Phys. Chem. Solids **112**, 127 (2018)
77. A.S. Rad, K. Ayub, Comput. Theor. Chem. **1121**, 68 (2017)
78. J. Ye, G. Liu, M. Yan, Q. Zhu, L. Zhu, J. Huang, X. Yang, Anal. Chem. **91**(20), 13237 (2019)
79. R. Sahraei, F. Mohammadi, E. Soheyli, M. Roushani, J. Lumin. **187**, 421 (2017)
80. S.K. Mandal, D. Karmakar, S. Ghosal, S. Paul, D. Jana, ChemistrySelect **4**(31), 9102 (2019)
81. S. Ghosal, H. Luitel, S.K. Mandal, D. Sanyal, D. Jana, J. Phys. Chem. Solids **136**, 109175 (2020)
82. L. Wang, P. Wang, B. Huang, X. Ma, G. Wang, Y. Dai, X. Zhang, X. Qin, Appl. Surf. Sci. **391**, 557 (2017)
83. D.A. Reddy, G. Murali, R. Vijayalakshmi, B. Reddy, Appl. Phys. A **105**(1), 119 (2011)
84. N. Prasad, K. Balasubramanian, Spectrochim. Acta Part A Mol. Biomol. Spectrosc. **173**, 687 (2017)
85. O. Brafman, S. Mitra, Phys. Rev. **171**(3), 931 (1968)
86. M. Dimitrievska, H. Xie, A. Jackson, X. Fontané, M. Espíndola-Rodríguez, E. Saucedo, A. Pérez-Rodríguez, A. Walsh, V. Izquierdo-Roca, Phys. Chem. Chem. Phys. **18**(11), 7632 (2016)
87. Y. Cheng, C. Jin, F. Gao, X. Wu, W. Zhong, S. Li, P.K. Chu, J. Appl. Phys. **106**(12), 123505 (2009)

88. M. Abdulkhadar, B. Thomas, Nanostruct. Mater. **5**(3), 289 (1995)
89. S. Radhu, C. Vijayan, Mater. Chem. Phys. **129**(3), 1132 (2011)
90. J. Kim, H. Rho, J. Kim, Y.J. Choi, J.G. Park, J. Raman Spectrosc. **43**(7), 906 (2012)
91. A. Sadeghnejad, L. Lu, C.J. Kiely, B.W. Berger, S. McIntosh, RSC Adv. **7**(61), 38490 (2017)
92. N. Eryong, L. Donglai, Z. Yunsen, B. Xue, Y. Liang, J. Yong, J. Zhifeng, S. Xiaosong, Appl. Surf. Sci. **257**(21), 8762 (2011)
93. J. Choi, S. Yoon, F.S. Kim, N. Kim, J. Alloy. Compd. **671**, 360 (2016)
94. N. Karar, F. Singh, B. Mehta, J. Appl. Phys. **95**(2), 656 (2004)
95. M. Osman, A. Othman, W.A. El-Said, A. Abd-Elrahim, A. Abu-Sehly, J. Phys. D Appl. Phys. **49**(5), 055304 (2015)
96. W. Peng, G. Cong, S. Qu, Z. Wang, Opt. Mater. **29**, 313 (2006)
97. K. Jeyasubramanian, M. Nisanthi, V. Benitha, N. Selvakumar, Acta Metall. Sinica (English Lett.) **28**(1), 103 (2015)
98. M.V. Limaye, S.B. Singh, R. Das, P. Poddar, S.K. Kulkarni, J. Solid State Chem. **184**(2), 391 (2011)
99. D.Z. Zegeye, Int. J. Sci. Res. **5**, 214 (2016)
100. X. Tian, P. Gao, Y. Nie, C. Yang, Z. Zhou, Y. Li, Y. Wang, Chem. Commun. **53**(49), 6589 (2017)
101. Y. Li, Y. Zhang, G. Wang, S. Li, R. Han, W. Wei, J. Mol. Liq. **263**, 53 (2018)
102. G.F. Malash, M.I. El-Khaiary, J. Colloid Interface Sci. **348**(2), 537 (2010)
103. C.H. Nguyen, C.C. Fu, R.S. Juang, J. Clean. Prod. **202**, 413 (2018)
104. A. Mills, D. Hazafy, J. Parkinson, T. Tuttle, M.G. Hutchings, Dyes Pigm. **88**(2), 149 (2011)
105. O. Kazak, Y.R. Eker, I. Akin, H. Bingol, A. Tor, J. Environ. Chem. Eng. **5**(3), 2639 (2017)
106. S. Paul, S. Ghosh, S.K. De, Langmuir **34**(14), 4324 (2018)
107. K. Ghezali, L. Mentar, B. Boudine, A. Azizi, J. Electroanal. Chem. **794**, 212 (2017)
108. L. Assali, W. Machado, J. Justo, Phys. Rev. B **69**(15), 155212 (2004)
109. S. Lany, A. Zunger, Phys. Rev. B **78**(23), 235104 (2008)
110. K. Sato, L. Bergqvist, J. Kudrnovsky, P.H. Dederichs, O. Eriksson, I. Turek, B. Sanyal, G. Bouzerar, H. Katayama-Yoshida, V. Dinh, et al., Rev. Modern Phys. **82**(2), 1633 (2010)
111. B. Zhang, L. Zhu, L. Lin, W. Yu, H. Tao, Y. Xu, F. Guo, L. Li, J. Huang, Vacuum **167**, 59 (2019)
112. F. Giustino, *Materials Modelling Using Density Functional Theory: properties and Predictions* (Oxford University Press, 2014), pp. 197–203
113. M. Dressel, G. Grüner, *Electrodynamics of Solids: optical Properties of Electrons in Matter* (American Association of Physics Teachers, 2002), pp. 61–68
114. J. Tauc, R. Grigorovici, A. Vancu, Phys. Status Solidi (b) **15**(2), 627 (1966)
115. R. Radwanski, Z. Ropka, Solid State Commun. **112**(11), 621 (1999)
116. M. Jothibas, C. Manoharan, S.J. Jeyakumar, P. Praveen, I.K. Punithavathy, J.P. Richard, Sol. Energy **159**, 434 (2018)
117. R. Wang, H. Liang, J. Hong, Z. Wang, J. Photochem. Photobiol. A **325**, 62 (2016)
118. J. Sharma, A. Gupta, O. Pandey, Ceram. Int. **45**(11), 13671 (2019)
119. R.S. Kumar, V. Veeravazhuthi, N. Muthukumarasamy, M. Thambidurai, M. Elango, A. Gnanaprakasam, G. Rajesh, SN Appl. Sci. **1**(3), 268 (2019)
120. S. Ham, D. Choi, D.J. Jang, Mater. Res. Bull. **116**, 32 (2019)
121. P. Sivakumar, K.G. Kumar, P. Sivakumar, S. Renganathan, J. Nanostructure Chem. **4**, 107 (2014)
122. A. Iwase, K. Ii, A. Kudo, Chem. Commun. **54**, 6117 (2018)
123. X. Wang, Z. Cao, Y. Zhang, H. Xu, S. Cao, R. Zhang, Chem. Eng. J. **385**, 123782 (2020)
124. H. Li, T. Sun, L. Zhang, Y. Cao, Nanoscale **12**, 18180 (2020)

Recent Advances and Applications of Modified-Semiconductor Photocatalyst in Pollutant Degradation

Pin Chen, Yixin Zhai, Yue Bao, and Shukui Zhu

Abstract Ensuring the cleanliness and non-toxicity of water resources is crucial as they are the most valuable assets for human beings. Water pollution has emerged as a significant worldwide concern in recent times, particularly due to the contamination of diverse organic substances such as pharmaceuticals and personal care products (PPCPs), persistent organic pollutants (POPs), and organic dyes. Recently, the use of reactive oxidative radicals or species in photocatalysis, which is an advanced oxidation process, has garnered significant interest for the remediation of organic pollution. The utilization of photocatalysis in solar-powered reactions holds great potential in tackling energy and environmental issues. Effective light absorption, enhanced charge separation and mobility, and expedited surface reactions are crucial factors that greatly impact the effectiveness of photocatalysis, and these can be achieved through the thoughtful design of photocatalysts. Numerous global endeavors have been undertaken thus far to create and pursue high-performance materials, encompassing techniques like doping, amalgamating with quantum dots, regulating heterojunctions, optimizing exposed facets. The promising candidates consist of inorganic metal alloy/metal oxide/metal sulfide, organic–inorganic hybrid materials such as metal–organic frameworks (MOFs), and organic semiconductors like covalent organic frameworks (COFs). In the final part of the analysis, there are also discussions about the main obstacles and viewpoints regarding photocatalysts, aiming to contribute to the advancement of this vibrant area of research.

Keywords Advanced materials · Photocatalysis · Pollutants degradation · Organic materials · Metal–organic frameworks (MOFs) · Covalent organic frameworks (COFs)

P. Chen · Y. Zhai · Y. Bao · S. Zhu (✉)
State Key Laboratory of Biogeology and Environmental Geology, China University of Geosciences, Wuhan 430074, China
e-mail: shukuizhu@126.com

S. J. Ikhmayies (ed.), *Advances in Catalysts Research*, Advances in Material Research and Technology, https://doi.org/10.1007/978-3-031-49108-5_6

1 Introduction

Over the past few decades, the excessive utilization of fossil fuels due to industrialization and urbanization has led to significant concerns regarding the energy crisis and environmental pollution. Hence, it is urgent to discover ecofriendly techniques for environmental restoration and investigate hopeful approaches for the long-term utilization of renewable energy [1, 2]. Solar energy, being the most abundant energy resource on the planet, is limitless and widely spread, demonstrating economic viability and substantial possibilities for practical use [3–5]. Photocatalysis has garnered significant interest since Fujishima and Honda first unveiled the photoelectrochemical (PEC) H_2-evolution on the titanium dioxide (TiO_2) electrode in 1972 [6]. This field has been recognized for its ability to address environmental issues (Fig. 1) [7] over the past few decades. Additionally, it has led to advancements in PEC water splitting, CO_2 reduction, electricity generation, and the production of value-added chemicals. The progress made in other areas of study besides pollutant degradation is not covered in this chapter, and readers can consult recent papers [4, 8–12] for further details.

The photocatalytic system has garnered significant interest from researchers, with semiconductor photocatalysts being the primary focus [5]. Thus far, the candidates encompass inorganic substances [13–16], organic substances [17–19], hybrids [20–22], metal–organic frameworks (MOFs) [20, 23–25], together with covalent organic frameworks (COFs) [26–28]. Furthermore, drawing inspiration from the process of natural photosynthesis, there has been a strong emphasis on advancing the photochemical process through the utilization of photoactive semiconductors. Under light irradiation, the semiconductor photocatalyst undergoes an advanced oxidation–reduction process known as photocatalysis, which takes place on its surface. It is essential for seeking out an efficient semiconductor photocatalyst to ultimately realize the highly efficient degradation of pollutants. Nevertheless, when exposed to

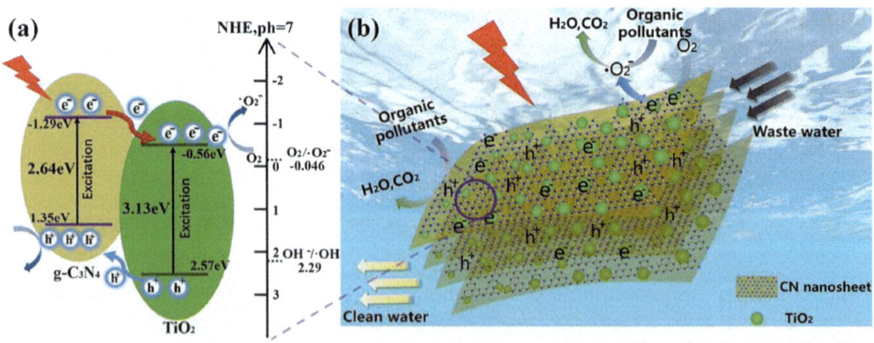

Fig. 1 **a** Diagram showing the separation of carriers induced by light in g-C_3N_4/TiO_2; **b** Diagram illustrating the potential degradation mechanism. Reproduced from Sheng et al. [7], with from Elsevier. Copyright (2019)

visible light, pure P25 (TiO_2) experiences a 30% degradation rate in rhodamine B (RhB) degradation [14]. The low utilization efficiency of sunlight, charge separation and mobility, and surface reaction caused a very low degradation rate. Therefore, it continues to be a major obstacle to create and produce effective sunlight-powered photocatalysts for the breakdown of pollutants.

Various techniques have been proposed to enhance the reactivity of photocatalysts, which are vital in achieving efficient solar-energy conversion [5]. These methods include doping [29, 30], loading cocatalysts [31], creating heterojunction [32], and controlling morphology [33]. Doping involves deliberately introducing appropriate anions or cations into the host lattices so as to modify the electrical properties of semiconductors by enhancing the densities of electrons or holes. Hence, the utilization of doping can effectively modify the constitution of semiconductor photocatalysts, thereby enhancing the segregation of electron-hole pairs generated by light exposure [29]. Cocatalysts have the ability to act as the active locations for photocatalytic reactions, enhancing the efficiency of separating the charges generated by light exposure [34]. Combining various semiconductors with suitable band edges in a heterojunction structure has emerged as an additional effective method to inhibit the recombination of electron-hole pairs [32]. Enhancing light absorption and improving photocatalytic properties are well-recognized outcomes of nanostructure manipulation in photocatalysts [3, 5]. Hence, it is a hot topic in current research to discover an efficient approach for synthesizing semiconductor photocatalysts that possess excellent light absorption capabilities and effective charge separation.

The objective of this chapter is to introduce the basics regarding the degradation of pollutants through photocatalysis, with a specific emphasis on the development and creation of photocatalysts. Over the last few years, there has been notable advancement in understanding the process of contaminant degradation and creating innovative materials to aid in its implementation. In addition, some excellent researches involving efficient photocatalytic degradation designs are summarized here. Several approaches have been suggested to fulfill enhanced performance criteria, such as enhancing light absorption, facilitating separation and transportation of photogenerated charges. In conclusion, the chapter presents a summary of the latest progress in photocatalytic degradation to demonstrate various common and practical approaches that are anticipated to address multiple specific challenges related to environmental remediation.

2 Charge Separation and Transport

At present, there is a tendency to elucidate the intricate physicochemical mechanisms of semiconductor photocatalysis using the theory of energy bands [2]. The photocatalysis process begins with photon absorption. When the semiconductor absorbs energy ($h\nu$) that is identical or greater than the bandgap energy (E_g), electrons (e^-) in the valence band (VB) are excited to the conduction band (CB) within an extremely short time, leaving behind holes (h^+) in the VB. As a result, the CB and VB of the semiconductor are occupied by e^- and h^+, respectively [35, 36]. Subsequently, h^+ and e^- take part in the redox reaction on the surface of the nanomaterial. To degrade the desired pollutants, h^+ has the ability to directly oxidize the contaminants or interact with H_2O/OH^- to produce hydroxyl radical (•OH).Typically, the e^- seizes the adsorbed oxygen (O_2) to produce superoxide radicals ($\bullet O_2^-$), which in turn contribute to the breakdown of pollutants along with the generated •OH. During the process of converting energy, it is possible to reduce carbon dioxide (CO_2)/H^+ to hydrocarbon/hydrogen (H_2), while H_2O/OH^- can be transformed into O_2 through oxidation. One of the most important steps is the trapping of electrons produced by light by dioxygen to produce $\bullet O_2^-$. This process not only prevents the recombination of light-generated carriers but also extends the lifespan of the positive charges [2, 19, 37–40], as depicted in Fig. 2. In organic semiconductors and polymers, it is possible for charge carriers to be produced as they transit from the HOMO (highest occupied molecular orbital) to the LUMO (lowest unoccupied molecular orbital). Semiconductors give rise to two kinds of bandgaps, viz., direct and indirect. Figure 3 illustrates that in the scenario of a direct band gap, the momentum (i.e., wave vector) of the electron in the VBM is identical to that of the electron in the CBM. Therefore, when exposed to a high-energy photon that exceeds the material's band gap, an electron can easily transition from valence band to conduction band, while experiencing minimal changes in its momentum. Besides, in an indirect band gap semiconductor, the momenta of the states in the CB's minimum and the VB's maximum are not equal. Therefore, in order to preserve momentum, an electron needs to experience a substantial alteration in the momentum to become stimulated across the band gap and generate the electron-hole pair within this substance. This happens owing to the fact that the electron not only interacts with the initial photon to acquire energy, but also with a quantized lattice vibration called phonon, which can either increase or decrease the momentum. Due to the participation of the electron, photon, together with phonon in this particular process of absorption, its occurrence is less likely and takes place at a significantly reduced pace compared to direct absorption [4].

Ultimately, the quantum efficiency (QE) of the photocatalytic process is dictated by the photo-excited carriers. This encompasses a kinetic competition, wherein these carriers either partake in the desired reactions or contribute to the recombination of the photoexcited electron-hole pairs. In order to promote the photocatalytic reaction, it is necessary for the photoexcited electron-hole pair to move toward the redox-active sites located on the surface of the photocatalyst. Additionally, there should be interfacial charge transfer in order to extract and separate the charges [4]. Prompt action is necessary for the catalytic reactions as the lifespan of semiconductor photocatalysts is usually brief, varying from picoseconds to nanoseconds, and aligns with the recombination time scale. An effective photocatalyst can be determined by the carrier lifetime, which is a measure of the semiconductor's capability [4].

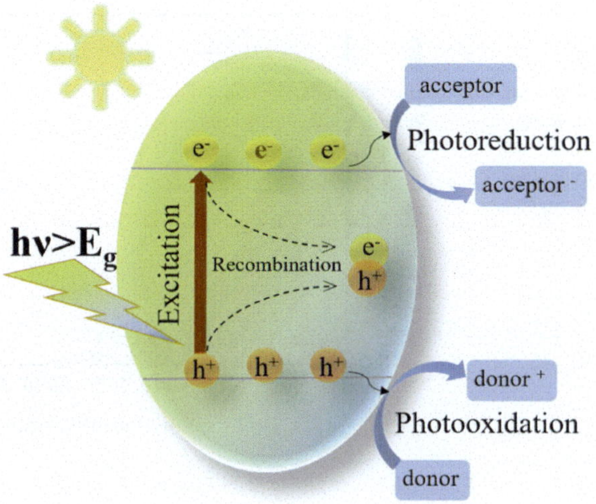

Fig. 2 Schematic illustration of semiconductor photocatalytic processes. Reproduced from Huang et al. [2], with permission from Elsevier. Copyright (2019)

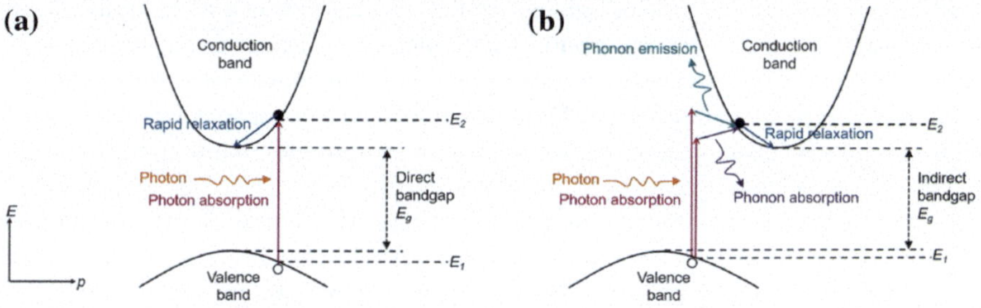

Fig. 3 Photon absorption phenomena are delineated for two distinct scenarios: **a** a direct band gap semiconductor, where an incident photon with energy denoted by $h\nu = E_2 - E_1$ exceeds the energy of the band gap (E_g), and **b** an indirect band gap semiconductor, categorized by two sub-cases of photon energy: one (for a photon) where $h\nu < E_2 - E_1$ and another (for a photon) where $h\nu > E_2 - E_1$. Reproduced from Wang and Domen [4], with permission from American Chemical Society. Copyright (2020)

2.1 *Challenges*

2.1.1 Visible-Light Harvesting

The process of semiconductor photocatalysis for transforming solar energy into chemical fuels initiates with the absorption of photons. When semiconductor nanomaterials are exposed to radiation, they become optically excited if the incoming light exhibits an energy that is equal to or greater than the band gap of the nanomaterial.

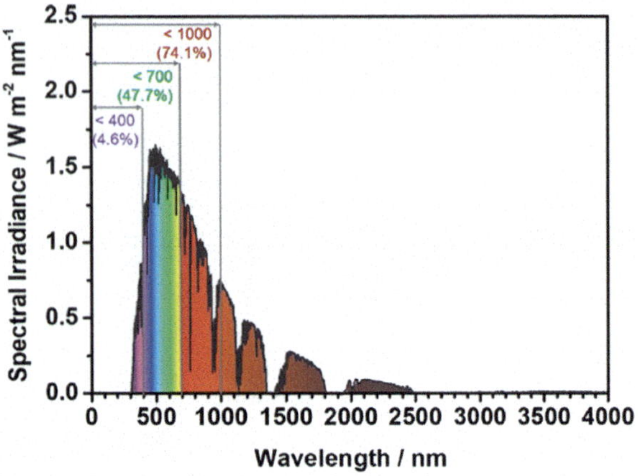

Fig. 4 AM 1.5 G solar spectrum based on the ASTM G173-03 reference spectrum. Reproduced with the permission from Wang and Domen [4], with permission from American Chemical Society. Copyright (2020)

This leads to the excitation of electrons from VB to CB within a femtosecond timeframe. In organic semiconductors, including dyes and polymers, the movement from the HOMO to the LUMO can also generate charge carriers. After excitation, electrons and holes quickly relax to the bottom of CB and the top of VB in a comparable timeframe. The quantification of light absorption by a particle in a semiconductor can be decided by the material's absorption coefficient, $\alpha(\lambda)$. The difference between E_2 and E_1 for a given photon energy is equal to $h\nu$. The sun emits a variety of radiation, spanning from X-ray to radio wave. However, the highest concentration of solar radiation (43%) that reaches the Earth's surface falls within the visible light spectrum, specifically wavelengths ranging from 400 to 700 nm (Fig. 4) [4]. Due to their excellent effectiveness, affordability, and lack of toxicity, zinc oxide (ZnO) and titanium dioxide (TiO_2) are widely favored as photocatalysts for the elimination of contaminants. Nevertheless, the photocatalysts possess a significant energy difference of 3.2 eV and therefore solely capture ultraviolet (UV) radiation with wavelengths less than 387 nm, constituting a mere 3–5% of the total solar energy [41, 42]. Hence, in order to enhance the breakdown of pollutants through photocatalysis, it is imperative to concentrate on the photocatalysts that can harness visible light.

2.1.2 Recombination

Following irradiation, the electron-hole pairs produced by light return to their original equilibrium states. In other words, the electrons generated by light move from CB to VB, where they combine with holes in VB, resulting in a decline in the number of carriers. Various materials can display significantly distinct bulk recombination kinetics. For example, the recombination rate of hematite (α-Fe_2O_3) is approximately 103 times higher than that of TiO_2, potentially because of the presence of the space charge layer (SCL) in optical transition of hematite [43]. On the one hand, band-to-band recombination is attributed to the high concentration of holes and electrons in direct band gap semiconductors. On the flip side, the process of recombination within indirect band gap semiconductors is propelled through the interchange of energy with a phonon, notably through the mechanism of Shockley-Read-Hall recombination [4]. Surface states can also undergo recombination, and the semiconductors frequently occupy extensive surface regions [44]. It is essential to enhance efficient photoexcited carriers' separation for many semiconductor photocatalysts.

2.2 *Strategies*

Numerous global endeavors have been undertaken thus far to develop and pursue high-performance nanomaterials for photocatalytic reactions driven by visible light. These endeavors encompass techniques such as doping, compositing with quantum dots, controlling heterojunctions, and optimizing exposed facets. The modified photocatalysts for removing contaminants under visible light are summarized in Table 1, allowing for the utilization of their complementary benefits and overcoming the drawbacks they may have. This section presents various approaches to manipulate the photophysical and other physicochemical characteristics of photocatalysts.

2.2.1 Doping

Regulating the photophysical and other physicochemical properties of photocatalysts can be effectively achieved through the introduction of foreign elements, which is known as doping. The characteristics that can be altered using this approach encompass optical absorption, p/n (where p represents the number of holes and n represents the number of free electrons) properties, charge density/mobility/ separation, particle size, hydrophilicity, crystallinity, defect density, and surface structure. Every single one of these alterations has the potential to enhance the material's photocatalytic effectiveness when it comes to breaking down pollutants.

Several studies have suggested that incorporating metals could be a meaningful approach to enhance the light absorption of photocatalysts with wide band gaps. According to Zhang and colleagues [62], $Bi_4O_5I_2$ nanosheets doped with Er^{3+}/Yb^{3+} were created synthetically and have shown remarkable efficiency in degrading RhB and imidacloprid (IM) when exposed to visible or near-infrared (NIR) light. The study conducted by Wang and colleagues [45] obtained the Co-BiOCl photocatalyst which could respond to visible light. Shi et al. constructed photocatalysts of TiO_2-x, which were self-doped with Ti^{3+} [51]. The photocatalysts exhibited excellent performance in terms of photocatalysis when exposed to visible light. According to Reddy et al. [63], a superb photocatalyst with (Al, Ni) co-doped ZnO structures was successfully synthesized, which exhibits excellent performance under visible light irradiation. The study conducted by Wang and colleagues claimed that a photocatalyst with an expanded photoabsorption region was created by doping g-C_3N_4 with potassium (K). The addition of metal dopants not only increased the optical absorption, but also

Table 1 The modified photocatalysts for the removal of organic pollutants

Composites	Preparation method	Applications	References
PO_4-doped Bi_2WO_6	PO_4-doped Bi_2WO_6 was prepared via the urea-precipitation method	Cr (VI), Rhodamine B	[29]
Co-doped BiOCl	Co-BiOCl nanosheets were synthesized via a simple hydrothermal route	Bisphenol A	[45]
Fe-CeO_2	Fe-doped CeO_2 nanosheets were synthesized by the solvothermal method	Salicylic acid	[46]
SKA-CN(1)	S, K-doped/alkalized g-C_3N_4 nanomaterial was fabricated via a thermal polymerization process	Bisphenol A	[47]
g-C_3N_4/TiO_2	C_3N_4/TiO_2 was prepared via in situ hydrothermal synthesis followed by calcination	Rhodamine B	[37]
g-C_3N_4/ZnO	g-C_3N_4/ZnO was prepared via heating the precursor obtained by the deposition–precipitation process	Rhodamine B, Cr^{6+}	[41]
CTFNS (2)/CNNS (3)	CTFNS/CNNS heterojunctions were successfully synthesized through an electrostatic force driven self-assembly utilizing amine-functionalized CNNS and carboxyl-rich CTFNS	Sulfamethazine	[48]
AgBr/$BiPO_4$/g-C_3N_4	AgBr/$BiPO_4$/g-C_3N_4 photocatalysts were acquired by deposition-precipitation method and they were turned into stable Ag/AgBr/$BiPO_4$/g-C_3N_4 system in the process of RB19 degradation	Reactive Blue 19 (RB19)	[49]
Z-scheme AgI/WO_3	AgI/WO_3 nanocomposites were synthesized via a simple precipitation method	Tetracycline	[50]
TiO_{2-x}/CDs(4)	TiO_{2-x}/CDs photocatalytic system was established by a low-temperature hydrothermal process	Rhodamine B, Cr^{6+}	[51]
NCDs (5)/g-C_3N_4	NCDs/g-C_3N_4 composite was constructed through loading NCD nanoparticles (NPs) onto the interlayers as well as surfaces of g-C_3N_4 by a simple polymerized process	Indomethacin	[52]
g-C_3N_4/TiO_2 {001}	g-C_3N_4/TiO_2 {001} was hybridized with g-C_3N_4 utilizing a solvent evaporation procedure	Organic molecules	[53]

(continued)

Table 1 (continued)

Composites	Preparation method	Applications	References
Mesoporous BiOBr	BiOBr photocatalysts were obtained by solvothermal process with diverse synthetic conditions	Rhodamine B	[54]
Exfoliate LCN [(6)]	The modified carbon nitride (LCN) was successfully synthesized via facile thermal copolymerization	Sulfamethazine	[55]

[(1)]SKA-CN: S, K-doped alkalized-carbon nitride
[(2)]CTFNS: Covalent triazine-based framework nanosheets
[(3)]CNNS: Carbon nitride nanosheets
[(4)]CDs: Carbon dots
[(5)]NCDs: N-doped carbon dots
[(6)]LCN: L-cysteine modified carbon nitride

Table 2 The photocatalytic performances of typical MOF-based nanocomposites

Composites	Preparation method	Application	References
g-C_3N_4@ZIF-8	In-situ deposition of ZIF-8 from its precursors on the g-C_3N_4	Tetracycline degradation	[56]
Z-scheme g-C_3N_4/UiO-66	Facilely fabricated using ball-milling method	Cr(VI) reduction	[57]
g-C_3N_4/MIL-53(Al)	Solvothermal deposition of MIL-53(Al) on the g-C_3N_4	RhB degradation	[58]
g-C_3N_4/CuBTC	In-situ deposition of CuBTC from its precursors on the g-C_3N_4	Dimethyl chlorophosphate degradation	[59]
g-C_3N_4/NH_2-MIL-125	By ultrasonication	Diclofenac degradation	[60]
CdS/g-C_3N_4/MOF	Using a facile solvothermal method	Dye degradation	[61]
g-C_3N_4/PDI@MOF	In situ growth of NH_2-MIL-53(Fe) onto g-C_3N_4/PDI layer	Pharmaceutical, phenolic micropollutants	[20]
MWCNT@MOF-derived In_2S_3	Via facile MOF sulfidation process	Tetracycline degradation	[25]
Fe_3O_4@MIL-100(Fe)	Solvothermal deposition of MIL-100(Fe) on Fe_3O_4	Methylene blue degradation	[24]

reduced the band gap and enhanced the efficiency of separating e^{-}-h^{+} pairs. Active radicals can be generated by the doped metal ions, which function as the redox center [64].

Strengthening the separation of photogenerated electron holes and expanding optical absorption can be achieved through nonmetallic doping. Huang and colleagues [65] reported that N-TiO_2@C photocatalyst was acquired, exhibiting outstanding performance in the decomposition of phenol using visible-light irradiation. The study conducted by Jiang and colleagues [66] disclosed that novel nanosheets of g-C_3N_4 containing self-doped nitrogen could be successfully prepared. The photocatalyst improved the absorption and utilization of visible light and extended the lifespan of charge carriers generated by light. Hu et al. [67] enhanced the performance of metal-free photocatalysts by codoping g-C_3N_4 with phosphorus and sulfur. Similarly, according to Jiang et al. [68], the carbon nitride doped with phosphorus and sulfur (PSCN) exhibited diminished photoluminescence (PL) emission intensity, indicating the inhibition of recombination of the photogenerated charges. Specifically, the substitution of C with P and the addition of S as an interstitial dopant significantly improved the charge transfer along the N-S-N-C-N-P channel, facilitating the photocatalytic reaction across the heptazine units. As a result, PSCN emerged as a remarkably active and efficient metal-free photocatalyst. Guo et al. [69] synthesized Cl-doped porous g-C_3N_4 (CN-Cl), the Cl played an essential role in the photocatalyst. The introduction of chlorine-doped material controlled the electronic configuration of CN, leading to an increase in specific surface area which supplied a larger number of reactive sites and restrained the recombination of photoinduced holes and electrons in CN. Huang et al. [70] reported a one-step methodology to produce P and O co-doped g-C_3N_4 coated anatase TiO_2 nanoparticles. The addition of phosphorus (P) and oxygen (O) during doping increased the performance of g-C_3N_4 to absorb visible light, while the presence of anatase TiO_2 nanoparticles enhanced the adsorption characteristics of enrofloxacin (ENFX) and facilitated the separation of the photoinduced carriers in POCN/anatase TiO_2 system.

2.2.2 Quantum Confinement Effect

Quantum dots (QDs) are nanoscale semiconductor particles with sizes ranging from a few to several tens of nanometers. QDs display quantum confinement effects due to their small size in comparison to the visible light wavelength, resulting in the emergence of distinctive electronic properties and high absorption coefficients. The band gap energy of QDs exhibits significant variations based on their size, which is a remarkable characteristic of QDs. This property enables easy adjustment of the CB and VB positions by simply altering the size of the quantum dots. Moreover, the necessary photon energy threshold regarding multiple electron generation (MEG) in QDs can occur within the visible or near-infrared (NIR) spectral range. This enables the production of multiple electron-hole pairs from a single absorbed photon.

Carbon dots (CDs), belonging to the category of QDs, are a novel kind of carbon nanomaterials that consist of a combination of sp^2/sp^3 carbon atoms and possess diverse surface functional groups. Their distinct characteristics, which include excellent water solubility, up-conversion PL (UPPL) behavior, and remarkable photoelectron transfer, have generated significant interest. Moreover, the CD's absorption range coincides with the majority of the solar spectrum, which is advantageous in enhancing the light absorption capabilities of composites modified with CDs.

Jamila et al. [71] developed a highly efficient ternary photocatalyst using nitrogen-doped carbon quantum dots (NCQDs). Li et al. [72] successfully prepared a novel photocatalyst called CDs/g-C_3N_4/SnO_2 by modifying g-C_3N_4/SnO_2 with CDs. The highest degradation rate of indomethacin (IDM) was observed when the CDs loading content was 0.5%, which was 5.62 times greater than the degradation performance of pristine g-C_3N_4. Xie and colleagues [73] demonstrated that the CDs/g-C_3N_4/MoO_3 nanocomposite could be effectively synthesized, resulting in a novel Z-scheme photocatalyst with modified CDs. According to systematic research, the photocatalyst named CDs/g-C_3N_4/MoO_3 showed dramatically enhanced photocatalytic activity under visible light for breaking down tetracycline (TC) in comparison to the original g-C_3N_4 as well as g-C_3N_4/MoO_3 mixture. The highest degradation rate of TC was achieved by doping 0.5% CDs, which was 3.5 or 46.2 times greater than that of g-C_3N_4/MoO_3 or g-C_3N_4, respectively. The improved photocatalytic efficiency of the CDs/g-C_3N_4/MoO_3 can be credited to the combined impact of CD characteristics (such as outstanding UPPL functionality and strong charge separation capability) and the structure of Z-scheme heterojunction.

Other QDs such as CdS, PbSe, PbS, and ZnS were introduced to photocatalysts to reform the photocatalytic performance. These QDs possess the ability to harvest light effectively, generate multiple excitons, and efficiently separate photoinduced charge carriers from other materials, leading to their advantageous photocatalytic characteristics. Ji et al. [74] synthesized a hierarchical nanoflower heterojunction photocatalyst $Bi_2O_2CO_3$/CdS, where CdS QDs were used to decorate the 3D structure. The purpose of this synthesis was to study the removal effect of rhodamine B (RhB) through photocatalysis. The photocatalytic degradation performance of $Bi_2O_2CO_3$/CdS was greatly improved in comparison to the original $Bi_2O_2CO_3$ and CdS QDs. The improved photocatalytic efficiency was credited to the combined influence of a hierarchical arrangement as well as heterojunction, leading to a significant augmentation in both the reaction active sites and the transfer of photogenerated carriers. CdS-Bi_2WO_6 photocatalysts were synthesized by Ge and Liu [75] using QDs as sensitizers. The Bi_2WO_6 samples unfolded a red shift and enhanced absorption in the visible light region after being sensitized by CdS QDs, resulting in high efficiency for degrading methyl orange. The improved photocatalytic efficiency can be credited to the collaborative impact of CdS QDs and Bi_2WO_6, which enhanced the migration efficiency of photogenerated carriers. Vanashi and Ghasemzadeh [76] synthesized a hydrogel called PbSe-QDs/NCH, which is a nanocomposite containing novel PbSe quantum dots. The photocatalyst exhibited favorable photocatalytic performance when exposed to visible light. Zhang et al. constructed a heterojunction by rationally combining 0D QDs (PCZ) with 2D g-C_3N_4 nanosheet, which are responsive to near-infrared (NIR) light and have a core@shell@shell structure of PbS@CdS@ZnS [77]. Efficient charge transfer between high-quality PCZ QDs highly dispersed on the g-C_3N_4 nanosheet and their strong interactions result in the broadband optical absorption. The interaction also endows PCZ QDs/g-C_3N_4 with extraordinary photocatalytic activity across the ultraviolet (UV) to NIR regions. The normalized rate constant achieved with the optimum QDs loading level surpasses the highest reported value regarding NIR-driven photocatalysis.

2.2.3 Z-Scheme Configurations

For the heterojunction-type photocatalytic system (Fig. 5) [78], it is demonstrated that the electrons produced by light in the conduction band of photocatalyst I (PC I) move to the conduction band of photocatalyst II (PC II), whereas the holes generated in the valence band of PC II shift to the valence band of PC I. Consequently, the holes and electrons generated by light are physically separated, effectively preventing their unwanted recombination. Nonetheless, the drawback lies in the fact that the redox capacity of holes and electrons generated by light is diminished following charge transfer due to the less positive potential of the top of the valence band (TVB) in PC I compared to PC II, and the less negative potential of the bottom of the conduction band (BCB) in PC II compared to PC I. Therefore, the current heterojunction-based photocatalytic system faces challenges in achieving both efficient charge separation and robust redox capability. Consequently, it is imperative to investigate novel photocatalytic systems in order to address the previously mentioned issues.

For artificial Z-scheme photocatalytic systems, they are inspired by natural photosynthesis. Hence, these systems also incorporate the spatial separation regarding photogenerated electrons as well as holes, thus effectively minimizing the recombination of the electron-hole pair within the system. Additionally, the accumulation of light-induced electrons on the conduction band of photosystem (PS) I creates an area abundant in electrons, effectively inhibiting the photooxidation process of PS I. Likewise, the accumulation of holes generated by light on the valence band of PS II creates a region abundant in holes, serving as a shield against photoreduction for PS II. Nonetheless, this also indicates that PS II and PS I are susceptible to photooxidation and photoreduction, respectively. Therefore, a photocatalyst lacking sufficient resistance to photooxidation is unsuitable for use as PS II. Similarly, a photocatalyst lacking strong resistance to photoreduction is unsuitable for use as PS I. Additionally, due to electron-hole recombination, the Z-scheme photocatalytic systems generate only half the amount of photogenerated electrons together with holes compared to the heterojunction-type photocatalytic systems under the same conditions. Despite this, the Z-scheme photocatalytic systems continue to garner significant interest because of their distinctive mechanism regarding electron transfer.

The A/D pair serves as a prevalent electron mediator in artificial Z-scheme photocatalytic systems. The Z-scheme system, named PS-A/D-PS, is composed of an A/D pair, PS I, and PS II, as depicted in Fig. 6 [78]. There is no physical connection between PS I and II, while the transfer of electrons from the conduction band of PS II to the valence band of PS I relies entirely on the under mentioned redox reactions of the A/D pair.

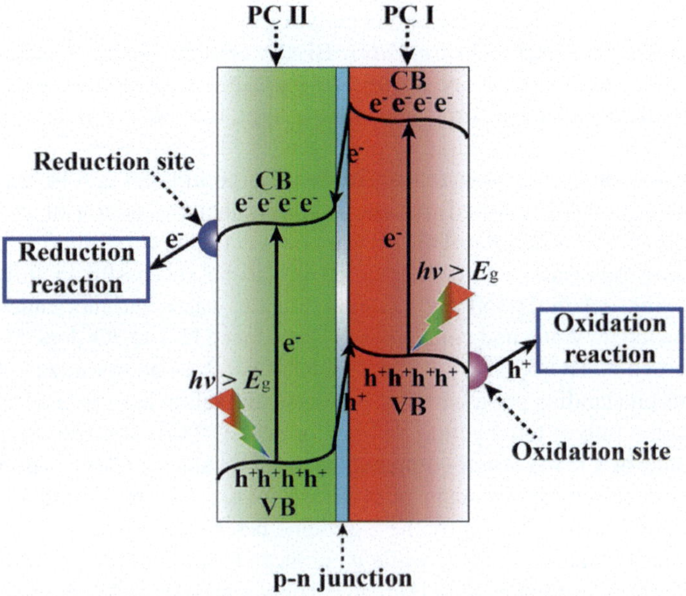

Fig. 5 Charge transfer in a heterojunction-type photocatalytic system. Reproduced from Zhou et al. [78], with permission from John Wiley and Sons. Copyright (2014)

$$A + e^- \rightarrow D(\text{CBofPSII}) \tag{1}$$

$$D + h^+ \rightarrow A(\text{VBofPSI}) \tag{2}$$

Jiang et al. synthesized a WO_3/g-C_3N_4/Bi_2O_3 photocatalyst, which operates through a direct solid-state dual Z-scheme [79]. Under visible light irradiation,

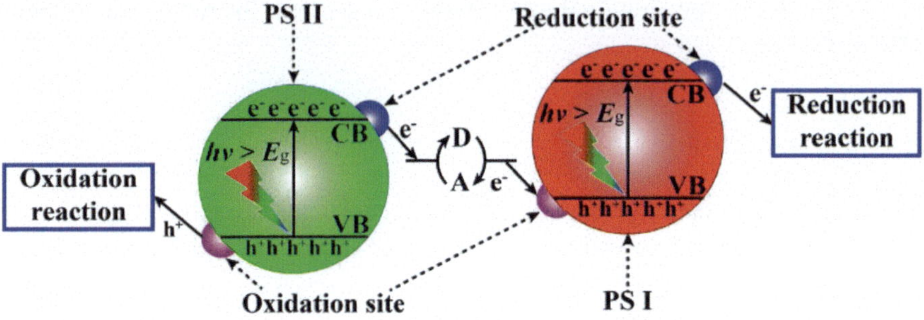

Fig. 6 The Z-scheme photocatalytic mechanism presented based on the system of natural photosynthesis. Reproduced from Zhou et al. [78], with permission from John Wiley and Sons. Copyright (2014)

the photocatalyst unfolded better photocatalytic performance in degrading tetracycline compared with pure WO_3, g-C_3N_4, Bi_2O_3, or their binary composites. The improved photocatalytic activity of the WO_3/g-C_3N_4/Bi_2O_3 photocatalyst can be attributed to the enhanced absorption of visible light, increased surface area, as well as improved efficiency in separating photogenerated electron-hole pairs. Furthermore, the photocatalyst demonstrates excellent durability and can be used repeatedly. Hasija et al. [80] reported that the Z-scheme heterojunction of phosphorus-doped g-C_3N_4/AgI/ZnO/CQDs (PGCN) was constructed successfully. The combined action of adsorption and photocatalysis proved to be successful in fully decomposing 2,4-dinitrophenol (DNP) into carbon dioxide, water, and inorganic ions. The study conducted by Wang and colleagues [81] produced N-TiO_2/O-doped N vacancy g-C_3N_4 (named N-TiO_2/CNO_{NV}-2). N-TiO_2/CNO_{NV}-2, as an optimized photocatalyst, exhibits outstanding photocatalytic performance, achieving approximately triple the degradation rate of tetracycline hydrochloride (TC-HCl) and forty-one-fold the reduction rate of Cr(VI) when compared with N vacancy g-C_3N_4. The improved photocatalytic activity mechanism of N-TiO_2/CNO_{NV}-2 is due to the acceleration of electron and hole separation by the Z-scheme heterojunction. Additionally, $\bullet O_2^-$ and h^+ primarily attack TC-HCl, while e^- reduces Cr (VI) in the presence of both TC-HCl and Cr (VI). Tang et al. [82] created a photocatalyst (called AAC) consisting of AgI, Ag_3PO_4, and g-C_3N_4, which was synthesized into a dual Z-scheme structure. The photocatalyst with a dual Z-scheme demonstrates superior efficiency in decomposing nitenpyram (NTP) compared with pure g-C_3N_4, Ag_3PO_4, AgI, and their combined composites. The findings from photoluminescence spectroscopy together with transient photocurrent response indicate a notable enhancement in the photogenerated electrons and holes separation efficiency for AAC, leading to an advantageous improvement in its photocatalytic performance.

Thankfully, numerous findings have shown that diverse photocatalysts offer various benefits. Firstly, the light absorption of materials with a broad range of energy gaps can be significantly improved by incorporating narrow band gap semiconductors or molecules. Secondly, the build-in electric field between the semiconductors can effectively enhance the separation and movement of electrons and holes generated by light. Lastly, integrating the co-catalyst can decrease the redox overpotential of the active site.

2.2.4 Facet Control

The photocatalytic activity of particulate semiconductors is influenced by their crystal facets. Reactions occurring at the materials' surfaces or interfaces are closely tied to the atomic configurations of the exposed surfaces and their associated physicochemical characteristics [4].

Shahzad et al. [83] developed a hybrid photocatalyst using $Ti_3C_2T_x$ (MXene) nanosheets. Controlled oxidation action led to the formation of a heterostructure in the as-prepared photocatalyst, achieved by using the Schottky junction between TiO_2-MXene interfaces. The abilities of the original MXene and the newly fabricated

001-T/MX composite to adsorb and degrade carbamazepine (CBZ) were examined. The calculated value of the observed rate constant ((K_{app}) for CBZ in the presence of ultraviolet light was 0.0304 min^{-1}, which exceeded the value observed under natural sunlight. The degradation ability was significantly influenced by acidic conditions (pH ranging from 3.0 to 5.0). Gu et al. [53] combined g-C_3N_4 with anatase TiO_2 nanosheets that have prominent {001} facets. The photocatalytic procedure suggests the existence of a successful charge separation between TiO_2 and g-C_3N_4, specifically the movement of photogenerated holes from the VB of TiO_2 to the HOMO of g-C_3N_4, and the introduction of electrons from the LUMO of g-C_3N_4 to the CB of TiO_2. The hybrid achieves an enhancement in photoactivity with the aid of either UV or visible light because of this collaborative effect. Tian et al. [84] found that a p-n junction was formed by creating a BiOI/g-C_3N_4 composite. Upon treatment with cetyltrimethylammonium-bromide (CTAB), the {002} surface of g-C_3N_4 exhibits a positive charge (g-$C_3N_4^+$) and the BiOI nanosheets are subsequently arranged perpendicularly on top of g-$C_3N_4^+$. The findings from the study on the photodegradation of various industrial pollutants and antibiotics indicate that B001/CN002 exhibits significantly greater photoactivity compared to g-C_3N_4, g-$C_3N_4^+$, BiOI, and B110/CN002$^+$. The improved separation and transport of charges at the B001/CN002 interface significantly contribute to the production of $^1O_2$ and $•O_2^-$, which is responsible for the exceptional photocatalytic performance. Furthermore, Ag_3PO_4 is deposited on the {040} facet of $BiVO_4$ to create a novel photocatalyst, whose photodegradation performance of the contaminant is further enhanced by incorporating polyaniline (PANI) and facet engineering [85]. The usage of PANI in the ternary nanocomposite achieves outstanding photocatalytic effectiveness, resulting in an 85.92% degradation efficiency for ciprofloxacin and a maximum rate constant at 0.00894 L mg^{-1} min^{-1}. This is 10.1, 5.6, and 1.6 times higher than the degradation efficiency of $BiVO_4$, $BiVO_4$/PANI, and $BiVO_4/Ag_3PO_4$, respectively.

3 Different Kinds of Photocatalysis

The classification of this section of the manuscript is determined by the various forms of photocatalysis. Fujishima et al. [6] proposed the groundbreaking research that sparked the field of photocatalytic water splitting during the 1970s. Since then, there has been a strong focus on this field of study for many years, leading to notable progress. Semiconductor photocatalysts, being the main element of the photocatalytic system, have garnered the majority of research attention. So far, numerous candidates including TiO_2, WO_3, $BiVO_4$, ZnO:GaN, CdS, g-C_3N_4, black phosphorus, and metal halide perovskites have been extensively studied in order to discover remarkably effective semiconductor photocatalysts that can eventually achieve commercially viable photocatalytic degradation of pollutants. MXenes, referred to as a family of 2D transition metal carbides, nitrides, or carbonitrides, have also been authenticated to be promising co-catalysts for pollutant degradation. The primary issue causing a lack of efficiency in decomposing pollutants on photocatalysts is the substantial loss of energy during the subsequent stages of light absorption, separation, and transfer of charges, as well as surface reaction. Hence, the creation of effective solar-powered photocatalysts for the extensive elimination of pollutants continues to be a major obstacle. This section is classified as follows:

3.1 Metal-Organic Frameworks

Lately, metal-organic frameworks (MOFs) have been recognized as a flexible foundation for the creation of solid catalysts. MOFs are a novel type of crystalline molecular solids that are constructed by connecting organic ligands with metals/metal-clusters [86]. The structural features of MOFs were illustrated in Fig. 7. MOFs offer numerous benefits when compared with conventional metal oxide semiconductors, including high specific surface area, abundant topology, and a readily adjustable porous morphology. All molecular assemblies and solids can exhibit a certain level of flexibility, including those that are commonly regarded as "rigid" such as the classic MOF-5, because coordination polymers tend to demonstrate extensive dynamic behavior on a larger scale, often referred to as "flexibility". Unlike traditional inorganic semiconductors that possess a spread-out CB and VB, MOFs can be characterized as molecules organized in a crystalline structure [87]. The diverse applications benefit from the distinct capabilities of these MOFs with affluent properties. It includes luminescence, gas segregation, and uptake, preparation of magnetic devices and materials, chemical sensing and detection, proton conductance, energy preservation and transformation, as well as biomedical applications. Photocatalysis has also become a potential application for MOFs. Initially, MOFs have the ability to combine photosensitizers and catalytic constituents within a singular substance through the fixation of the active locations on metal nodes, organic connectors, or enclosed guest particles within the cavities. In addition, the swift movement and diffusion of substrates and products is facilitated by the considerable permeability of MOFs. The distinct and clearly defined crystalline structure of MOFs offers a singular opportunity to explore the mechanism regarding energy transfer in the photocatalytic process, which proves challenging to examine in alternative photocatalytic systems. In addition, unlike photocatalysts that are homogeneous, MOFs have the advantage of being easily separated from the solution and can be reused several times. Hence, they contribute to prolong the lifespan of the photocatalysts while minimizing waste and contamination [88].

The utilization of MOFs as photocatalysts has paralleled the advancement of photocatalysis and has been accompanied by the expansion of both the structural and functional intricacy of MOFs. By far, there have been many reports showcasing the distinct superiorities of photocatalysis using MOFs compared to solely organic or inorganic systems. Additionally, exceptional reviews are published nearly every year. MOF-derived photocatalysts can be categorized into three groups based on the inherent structural functions of the photocatalytic center (Fig. 8). With the 0D inorganic cluster nodes serving as the catalyst, the semiconductor dots photocatalysts in

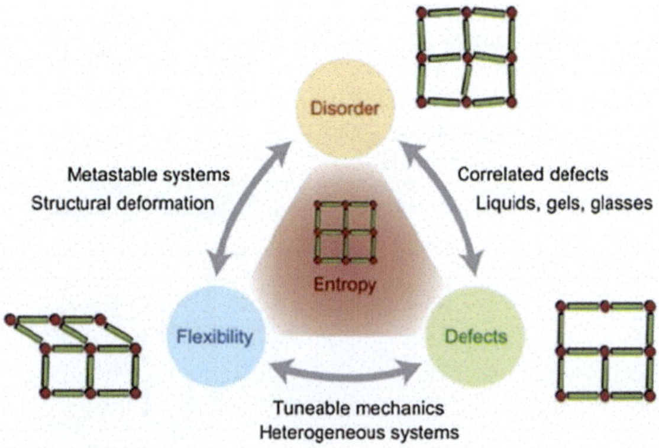

Fig. 7 Flexibility, defect, and disorder in MOFs. Reproduced from Bennett et al. [89], with permission from Springer Nature. Copyright (2016)

type I MOFs are effectively separated by the organic linkers and occasionally by the MOFs' pores. The MOFs are characterized as individual semiconductor nanoparticles that are evenly and uniformly distributed within the crystal lattice. In type II MOFs, the connectors consist of functional organic and metal-organic dye-centric photocatalysts, which are anchored and delineated by the metal junctions. These MOFs are perceived as dispersed dye-centric photocatalysts organized in a structured sequence in an isolated manner. In type III MOFs, photocatalytic entities with appropriate dimensions and configurations are contained within the MOFs' pores. These catalysts are viewed as consistently arrayed supramolecular frameworks, separated by the MOF scaffold. To prevent the clustering of these nanoscale sites, the type I MOFs possess an appropriate concentration of catalytic sites, leading to enhanced reactivity toward protons or carbon dioxide, and the breakdown of organic contaminants in both water and air. Furthermore, apart from the segregation and stabilization of photoactive locations, photocatalysts based on type II MOFs also offer a chance for the photoactive component to collaborate with other functionalities concurrently or in synergy, resulting in versatile catalysts that cover a wide range of spectra or enable the fusion of photocatalysis with metal- or organo-catalysis. This type of MOFs has been successfully used as photocatalysts for traditional organic synthesis, radical chemistry, and novel consecutive photoinduced electron transfer (con-PET) processes. The development of photocatalysis in relation to type III MOF is still lacking. Although the combination of MOF porosity and photocatalyst has resulted in the development of efficient photocatalysts [90], only metal nanoparticles (NPs), metallic alloys, polyoxometalates (POMs), and fullerene have been incorporated into the pores of MOFs for photocatalysis.

Different techniques used for synthesizing MOFs include solvothermal, heating with the assistance of microwaves, and mechanochemical methods [91]. The temperature at which MOFs are synthesized usually distinguishes between solvothermal and non-solvothermal, which determines the type of reactor that must be used. While there is no universally accepted explanation for solvothermal reactions, we adopt the definition provided by Rabenau [92], which refers to reactions occurring in sealed

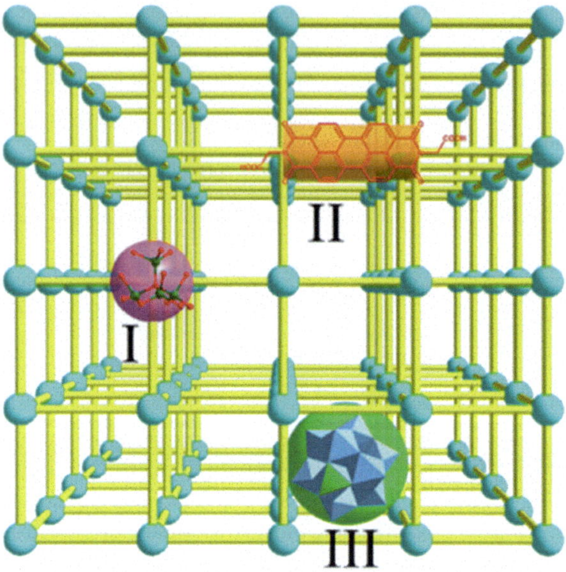

Fig. 8 A scheme illustrating the three categories of MOF-based photocatalysts. Reproduced from Zeng et al. [90], with permission from American Chemical Society. Copyright (2016)

containers under self-generated pressure exceeding the boiling point of the solvent. In contrast, non-solvothermal reactions occur at or below the boiling point under normal pressure, thereby distinguishing the synthetic conditions [93]. Currently, solvothermal synthesis remains the predominant method for laboratory preparation of MOFs, typically requiring numerous organic solvents, extended reaction durations, and elevated temperatures. Clearly, producing MOFs on a large scale [94] is not a cost-effective or environmentally friendly approach. Microwave-assisted techniques utilize waves with frequencies ranging from 300 MHz to 300 GHz to reduce the reaction time to just several hours or minutes, without compromising the quality of the MOF product. Microwave-assisted synthesis offers several benefits compared to alternative synthetic approaches, including superior phase purity, precise regulation of crystal growth and reaction conditions, and increased product yields [95]. The preparation of MOFs using mechanical force as the driving force for chemical transformation has proven to be a fast and effective method in mechanochemical reaction. Mechanochemical synthesis depends on the direct assimilation of mechanical energy by reagents, typically in the form of solids, while undergoing milling or grinding procedures [96]. The potential to carry out chemical reactions at ambient temperature (or marginally higher) accompanied by a notable decrease in solvent usage makes mechanochemical synthesis a highly appealing approach [97].

Adsorption and advanced oxidation processes (AOPs) have been pursued for applications of MOFs in water depollution. The detailed examination of MOFs has revealed their effectiveness in various environmental applications, particularly in the adsorption-based removal of pollutants [98]. MOFs exhibit semiconductor-like properties when exposed to light, as the absorption of light by organic ligands triggers the activation of metal sites through the charge transfer from ligands to metal clusters

[99]. To ensure environmental protection, it is crucial for the utilized MOF catalysts to exhibit exceptional stability, preventing any possibility of transition metal or even heavy metal leaching into the solution. The utilization of MOFs in photocatalytic treatments of organic contaminants also requires reusability as a fundamental requirement [100]. In the last few years, there has been an increasing urgency and importance in studying the utilization of MOFs for eliminating persistent pollutants, particularly in relation to environmental concerns [101]. The authors (He and others) [102] reported that a controlled calcination of the Ti-MOF precursor was used to create an N-TiO_2 nanostructure. The absorption coefficient can be significantly increased and the band gap can be narrowed through the successful doping of N elements. The research conducted by Araya and colleagues found that the synthesis of MIL-53(Fe) [103] was accomplished, and it was effectively combined with anionic resin (i.e., Amberlite IRA 200) and cationic resin (viz., Amberlite IRA 900) to produce composite photocatalysts, namely AMIL-53(Fe) and DMIL-53(Fe). This study provides a perspective on using resins as loaded agents instead of conventional supports in order to enhance the degradation selectivity, catalytic efficiency, stability, and optical characteristics of MOF-based photocatalysts. Ramezanalizadeh et al. [104] reported that a magnetic composite of bimetallic Ni/Co-based MOF was successfully prepared, incorporating the magnetic $BiFeO_3$. The improved photocatalytic behavior of the MOF/$BiFeO_3$ nanocomposite, as indicated by the acquired findings, can be ascribed to the efficient separation of photogenerated holes and electrons, presence of selective channels, extensive surface area, substantial visible light absorption, convenient transportation of guest molecules, and favorable properties regarding sorption. Under visible light, Fig. 9 depicted the elimination of organic contaminants on M88/GO-9 nanocomposite through a schematic illustration [105]. When exposed to visible light, M88/GO-9 can be stimulated to produce electrons and holes at conduction and valence bands. The photogenerated positive charges with a powerful ability to cause oxidation can directly oxidize the RhB. In the meantime, the presence of extremely thin GO sheets allows for the migration of photoexcited electrons to GO, enhancing the effective separation of photoinduced electron-hole pairs. Additionally, when hydrogen peroxide is present, H_2O_2 can capture the electrons in the CB. Simultaneously, M88/GO-9's surface generates highly active $\bullet O_2^-$ radicals that can further oxidize RhB.

3.2 *Conventional Inorganic Materials*

Hiroshi Fujimajima and Kenichi Honda published a significant study in 1972, demonstrating that the TiO_2 electrode has the ability to split H_2O into oxygen and hydrogen when exposed to ultraviolet light. This indicated a fresh approach to address the energy shortage, which promptly triggered a worldwide surge in scientific investigation [6]. Subsequently, Frank and Bard effectively converted CN- to OCN- by utilizing TiO_2 as a photocatalyst, a significant action that advanced and expedited the utilization of photocatalysts in the treatment of wastewater [106]. Ever since, TiO_2

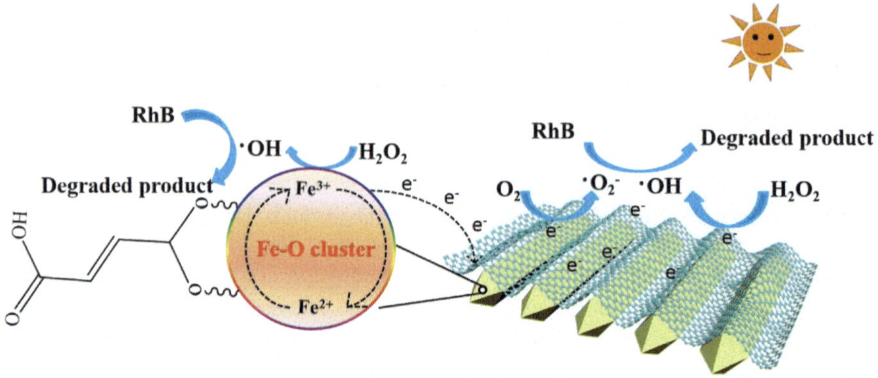

Fig. 9 Schematically illustrated mechanisms for the oxidation of RhB dye over M88/GO-9 nanocomposite. Reproduced from Liu et al. [105], with permission from Elsevier. Copyright (2018)

has been employed as a photocatalyst for various purposes within the realm of environmental control (Fig. 10). The extensive investigation of photocatalyst technology further expands the potential scope of photocatalysts' applications.

Semiconductor photochemistry involves the light-induced redox reaction of semiconductors in photocatalytic action. A semiconductor consists of a VB with low energy and a CB with high energy, and the gap between the CB and the VB is referred to as a forbidden band. When the incident light has energy higher than semiconductor's band gap, photons excite electrons from VB to CB, creating corresponding holes in VB. An electric field separates the electrons and holes produced by light and causes them to migrate toward the surfaces of semiconductor particles. The pores created by light have powerful oxidative characteristics and are capable

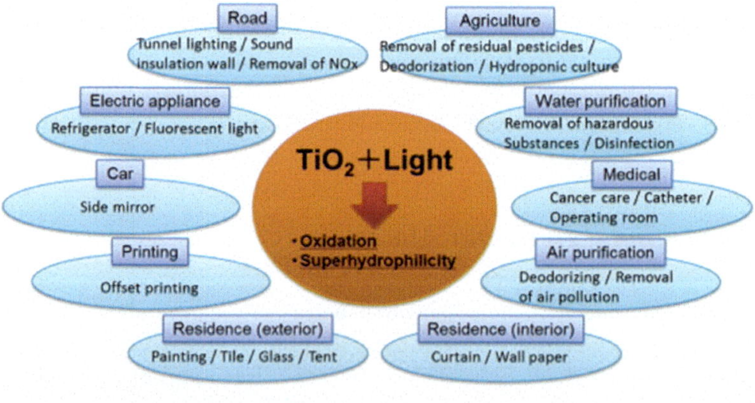

Fig. 10 Applications of TiO_2 in photocatalysis. Reproduced with the permission from Nakata et al. [107]. All rights reserved (2012), Elsevier

of oxidizing substances that are adsorbed on the surface or in the solution of the semiconductor [108].

The step-by-step procedure of a photocatalytic reaction is as outlined. Upon light absorption, electrons in VB become excited and move to CB, leading to the presence of a positive hole (h^+) in VB [109]. Reduction reactions occurred by the e^- in CB, while oxidation reactions can be induced via the h^+ in VB. Several different semiconductor materials, including TiO_2, zinc oxide (ZnO), g-C_3N_4, bismuth vanadate ($BiVO_4$), cadmium sulfide (CdS), and tungsten trioxide (WO_3), have been employed in photocatalytic processes.

Carey et al. [110] conducted a study in 1976. The researchers found that nano-TiO_2, when exposed to ultraviolet radiation, has the ability to break down polychlorinated biphenyls, which are organic compounds that are difficult to degrade [111]. This discovery marked a significant advancement in the fields of photocatalytic degradation of semiconductor nanomaterials. Bard et al. performed a study in 1980. They suggested a mechanism of photocatalysis by employing TiO_2 as a catalyst [112], which advanced the progress of TiO_2 in the realm of photocatalysis. Due to its elevated redox potential, increased reactivity, enhanced efficiency, chemical stability, lack of toxicity, simple synthesis, and affordability [113], TiO_2 stands out as a remarkable semiconductor material extensively investigated and widely utilized in important photocatalytic reactions.

In nature, the well-known minerals rutile, anatase, and brookite are the primary occurrences of TiO_2. Generally, the three types consist of TiO_6 octahedra with varying distortions (Fig. 11) [114]. Anatase is the most expected form among the various types of TiO_2 due to its thematic characteristics that enhance photocatalytic activity, including a substantial difference in energy levels, the existence of hydroxyl groups, a large surface area, and high porosity [115]. Rutile exhibits the highest thermodynamic stability under normal environmental conditions. However, anatase exhibits greater stability compared to rutile when the particle size decreases below 14 nm [116]. Degussa P25 titania is the name given to commercially available TiO_2, consisting of 75% anatase and 25% rutile. Anatase exhibits a higher photocatalytic efficiency under UV irradiation compared to the rutile phase due to its unique lattice structures.

The energy contained in electrons and holes generated by light is utilized to carry out thermodynamically favorable reactions in the process of pollutant degradation. TiO_2 stands out among other semiconductor materials due to its ability to perform both oxidation and reduction processes, which is a unique characteristic [117]. There are also various limitations associated with the utilization of TiO_2 as a technique for treating water through photocatalysis. Less than 5% of the visible and ultraviolet radiation that falls on its surface is absorbed, according to [118]. The capacity for adsorption, especially in the case of organic pollutants, is also quite restricted. The utilization of nanoparticles often leads to their aggregation and enlargement, resulting in a decrease in surface area and active sites. Additionally, they demonstrate significant scattering and both of these factors contribute to a decline in efficiency [119]. As a result, the catalysts made with doped TiO_2 were created. Extensive research has been conducted in the past on the utilization of doped TiO_2 catalysts, which has demonstrated their effectiveness and affordability in eliminating pollutants from

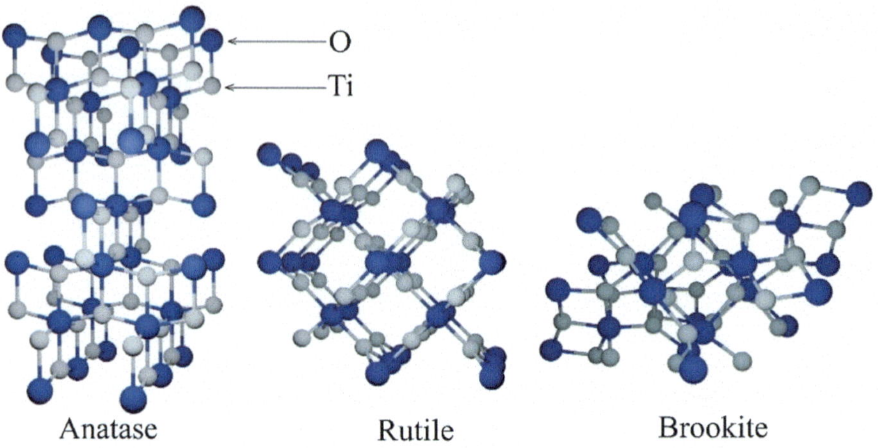

Fig. 11 The anatase, rutile, and brookite polymorphs of TiO_2 that have distinct crystal structures. Reproduced from Peiris et al. [114], with permission from John Wiley and Sons. Copyright (2021)

wastewater. Figure 12 shows the electron transfer mechanism in the photocatalytic process of gold-modified TiO_2 nanocomposite. According to the research conducted by Zhou and colleagues [120], a set of Pt nanocrystals (Pt NCs) were produced on a TiO_2 substrate, allowing for the adjustment of the layer's thickness. The photoreduction method was used to prepare well-dispersed platinum nuclei at ambient temperature. After being calcined in O_2, C-Pt/TiO_2-O_2-1 exhibits platinum nanoclusters (Pt NCs) of approximately 3 nm in size, featuring an active surface and remarkable resistance to sintering. Consequently, it demonstrates the most superior catalytic performance in the oxidation of CO.

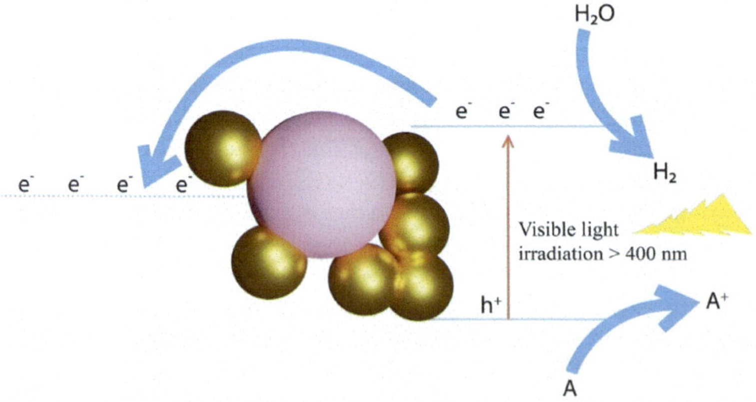

Fig. 12 When exposed to visible light, gold-modified TiO_2 nanocomposites promote the transfer of electron from gold to TiO_2 through an electron transfer mechanism. Reproduced from Peiris et al. [114], with the permission from John Wiley and Sons. Copyright (2021)

Zinc white, also referred to as zinc oxide, is a white hexagonal crystal or white powder. Similar to GaN, ZnO possesses a wurtzite arrangement; however, ZnO is a sizable single crystal in bulk form [121]. Zinc oxide exhibits clearly defined crystal structures, typically found in rock salt, wurtzite, or cubic (zinc blende) forms. ZnO can form a rocksalt structure when subjected to high pressure, making this structure of ZnO unique. Among the three structures, the ZnO wurtzite structure exhibits the greatest thermodynamic stability. ZnO [122] typically adopts this prevalent structure. Zinc oxide, with its favorable characteristics, finds applications in various sectors including catalysis, rubber and paint manufacturing, ceramic production, varistors, fertilizers, and cosmetics.

In today's field of photocatalysis, ZnO has become the forefront contender for an effective and promising role in green environmental management. This is due to its extensive and direct band gap at 3.37 eV at 300 K, along with a significant exciton binding energy at 60 meV at room temperature [123]. ZnO can be easily cultivated and maintains thermodynamic stability in the form of a hexagonal wurtzite structure [124]. ZnO is a safe substance for the environment since it is harmonious with living beings. This quality makes it suitable for a wide variety of everyday uses that pose no threats to human well-being or the environment [125]. Figure 13 illustrates the process of ZnO's photodegradation of organic pollutants. ZnO and TiO_2 exhibit nearly identical photocatalytic properties due to their similar band gap energy at 3.2 eV. However, compared with TiO_2, ZnO is relatively cheap and it is not economical to use TiO_2 in large-scale water treatment [126].

According to [127], the band gap energy width and photo etching are among the major drawbacks of conventional ZnO. The light absorption regarding ZnO is restricted to the ultraviolet regions owing to its wide band gap energy. This causes a rapid reorganization of the charge produced by the light, resulting in a low photocatalytic efficiency [128]. Therefore, ZnO was modified by introducing different kinds of

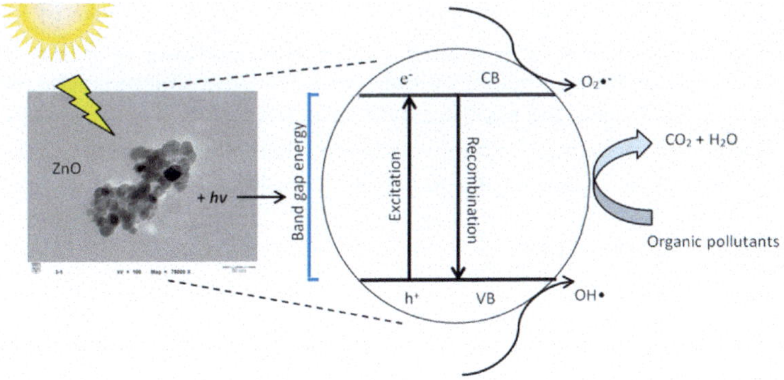

Fig. 13 ZnO facilitates the breakdown of organic pollutants when exposed to sunlight. Reproduced from Ong et al. [125], with the permission from Elsevier. Copyright (2018)

metal dopants into the semiconductor. The modification of the semiconductor significantly decreases the recombination ratio of the carriers and enhances the catalytic activity. To summarize, the future holds great promise for the practical use of altered ZnO photocatalysts in industries, thanks to its exceptional photocatalytic efficiency, stability under light exposure, and lack of toxicity.

g-C_3N_4 as an n-type non-metallic polymer semiconductor was first discovered in 2009 [129], and has been used in photocatalytic hydrogen production. g-C_3N_4 is formed by combining tri-s-triazine (C_6N_7) and s-triazine (C_3N_4) rings to create its structure. At room temperature [130], g-C_3N_4, which contains regular tri-s-triazine units, exhibits the highest stability among all C_3N_4 phases. g-C_3N_4 possesses a lamellar structure in two dimensions (2D) with a π conjugated system and a moderate energy gap at 2.7 eV. These characteristics result in distinctive photoelectric properties, excellent chemical stability, and efficient absorption of visible light, distinguishing it from conventional semiconductors like TiO_2 and ZnO.

In contrast to conventional photocatalysts, the precursor substances needed for the production of g-C_3N_4 are inexpensive, environmentally friendly, and readily accessible. The composition consists of carbon and nitrogen, two elements that are abundant in the earth, with a molar ratio of 0.75. The fabrications for it are incredibly straightforward, although this does not imply that the process is a single-step reaction. Various techniques, including high temperature calcination, molten salt, molecular self-assembly, microwave irradiation, ionic liquid methods, solvothermal strategies, chemical vapor deposition, and hydrothermal method, are commonly used for the preparation of g-C_3N_4-based photocatalyst [131]. The optical and electronic characteristics of g-C_3N_4 are greatly influenced by the materials utilized in its preparation, including dicyandiamide, cyanamide, and melamine. At present, the approaches of preparing g-C_3N_4 are very mature. The challenges that need to be addressed are the limited surface area and the lack of active sites.

Due to the appropriate band gap and position of CB and VB, g-C_3N_4 has recently posed a challenge to the dominance of the initial generation of semiconductors. Nevertheless, the unsatisfactory properties of g-C_3N_4 persist because of its inadequate absorption of visible light, limited surface area, poor conductivity, and high rate of recombination for photoelectron-hole pairs. These problems can be overcome by modifying g-C_3N_4, and the photocatalytic effectiveness can be improved [132].

For the sake of enhancing the photoelectronic characteristics of g-C_3N_4, diverse techniques like doping, creating heterojunctions, combining with carbon-based materials, and introducing defects have been employed to achieve diverse morphologies with enhanced properties [132].

Doping has been applied to improve the photoelectronic characteristics of g-C_3N_4. Due to the layered structure and the presence of heteroatoms in its cavity, the conduction band of doped g-C_3N_4 increases without being affected by the absorption of visible light [133]. The study conducted by Kamal and colleagues [134] reported that BCN/$NiFe_2O_4$ nanocomposites were effectively synthesized using boron-doped g-C_3N_4 as a photocatalyst for visible light-induced degradation of methylene blue. The findings indicate that BCN/$NiFe_2O_4$ nanocomposites exhibit a superior degradation efficiency (98%) compared to both BCN and $NiFe_2O_4$ nanocomposites. Liu and colleagues [135] claimed that sulfur was added to g-C_3N_4, resulting in a doped material that displayed a promising photoreactivity, as confirmed by both theoretical and experimental analysis. In addition, Yang and colleagues [136] reported that a comparable (3D) phosphorus (P)-doped flower-shaped g-C_3N_4 catalyst was synthesized and employed for the identical process of photocatalytic water splitting. Phosphorus atoms decelerate the recombination rate of holes as well as electrons generated by light, thereby enhancing the efficiency of the photocatalyst.

The drawbacks of g-C_3N_4 include inadequate separation of charges generated by light and limited absorption of light. The heterojunction has the ability to decrease the energy difference between bands and enhance both the absorption of light and the transfer of charges. Up to now, various heterojunctions have been fabricated including Z-scheme, type II, and Schottky junctions. Zhang and colleagues [137] reported that a Z-scheme system comprising of 2D/2D BiOBr/CDs/g-C_3N_4 was created. When the nanocomposite was utilized for the photodegradation of ciprofloxacin (CIP) as well as tetracycline (TC), it exhibited enhanced performance. The authors Zhang and colleagues [138] reported that a type II heterojunction was created using sulfur-mediated g-C_3N_4 (CNS) together with pristine g-C_3N_4 (CN). The utilization of the system led to the desired band alignment and enhanced the transfer of charges. The desired outcome is for the difference in energy levels between the CB of CN and CNS to facilitate the transfer of the photogenerated electron from CN to CNS. The hindrance of recombination of charges generated by light is considered essential in photoreactions. The CN/CNS heterojunctions exhibited significant enhancement in photocatalytic hydrogen generation. By combining the valuable metal with g-C_3N_4, it becomes possible to build Schottky junctions, which in turn create interfaces facilitating enhanced charge transfer by minimizing recombination. Typically, plasmonic metals like gold, silver, palladium, platinum, rhodium, and iridium [139] are commonly employed to build such a junction.

To enhance the optoelectronic capabilities, g-C_3N_4 has been combined with carbonaceous substances like carbon nanotubes (CNTs), graphene (G), and fullerenes (C_{60}). The specific surface area of these materials is high, and they possess thermal or chemical stability together with high conductivity. Li et al. [140] reported that a nanocomposite of carbon and carbon nitride was created, resulting in improved degradation of methylene blue (MB) through enhanced light absorption efficiency, adsorption, as well as charge separation due to the synergistic effects of both materials.

The morphologies and electronic band structures of g-C_3N_4 can be altered by introducing defects. The combined impact of these two alterations leads to improved separation of charges, absorption of light (shift towards red), and increased surface area, which serve as the foundation for light-induced reactions. Niu et al. reported that a rapid (5 min) thermal process was devised to alter graphitic carbon nitride by introducing significant quantities of imperfections. With the rise in the treating temperature, the CNQs exhibited augmented absorbance along with the visibly shifted intrinsic light absorption edge, indicating the emergence of a new light absorption edge. Different defects are believed to have unique impacts on the boost of light absorption, as indicated by both experimental and theoretical findings.

Essentially, g-C_3N_4 is a semiconductor without metals that has a customizable band gap, excellent thermal/chemical stability, and appealing electronic characteristics. UV radiation can be absorbed by g-C_3N_4 because of its band gap of 2.7 eV. The combination of g-C_3N_4 with other nanomaterials through exfoliation, doping, and amalgamation can boost the optoelectronic characteristics of both the g-C_3N_4 nanomaterial and the nanocomposites, showcasing synergistic properties.

3.3 Covalent Organic Frameworks

Covalent organic frameworks (COFs) are a novel category of crystalline organic porous materials. They are formed by linking various organic units together through covalent bonds. These frameworks possess a well-engineered pore structure, a large surface area, remarkable stability, and can be easily functionalized at the molecular level [141]. Hydrazine-based-COF has demonstrated a commendable efficiency in generating hydrogen under visible light, comparable to Pt-modified amorphous melon and crystalline poly (triazine imide), two well-known photocatalysts. The achievement has sparked the investigation of photocatalysts based on COF in the entire community (Fig. 14).

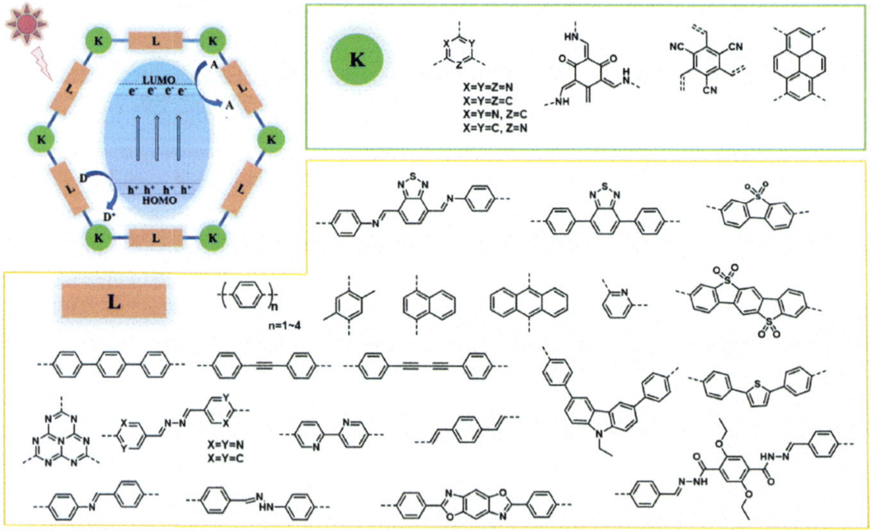

Fig. 14 The configurations of COF photocatalysts (K denotes knots; L denotes linkers) [142]. Reproduced from Wang et al. [139], with permission from Royal Society of Chemistry. Copyright (2020)

The use of COFs as catalysts has garnered significant interest because of their easily manageable framework architectures, a wide range of structures, and adaptability. Currently, several publications have already emerged, primarily focusing on the design, synthesis, structure, and utilization of COFs. COFs can be divided into 2D and three-dimensional 3D structures, relying on the size and dimension of the building blocks. The two of them possess a network structure that is not closed, leading to the presence of channels or nanopores that have consistent sizes ranging from angstroms to nanometers [143]. The distinctive arrangement of their components makes them appealing for various uses, such as storing and separating gases, delivering drugs, and converting and storing energy [144]. Furthermore, by incorporating functional groups, COFs can acquire distinctive optical and optoelectronic characteristics. COF materials [145] offer vast possibilities due to the wide range of structural and functional variations and diversity of the organic components.

3.3.1 Building Blocks of COFs

COFs are made up of π-frameworks as well as covalent bonds that contain multiple reactive sites. COFs are categorized into 3 types, namely boron-containing, triazine-based, as well as nitrogen-based COFs, depending on their linkages together with building blocks [146].

The synthesis of boroxine linkages/boronate ester linkages can be achieved by either self-condensing a single building unit (boronic acid) or co-condensing building units (boronic acid and catechol), respectively [147]. The formation of COF-1 relies on a molecular dehydration process, where three boronic acid molecules come together to create a flat six-membered B_3O_3 (boroxine) ring, resulting in the removal of three water molecules. Boronate-ester-linked COF-5 was obtained through the co-condensation of the hexahydroxytriphenylene knot as well as 1,4-benzenediboronic acid edge unit [148]. COFs that contain boron exhibit excellent thermal stability, crystalline structure, photoconductivity, and large surface areas [149].

CTFs, which are a unique type of COFs, are formed through cyclotrimerization of nitriles under ionothermal condition, amidine-based polycondensation, Friedel-Crafts reaction, or strong Brønsted acidic condition to create ultra-stable triazine linkages [150]. Compared to COFs containing boron, CTFs exhibit higher stability, a greater amount of nitrogen atoms, adjustable structure, and a significant level of conjugation [151]. They also have lower crystallinity. The emerging field involves the development of novel linkages in highly functionalized fully-conjugated COFs. In a study, a completely sp^2-carbon 2D COF (TTO-COF) was created and produced through acid-catalyzed Aldol reaction, utilizing the uniform assembly of triazine unit into COF. Figure 15 demonstrates that TTO-COF exhibited superior performance and reusability in the photocatalytic elimination of dyes and C-H functionalization of arenes and heteroarenes compared with imine COFs as well as $g\text{-}C_3N_4$. This photosensitive semiconductor displayed good chemical stability under harsh condition (such as strong acid and strong base) and demonstrated excellent conduction efficiency of the carrier, with an optical bandgap of 2.46 eV.

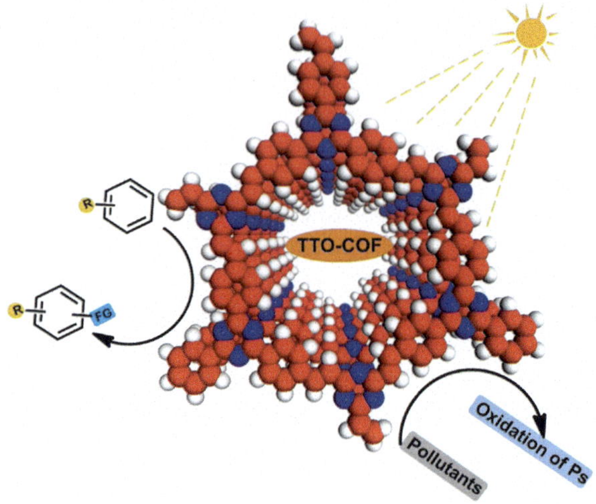

Fig. 15 Triazine-based COF. Reproduced from Yang et al. [152], with permission from Elsevier. Copyright (2020)

The COFs made from imines are created through the co-condensation of aldehydes with amines [153]. In comparison to boron-based COF, imine-linked COF exhibit greater stability in water, acidic or basic environments, and organic solvent, although their crystallinity may be inferior or comparable [154]. A potential heterogeneous photocatalyst for photoinduced atom transfer radical polymerization (ATRP) called TAPPy-TPA-COF, which is an imine-based covalent organic framework (Fig. 16), has been synthesized. Polymerizations of methyl methacrylate (MMA) are controlled well by maintaining a balance between Cu(I) activation and Cu(II) inactivation. The experiments on chain extension have additionally shown the accuracy of polymer chain ends. In the meantime, the experiments on catalyst recycling have demonstrated the durability of TAPPy-TPA-COF in ATRP reactions. The findings affirm the possibility of employing COFs as heterogeneous photocatalysts for copper-mediated ATRP when exposed to visible light.

In addition to the linkages mentioned above, COF fabrication has also utilized other types of linkages including carbamate, borosilicate, phenazine, and squaraine linkages. As previously mentioned, different connection patterns have been developed in relation to the formation of COF. Various connections result in diverse formations and characteristics, typically associated with stability [156].

3.3.2 Synthesis and Properties of COFs

COF can be synthesized using solvothermal, ionothermal, and either ambient temperature or microwave methods. Typically, a solvothermal technique is employed to synthesize the majority of COFs, wherein the reaction is conducted in a sealed container at temperatures ranging from 80 to 120 °C for a duration of three to

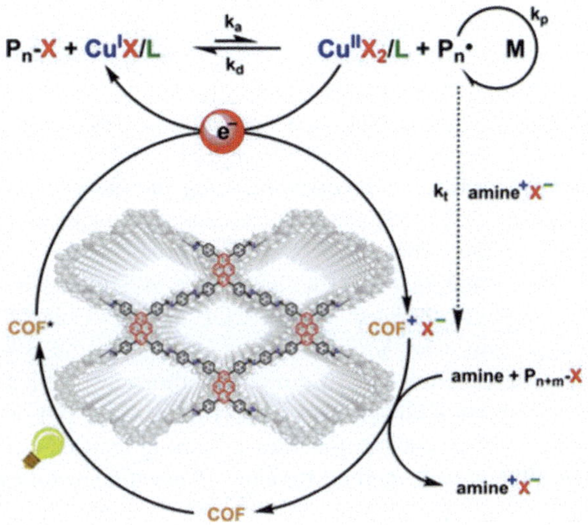

Fig. 16 COF derived from imine. Reproduced from Fu et al. [155], with the permission from John Wiley and Sons. Copyright (2021)

seven days. Microwave heating offers numerous benefits in comparison to traditional heating method. These include faster reaction rate, increased product yield, reduced energy consumption, gentler reaction conditions, and enhanced selectivity in chemical reactions [149]. The main methods for conducting syntheses at room temperature are mechanochemical grinding as well as the solvent method. Compared to the solvent thermal method, the room-temperature synthesis of high crystallinity COFs results in elevated thermal stability and surface area. Nevertheless, the high-priced chemicals employed in the production complicate the scaling up of COF applications [154].

COFs have several advantages over traditional semiconductors in terms of photocatalysis. Firstly, the designability of COFs allows for targeted structures and special properties related to photocatalytic reactions, like excellent absorption of visible light and efficient separation and transfer of electron-hole pair. Secondly, the large surface area of COFs provides numerous catalytic sites, while their crystalline and porous structure facilitates rapid charge transport and reduces the likelihood of charge trapping due to defects, thus minimizing electron-hole recombination. Thirdly, COFs with strong covalent bond exhibit high chemical as well as thermal stability, and the incorporation of photoactive units within the robust framework prevents photocorrosion and enhances the excited state's lifetime. Lastly, the extended π-conjugated structure in both the plane and stacking direction guarantees high mobility of charge carriers. The remarkable intrinsic characteristics give COFs immense potential in the conversion of solar energy and the restoration of the environment. They are considered to be on par with or even surpass MOFs and traditional photocatalytic semiconductor in terms of performance [142].

3.3.3 Applications

Promising are two distinct categories among the various applications of photocatalysis. Environmental applications include the breakdown of pollutants, the elimination of bacteria, and the facilitation of metal-free cross-coupling reactions. Another area of application involves utilizing the photocatalytic characteristics for energy purposes [157, 158]. Here, we list some applications of COF as a photocatalysts for pollutant degradation.

Photocatalytic properties have been investigated in certain 3D COFs. For instance, 3D porphyrin-derived 3D-Por-COF and 3D-CuPor-COF containing photoelectric components were prepared using [4 + 4] imine condensation and utilized in the photodegradation of 9,10-dimethylanthracene [159]. The authors Liu and colleagues reported that COF-PD/AgI was synthesized [160] using 2, 5-diaminopyridine and 1, 3, 5-triformylphloroglucinol as starting materials. The synthesis process involved the solvothermal method for COF-PD synthesis, followed by the co-precipitation of AgI with COF-PD. The addition of AgI resulted in a decrease in the recombination of charge carriers in the binary composites based on AgI. Therefore, it is anticipated that the incorporation of AgI into COFs can boost the segregation of photoinduced electrons and holes, consequently enhancing their photocatalytic efficiency when exposed to visible light.COF-PD/AgI has a good degradation effect on organic pollutants such as RhB and acetaminophen (ACTP). COF-PD/AgI can remove 5 mg/L RhB and ACTP under visible light irradiation within 80 and 160 min, respectively, through photocatalysis. The photocatalytic process was found to be influenced by $\bullet O_2^-$ and h^+, with h^+ making a relatively smaller contribution compared to $\bullet O_2^-$. The primary mechanism of RhB degradation via COF-PD/AgI is the dominance of $\bullet O_2^-$ radical-induced N-deethylation, while ACTP degradation is primarily driven by hydroxylation (Fig. 17). Furthermore, research has demonstrated that adjustable covalent organic structures featuring diverse positions of heterocyclic nitrogen are capable of efficiently breaking down ACTP when exposed to visible light [161]. Additionally, the covalent organic framework COF-JLU19, linked by amide bonds, has exhibited the ability to degrade RhB in water when exposed to sunlight [162].

Chemically stable TpMA COFs were prepared in the Lv [26] group through a mechanochemical reaction at room temperature using 1, 3, 5-tricarbonylresorcinol (Tp) and melamine (MA). TpMA exhibited effective elimination of both phenol and methyl orange. After 60 min of irradiation, TpMAC $_{(3\ \mathrm{mL})}$ successfully broke down phenol entirely, while TpMAC $_{(1\ \mathrm{mL})}$ catalyzed the decomposition of approximately 83.5% of phenol. In line with the breakdown of phenol through photocatalysis, TpMAC $_{(3\ \mathrm{mL})}$ and TpMAC $_{(1\ \mathrm{mL})}$ resulted in the photodegradation of 89% and 79% of methyl orange, respectively, following 60 min of exposure to visible light. The exfoliated slender strip-like shape of TpMA COFs demonstrated an increased rate for phenol degradation and maintained a catalytic efficiency of 87.6% even after being reused 4 times. This indicates that TpMA, synthesized through mechanochemical means, possesses satisfactory recyclability and stability. In addition, the triadic structure of the covalent organic framework, which consists of the C_3N_4 active

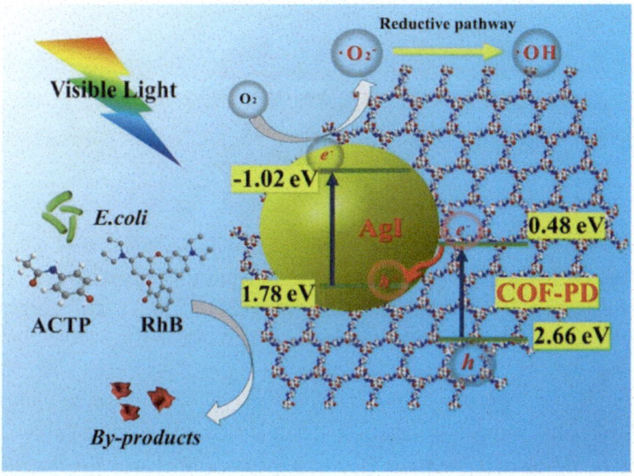

Fig. 17 Photocatalysis of COF-PD/AgI. Reproduced from Liu et al. [160], with permission from Elsevier. Copyright (2020)

center, photoelectron shift platform, and electron-withdrawing units, was found to be effective in the photodegradation of phenol together with methyl orange [163].

Titanium dioxide is extensively studied as a photocatalyst for various photocatalytic purposes. COFs acted as agents for interface engineering to produce Fe-doped 5Fe-TiO_2@COF nanoparticles that were ultrafine and well-dispersed, with a size of approximately 2.3 ± 0.9 nm [164]. Under the influence of surrounding light, a small amount of light-sensitive Fe-TiO_2@COF mixture decomposed methylene blue with excellent durability, and it was capable of being utilized repeatedly without impacting the rate of photodegradation. COF, acting as the support of the catalyst, enhanced the adsorption of organic compounds, while also stabilizing and regulating the growth of nanoparticles, ultimately leading to an augmentation in mass transport through porous channels [149].

3.4 Organic Materials

The structural diversity, synthetic modularity, as well as feasibility for precise tuning of optoelectronic and structural properties make organic semiconductors advantageous. Furthermore, organic semiconductors, commonly known as organic conjugated molecule, possess a distributed p-electron system, rendering them as potential components for structured nanoscale superstructures. Thermal deposition and facile solution processing enable the utilization of their structural adaptability for the fabrication of lightweight, large-scale, flexible, and stretchable devices. Figure 18 displayed the structures of various representative organic photocatalysts. Light absorption in the UV-V is range is the primary condition for a photocatalyst to be usable. Due to their robust absorption bands in the ultraviolet–visible–near-infrared (UV–Vis–NIR) range, organic semiconductor exhibits proficiency in this aspect. Various types of organic compounds, including porphyrin (Pors), phthalocyanine,

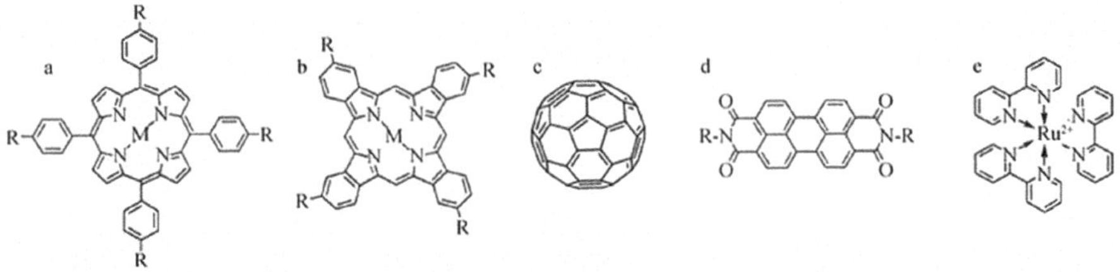

Fig. 18 Examples of various organic photocatalysts include: **a** Porphyrins, **b** Phthalocyanines, **c** Fullerene (C60), **d** Perylene diimides (PDIs), and **e** Ruthenium complexes with bipyridine ligands. Reproduced from Chen et al. [165], with permission from Springer Nature. Copyright (2017)

fullerene, perylenetetracarboxylic diimide derivative, and Ru polypyridyl complex, have been utilized for diverse photocatalytic applications [165].

Porphyrins consist of four pyrrolic units linked by methine bridges, creating fully conjugated aromatic macrocycles containing 18 π electrons. The porphyrin derivatives possess distinct characteristics that enable them to efficiently convert photo, electrical, and chemical energy, making them crucial components in energy conversion systems. In addition, they function as photocatalysts that transform light into chemical energy [166].

In aerated neutral aqueous solutions, the herbicides atrazine and ametryn can be degraded by the sensitizers 5, 10, 15, 20-tetrakis (2,6-dichloro-3-sulfophenyl) porphyrin and 5, 10, 15, 20-tetrakis (4-sulfophenyl) porphyrin using visible light. The degradation rate for atrazine was 30%, while for ametryn, it reached 63%. Characterization of the end products revealed that they were dealkylated s-triazines [167]. Metalloporphyrin complexes can be formed by the binding of porphyrins with nearly all metal ions mentioned in the periodic table, leading to the coordination number ranging from four to six. For example, copper (II) porphyrin can photodegrade 4-nitrophenol [168]. The water-soluble TDCPPS and its iron complex ZnTDCPPS have been effectively employed in the oxidation of 4-CP and 2,6-DMP [169]. Furthermore, the visible-light irradiation has shown the photocatalytic capabilities of the antimony porphyrin complex supported on silica gel, including the removal of chlorophenols, oxidation of cycloalkenes, and the ability to kill E. coli and Legionella species [170].

Various nanostructures, including nanowires, nanorods, nanotubes, and nanoparticles, have been created from Pors for the purpose of visible-light photocatalysis. As an example, Guo and colleagues [171] reported that the supramolecular photocatalytic efficiency of nanofibers and nanospheres based on porphyrin has been investigated in relation to the degradation of RhB pollutants when exposed to visible light. The nanofibers exhibit noticeable photocatalytic effectiveness in the photodegradation of RhB molecules, with the ability to carry out the reaction up to 8 consecutive times.

Phthalocyanines (Pcs) are a class of macrocyclic compounds that are structurally analogous to porphyrins. They consist of four isoindole units linked through nitrogen atoms to form a large, planar, conjugated system. Typically, these molecules can coordinate with a variety of metals at their center. The basic phthalocyanine structure is aromatic and exhibits deep blue-green coloration. This compound belongs to a highly potential group of catalysts for visible light due to its significant advancement in the visible region. Moreover, it is environmentally friendly, inexpensive, and readily available [172].

In applications where phthalocyanines serve as visible photocatalysts for organic pollutant degradation, they tend to aggregate upon dissolution in water, leading to dimer formation or even greater degrees of polymerization. Notably, their catalytic efficacy is optimal in their monomeric state. Consequently, numerous strategies have been explored to minimize the aggregation propensity of phthalocyanines in aqueous environments. Among these, anchoring phthalocyanines to carriers, such as clay, molecular sieves, or resins, has proven to be a straightforward and highly efficient approach. [173]. Tao et al. [174] reported that they discovered when exposed to visible light, iron sulfonate phthalocyanine loaded onto anion exchange resin efficiently decomposed orange II, while showing limited effectiveness in breaking down cationic dyes. This suggests that the photocatalyst can only interact with substrates that are adsorbed onto its surface. The Xiong [175] team utilized modified bentonite to immobilize palladium sulfonated phthalocyanine for the purpose of eliminating 2,4,6-trichlorophenol (TCP) using visible light. They also examined how the degradation efficiency was influenced by the frequency of recovery cycles. After seven recoveries, a slight decline in degradation efficiency was observed, which was attributed to the adsorption of intermediate substances onto the surface of bentonite, resulting in a reduction of the reaction area and the sampling of a lower concentration of catalyst. Additionally, Xiong et al. [176] reported that the effectiveness of multiphase palladium phthalocyanine photocatalysts in studying the degradation of TCP was demonstrated by loading them onto water talc and comparing them with homogeneous catalysts.

At present, phthalocyanines are employed for the remediation of recalcitrant organic contaminants in water, primarily encompassing phenolic compounds and organic pigments. The phthalocyanine derivatives from the existing research demonstrated excellent degradation of peacock green, reactive red, rhodamine B, and orange yellow II. They achieved a decolorization rate close to 100% and exhibited high mineralization. For example, Sorokin et al. [177] successfully degraded 2, 4-chlorophenol (2,4-DCP) in water using water-soluble iron phthalocyanine and obtained a fairly high mineralization rate. The authors Zhang and colleagues [178] reported that the study discovered that the presence of copper phthalocyanine (CuPc), an organic dye sensitizer, significantly increased the light absorption of Bi_2MoO_6. This enhancement resulted in excellent degradation of phenol and rhodamine B. In hydrothermal conditions, Wang's team [179] successfully produced anatase nanophotocatalyst using copper phthalocyanine tetrasulfonate (CuPcTs). In the presence of CuPcTs/TiO_2 catalyst, the degradation rate of the methyl orange solution is significantly higher compared with the unmodified TiO_2. After being exposed to visible light for 30 min ($\lambda > 450$ nm), the unmodified TiO_2 only achieves a degradation rate of 14.48%, whereas the CuPcTs/TiO_2 catalyst significantly increases it to 73.39%.

Carbon nanomaterials possess an extensive range of physical and chemical characteristics. Fullerenes, being a significant constituent of the carbon nanomaterial group, inherently possess numerous characteristics. At this point, Fullerene (C_{60}) is among the materials that have been extensively researched. C_{60} possesses numerous physical and chemical characteristics, including harmlessness and superconductivity. Undoubtedly, the unique characteristics of photocatalytic reactions have captured our interest [180].

The unique characteristics of C_{60} in photocatalytic reactions are primarily demonstrated in 3 main areas. The photocatalytic properties of this material are influenced by its distinctive morphological structure, which is characterized by a three-dimensional symmetrical arrangement and numerous internal micropores. The remarkable structural characteristic of C_{60} molecules contributes to their high quantum efficiency in the photocatalytic reaction. Additionally, it possesses favorable physical/chemical stability, which guarantees the enduring presence of C_{60} in composite material and enhances its effectiveness in the process of photocatalysis. After being stimulated by light, it exhibits distinctive optical characteristics, which involve processes of electron and energy transfer. The entire range of visible and ultraviolet light can be absorbed by them, and the photocatalytic reaction is effectively utilized within the visible spectrum. According to [180], functional fullerene is a successful method for modifying its electronic characteristics.

Despite the numerous benefits of fullerenes, their dispersion as well as solubility in solution is quite poor. Hence, there has been significant interest in the alteration of fullerene to acquire the derivative for the purpose of photocatalytic reactions. For instance, Bai and colleagues [181] reported that a hydroxyl group was added to the fullerene, resulting in a polyhydroxy fullerene. This was then mixed with TiO_2 to create a photocatalyst used for eliminating the organic dye. The study conducted by Djordjevic and colleagues claimed that a combination catalyst consisting of polyhydroxy fullerene and TiO_2 was created [182], who then investigated its effectiveness in breaking down the herbicide mesotrione.

New composite-based photocatalysts for the photocatalytic reaction have been created by combining fullerenes with different wide band gap semiconductor photocatalysts, as demonstrated by Yu et al. [183]. A nanocomposite consisting of fullerene and TiO_2 was created using a straightforward hydrothermal crystallization technique to degrade vapor phase acetone through photocatalysis. Ma and colleagues [184] reported that a composite catalyst of C_{60}/BiOCl was created by in-situ preparation, enabling the degradation of RhB and phenol in simulated sunlight. Park et al. [185] claimed that a composite catalyst was prepared by utilizing water-soluble fullerol (i.e., $C_{60}(OH)_x$) to activate TiO_2 in the presence of visible light. This catalyst exhibited a positive impact on the conversion of harmful Cr(VI) into less harmful Cr(III) in water.

PDI, also known as perylene-3, 4, 9, 10-tetracarboxylic diimide, possesses distinctive characteristics in terms of its optical and electronic attributes. These include a notable oxidation potential, a high molar extinction coefficient, and excellent thermal and photochemical stability. Consequently, PDI has found extensive utilization as a fundamental building block for the production of organic photofunctional materials, particularly in the realm of fluorescent sensors. Additionally, PDI has been employed as a photocatalyst for achieving photoinduced chemical conversion [186].

In the photocatalytic elimination of phenol [187], the hybrid PDI/BiOCl photocatalyst displayed 2.2 as well as 1.6 times higher photocatalytic activity than BiOCl and PDI, respectively. Zeng and colleagues [188] reported that the aim of their study was to improve the absorption of visible light and enhance the photocatalytic effectiveness through the creation of the organic-inorganic Z-scheme system known as WO_3@Cu@PDI. It appears that WO_3@Cu@PDI exhibits remarkable capability in absorbing visible light and degrading TC through photocatalysis. When exposed to visible light, the composites degrade at a rate that is 40 times higher than WO_3 and 5 times faster than PDI. After being reused three times, the photocatalytic efficiency of WO3@Cu@PDI towards TC remained at 85%.

3D-PDI demonstrated remarkable degradation efficiency and durability towards various tetracycline antibiotics (viz., tetracycline, chlortetracycline, and oxytetracycline) under exposure to visible light. The improved adsorption and breakdown capabilities of TC by 3D-PDI were primarily attributed to the greater specific surface area for surface interaction and the π-electron conjugation within the supramolecular system of the 3D network. TC experiences hydroxylation, demethylation, aromatization, together with ring-opening reactions, ultimately resulting in full mineralization to CO_2 and H_2O [189].

Extensive research has been conducted on Ru complexes, particularly Ru polypyridyl complexes or their counterparts, due to their enduring excited states, remarkable efficiency in formation, prominent absorption bands in the visible spectrum, and exceptional stability in photochemical reactions. Currently, they remain the most widely used molecules in dye-sensitized photocatalysis. An example is the absorption spectrum of tris(2,2′-bipyridyl)ruthenium(II), viz., $[Ru(bpy)_3]^{2+}$, which displays a powerful wide range concentrated at approximately 450 nm, demonstrating a high quantum efficiency (Φ) of about 1 for the photoexcited state. Furthermore, this activated state is able to function as both an electron donor and an electron acceptor in intermolecular oxidative-reductive processes. This capability to facilitate effective electron-transfer interactions between electron-abundant donors and electron-deficient acceptors underscores its significant potential in solar energy utilization [190].

To enhance the light-capturing and quantum effectiveness of TiO_2, a fresh polypyridyl complex of Ru(II), known as [RuII(tptz)(CH_3CN)Cl_2], was created and employed as a photosensitizer dye responsive to visible light. The photocatalytic activity for the elimination of ATZ was greatly improved in the Ru-CMP-TiO_2/rGO hybrid compared with pure TiO_2. The degradation rate of ATZ in the Ru-CMP-TiO_2/rGO hybrid photocatalytic process was 9 times higher compared to the degradation rate with commercial TiO_2. The improved photocatalytic performance of the catalyst that was prepared can be credited to its superior light absorption and effective electron transfer, which is a result of its LUMO position being more compatible with the CB of TiO_2. Furthermore, the remarkable conductive properties and adsorptive capability of graphene played a role in enhancing the photocatalytic behavior. Therefore, the Ru-CMP-TiO_2/rGO hybrid nanomaterial possesses these characteristics that render it a highly suitable option for the treatment of polluted water using photocatalysis [191].

The cationic ion-exchange resin (Amberlite IRA 200) was used to support the Fe(II) complex of 2,2′-bipyridine ($[FeII(bpy)_3]^{2+}$). The efficient visible-light-responsive catalyst, which was proven to be effective, degraded organic contaminants like RhB, malachite green (MG), as well as N, N-dimethylaniline (DMA). The removal yield of total organic carbon (TOC) was 60%, 58% and 75% for RhB, MG and DMA, respectively. Furthermore, the system exhibited consistent catalytic performance throughout 22 consecutive cycles (~100 h) in the degradation of RhB, without any notable decline. Additionally, it could be easily retrieved from the reaction mixture for future use by means of filtration [192].

There are various reaction pathways that organic photocatalysts can utilize. It appears more straightforward to elucidate the reaction mechanism in homogeneous media due to the diversity of pathways involved. Marin et al. [193] have summarized potential mechanistic pathways. Essentially, the reaction process of organic catalysts occurs via the under mentioned pathways (Fig. 19). (i) Firstly, a photocatalyst (referred to as P) absorbs light with a specific wavelength and transitions to the initial singlet excited state $^1P^*$. $^1P^*$ has the ability to engage in electron-transfer reactions with a representative pollutant (referred to as Q), resulting in the formation of the oxidized state of pollutant ($Q^{\bullet+}$) and the reduced state of photocatalyst ($P^{\bullet-}$). The formation of the triplet state $^3P^*$ by intersystem crossing (isc) from $^1P^*$ or (ii, iii) $^1P^*$ can result in the reaction with water, producing hydroxyl radical OH• and the radical anion of photocatalyst ($P^{\bullet-}$). Subsequently, OH• participates in the oxidation of the pollutants. (iv) Additionally, electron-transfer reaction can occur for $^3P^*$, resulting in the creation of $Q^{\bullet+}$ and $P^{\bullet-}$. (v) The energy transfer process between $^3P^*$ and O_2 leads to the formation of singlet oxygen ($^1O_2$). (vi) The formation of a P-Q complex is possible between the photocatalyst's ground state and the pollutant. Light irradiation causes charge separation in the complex, resulting in $Q^{\bullet+}$ as well as $P^{\bullet-}$. (vii) Another route includes the interaction between $P^{\bullet-}$ and molecular oxygen, leading to the formation of superoxide radical anion $\bullet O_2^-$. The diverse mechanistic routes outlined in this document offer an opportunity to develop a treatment tailored to the specific disposal of various pollutants.

4 Conclusions and Future Perspectives

Photocatalysis, being a technology with great potential, has been extensively studied for its ability to efficiently break down different contaminants and transform solar energy to sustainable chemical energy. Its applications have been widely investigated in order to address environmental contamination and the global energy crisis. So far, photocatalysts have achieved a flourishing phase and demonstrated a hopeful performance in the aforementioned applications. Research attention will persist until significant advancements are achieved for widespread implementation.

Although significant photocatalytic outcomes have been achieved, there remain several crucial and thought-provoking subjects that require further investigation. For large-scale applications, it is crucial to have photocatalysts that have efficient

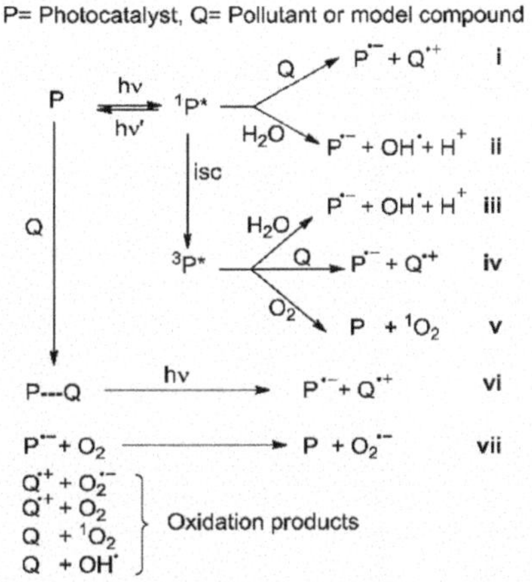

Fig. 19 Proposed mechanistic pathways. Reproduced from Marin et al. [193], with permission from the American Chemical Society. Copyright (2012)

absorption of visible light, minimal resistance in charge carrier transfer, and excellent physical and chemical stability. In addition, a highly effective photocatalyst typically needs a relatively limited bandgap to broaden the range of light it can utilize. At the same time, the produced photocatalysts should generate a maximum number of charge carriers with suitable energy level so as to carry out a particular photocatalytic redox reaction. Furthermore, the migration of charge carrier on the surface of photocatalyst is highly influenced by the energy band bending of semiconductor. Hence, it is crucial to regulate the energy band structure. Generally speaking, the presence of inherent electric fields at the boundary of semiconductors aids in the segregation of the electron-hole pair. Nevertheless, the enhancement of carriers' dynamics can only occur when the migration of photogenerated charge carriers aligns with the built-in electric field's direction. Hence, it is a thrilling revelation to incorporate the external electric field generated by polarization for altering the semiconductor's band structure. The movement of charge carriers on the surface of two semiconductors is also influenced by factors like crystal structure, including imperfections, lattice dimensions, and other related characteristics. Improving photocatalytic performance of the target photocatalyst can also be achieved via optimizing the interface characteristics of semiconductors.

The current review aims to offer guidance for the development of future design strategies to improve the specific characteristics of photocatalytic devices, ultimately resulting in the efficient degradation of contaminants through the use of photocatalytic technology.

References

1. V. Hasija, P. Raizada, A. Sudhaik, K. Sharma, A. Kumar, P. Singh, S.B. Jonnalagadda, V.K. Thakur, Recent advances in noble metal free doped graphitic carbon nitride based nanohybrids

for photocatalysis of organic contaminants in water: a review. Appl. Mater. Today **15**, 494–524 (2019). https://doi.org/10.1016/j.apmt.2019.04.003

2. D. Huang, S. Chen, G. Zeng, X. Gong, C. Zhou, M. Cheng, W. Xue, X. Yan, J. Li, Artificial Z-scheme photocatalytic system: What have been done and where to go?, Coordin. Chem. Rev. **385**, 44–80 (2019). https://doi.org/10.1016/j.ccr.2018.12.013
3. Y. Ding, I.S. Yang, Z. Li, X. Xia, W.I. Lee, S. Dai, D.W. Bahnemann, J.H. Pan, Nanoporous TiO_2 spheres with tailored textural properties: Controllable synthesis, formation mechanism, and photochemical applications. Prog. Mater. Sci. **109**, 100620 (2020). https://doi.org/10.1016/j.pmatsci.2019.100620
4. Q. Wang, K. Domen, Particulate photocatalysts for light-driven water splitting: Mechanisms, challenges, and design strategies. Chem. Rev. **120**, 919–985 (2020). https://doi.org/10.1021/acs.chemrev.9b00201
5. M. Xiao, Z. Wang, M. Lyu, B. Luo, S. Wang, G. Liu, H.M. Cheng, L. Wang, Hollow nanostructures for photocatalysis: advantages and challenges. Adv. Mater. **31**, e1801369 (2019). https://doi.org/10.1002/adma.201801369
6. A. Fujishima, K. Honda, Electrochemical photolysis of water at a semiconductor electrode. Nature **238**, 37–38 (1972)
7. Y. Sheng, Z. Wei, H. Miao, W. Yao, H. Li, Y. Zhu, Enhanced organic pollutant photodegradation via adsorption/photocatalysis synergy using a 3D g-C_3N_4/TiO_2 free-separation photocatalyst. Chem. Eng. J. **370**, 287–294 (2019). https://doi.org/10.1016/j.cej.2019.03.197
8. S. Wang, G. Liu, L. Wang, Crystal facet engineering of photoelectrodes for photoelectrochemical water splitting. Chem. Rev. **119**, 5192–5247 (2019). https://doi.org/10.1021/acs.chemrev.8b00584
9. A. Dhakshinamoorthy, Z. Li, H. Garcia, Catalysis and photocatalysis by metal organic frameworks. Chem. Soc. Rev. **47**, 8134–8172 (2018). https://doi.org/10.1039/c8cs00256h
10. Y. Liu, Z. Liu, D. Huang, M. Cheng, G. Zeng, C. Lai, C. Zhang, C. Zhou, W. Wang, D. Jiang, H. Wang, B. Shao, Metal or metal-containing nanoparticle@MOF nanocomposites as a promising type of photocatalyst, Coordin. Chem. Rev. **388**, 63–78 (2019). https://doi.org/10.1016/j.ccr.2019.02.031
11. X. Li, J.G. Yu, M. Jaroniec, Hierarchical photocatalysts. Chem. Soc. Rev. **45**, 2603–2636 (2016). https://doi.org/10.1039/c5cs00838g
12. M. Bhadra, S. Kandambeth, M.K. Sahoo, M. Addicoat, E. Balaraman, R. Banerjee, Triazine functionalized porous covalent organic framework for photo-organocatalytic E-Z isomerization of olefins. J. Am. Chem. Soc. **141**, 6152–6156 (2019). https://doi.org/10.1021/jacs.9b01891
13. Y.Z. Li, Z. Li, Y.S. Xia, H.F. Li, J.H. Shi, A.M. Zhang, H.H. Huo, S.Y. Tan, L.Z. Gao, Fabrication of ternary AgBr/$BiPO_4$/g-C_3N_4 heterostructure with dual Z-scheme and its visible light photocatalytic activity for Reactive Blue 19. Environ. Res. **192** (2021). https://doi.org/10.1016/j.envres.2020.110260
14. G. Liu, F. He, J. Zhang, L. Li, F. Li, L. Chen, Y. Huang, Yolk–shell structured Fe_3O_4@C@F-TiO_2 microspheres with surface fluorinated as recyclable visible-light driven photocatalysts. Appl. Catal. B Environ. **150–151**, 515–522 (2014). https://doi.org/10.1016/j.apcatb.2013.12.050
15. Y. Wang, W. Jiang, W. Luo, X. Chen, Y. Zhu, Ultrathin nanosheets g-C_3N_4@Bi_2WO_6 core-shell structure via low temperature reassembled strategy to promote photocatalytic activity. Appl. Catal. B Environ. **237**, 633–640 (2018). https://doi.org/10.1016/j.apcatb.2018.06.013
16. X. Yuan, D. Shen, Q. Zhang, H. Zou, Z. Liu, F. Peng, Z-scheme Bi_2WO_6/$CuBi_2O_4$ heterojunction mediated by interfacial electric field for efficient visible-light photocatalytic degradation of tetracycline. Chem. Eng. J. **369**, 292–301 (2019). https://doi.org/10.1016/j.cej.2019.03.082
17. T. Cai, W. Zeng, Y. Liu, L. Wang, W. Dong, H. Chen, X. Xia, A promising inorganic-organic Z-scheme photocatalyst Ag_3PO_4/PDI supermolecule with enhanced photoactivity and photostability for environmental remediation. Appl. Catal. B Environ. **263**, 118327 (2020). https://doi.org/10.1016/j.apcatb.2019.118327

18. P. Chen, L. Blaney, G. Cagnetta, J. Huang, B. Wang, Y. Wang, S. Deng, G. Yu, Degradation of ofloxacin by perylene diimide supramolecular nanofiber sunlight-driven photocatalysis. Environ. Sci. Technol. **53**, 1564–1575 (2019). https://doi.org/10.1021/acs.est.8b05827
19. H. Zhang, X. Chen, Z. Zhang, K. Yu, W. Zhu, Y. Zhu, Highly-crystalline triazine-PDI polymer with an enhanced built-in electric field for full-spectrum photocatalytic phenol mineralization. Appl. Catal. B Environ. **287**, 119957 (2021). https://doi.org/10.1016/j.apcatb.2021.119957
20. Y. Li, Y. Fang, Z. Cao, N. Li, D. Chen, Q. Xu, J. Lu, Construction of g-C_3N_4/PDI@MOF heterojunctions for the highly efficient visible light-driven degradation of pharmaceutical and phenolic micropollutants. Appl. Catal. B Environ. **250**, 150–162 (2019). https://doi.org/10.1016/j.apcatb.2019.03.024
21. J. Zhang, H. Yang, S. Xu, L. Yang, Y. Song, L. Jiang, Y. Dan, Dramatic enhancement of visible light photocatalysis due to strong interaction between TiO_2 and end-group functionalized P3HT. Appl. Catal. B Environ. **174–175**, 193–202 (2015). https://doi.org/10.1016/j.apcatb.2015.02.034
22. X.J. Bai, H.Y. Li, Z.Y. Zhang, X.R. Zhang, C. Wang, J. Xu, Y.F. Zhu, Carbon nitride nested tubes with graphene as a dual electron mediator in Z-scheme photocatalytic deoxynivalenol degradation, Catal. Sci. Technol. **9**, 1680–1690 (2019). https://doi.org/10.1039/c9cy00209j
23. Z. Wu, Y. Wang, Z. Xiong, Z. Ao, S. Pu, G. Yao, B. Lai, Core-shell magnetic Fe_3O_4@Zn/Co-ZIFs to activate peroxymonosulfate for highly efficient degradation of carbamazepine. Appl. Catal. B Environ. **277**, 119136 (2020). https://doi.org/10.1016/j.apcatb.2020.119136
24. C.F. Zhang, L.G. Qiu, F. Ke, Y.J. Zhu, Y.P. Yuan, G.-S. Xu, X. Jiang, A novel magnetic recyclable photocatalyst based on a core–shell metal–organic framework Fe_3O_4@MIL-100(Fe) for the decolorization of methylene blue dye. J. Mater. Chem. A **1**, 14329 (2013). https://doi.org/10.1039/c3ta13030d
25. G. Capilli, M. Costamagna, F. Sordello, C. Minero, Synthesis, characterization and photocatalytic performance of p-type carbon nitride. Appl. Catal. B Environ. **242**, 121–131 (2019). https://doi.org/10.1016/j.apcatb.2018.09.057
26. H. Lv, X. Zhao, H. Niu, S. He, Z. Tang, F. Wu, J.P. Giesy, Ball milling synthesis of covalent organic framework as a highly active photocatalyst for degradation of organic contaminants. J. Hazard. Mater. **369**, 494–502 (2019). https://doi.org/10.1016/j.jhazmat.2019.02.046
27. Q. Liao, D. Wang, C. Ke, Y. Zhang, Q. Han, Y. Zhang, K. Xi, Metal-free Fenton-like photocatalysts based on covalent organic frameworks. Appl. Catal. B Environ. **298**, 120548 (2021). https://doi.org/10.1016/j.apcatb.2021.120548
28. S.W. Lv, J.M. Liu, F.E. Yang, C.Y. Li, S. Wang, A novel photocatalytic platform based on the newly-constructed ternary composites with a double p-n heterojunction for contaminants degradation and bacteria inactivation. Chem. Eng. J. **409**, 128269 (2021). https://doi.org/10.1016/j.cej.2020.128269
29. C. Li, G. Chen, J. Sun, J. Rao, Z. Han, Y. Hu, W. Xing, C. Zhang, Doping effect of phosphate in Bi_2WO_6 and universal improved photocatalytic activity for removing various pollutants in water. Appl. Catal. B Environ. **188**, 39–47 (2016). https://doi.org/10.1016/j.apcatb.2016.01.054
30. P. Chen, S. Di, X. Qiu, S. Zhu, One-step synthesis of F-TiO_2/g-C_3N_4 heterojunction as highly efficient visible-light-active catalysts for tetrabromobisphenol A and sulfamethazine degradation. Appl. Surf. Sci. **587**, 152889 (2022). https://doi.org/10.1016/j.apsusc.2022.152889
31. J. Qu, D. Chen, N. Li, Q. Xu, H. Li, J. He, J. Lu, Ternary photocatalyst of atomic-scale Pt coupled with MoS_2 co-loaded on TiO_2 surface for highly efficient degradation of gaseous toluene. Appl. Catal. B Environ. **256**, 117877 (2019). https://doi.org/10.1016/j.apcatb.2019.117877
32. X. Jin, Q. Guan, T. Tian, H. Li, Y. Han, F. Hao, Y. Cui, W. Li, Y. Zhu, Y. Zhang, In_2O_3/boron doped g-C_3N_4 heterojunction catalysts with remarkably enhanced visible-light photocatalytic efficiencies. Appl. Surf. Sci. **504**, 144241 (2020). https://doi.org/10.1016/j.apsusc.2019.144241

33. M. Zhang, J. Xu, R. Zong, Y. Zhu, Enhancement of visible light photocatalytic activities via porous structure of g-C_3N_4. Appl. Catal. B Environ. **147**, 229–235 (2014). https://doi.org/10.1016/j.apcatb.2013.09.002
34. B.B. Wang, P. Li, C.L. Du, Y. Wang, D.X. Gao, S.T. Li, L.Y. Zhang, F.Y. Wen, Synergetic effect of dual co-catalysts on the activity of $BiVO_4$ for photocatalytic carbamazepine degradation. RSC Adv. **9**, 41977–41983 (2019). https://doi.org/10.1039/c9ra07152k
35. S. Ida, A. Takashiba, S. Koga, H. Hagiwara, T. Ishihara, Potential gradient and photocatalytic activity of an ultrathin p-n junction surface prepared with two-dimensional semiconducting nanocrystals. J. Am. Chem. Soc. **136**, 1872–1878 (2014). https://doi.org/10.1021/ja409465k
36. D. Huang, Z. Li, G. Zeng, C. Zhou, W. Xue, X. Gong, X. Yan, S. Chen, W. Wang, M. Cheng, Megamerger in photocatalytic field: 2D g-C_3N_4 nanosheets serve as support of 0D nanomaterials for improving photocatalytic performance. Appl. Catal. B Environ. **240**, 153–173 (2019). https://doi.org/10.1016/j.apcatb.2018.08.071
37. R. Hao, G. Wang, C. Jiang, H. Tang, Q. Xu, In situ hydrothermal synthesis of g-C_3N_4/TiO_2 heterojunction photocatalysts with high specific surface area for Rhodamine B degradation. Appl. Surf. Sci. **411**, 400–410 (2017). https://doi.org/10.1016/j.apsusc.2017.03.197
38. J. Jin, J. Yu, D. Guo, C. Cui, W. Ho, A hierarchical Z-Scheme CdS-WO_3 photocatalyst with enhanced CO_2 reduction activity. Small **11**, 5262–5271 (2015). https://doi.org/10.1002/smll.201500926
39. J. Yu, J. Low, W. Xiao, P. Zhou, M. Jaroniec, Enhanced photocatalytic CO_2 reduction activity of anatase TiO_2 by coexposed 001 and 101 facets. J. Am. Chem. Soc. **136**, 8839–8842 (2014). https://doi.org/10.1021/ja5044787
40. X.B. Li, J. Xiong, X.M. Gao, J.T. Huang, Z.J. Feng, Z. Chen, Y.F. Zhu, Recent advances in 3D g-C_3N_4 composite photocatalysts for photocatalytic water splitting, degradation of pollutants and CO_2 reduction. J. Alloy. Compd. **802**, 196–209 (2019). https://doi.org/10.1016/j.jallcom.2019.06.185
41. W. Liu, M. Wang, C. Xu, S. Chen, Facile synthesis of g-C_3N_4/ZnO composite with enhanced visible light photooxidation and photoreduction properties. Chem. Eng. J. **209**, 386–393 (2012). https://doi.org/10.1016/j.cej.2012.08.033
42. S. Zhang, J. Yi, J. Chen, Z. Yin, T. Tang, W. Wei, S. Cao, H. Xu, Spatially confined Fe_2O_3 in hierarchical SiO_2@TiO_2 hollow sphere exhibiting superior photocatalytic efficiency for degrading antibiotics. Chem. Eng. J. **380**, 122583 (2020). https://doi.org/10.1016/j.cej.2019.122583
43. S.R. Pendlebury, X. Wang, F. Le Formal, M. Cornuz, A. Kafizas, S.D. Tilley, M. Gratzel, J.R. Durrant, Ultrafast charge carrier recombination and trapping in hematite photoanodes under applied bias. J. Am. Chem. Soc. **136**, 9854–9857 (2014). https://doi.org/10.1021/ja504473e
44. C. Zachaus, F.F. Abdi, L.M. Peter, R. van de Krol, Photocurrent of $BiVO_4$ is limited by surface recombination, not surface catalysis. Chem. Sci. **8**, 3712–3719 (2017). https://doi.org/10.1039/c7sc00363c
45. C.Y. Wang, Y.J. Zhang, W.K. Wang, D.N. Pei, G.X. Huang, J.J. Chen, X. Zhang, H.Q. Yu, Enhanced photocatalytic degradation of bisphenol A by Co-doped BiOCl nanosheets under visible light irradiation. Appl. Catal. B Environ. **221**, 320–328 (2018). https://doi.org/10.1016/j.apcatb.2017.09.036
46. W. Wang, Q. Zhu, F. Qin, Q. Dai, X. Wang, Fe doped CeO_2 nanosheets as Fenton-like heterogeneous catalysts for degradation of salicylic acid. Chem. Eng. J. **333**, 226–239 (2018). https://doi.org/10.1016/j.cej.2017.08.065
47. L. Xu, L. Li, L. Yu, J.C. Yu, Efficient generation of singlet oxygen on modified g-C_3N_4 photocatalyst for preferential oxidation of targeted organic pollutants. Chem. Eng. J. **431**, 134241 (2022). https://doi.org/10.1016/j.cej.2021.134241
48. S. Cao, Y. Zhang, N. He, J. Wang, H. Chen, F. Jiang, Metal-free 2D/2D heterojunction of covalent triazine-based frameworks/graphitic carbon nitride with enhanced interfacial charge separation for highly efficient photocatalytic elimination of antibiotic pollutants. J. Hazard. Mater. **391**, 122204 (2020). https://doi.org/10.1016/j.jhazmat.2020.122204

49. Y. Li, Z. Li, Y. Xia, H. Li, J. Shi, A. Zhang, H. Huo, S. Tan, L. Gao, Fabrication of ternary $AgBr/BiPO_4/g-C_3N_4$ heterostructure with dual Z-scheme and its visible light photocatalytic activity for Reactive Blue 19. Environ. Res. **192**, 110260 (2021). https://doi.org/10.1016/j.envres.2020.110260
50. T. Wang, W. Quan, D. Jiang, L. Chen, D. Li, S. Meng, M. Chen, Synthesis of redox-mediator-free direct Z-scheme AgI/WO_3 nanocomposite photocatalysts for the degradation of tetracycline with enhanced photocatalytic activity. Chem. Eng. J. **300**, 280–290 (2016). https://doi.org/10.1016/j.cej.2016.04.128
51. C. Shi, H. Qi, Z. Sun, K. Qu, Z. Huang, J. Li, M. Dong, Z. Guo, Carbon dot-sensitized urchin-like Ti^{3+} self-doped TiO_2 photocatalysts with enhanced photoredox ability for highly efficient removal of Cr^{6+} and RhB. J. Mater. Chem. C **8**, 2238–2247 (2020). https://doi.org/10.1039/c9tc05513d
52. F. Wang, P. Chen, Y. Feng, Z. Xie, Y. Liu, Y. Su, Q. Zhang, Y. Wang, K. Yao, W. Lv, G. Liu, Facile synthesis of N-doped carbon dots/$g-C_3N_4$ photocatalyst with enhanced visible-light photocatalytic activity for the degradation of indomethacin. Appl. Catal. B Environ. **207**, 103–113 (2017). https://doi.org/10.1016/j.apcatb.2017.02.024
53. L. Gu, J. Wang, Z. Zou, X. Han, Graphitic-C3N4-hybridized TiO2 nanosheets with reactive 001 facets to enhance the UV- and visible-light photocatalytic activity. J. Hazard. Mater. **268**, 216–223 (2014). https://doi.org/10.1016/j.jhazmat.2014.01.021
54. J. Xu, W. Meng, Y. Zhang, L. Li, C. Guo, Photocatalytic degradation of tetrabromobisphenol A by mesoporous BiOBr: efficacy, products and pathway. Appl. Catal. B Environ. **107**, 355–362 (2011). https://doi.org/10.1016/j.apcatb.2011.07.036
55. C. Zhou, Z. Zeng, G. Zeng, D. Huang, R. Xiao, M. Cheng, C. Zhang, W. Xiong, C. Lai, Y. Yang, W. Wang, H. Yi, B. Li, Visible-light-driven photocatalytic degradation of sulfamethazine by surface engineering of carbon nitride Properties, degradation pathway and mechanisms. J. Hazard. Mater. **380**, 120815 (2019). https://doi.org/10.1016/j.jhazmat.2019.120815
56. X. Yuan, S. Qu, X. Huang, X. Xue, C. Yuan, S. Wang, L. Wei, P. Cai, Design of core-shelled $g-C_3N_4$@ZIF-8 photocatalyst with enhanced tetracycline adsorption for boosting photocatalytic degradation. Chem. Eng. J. **416**, 129148 (2021). https://doi.org/10.1016/j.cej.2021.129148
57. X.H. Yi, S.Q. Ma, X.D. Du, C. Zhao, H. Fu, P. Wang, C.C. Wang, The facile fabrication of 2D/3D Z-scheme $g-C_3N_4$/UiO-66 heterojunction with enhanced photocatalytic Cr(VI) reduction performance under white light. Chem. Eng. J. **375**, 121944 (2019). https://doi.org/10.1016/j.cej.2019.121944
58. D. Guo, R.Y. Wen, M.M. Liu, H.X. Guo, J.H. Chen, W. Weng, Facile fabrication of $g-C_3N_4$/MIL-53(Al) composite with enhanced photocatalytic activities under visible-light irradiation. Appl. Organomet. Chem. **29**, 690–697 (2015). https://doi.org/10.1002/aoc.3352
59. D.A. Giannakoudakis, N.A. Travlou, J. Secor, T.J. Bandosz, Oxidized $g-C_3N_4$ nanospheres as catalytically photoactive linkers in MOF/$g-C_3N_4$ composite of hierarchical pore structure. Small **13** (2017). https://doi.org/10.1002/smll.201601758
60. V. Muelas-Ramos, M.J. Sampaio, C.G. Silva, J. Bedia, J.J. Rodriguez, J.L. Faria, C. Belver, Degradation of diclofenac in water under LED irradiation using combined $g-C_3N_4/NH_2$-MIL-125 photocatalysts. J. Hazard. Mater. **416**, 126199 (2021). https://doi.org/10.1016/j.jhazmat.2021.126199
61. Y. Chen, B. Zhai, Y. Liang, Y. Li, J. Li, Preparation of CdS/ $g-C_3N_4$/ MOF composite with enhanced visible-light photocatalytic activity for dye degradation. J. Solid State Chem. **274**, 32–39 (2019). https://doi.org/10.1016/j.jssc.2019.01.038
62. D. Zhang, S. Liang, S. Yao, H. Li, J. Liu, Y. Geng, X. Pu, Highly efficient visible/NIR photocatalytic activity and mechanism of Yb^{3+}/Er^{3+} co-doped $Bi_4O_5I_2$ up-conversion photocatalyst. Sep. Purif. Technol. **248**, 117040 (2020). https://doi.org/10.1016/j.seppur.2020.117040
63. I.N. Reddy, C.V. Reddy, J. Shim, B. Akkinepally, M. Cho, K. Yoo, D. Kim, Excellent visible-light driven photocatalyst of (Al, Ni) co-doped ZnO structures for organic dye degradation. Catal. Today **340**, 277–285 (2020). https://doi.org/10.1016/j.cattod.2018.07.030
64. Z. Zhu, C. Ma, K. Yu, Z. Lu, Z. Liu, P. Huo, X. Tang, Y. Yan, Synthesis Ce-doped biomass carbon-based $g-C_3N_4$ via plant growing guide and temperature-programmed technique for

degrading 2-Mercaptobenzothiazole. Appl. Catal. B Environ. **268**, 118432 (2020). https://doi.org/10.1016/j.apcatb.2019.118432

65. H. Huang, Y. Song, N. Li, D. Chen, Q. Xu, H. Li, J. He, J. Lu, One-step in-situ preparation of N-doped TiO_2@C derived from Ti_3C_2 MXene for enhanced visible-light driven photodegradation. Appl. Catal. B Environ. **251**, 154–161 (2019). https://doi.org/10.1016/j.apcatb.2019.03.066
66. L. Jiang, X. Yuan, G. Zeng, J. Liang, Z. Wu, H. Yu, D. Mo, H. Wang, Z. Xiao, C. Zhou, Nitrogen self-doped g-C_3N_4 nanosheets with tunable band structures for enhanced photocatalytic tetracycline degradation. J. Colloid Interface Sci. **536**, 17–29 (2019). https://doi.org/10.1016/j.jcis.2018.10.033
67. C. Hu, W.-Z. Hung, M.-S. Wang, P.-J. Lu, Phosphorus and sulfur codoped g-C_3N_4 as an efficient metal-free photocatalyst. Carbon **127**, 374–383 (2018). https://doi.org/10.1016/j.carbon.2017.11.019
68. L. Jiang, X. Yuan, G. Zeng, X. Chen, Z. Wu, J. Liang, J. Zhang, H. Wang, H. Wang, Phosphorus- and sulfur-codoped g-C_3N_4: Facile preparation, mechanism insight, and application as efficient photocatalyst for tetracycline and methyl orange degradation under visible light irradiation. ACS Sustain. Chem. Eng. **5**, 5831–5841 (2017). https://doi.org/10.1021/acssuschemeng.7b00559
69. F. Guo, M. Li, H. Ren, X. Huang, K. Shu, W. Shi, C. Lu, Facile bottom-up preparation of Cl-doped porous g-C_3N_4 nanosheets for enhanced photocatalytic degradation of tetracycline under visible light. Sep. Purif. Technol. **228**, 115770 (2019). https://doi.org/10.1016/j.seppur.2019.115770
70. J. Huang, D. Li, R. Li, P. Chen, Q. Zhang, H. Liu, W. Lv, G. Liu, Y. Feng, One-step synthesis of phosphorus/oxygen co-doped g-C_3N_4/anatase TiO_2 Z-scheme photocatalyst for significantly enhanced visible-light photocatalysis degradation of enrofloxacin. J. Hazard. Mater. **386**, 121634 (2020). https://doi.org/10.1016/j.jhazmat.2019.121634
71. G.S. Jamila, S. Sajjad, S.A.K. Leghari, M. Long, Nitrogen doped carbon quantum dots and GO modified WO_3 nanosheets combination as an effective visible photo catalyst. J. Hazard. Mater. **382**, 121087 (2020). https://doi.org/10.1016/j.jhazmat.2019.121087
72. D. Li, J. Huang, R. Li, P. Chen, D. Chen, M. Cai, H. Liu, Y. Feng, W. Lv, G. Liu, Synthesis of a carbon dots modified g-C_3N_4/SnO_2 Z-scheme photocatalyst with superior photocatalytic activity for PPCPs degradation under visible light irradiation. J. Hazard. Mater. **401**, 123257 (2021). https://doi.org/10.1016/j.jhazmat.2020.123257
73. Z. Xie, Y. Feng, F. Wang, D. Chen, Q. Zhang, Y. Zeng, W. Lv, G. Liu, Construction of carbon dots modified MoO_3/g-C_3N_4 Z-scheme photocatalyst with enhanced visible-light photocatalytic activity for the degradation of tetracycline. Appl. Catal. B Environ. **229**, 96–104 (2018). https://doi.org/10.1016/j.apcatb.2018.02.011
74. N. Cao, Y. Quan, A. Guan, C. Yang, Y. Ji, L. Zhang, G. Zheng, Oxygen vacancies enhanced cooperative electrocatalytic reduction of carbon dioxide and nitrite ions to urea, J. Colloid. Interface. Sci. **577**, 109–114 (2020). https://doi.org/10.1016/j.jcis.2020.05.014
75. L. Ge, J. Liu, Efficient visible light-induced photocatalytic degradation of methyl orange by QDs sensitized CdS-Bi_2WO_6. Appl. Catal. B Environ. **105**, 289–297 (2011). https://doi.org/10.1016/j.apcatb.2011.04.016
76. A. Keshtkar Vanashi, H. Ghasemzadeh, Photocatalytic production of hydroxyl radical by PbSe quantum dot nanocomposite hydrogel. Appl. Surf. Sci. **564**, 150467 (2021). https://doi.org/10.1016/j.apsusc.2021.150467
77. Q. Zhang, F. Yang, S. Zhou, N. Bao, Z. Xu, M. Chaker, D. Ma, Broadband photocatalysts enabled by 0D/2D heterojunctions of near-infrared quantum dots/graphitic carbon nitride nanosheets. Appl. Catal. B Environ. **270**, 118879 (2020). https://doi.org/10.1016/j.apcatb.2020.118879
78. P. Zhou, J. Yu, M. Jaroniec, All-solid-state Z-scheme photocatalytic systems. Adv. Mater. **26**, 4920–4935 (2014). https://doi.org/10.1002/adma.201400288
79. L. Jiang, X. Yuan, G. Zeng, J. Liang, X. Chen, H. Yu, H. Wang, Z. Wu, J. Zhang, T. Xiong, In-situ synthesis of direct solid-state dual Z-scheme WO_3/g-C_3N_4/Bi_2O_3 photocatalyst for

the degradation of refractory pollutant. Appl. Catal. B Environ. **227**, 376–385 (2018). https://doi.org/10.1016/j.apcatb.2018.01.042
80. V. Hasija, A. Sudhaik, P. Raizada, A. Hosseini-Bandegharaei, P. Singh, Carbon quantum dots supported AgI /ZnO/phosphorus doped graphitic carbon nitride as Z-scheme photocatalyst for efficient photodegradation of 2, 4-dinitrophenol. J. Environ. Chem. Eng. **7**, 103272 (2019). https://doi.org/10.1016/j.jece.2019.103272
81. Y. Wang, L. Rao, P. Wang, Z. Shi, L. Zhang, Photocatalytic activity of N-TiO_2/O-doped N vacancy g-C_3N_4 and the intermediates toxicity evaluation under tetracycline hydrochloride and Cr(VI) coexistence environment. Appl. Catal. B Environ. **262**, 118308 (2020). https://doi.org/10.1016/j.apcatb.2019.118308
82. M. Tang, Y. Ao, C. Wang, P. Wang, Facile synthesis of dual Z-scheme g-C_3N_4/Ag_3PO_4/AgI composite photocatalysts with enhanced performance for the degradation of a typical neonicotinoid pesticide. Appl. Catal. B Environ. **268**, 118395 (2020). https://doi.org/10.1016/j.apcatb.2019.118395
83. A. Shahzad, K. Rasool, M. Nawaz, W. Miran, J. Jang, M. Moztahida, K.A. Mahmoud, D.S. Lee, Heterostructural TiO_2/$Ti_3C_2T_x$ (MXene) for photocatalytic degradation of antiepileptic drug carbamazepine. Chem. Eng. J. **349**, 748–755 (2018). https://doi.org/10.1016/j.cej.2018.05.148
84. N. Tian, H. Huang, S. Wang, T. Zhang, X. Du, Y. Zhang, Facet-charge-induced coupling dependent interfacial photocharge separation: a case of BiOI/g-C_3N_4 p-n junction. Appl. Catal. B Environ. **267**, 118697 (2020). https://doi.org/10.1016/j.apcatb.2020.118697
85. S. Chen, D. Huang, G. Zeng, W. Xue, L. Lei, P. Xu, R. Deng, J. Li, M. Cheng, In-situ synthesis of facet-dependent $BiVO_4$/Ag_3PO_4/PANI photocatalyst with enhanced visible-light-induced photocatalytic degradation performance: synergism of interfacial coupling and hole-transfer. Chem. Eng. J. **382**, 122840 (2020). https://doi.org/10.1016/j.cej.2019.122840
86. J.L. Wang, C. Wang, W. Lin, Metal–organic frameworks for light harvesting and photocatalysis. ACS Catal. **2**, 2630–2640 (2012). https://doi.org/10.1021/cs3005874
87. Q. Wang, Q. Gao, A.M. Al-Enizi, A. Nafady, S. Ma, Recent advances in MOF-based photocatalysis: environmental remediation under visible light. Inorg. Chem. Front. **7**, 300–339 (2020). https://doi.org/10.1039/c9qi01120j
88. W. Zhao, B. Wang, G. Yu, Antibiotic resistance genes in China: occurrence, risk, and correlation among different parameters, Environ. Sci. Pollut. Res. Int. **25**, 21467–21482 (2018). https://doi.org/10.1007/s11356-018-2507-z
89. T.D. Bennett, A.K. Cheetham, A.H. Fuchs, F.X. Coudert, Interplay between defects, disorder and flexibility in metal-organic frameworks. Nat. Chem. **9**, 11–16 (2017). https://doi.org/10.1038/nchem.2691
90. L. Zeng, X. Guo, C. He, C. Duan, Metal–organic frameworks: Versatile materials for heterogeneous photocatalysis. ACS Catal. **6**, 7935–7947 (2016). https://doi.org/10.1021/acscatal.6b02228
91. N. Stock, S. Biswas, Synthesis of metal-organic frameworks (MOFs): routes to various MOF topologies, morphologies, and composites. Chem. Rev. **112**, 933–969 (2012). https://doi.org/10.1021/cr200304e
92. A. Rabenau, The role of hydrothermal synthesis in preparative chemistry. Angew. Chem. Int. Ed. **24**, 1026–1040 (1985). https://doi.org/10.1002/anie.198510261
93. K. Byrappa, M. Yoshimura, Hydrothermal technology—principles and applications, pp. 1–49 (2013). https://doi.org/10.1016/b978-0-12-375090-7.00001-3
94. Y.H. Huang, W.S. Lo, Y.W. Kuo, W.J. Chen, C.H. Lin, F.K. Shieh, Green and rapid synthesis of zirconium metal-organic frameworks via mechanochemistry: UiO-66 analog nanocrystals obtained in one hundred seconds. Chem. Commun. **53**, 5818–5821 (2017). https://doi.org/10.1039/c7cc03105j
95. J.F. Kurisingal, R. Babu, S.-H. Kim, Y.X. Li, J.-S. Chang, S.J. Cho, D.-W. Park, Microwave-induced synthesis of a bimetallic charge-transfer metal organic framework: a promising host for the chemical fixation of CO_2, Catal. Sci. Technol. **8**, 591–600 (2018). https://doi.org/10.1039/c7cy02063e

96. P. Balaz, M. Achimovicova, M. Balaz, P. Billik, Z. Cherkezova-Zheleva, J.M. Criado, F. Delogu, E. Dutkova, E. Gaffet, F.J. Gotor, R. Kumar, I. Mitov, T. Rojac, M. Senna, A. Streletskii, K. Wieczorek-Ciurowa, Hallmarks of mechanochemistry: from nanoparticles to technology. Chem. Soc. Rev. **42**, 7571–7637 (2013). https://doi.org/10.1039/c3cs35468g
97. S. Glowniak, B. Szczesniak, J. Choma, M. Jaroniec, Mechanochemistry: toward green synthesis of metal-organic frameworks. Mater. Today **46**, 109–124 (2021). https://doi.org/10.1016/j.mattod.2021.01.008
98. C.C. Wang, J.R. Li, X.L. Lv, Y.Q. Zhang, G.S. Guo, Photocatalytic organic pollutants degradation in metal-organic frameworks. Energ Environ. Sci. **7**, 2831–2867 (2014). https://doi.org/10.1039/c4ee01299b
99. T. Zhang, W. Lin, Metal-organic frameworks for artificial photosynthesis and photocatalysis. Chem. Soc. Rev. **43**, 5982–5993 (2014). https://doi.org/10.1039/c4cs00103f
100. M.M. Zhou, Y.N. Wu, J.L. Qiao, J. Zhang, A. McDonald, G.T. Li, F.T. Li, The removal of bisphenol A from aqueous solutions by MIL-53(Al) and mesostructured MIL-53(Al). J. Colloid Interface Sci. **405**, 157–163 (2013). https://doi.org/10.1016/j.jcis.2013.05.024
101. X. Li, W. Guo, Z. Liu, R. Wang, H. Liu, Quinone-modified NH_2-MIL-101(Fe) composite as a redox mediator for improved degradation of bisphenol A. J. Hazard. Mater. **324**, 665–672 (2017). https://doi.org/10.1016/j.jhazmat.2016.11.040
102. Y. He, X. Zhang, Y. Wei, X. Chen, Z. Wang, R. Yu, Ti-MOF derived N-doped TiO_2 nanostructure as visible-light-driven photocatalyst. Chem. Res. Chin. U. **36**, 447–452 (2020). https://doi.org/10.1007/s40242-020-0106-2
103. T. Araya, M. Jia, J. Yang, P. Zhao, K. Cai, W. Ma, Y. Huang, Resin modified MIL-53 (Fe) MOF for improvement of photocatalytic performance. Appl. Catal. B Environ. **203**, 768–777 (2017). https://doi.org/10.1016/j.apcatb.2016.10.072
104. H. Ramezanalizadeh, F. Manteghi, Immobilization of mixed cobalt/nickel metal-organic framework on a magnetic BiFeO3: a highly efficient separable photocatalyst for degradation of water pollutions. J. Photochem. Photobio. A **346**, 89–104 (2017). https://doi.org/10.1016/j.jphotochem.2017.05.041
105. N. Liu, W. Huang, X. Zhang, L. Tang, L. Wang, Y. Wang, M. Wu, Ultrathin graphene oxide encapsulated in uniform MIL-88A(Fe) for enhanced visible light-driven photodegradation of RhB. Appl. Catal. B Environ. **221**, 119–128 (2018). https://doi.org/10.1016/j.apcatb.2017.09.020
106. S.N. Frank, A.J. Bard, Semiconductor electrodes. II. Electrochemistry at n-type titanium dioxide electrodes in acetonitrile solutions. J. Am. Chem. Soc. **97**, 7427–7433 (1975)
107. K. Nakata, T. Ochiai, T. Murakami, A. Fujishima, Photoenergy conversion with TiO_2 photocatalysis: new materials and recent applications. Electrochim. Acta **84**, 103–111 (2012). https://doi.org/10.1016/j.electacta.2012.03.035
108. E.I. Kapinus, T.A. Khalyavka, V.V. Shimanovskaya, T.I. Viktorova, V.V. Strelko, Photocatalytic activity of spectro-pure titanium dioxide: effects of crystalline structure, specific surface area and sorption properties. Int. J. Photoenergy **5**, 159–166 (2003). https://doi.org/10.1155/s1110662x0300028x
109. P.V.L. Reddy, K.-H. Kim, H. Song, Emerging green chemical technologies for the conversion of CH4 to value added products. Renew. Sustain. Energy Rev. **24**, 578–585 (2013). https://doi.org/10.1016/j.rser.2013.03.035
110. J.H. Carey, J. Lawrence, H.M. Tosine, Photodechlorination of PCB's in the presence of titanium dioxide in aqueous suspensions, Bull. Environ. Contam. Toxicol. **16**, 697–701 (1976). https://doi.org/10.1007/bf01685575
111. H. Choi, E. Stathatos, D.D. Dionysiou, Photocatalytic TiO_2 films and membranes for the development of efficient wastewater treatment and reuse systems. Desalination **202**, 199–206 (2007). https://doi.org/10.1016/j.desal.2005.12.055
112. I. Iruml, W.W. Dunn, K.O. Wllbourn, F.-R.F. Fan, A.J. Bard, Heterogeneous photocatalytic oxidation of hydrocarbons on platinized titanium dioxide powders. J. Phys. Chem. **84**, 3207–3210 (1980). https://doi.org/10.1021/j100461a015

113. J. Schneider, M. Matsuoka, M. Takeuchi, J.L. Zhang, Y. Horiuchi, M. Anpo, D.W. Bahnemann, Understanding TiO_2 photocatalysis: mechanisms and materials. Chem. Rev. **114**, 9919–9986 (2014). https://doi.org/10.1021/cr5001892
114. S. Peiris, H.B. Silva, K.N. Ranasinghe, S.V. Bandara, I.R. Perera, Recent development and future prospects of TiO_2 photocatalysis. J. Chin. Chem. Soc. **68**, 738–769 (2021). https://doi.org/10.1002/jccs.202000465
115. T. Ohno, K. Sarukawa, K. Tokieda, M. Matsumura, Morphology of a TiO_2 photocatalyst (Degussa, P25) consisting of anatase and rutile crystalline phases. J. Catal. **203**, 82–86 (2001). https://doi.org/10.1006/jcat.2001.3316
116. V. Chandrasekhar, C. Mohapatra, R. Banerjee, A. Mallick, Synthesis, structure, and H_2/CO_2 adsorption in a three-dimensional 4-connected triorganotin coordination polymer with a sqc topology. Inorg. Chem. **52**, 3579–3581 (2013). https://doi.org/10.1021/ic302528b
117. S. Yin, Q. Zhang, F. Saito, T. Sato, Preparation of visible light-activated titania photocatalyst by mechanochemical method. Chem. Lett. **32**, 358–359 (2003). https://doi.org/10.1246/cl.2003.358
118. B. Gao, P.S. Yap, T.M. Lim, T.-T. Lim, Adsorption-photocatalytic degradation of Acid Red 88 by supported TiO_2: effect of activated carbon support and aqueous anions. Chem. Eng. J. **171**, 1098–1107 (2011). https://doi.org/10.1016/j.cej.2011.05.006
119. M. Khairy, W. Zakaria, Effect of metal-doping of TiO_2 nanoparticles on their photocatalytic activities toward removal of organic dyes. Egy. J. Petrol. **23**, 419–426 (2014). https://doi.org/10.1016/j.ejpe.2014.09.010
120. M. Zhou, M. Li, C. Hou, Z. Li, Y. Wang, K. Xiang, X. Guo, Pt nanocrystallines/TiO_2 with thickness-controlled carbon layers: preparation and activities in CO oxidation. Chinese Chem. Lett. **29**, 787–790 (2018). https://doi.org/10.1016/j.cclet.2018.03.010
121. S. Dubbaka, Branched zinc oxide nanostructures: synthesis and photo catalysis study for application in dye sensitized solar cells. University of Arkansas (2008). https://doi.org/10.1201/b11518-10
122. A. Janotti, C.G. Van de Walle, Fundamentals of zinc oxide as a semiconductor. Rep. Prog. Phys. **72** (2009). https://doi.org/10.1088/0034-4885/72/12/126501
123. Z.L. Wang, Nanostructures of zinc oxide. Mater. Today **7**, 26–33 (2004). https://doi.org/10.1016/s1369-7021(04)00286-x
124. L. Schmidt-Mende, J.L. MacManus-Driscoll, ZnO—nanostructures, defects, and devices. Mater. Today **10**, 40–48 (2007). https://doi.org/10.1016/s1369-7021(07)70078-0
125. C.B. Ong, L.Y. Ng, A.W. Mohammad, A review of ZnO nanoparticles as solar photocatalysts: synthesis, mechanisms and applications. Renew. Sust. Energy Rev. **81**, 536–551 (2018). https://doi.org/10.1016/j.rser.2017.08.020
126. N. Daneshvar, D. Salari, A.R. Khataee, Photocatalytic degradation of azo dye acid red 14 in water on ZnO as an alternative catalyst to TiO_2. J. Photochem. Photobio. A **162**, 317–322 (2004). https://doi.org/10.1016/s1010-6030(03)00378-2
127. R.K. Dutta, B.P. Nenavathu, S. Talukdar, Anomalous antibacterial activity and dye degradation by selenium doped ZnO nanoparticles. Colloid. Surfaces. B **114**, 218–224 (2014). https://doi.org/10.1016/j.colsurfb.2013.10.007
128. X.C. Wang, K. Maeda, A. Thomas, K. Takanabe, G. Xin, J.M. Carlsson, K. Domen, M. Antonietti, A metal-free polymeric photocatalyst for hydrogen production from water under visible light. Nat. Mater. **8**, 76–80 (2009). https://doi.org/10.1038/nmat2317
129. Y. Xu, S.P. Gao, Band gap of C_3N_4 in the GW approximation. Int. J. Hydrogen Energ. **37**, 11072–11080 (2012). https://doi.org/10.1016/j.ijhydene.2012.04.138
130. F.M. Pinto, F.A. La Porta, Current perspective on synthesis, properties, and application of graphitic carbon nitride related-compounds. Emerging research in science and engineering based on advanced experimental and computational strategies, pp. 413–432 (2020)
131. X.L. Liu, R. Ma, L. Zhuang, B.W. Hu, J.R. Chen, X.Y. Liu, X.K. Wang, Recent developments of doped g-C_3N_4 photocatalysts for the degradation of organic pollutants. Crit. Rev. Environ. Sci. Technol. **51**, 751–790 (2021). https://doi.org/10.1080/10643389.2020.1734433

132. T. Kar, S. Godavarthi, S.K. Pasha, K. Deshmukh, L. Martinez-Gomez, M.K. Kesarla, Layered materials and their heterojunctions for supercapacitor applications: a review. Crit. Rev. Solid State Mater. Sci. https://doi.org/10.1080/10408436.2021.1886048
133. Z.H. Sheng, L. Shao, J.J. Chen, W.J. Bao, F.B. Wang, X.H. Xia, Catalyst-free synthesis of nitrogen-doped graphene via thermal annealing graphite oxide with melamine and its excellent electrocatalysis. ACS Nano **5**, 4350–4358 (2011). https://doi.org/10.1021/nn103584t
134. S. Kamal, S. Balu, S. Palanisamy, K. Uma, V. Velusamy, T.C.K. Yang, Synthesis of boron doped $C_3N_4/NiFe_2O_4$ nanocomposite: an enhanced visible light photocatalyst for the degradation of methylene blue. Results Phys. **12**, 1238–1244 (2019). https://doi.org/10.1016/j.rinp.2019.01.004
135. G. Liu, P. Niu, C.H. Sun, S.C. Smith, Z.G. Chen, G.Q. Lu, H.M. Cheng, Unique electronic structure induced high photoreactivity of sulfur-doped graphitic C_3N_4. J. Am. Chem. Soc. **132**, 11642–11648 (2010). https://doi.org/10.1021/ja103798k
136. H. Yang, Y. Zhou, Y. Wang, S. Hu, B. Wang, Q. Liao, H. Li, J. Bao, G. Ge, S. Jia, Three-dimensional flower-like phosphorus-doped g-C_3N_4 with a high surface area for visible-light photocatalytic hydrogen evolution. J. Mater. Chem. A. **6**, 16485–16494 (2018). https://doi.org/10.1039/c8ta05723k
137. M.M. Zhang, C. Lai, B.S. Li, D.L. Huang, G.M. Zeng, P. Xu, L. Qin, S.Y. Liu, X.G. Liu, H. Yi, M.F. Li, C.C. Chu, Z. Chen, Rational design 2D/2D BiOBr/CDs/g-C_3N_4 Z-scheme heterojunction photocatalyst with carbon dots as solid-state electron mediators for enhanced visible and NIR photocatalytic activity: kinetics, intermediates, and mechanism insight. J. Catal. **369**, 469–481 (2019). https://doi.org/10.1016/j.jcat.2018.11.029
138. J.S. Zhang, M.W. Zhang, R.Q. Sun, X.C. Wang, A facile band alignment of polymeric carbon nitride semiconductors to construct isotype heterojunctions. Angew. Chem. Int. Edit. **51**, 10145–10149 (2012). https://doi.org/10.1002/anie.201205333
139. D.L. Huang, X.L. Yan, M. Yan, G.M. Zeng, C.Y. Zhou, J. Wan, M. Cheng, W.J. Xue, Graphitic carbon nitride-based heterojunction photoactive nanocomposites: applications and mechanism insight. ACS Appl. Mater. Inter. **10**, 21035–21055 (2018). https://doi.org/10.1021/acsami.8b03620
140. Y. Li, M. Meng, C.N. Ji, S. Teng, L.Y. Gao, R.J. Qu, Z.L. Yang, C.M. Sun, Y. Zhang, Soft-template synthesis of hybrid carbon and carbon nitride composites with enhanced photocatalytic activity for the degradation of methylene blue under visible light. Environ. Prog. Sustain. Energ. **38** (2019). https://doi.org/10.1002/ep.13186
141. P. Niu, M. Qiao, Y.F. Li, L. Huang, T.Y. Zhai, Distinctive defects engineering in graphitic carbon nitride for greatly extended visible light photocatalytic hydrogen evolution. Nano Energ. **44**, 73–81 (2018). https://doi.org/10.1016/j.nanoen.2017.11.059
142. H. Wang, H. Wang, Z.W. Wang, L. Tang, G.M. Zeng, P. Xu, M. Chen, T. Xiong, C.Y. Zhou, X.Y. Li, D.N. Huang, Y. Zhu, Z.X. Wang, J.W. Tang, Covalent organic framework photocatalysts: structures and applications. Chem. Soc. Rev. **49**, 4135–4165 (2020). https://doi.org/10.1039/d0cs00278j
143. S.S. Yuan, X. Li, J.Y. Zhu, G. Zhang, P. Van Puyvelde, B. Van der Bruggen, Covalent organic frameworks for membrane separation. Chem. Soc. Rev. **48**, 2665–2681 (2019). https://doi.org/10.1039/c8cs00919h
144. X.H. Ren, G.C. Liao, Z.J. Li, H. Qiao, Y. Zhang, X. Yu, B. Wang, H. Tan, L. Shi, X. Qi, H. Zhang, Two-dimensional MOF and COF nanosheets for next-generation optoelectronic applications. Coordin. Chem. Rev. **435** (2021). https://doi.org/10.1016/j.ccr.2021.213781
145. L.P. Guo, S.B. Jin, Stable covalent organic frameworks for photochemical applications. Chemphotochem **3**, 973–983 (2019). https://doi.org/10.1002/cptc.201900089
146. R.W. Tilford, W.R. Gemmill, H.C. zur Loye, J.J. Lavigne, Facile synthesis of a highly crystalline, covalently linked porous boronate network. Chem. Mater. **18**, 5296–5301 (2006). https://doi.org/10.1021/cm061177g
147. A.P. Cote, A.I. Benin, N.W. Ockwig, M. O'Keeffe, A.J. Matzger, O.M. Yaghi, Porous, crystalline, covalent organic frameworks. Science **310**, 1166–1170 (2005). https://doi.org/10.1126/science.1120411

148. Y. Du, K.M. Mao, P. Kamakoti, B. Wooler, S. Cundy, Q.C. Li, P. Ravikovitch, D. Calabro, The effects of pyridine on the structure of B-COFs and the underlying mechanism. J. Mater. Chem. A **1**, 13171–13178 (2013). https://doi.org/10.1039/c3ta12515g
149. J. You, Y. Zhao, L. Wang, W. Bao, Recent developments in the photocatalytic applications of covalent organic frameworks: a review. J. Clean. Prod. **291**, 125822 (2021). https://doi.org/10.1016/j.jclepro.2021.125822
150. P. Katekomol, J. Roeser, M. Bojdys, J. Weber, A. Thomas, Covalent triazine frameworks prepared from 1, 3, 5-tricyanobenzene. Chem. Mater. **25**, 1542–1548 (2013). https://doi.org/10.1021/cm303751n
151. F. Haase, E. Troschke, G. Savasci, T. Banerjee, V. Duppel, S. Dorfler, M.M.J. Grundei, A.M. Burow, C. Ochsenfeld, S. Kaskel, B.V. Lotsch, Topochemical conversion of an imine- into a thiazole-linked covalent organic framework enabling real structure analysis. Nat. Commun. **9**, 2600 (2018). https://doi.org/10.1038/s41467-018-04979-y
152. Y. Yang, H. Niu, L. Xu, H. Zhang, Y. Cai, Triazine functionalized fully conjugated covalent organic framework for efficient photocatalysis. Appl. Catal. B Environ. **269**, 118799 (2020). https://doi.org/10.1016/j.apcatb.2020.118799
153. F.J. Uribe-Romo, J.R. Hunt, H. Furukawa, C. Klock, M. O'Keeffe, O.M. Yaghi, A crystalline imine-linked 3-D porous covalent organic framework. J. Am. Chem. Soc. **131**, 4570 (2009). https://doi.org/10.1021/ja8096256
154. W. Ma, Q. Zheng, Y. He, G. Li, W. Guo, Z. Lin, L. Zhang, Size-controllable synthesis of uniform spherical covalent organic frameworks at room temperature for highly efficient and selective enrichment of hydrophobic peptides. J. Am. Chem. Soc. **141**, 18271–18277 (2019). https://doi.org/10.1021/jacs.9b09189
155. X.L. Fu, Z. Lu, H.J. Yang, X.Y. Yin, L.Q. Xiao, L.X. Hou, Imine-based covalent organic framework as photocatalyst for visible-light-induced atom transfer radical polymerization. J. Polymer Sci. **59**, 2036–2044 (2021). https://doi.org/10.1002/pol.20210261
156. H. Hu, Q.Q. Yan, R.L. Ge, Y.A. Gao, Covalent organic frameworks as heterogeneous catalysts. Chin. J. Catal. **39**, 1167–1179 (2018). https://doi.org/10.1016/s1872-2067(18)63057-8
157. D.N. Bunck, W.R. Dichtel, Bulk synthesis of exfoliated two-dimensional polymers using hydrazone-linked covalent organic frameworks. J. Am. Chem. Soc. **135**, 14952–14955 (2013). https://doi.org/10.1021/ja408243n
158. X. Li, K. Kawai, M. Fujitsuka, Y. Osakada, COF-based photocatalyst for energy and environment applications. Surf. Interfaces **25** (2021). https://doi.org/10.1016/j.surfin.2021.101249
159. G. Lin, H. Ding, R. Chen, Z. Peng, B. Wang, C. Wang, 3D porphyrin-based covalent organic frameworks. J. Am. Chem. Soc. **139**, 8705–8709 (2017). https://doi.org/10.1021/jacs.7b04141
160. F.Y. Liu, C.Y. Nie, Q.Q. Dong, Z.Y. Ma, W. Liu, M.P. Tong, AgI modified covalent organic frameworks for effective bacterial disinfection and organic pollutant degradation under visible light irradiation. J. Hazard. Mater. **398** (2020). https://doi.org/10.1016/j.jhazmat.2020.122865
161. F. Liu, Z. Ma, Y. Deng, M. Wang, P. Zhou, W. Liu, S. Guo, M. Tong, D. Ma, Tunable covalent organic frameworks with different heterocyclic nitrogen locations for efficient Cr(VI) reduction, escherichia coli disinfection, and paracetamol degradation under visible-light irradiation. Environ. Sci. Technol. **55**, 5371–5381 (2021). https://doi.org/10.1021/acs.est.0c07857
162. X. Han, J. Huang, C. Yuan, Y. Liu, Y. Cui, Chiral 3D covalent organic frameworks for high performance liquid chromatographic enantioseparation. J. Am. Chem. Soc. **140**, 892–895 (2018). https://doi.org/10.1021/jacs.7b12110
163. S.J. He, Q.F. Rong, H.Y. Niu, Y.Q. Cai, Construction of a superior visible-light-driven photocatalyst based on a C_3N_4 active centre-photoelectron shift platform-electron withdrawing unit triadic structure covalent organic framework. Chem. Commun. **53**, 9636–9639 (2017). https://doi.org/10.1039/c7cc04515h
164. Y.M. Zhang, Y.M. Hu, J.H. Zhao, E. Park, Y.H. Jin, Q.J. Liu, W. Zhang, Covalent organic framework-supported Fe-TiO_2 nanoparticles as ambient-light-active photocatalysts. J. Mater. Chem. A **7**, 16364–16371 (2019). https://doi.org/10.1039/c9ta03649k

165. Y.Z. Chen, W.H. Li, L. Li, L.N. Wang, Progress in organic photocatalysts. Rare Met. **37**, 1–12 (2018). https://doi.org/10.1007/s12598-017-0953-2
166. J. Min Park, J.H. Lee, W.-D. Jang, Applications of porphyrins in emerging energy conversion technologies. Coordin. Chem. Rev. **407**, 213157 (2020). https://doi.org/10.1016/j.ccr.2019.213157
167. S.L.H. Rebelo, A. Melo, R. Coimbra, M.E. Azenha, M.M. Pereira, H.D. Burrows, M. Sarakha, Photodegradation of atrazine and ametryn with visible light using water soluble porphyrins as sensitizers. Environ. Chem. Lett. **5**, 29–33 (2007). https://doi.org/10.1007/s10311-006-0072-z
168. G.P. Yao, J. Li, Y. Luo, W.J. Sun, Efficient visible photodegradation of 4-nitrophenol in the presence of H_2O_2 by using a new copper(II) porphyrin-TiO_2 photocatalyst. J. Mol. Catal. A: Chem. **361**, 29–35 (2012). https://doi.org/10.1016/j.molcata.2012.04.013
169. C.J. Monteiro, M.M. Pereira, M.E. Azenha, H.D. Burrows, C. Serpa, L.G. Arnaut, M.J. Tapia, M. Sarakha, P. Wong-Wah-Chung, S. Navaratnam, A comparative study of water soluble 5,10,15,20-tetrakis(2,6-dichloro-3-sulfophenyl)porphyrin and its metal complexes as efficient sensitizers for photodegradation of phenols. Photochem. Photobio. Sci. **4**, 617–624 (2005). https://doi.org/10.1039/b507597a
170. T. Shiragami, J. Matsumoto, H. Inoue, M. Yasuda, Antimony porphyrin complexes as visible-light driven photocatalyst. J. Photochem. Photobio. C **6**, 227–248 (2005). https://doi.org/10.1016/j.jphotochemrev.2005.12.001
171. P.P. Guo, P.L. Chen, W.H. Ma, M.H. Liu, Morphology-dependent supramolecular photocatalytic performance of porphyrin nanoassemblies: from molecule to artificial supramolecular nanoantenna. J. Mater. Chem. **22**, 20243–20249 (2012). https://doi.org/10.1039/c2jm33253a
172. J. Chen, C.J. Zhu, Y. Xu, P.W. Zhang, T.X. Liang, Advances in phthalocyanine compounds and their photochemical and electrochemical properties. Curr. Org. Chem. **22**, 485–504 (2018). https://doi.org/10.2174/1385272821666171002122055
173. V. Iliev, V. Alexiev, L. Bilyarska, Effect of metal phthalocyanine complex aggregation on the catalytic and photocatalytic oxidation of sulfur containing compounds. J. Mol. Catal. A: Chem. **137**, 15–22 (1999). https://doi.org/10.1016/s1381-1169(98)00069-7
174. X. Tao, W. Ma, J. Li, Y. Huang, J. Zhao, J.C. Yu, Efficient degradation of organic pollutants mediated by immobilized iron tetrasulfophthalocyanine under visible light irradiation. Chem. Commun. 80–81 (2003). https://doi.org/10.1039/b209083j
175. Z.G. Xiong, Y.M. Xu, L.Z. Zhu, J.C. Zhao, Enhanced photodegradation of 2,4,6-trichlorophenol over palladium phthalocyaninesulfonate modified organobentonite. Langmuir **21**, 10602–10607 (2005). https://doi.org/10.1021/la051714w
176. Z.G. Xiong, Y.M. Xu, Immobilization of palladium phthalocyaninesulfonate onto anionic clay for sorption and oxidation of 2,4,6-trichlorophenol under visible light irradiation. Chem. Mater. **19**, 1452–1458 (2007). https://doi.org/10.1021/cm062437x
177. A. Sorokin, S. DeSuzzoniDezard, D. Poullain, J.P. Noel, B. Meunier, CO_2 as the ultimate degradation product in the H_2O_2 oxidation of 2,4,6-trichlorophenol catalyzed by iron tetrasulfophthalocyanine. J. Am. Chem. Soc. **118**, 7410–7411 (1996). https://doi.org/10.1021/ja960177m
178. Z. Zhang, W. Wang, D. Jiang, J. Xu, CuPc sensitized Bi_2MoO_6 with remarkable photoresponse and enhanced photocatalytic activity. Catal. Commun. **55**, 15–18 (2014). https://doi.org/10.1016/j.catcom.2014.06.004
179. Z. Wang, H. Chen, P. Tang, W. Mao, F. Zhang, G. Qian, X. Fan, Hydrothermal in situ preparation of the copper phthalocyanine tetrasulfonate modified titanium dioxide photocatalyst. Colloid. Surface. A **289**, 207–211 (2006). https://doi.org/10.1016/j.colsurfa.2006.04.049
180. Y. Pan, X. Liu, W. Zhang, Z. Liu, G. Zeng, B. Shao, Q. Liang, Q. He, X. Yuan, D. Huang, M. Chen, Advances in photocatalysis based on fullerene C_{60} and its derivatives: properties, mechanism, synthesis, and applications. Appl. Catal. B Environ. **265**, 118579 (2020). https://doi.org/10.1016/j.apcatb.2019.118579
181. W. Bai, V. Krishna, J. Wang, B. Moudgil, B. Koopman, Enhancement of nano titanium dioxide photocatalysis in transparent coatings by polyhydroxy fullerene. Appl. Catal. B Environ. **125**, 128–135 (2012). https://doi.org/10.1016/j.apcatb.2012.05.026

182. A. Djordjevic, D. Sojic Merkulov, M. Lazarevic, I. Borisev, I. Medic, V. Pavlovic, B. Miljevic, B. Abramovic, Enhancement of nano titanium dioxide coatings by fullerene and polyhydroxy fullerene in the photocatalytic degradation of the herbicide mesotrione. Chemosphere **196**, 145–152 (2018). https://doi.org/10.1016/j.chemosphere.2017.12.160
183. J.G. Yu, T.T. Ma, G. Liu, B. Cheng, Enhanced photocatalytic activity of bimodal mesoporous titania powders by C_{60} modification. Dalton T. **40**, 6635–6644 (2011). https://doi.org/10.1039/c1dt10274e
184. D. Ma, J. Zhong, R. Peng, J. Li, R. Duan, Effective photoinduced charge separation and photocatalytic activity of hierarchical microsphere-like C_{60}/BiOCl. Appl. Surf. Sci. **465**, 249–258 (2019). https://doi.org/10.1016/j.apsusc.2018.09.192
185. Y. Park, N.J. Singh, K.S. Kim, T. Tachikawa, T. Majima, W. Choi, Fullerol-titania charge-transfer-mediated photocatalysis working under visible light. Chem. Eur. J. **15**, 10843–10850 (2009). https://doi.org/10.1002/chem.200901704
186. J.T. Shang, H.Y. Tang, H.W. Ji, W.H. Ma, C.C. Chen, J.C. Zhao, Synthesis, characterization, and activity of a covalently anchored heterogeneous perylene diimide photocatalyst. Chin. J. Catal. **38**, 2094–2101 (2017). https://doi.org/10.1016/s1872-2067(17)62960-7
187. X.M. Gao, K.L. Gao, X.B. Li, Y.Y. Shang, F. Fu, Hybrid PDI/BiOCl heterojunction with enhanced interfacial charge transfer for a full-spectrum photocatalytic degradation of pollutants. Catal. Sci. Technol. **10**, 372–381 (2020). https://doi.org/10.1039/c9cy01722d
188. W.G. Zeng, T. Cai, Y.T. Liu, L.L. Wang, W.Y. Dong, H. Chen, X.N. Xia, An artificial organic-inorganic Z-scheme photocatalyst WO_3@Cu@PDI supramolecular with excellent visible light absorption and photocatalytic activity. Chem. Eng. J. **381** (2020). https://doi.org/10.1016/j.cej.2019.122691
189. Q.C. Zhang, L. Jiang, J. Wang, Y.F. Zhu, Y.J. Pu, W.D. Dai, Photocatalytic degradation of tetracycline antibiotics using three-dimensional network structure perylene diimide supramolecular organic photocatalyst under visible-light irradiation. Appl. Catal. B: Environ. **277** (2020). https://doi.org/10.1016/j.apcatb.2020.119122
190. Y. Xu, L. Duan, L. Tong, B. Akermark, L. Sun, Visible light-driven water oxidation catalyzed by a highly efficient dinuclear ruthenium complex. Chem. Commun. **46**, 6506–6508 (2010). https://doi.org/10.1039/c0cc01250e
191. A.H.C. Khavar, G. Moussavi, K. Yaghmaeian, A.R. Mahjoub, N. Khedri, M. Dusek, T. Vaclavu, M. Hosseini, A new Ru(ii) polypyridyl complex as an efficient photosensitizer for enhancing the visible-light-driven photocatalytic activity of a TiO_2/reduced graphene oxide nanocomposite for the degradation of atrazine: DFT and mechanism insights. RSC Adv. **10**, 22500–22514 (2020). https://doi.org/10.1039/c9ra06704c
192. W.H. Ma, J. Li, X. Tao, J. He, Y.M. Xu, J.C. Yu, J.C. Zhao, Efficient degradation of organic pollutants by using dioxygen activated by resin-exchanged Iron(II) bipyridine under visible irradiation. Angew. Chem. Int. Edit. **42**, 1029-+ (2003). https://doi.org/10.1002/anie.200390264
193. M.L. Marin, L. Santos-Juanes, A. Arques, A.M. Amat, M.A. Miranda, Organic photocatalysts for the oxidation of pollutants and model compounds. Chem. Rev. **112**, 1710–1750 (2012). https://doi.org/10.1021/cr2000543
194. W. Wang, P. Xu, M. Chen, G. Zeng, C. Zhang, C. Zhou, Y. Yang, D. Huang, C. Lai, M. Cheng, L. Hu, W. Xiong, H. Guo, M. Zhou, Alkali metal-assisted synthesis of graphite carbon nitride with tunable band-gap for enhanced visible-light-driven photocatalytic performance. ACS Sustain. Chem. Eng. **6**, 15503–15516 (2018). https://doi.org/10.1021/acssuschemeng.8b03965

Past, Present, and Future in the Development of Medium and High-Temperature Catalytic Processes for N_2O Decomposition

Yihao Wu, Yuanshuang Zheng, and Pascal Granger

Abstract Trace N_2O emissions in ammonia burner during the nitric acid production process have received special attention as N_2O is recognized as a potent greenhouse gas that actively participates in global warming. The abatement of N_2O complies to international commitments aiming to reduce the greenhouse gas emissions through the installation of appropriate technologies. Different strategies have been categorized, the most efficient one involving the implementation of heterogeneous catalytic reactor in various positions of nitric acid plant. The catalyst composition strongly depends on the running temperature privileging mixed-metal oxides for N_2O decomposition at medium and high-temperature whereas Platinum Group Metal-supported catalysts can be preferred for end-of-pipe technologies running at low temperature with the help of reducing agent. This present chapter focusses on medium and high-temperature (350–900 °C) application. Different theoretical and experimental approaches will be discussed in order to get more insights into the design of active sites, especially under more realistic running conditions to improve the understanding of kinetics and get more relevant reaction mechanisms that could further provide guidelines for the preparation of more stable and selective catalysts.

Keywords N_2O decomposition · Kinetics · Perovskite · Spinel structure · Nitric acid plant · Reaction mechanisms

---

Y. Wu · Y. Zheng · P. Granger (✉)
UMR 8181 - UCCS - Unité de Catalyse Et Chimie du Solide, Univ. Lille, CNRS, Centrale Lille, Univ. Artois, 59000 Lille, France
e-mail: pascal.granger@univ-lille.fr

Y. Wu
China Energy Longyuan Environmental Protection Co., Ltd, Building 1, Yard 16, W. 4th Ring Road, Beijing 100039, Haidian District, China

S. J. Ikhmayies (ed.), *Advances in Catalysts Research*, Advances in Material Research and Technology, https://doi.org/10.1007/978-3-031-49108-5_7

1 General Context and Specific Issues to the Development of Efficient and Stable Catalytic Systems for the Abatement of N_2O

The general characteristic of catalysis is mainly linked to substantial savings in energy, thanks to a strong reduction in the activation barrier and improved yields. But in some cases, the temperature can be a constraint either imposed or necessary. At the same time, a high selectivity and durability towards the target product is still an outstanding challenge in the elaboration of cost-efficient catalytic technology. By way of illustration, three-way catalysts, selective activation of light alkanes into unsaturated alkanes or N_2O decomposition run at high temperature (above 350 °C). In the first case, the high temperature of inlet exhaust gas is a constraint and can affect the catalyst durability [1]. In the particular case of the selective oxidation of hydrocarbons, the gaseous oxygen is non-selective unlike the more selective bulk oxygen species of the catalyst. Their activation requires a high temperature to ease their mobility and improve their reactivity [2]. The final example is not trivial and will be illustrated in this review. Taking nitric acid plants, for example, undesired N_2O is formed due to the incomplete catalytic oxidation of ammonia on deactivated Pt–Rh gauzes at the expense of NO_x yield [3]. Due to a global warming potential ~ 310 times higher than CO_2 and a longer lifetime, different commitments have been accepted by industrialized countries greenhouse gas emitters. For instance, the Kyoto protocol stipulated a reduction of the greenhouse gas by at least 5% during the period 2008–2012. More recently, the Paris Agreement on Climate Change entered into force on 4 November 2016 aiming to limit global warming to well below 2 °C, preferably to 1.5 °C compared to pre-industrial levels, also stipulated the reaching of global peaking of greenhouse gas emissions as soon as possible to achieve a climate neutral world by mid-century. In previous review papers, Perez-Ramirez [3, 4] pointed out the strategic importance in the reduction of N_2O emissions by treating a finite number of industrial sources due to its cost efficiency.

The first attempt has been achieved in adipic acid production where N_2O is a primary side product with a content of 30–50 vol.% in the exhaust gas [4]. Presently, particular attention is paid to nitric acid plant with exhaust gas containing much lower concentration (below 0.1–0.5 vol.%) which strongly affects the effectiveness of the catalyst and may require the use of intrinsically very active precious metals to promote transformation kinetics. Different strategies for the abatement of N_2O have been already discussed and classified in the nitric acid production chain [5] taking the development of more stable Pt–Rh gauze into account to avoid the side production of N_2O, until the implementation of catalytic end-of-pipe technology to treat tail gas. Intermediate options emerged [5] concerning the direct N_2O decomposition inside the ammonia burner (secondary option) or by treating the exhaust gas at medium temperature (tertiary option) to avoid the extra use of reducing agent as exemplified in Fig. 1. The operating temperature needed for the secondary option is imposed by the process reaching up 800 °C. Under these operating conditions, the parallel decomposition of NO_x must be avoided which would have a detrimental impact on the economy of nitric acid plant. In principle, no retrofit of ammonia burner is needed for recent installations and the use of reducing agent is not required compared with end-of-pipe technologies. However, some prerequisites related to catalytic properties in terms of thermal stability and selectivity must be considered to prevent any competitive NO_x conversion. Among the different strategies, an alternative consists in N_2O decomposition at an intermediate temperature downstream the absorption column and upstream of the tail-gas expander (see Fig. 1). This strategy does not induce any perturbation on the ammonia burner. The typical temperature of the outlet gas subjected to that treatment varies in the range 400–500 °C [6]. The size of the reactor and residence time can be adjusted to optimize the rate of N_2O decomposition. An excellent review is subjected to the state of the art regarding the most appropriate catalysts according to the operating conditions [7]. Note that the prerequisites differ according to the position of the catalytic system in the scheme of nitric acid process. Indeed, residual trace amount of N_2O and NO_x downstream the absorption column should be removed simultaneously, while the catalyst inside the ammonia burner must selectively convert N_2O and prevent any NO_x decomposition.

Different catalytic formulations have been previously developed for medium and high-temperature applications, *i.e.* Co_2AlO_4/CeO_2 by Yara International, CuO/Al_2O_3 by BASF, and Fe/ZSM-5 by Uhde [8]. Alternative catalytic formulations were evaluated at the lab-scale such as yttrium stabilized zirconia [9], Fe–Ce mixed oxides [10], ceria-zirconia [11], Fe_2O_3/Al_2O_3 [12], spinel structures [13] and K/Zn–Co_3O_4 spinel catalysts [14], mixed metal oxides from hydrotalcite precursors [15–19] and metal-substituted hexaaluminates [20–24]. Perovskite-type oxides (ABO_3) have also been investigated. Porzellanwerk Kloster Veilsdor has already patented supported $LaCoO_3$ perovskite as an active phase for high-temperature N_2O decomposition

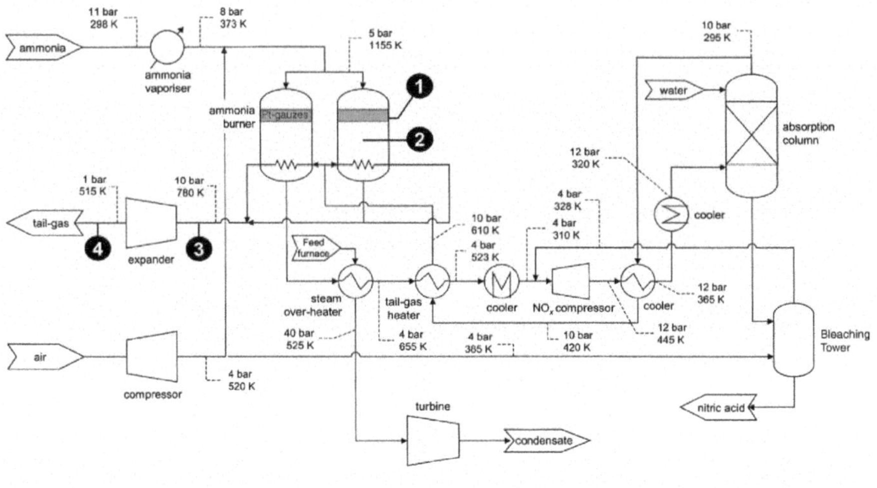

Fig. 1 Scheme of industrial plant for nitric acid production. Reprinted from Pérez-Ramiréz et al. [5], with permission from Elsevier. Copyright (2003)

[25]. Umicore also developed various perovskite-based formulations [26]. However, most of those studies were carried out under moderate conditions with an inlet gas mixture deviating from the real exhaust gas. In practice, the operating conditions according to the position of the catalytic process, i.e. medium temperature (MT) or high-temperature (HT) differ. In principle, the reaction mixture containing a high concentration of NO_x of the order of 5 vol.% against a concentration of the order of 0.1 vol.% for N_2O is a prerequisite for the study at high temperature. On the other hand, the concentration of NO_x should be much lower downstream of the absorption column. A careful examination of operating conditions in Table 1 underlines that these prerequisites are not fully respected.

Finally, we have to briefly consider low-temperature catalytic abatement processes (quaternary option). End-of-pipe technologies can be implemented in order to remove simultaneous N_2O and NO_x. In general, the low residual temperature of the tail gas prevents their decomposition and the use of a reducing agent is needed to clean the catalyst surface through the removal of strongly chemisorbed oxygen species by reacting with the reducing agent. Efficient selective-catalytic-reduction processes for the removal of NO_x exist working below 250 °C and can be a priori developed for this application. However, the simultaneous reduction of NO_x and N_2O at low temperature is a challenge because the exhaust gases must be reheated to ensure good performance which can, to a certain extent, considerably limit the applicability due to low profitability. The use of noble metals exhibiting exceptional turn-over-number for the removal of NO_x can also be considered for both low and medium-temperature application [27]. All these limits lead one to wonder if it is not more profitable to eliminate simultaneously NO_x and N_2O at medium temperatures, thanks to their decomposition by no longer resorting to the use of reducing agents, precious metals and without penalties on energy consumption for gas preheating. To this end, CuO/CeO_2 was found more competitive than precious metals exhibiting superior activity at moderate temperature and a long-term stability [7, 28, 29]. Some pecularities also appear on the use of Cu-doped zeolite, which also support this option based on the fact that NO_x can assist in the decomposition of N_2O [30].

2 Heterogeneous Catalysts for Medium and High-Temperature N_2O Decomposition

2.1 *Supported Versus Unsupported Catalysts: Impact of Support Effect, Additive, and Method of Preparation*

2.1.1 *Metal Oxides and Mixed Metal Oxides*

Metal oxides are suitable for MT and HT catalytic processes. Probably, the more the scarcity of high-temperature processes is likely due to high residual temperature and the importance to get high thermal stability in wet conditions [40, 41]. For MT

Table 1 Performance of various supported and bulk catalysts for N_2O decomposition running at medium (MT) and high (HT) temperatures

Catalyst	Inlet feed gas composition	$T_{(reaction)}$ (°C)	W/F_0 (g h L^{-1})	N_2O^a conv. (%)	Temperature range (°C)	Range of application	Ref
$Cu_{0.25}Co_{2.75}O_4$	1000 ppm N_2O 1000 ppm N_2O, 200 ppm NO, 2%O_2, 0.5%H_2O	370 370	5.5×10^{-2} 5.5×10^{-2}	100 5.0	300–650	MT	[31]
$Fe_{0.5}Ce_{0.5}O_2$	0.45 vol.% N_2O	500	1.7×10^{-2}	60	200–600	MT	[10]
K-doped Co_3O_4	0.1% N_2O	360	5.5×10^{-2}	30	300–450	MT	[32]
Co_3O_4-Cs on ceramic foam	0.1% N_2O	360	1.7×10^{-2}	65	300–450	MT	[33]
Fe-SSZ-13	1000 ppm N_2O, 4% O_2 1000 ppm N_2O, 4% O_2, 10% H_2O	477 537	1.5×10^{-2} 1.5×10^{-2}	50 50	377–550	MT	[34]
Ni-(Mg)-Al	0.1 vol.% N_2O	390	1.7×10^{-2}	20	300–450	MT	[15]
K/Zn-Co_3O_4	15 vol.% N_2O	240	0.17	40	200–600	MT	[14]
CoZSM-11	0.5 vol.% N_2O	375	5.5×10^{-2}	40	300–550	MT	[35]
Ce-Co–O binary system	1000 ppm N_2O	304	1.1×10^{-2}	50	250–450	MT	[36]
Mesoporous Co_3O_4 23.5 wt.% Co_3O_4/SiO_2	500 ppm N_2O, 3 vol.% O_2	400 400	1.25×10^{-2} 1.25×10^{-2}	5.0 23	300–500	MT	[37]
6 wt.% Cu/CeO_2	2500 ppm N_2O	400	1.7×10^{-2}	54	300–550	MT	[38]
Fe-ZSM-11	0.5% N_2O, 5%O_2, 2% H_2O	400	5.5×10^{-2}	60	300–550	MT	[39]
$LaMn_xAl_{12-x}O_{19}$	30 vol.% N_2O	700 500	3.3×10^{-2} 3.3×10^{-2}	90 85	500–800 500–800	HT	[21]

(continued)

Table 1 (continued)

Catalyst	Inlet feed gas composition	$T_{(reaction)}$ (°C)	W/F_0 ($g\ h\ L^{-1}$)	N_2O^a conv. (%)	Temperature range (°C)	Range of application	Ref
$BaRu_{0.2}FeAl_{10.8}O_{19}$	30 vol.% N_2O	700 500	3.3×10^{-2} 3.3×10^{-2}	100 70	500–800 500–800	HT	[21]
$BaFeAl_{11}O_{19}$	0.15 vol.%	700 550	1.7×10^{-2} 1.7×10^{-2}	75 12	500–800 500–800	HT	[23]
Fe-ZSM5	0.15 vol.%	550	1.7×10^{-2}	90	500–800	HT	[23]
$LaFeAl_{11}O_{19}$	0.15 vol.% 0.15 vol.%, 1 vol.% NO, 10 vol. O_2, 10 vol.% H_2O	750 750	1.7×10^{-2} 1.7×10^{-2}	85 40	750	HT	[24]
$Mn_{0.1725}Co_3AlO_x$ from hydrotalcite	0.1 vol.% N_2O, 0.03% vol.% NO, 4.5 vol.% O_2	300	5.8×10^{-2}	43	200–500	MT	[18]
K/Co-Al-500 HT (K/Co = 0.08)	500 ppm N_2O	250	3.3×10^{-2}	20	200–500	MT	[19]
Y-doped ZrO_2 (ZrO_2-0.05_Y)	200 ppm N_2O, 1400 ppm NO, 1 vol.% O_2, 15 vol.% H_2O	750	3.0×10^{-2}	59	600–900	HT	[9]
Fe_2O_3/Al_2O_3 (Fe/Al = 1)	0.5 vol.% N_2O, 3 vol.% NO, 0.5 vol.% O_2, 6 vol.% H_2O	750	$(1.7–3.3) \times 10^{-2}$	81	500–850	HT	[12]
$Ce_{0.52}Zr_{0.48}O_2$	2000 ppm N_2O, 1.4 vol.% NO, 1 vol.% O_2, 15 vol.% H_2O	660	6.0×10^{-2}	50	550–900	HT	[11]

(continued)

Table 1 (continued)

Catalyst	Inlet feed gas composition	$T_{(reaction)}$ (°C)	W/F_0 ($g\ h\ L^{-1}$)	N_2O^a conv. (%)	Temperature range (°C)	Range of application	Ref
Fresh $LaCo_{0.6}Fe_{0.4}O_3$	vol.% N_2O, 5 vol.% NO, 6 vol.% O_2, 15 vol% H_2O	588	4.7×10^{-2}	50	500–900	HT	[40]
Aged $LaCo_{0.6}Fe_{0.4}O_3$	vol.% N_2O, 5 vol.% NO, 6 vol.% O_2, 15 vol% H_2O	635	4.7×10^{-2}	50	500–900	HT	[40]
Fresh $La_{0.8}CoFeO_3$	vol.% N_2O, 5 vol.% NO, 6 vol.% O_2, 15 vol% H_2O	600	4.7×10^{-2}	60	500–900	HT	[41]
Aged $La_{0.8}CoFeO_3$	vol.% N_2O, 5 vol.% NO, 6 vol.% O_2, 15 vol% H_2O	600	4.7×10^{-2}	30	500–900	HT	[41]

[a] conversation corresponding to the selected reaction temperature

catalytic processes, the comparison between supported and unsupported heterogeneous catalysts is relevant showing the beneficial effect of specific surface area as well as the dispersion of the active phase [37]. On the other hand, more thermally stable bulk catalysts can be preferred for high-temperature applications with materials having low specific surface areas as this parameter is no longer crucial in this range of temperature. This assertion has been previously verified on Mn-substituted hexaaluminates revealing higher performance on the sample calcined at 1200 °C than those calcined at 500 °C with a much higher specific surface area [21]. Spinel structures have been widely studied to check the impact of the methods of preparation for optimizing the cobalt dispersion [42–46], the impact of the support materials [32, 33, 47–49], the impact of dopants especially potassium [31, 50, 51], the impact of the pre-treatment [36]. Different preparation methods have been compared in order to establish a classification and also identify the most important parameters, which influence their catalytic features.

Among all this literature some guidelines can be given. Firstly, the specific surface area cannot be reasonably considered as the most important parameter [52] as illustrated for high-temperature application and in some extent at medium temperatures. In many cases, the redox properties of transition metals exhibiting mixed valence are usually used as a main component in active phases (see Table 1) and this property outperforms the specific surface area. The redox properties are commonly enhanced through doping. For instance, such strategy can improve the reducibility of Co^{3+} to Co^{2+}, as Co^{2+} in tetrahedral coordination is expected to act as an active element [52]. The introduction of copper can promote the reduction of Co^{3+} to Co^{2+} in $Cu_xCo_{3-x}O_4$ for $x \leq 0.25$. Subsequent incorporation of potassium would stabilize Co^{2+} which benefits catalytic activity [31]. The incorporation of cerium in cobalt oxide can lead to cooperative effects in the binary system [36]. The oxygen storage properties of cerium oxide can be profitably used to improve the performance of cobalt oxide. CeO_2 itself is inactive but can act as promoter once combined with cobalt. As illustrated in Fig. 2, a second reaction pathway in the binary Ce-Co–O system can be envisioned to refill anionic vacancies Ce^{III}-O^- into $Ce^{IV}O^-$.

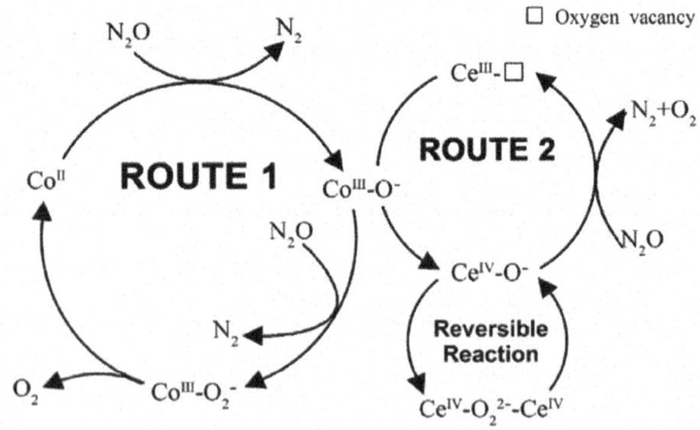

Fig. 2 Proposed scheme of N_2O decomposition catalysed by Ce-Co–O catalysts. Reprinted from You et al. [36], with permission from Elsevier. Copyright (2018)

In many cases, realistic inlet feed gas with catalyst working in wet conditions induced strong inhibiting effect for moderate reaction temperature. This detrimental effect, mostly observed for medium temperature catalytic processes, is generally weakly described and a question can arise regarding the impact of water on the surface properties of the catalyst in case of reconstructions with subsequent changes in the distribution of typical facets. Indeed, recent investigation showed that cobalt spinel nano-cube clusters with reactive oxygen species on (100) surface are more active than their nano-octahedral counterpart with reactive oxygen on (111) surface. Consequently, surface reconstructions could lead to a change in the surface reactivity. Zasada et al. [53] found that the redox changes vs. temperature from oxygen-precovered surface at low temperature to the formation of surface defects at high temperature are delayed on the (111) surface. Changes in morphology can be also induced by the selected preparation method [32]. Microwave-assisted coprecipitation method with calcination at 500 °C leads to $Mn_xCo_{1-x}Co_2O_4$ particles exhibiting a nanorod morphology. Mn substitution was found to improve both textural and structural properties. An optimum in activity is obtained for $x = 0.75$. At higher potassium content, the loss of specific surface area jointly with surface potassium enrichment and accumulation of oxygen at the surface would cause a decrease in catalytic activity.

2.2 *Zeolite Based Catalysts*

Metal-doped zeolites are conventionally prepared by ion-exhange method, which provides excellent metal dispersion in comparison to wet impregnation. Copper, cobalt, and iron have been mostly investigated for N_2O decomposition [30, 54, 55] and a large panel of zeolites has been explored, which differ from their porous structure, *i.e.* small vs. large channels [56–58]. Their hydrothermal stability is related to the Si/Al ratio. By way of illustration, the activity of Co/ZSM-11 in N_2O decomposition is improved with a decrease in the Si/Al ratio [35]. The pore topologies as well as the presence of strong Brønsted acid sites have a significant effect on the nature of extra framework iron species and their distribution [34]. In severe operating conditions dealumination process and aggregation of exchanged cations to form extra framework metal oxide clusters have usually a detrimental effect on the catalytic properties because it is commonly recognized that isolated exchanged cations bind to the zeolite framework are more active than extra framework metal oxide aggregates. However, some controversies can arise because of non-uniform catalytic behaviour according to the nature of the transition metal as well as the operating conditions. For instance, polynuclear FeO_x species, characterized by a lengthening of the Fe–O bond, in the extra framework of Fe/ZSM5 have been found to play a dominant role in N_2O decomposition. [59]. Such explanation differs from previous ones which stipulated that highly dispersed metal species are responsible for the catalytic activity. As a matter of fact, particular attention must be paid to the temperature conditions and likely the presence of steam. Among the multiple metal oxide species stabilized inside the porous structure, their contribution to the overall catalytic activity can change as a function of temperature switching from inactive to active species with a rise in temperature.

The diameter size of the channels of zeolite structure also appears as a critical parameter because zeolites having high framework density and low diameters of the channels, i.e. MFI, BEA, and FER, are more active than those exhibiting more open frameworks with larger diameter channels, *i.e.* FAU and MOR zeolites. This latter point is important because according to the diameter size of the channel, some perturbations due to mass transfer can occur. Accordingly, at increasing crystal size, the sensibility to transfer limitation can accentuate and then alter the catalyst effectiveness. Xiao et al. [60] reported that these limitations cannot be no-longer neglected for crystal size $\geq 5\ \mu m$. Interestingly, the significance of oxygen desorption as a critical parameter for establishing relevant comparisons with catalytic activity could be highly criticizable [61]. Indeed, the desorption rate would no longer reflect the strength of interaction between oxygen and the substrate in case of a slow diffusion process in the small channels. In some extent, Fe_xO_y inside the channel of the zeolite structure can preserve a significant activity depending on their nuclearity as reported elsewhere [62].

Some attempts combined micro-mesoporosity in order to improve the accessibility of ion-exchange positions [63]. The performance of micro-mesoporous Fe/Beta with those obtained on Fe/Beta having a single type of mesoporous structure has been compared. Despite the much lower iron content, the hierarchical mesoporous structure leads to a comparable activity compared to the single mesoporous structure. The formation of monomeric and small $Fe_x^{3+}O_y$ oligomers prevails in the former sample and would originate the catalytic activity. Let us note that this observation diverges from that earlier discussed for Fe/ZSM5 where the authors privilege the involvement of large clusters. But, there is no indication regarding their size and nuclearity. New conceptual ideas emerged that can summarize the complexity in determining the origin of the catalytic activity and the identification of the active sites. Santhosh Kumar et al. [58] concluded on Fe-SBA-15 that the confinement of iron species in the pore structures with suitable pore size and geometry could be the driving force in determining the catalytic properties. In principle, large mesopores are not suited to promote an intimate contact between the molecules and the iron active site. This complexity seems to be illustrated in this concept pointing out an optimal pore size structure for monitoring the dispersion of the active element, mass transfer limitations, and promoting confinement effects.

This complexity is also related to the fact that many investigations differ from their operating conditions, *i.e.* in the presence or in the absence of water. The water composition can also vary significantly in the inlet gas mixture. For instance, the presence of water can drastically alter the kinetic behaviour because of strong inhibiting effects hindering the adsorption of N_2O. This effect accentuates according to the partial pressure of water. In addition, structural changes through dealumination and/or Cu aggregation can also occur inducing a deterioration of the catalyst performances [64]

2.3 Design of Active Sites

N_2O decomposition is not thermodynamically limited but suffers from kinetic limitations because the desorption of chemisorbed oxygen coming from O_2 adsorption or N_2O dissociation is slow, which inhibits the reaction rate. Based on this, once dissociated, chemisorbed oxygen species should release the surface instantly through desorption, which would qualify the best heterogeneous catalysts. In practice, desorption is slow and requires high temperature to occur. Numerous investigations have focussed on the composition of various catalysts for designing active sites with optimal geometric and electronic properties, which most effectively would promote the desorption of oxygen. An important aspect is also the preferential development of bulk catalysts instead of supported catalysts as already discussed. Bulk catalysts have lower specific surface area, which means that most of the active component is not completely used. These can affect the economics of the process limiting rational approaches to reach maximum atom efficiency. In practice, better rationalization would concern the supported catalysts and more particularly for low and medium temperatures. However, they suffer from both high sensitivity to thermal sintering, and poisoning effects. As example, small CuO clusters dispersed on CeO_2 exposed to 2500 ppm N_2O, 1.5% NO, and 1.5% H_2O were found to be sensitive to the presence of water leading to surface reconstruction associated to the transformation of CuO to $CuO{\cdot}3H_2O$ at 400 °C [38]. Presently, the lack of understanding of the catalyst functionalities and the existing controversies on the identification of active sites has been proven as a serious limitation in the development of optimal systems.

Cobalt is a benchmark component stabilized as a single Co_3O_4 oxide exhibiting a spinel-type structure in which Co^{2+} and Co^{3+} coexist in different tetrahedral and octahedral coordinations respectively. Different methodologies have been implemented to identify the valence responsible for the catalytic activity. Eom et al. found a remarkable gain in conversion when Co_3O_4 is pre-reduced, which led the authors to conclude on the involvement of Co^{2+} [65]. The substitution of Co^{3+} by trivalent cations and Co^{2+} by divalent cations is an alternative method that can provide arguments for the identification of the valence of active Co species. Stelmachowski et al. found that the magnitude of the decline in catalytic activity was greater when Co^{3+} is substituted by Al^{3+}, which led the authors to the conclusion that Co^{3+} is more active with activation barrier (E_a) in the range 15–17 kJ/mol whereas Co^{2+} in octahedral position is much less active with $E_a = 27$–28 kJ/mol [66].

Co/hydroxyapatite (HAP) prepared by ion exchange, impregnation, and hydrothermal methods led to improved atom efficiency in comparison to Co_xO_y. This approach allows tuning the oxidation state of cobalt as well as coordination geometry [67]. The best compromise was obtained with ion exchange promoting the formation of Co^{2+} in octohedral position ($Co^{2+}\|Oh$) more active than Co^{2+} ($Co^{2+}\|Td$) and Co^{3+} ($Co^{3+}\|Oh$) respectively in tetrahedral and octahedral position. The authors concluded that the former coordination promotes the desorption of oxygen thanks to closer spacing distance between two $Co^{2+}\|Oh$.

Zeolite based materials are another variety of catalysts that led to promising catalytic properties. Again, the design of active site is still under debate [39]. As an example, a wide panel of iron species can form in zeolite structure such as: Fe^{3+} framework—isolated Fe^{3+} or Fe^{2+} bonded to the zeolite framework through Si–O-Fe or Al-O-Fe bridges—dinuclear Fe–O-Fe—oligomeric Fe oxo-species located in the large cavities—small FeO_x nanoparticles and—bulk FeO_x particles. The distribution of these iron species varies according to multiple preparation methods already reported. As a consequence, there is a strong non-uniformity in the definition of active sites. ZSM5 served as benchmark zeolite which led to the identification of a broad panel of active sites in N_2O decomposition such as Fe^{3+}-oxo oligomeric iron species, clustered iron species, or coordinatively unsaturated Fe^{3+} species [68–71]. Recent investigations revealed the superiority of chabazite structure, typically SAPO-34 and SSZ-13. Wang et al. [72] found better performance of Fe/SSZ-13 than Fe/Beta containing the same amount of iron, *i.e.* 0.63 wt.% in the temperature range 350–550 °C (see Figs. 3 and 4). However, they also observed a sharp loss of conversion in the presence 2.5% H_2O leading to the same apparent activation energy as seen in Fig. 4, which reflects comparable kinetic behaviour while they differ in dry conditions. The authors presume that the superior performances in dry conditions are related to a higher density of more active monomeric and dimeric iron species. Their stability in some extent could be questionable even if a better stability compared to Fe/FER is reported possibly due to a smaller pore structure. The discussion lies in many assumptions. Nonetheless, these assertions have not been questioned. Indeed Zhang et al. also claimed a remarkably high temperature hydrothermal stability of Fe/SAPO and Fe/SSZ-13 in comparison to the benchmark Fe/FER between 400 °C and 600 °C with inlet mixture composed of 500 ppm N_2O and 5% H_2O. Multiple iron species have been identified but the authors conclude that dinuclear $[HO\text{-}Fe\text{–}O\text{-}Fe\text{-}OH]^{2+}$ species are responsible for the catalytic activity in N_2O decomposition. The higher performance of Fe/SSZ-13 has been ascribed to a lower Fe–O binding energy, which favours oxygen desorption [73]. Let us note that site requirements can differ according to the nature of the transition metal and also the metal content. Indeed Lin et al. [74] found an optimal Fe/Al ratio corresponding to maximum catalytic activity. This has been explained by the formation of less active oligomeric Fe species than dimeric ones with increasing Fe loading. On the other hand, N_2O decomposition on Co-SSZ-13 zeolite obeys to a single-site route involving isolated Co^{2+} according to a first-order kinetic regime obtained at low partial pressure. The pecularities of the chabazite structure do not drastically differ from the comparison with FeZSM-11 which also emphasizes an excellent correlation between the normalized reaction rate and the amount of small FeO_x particles characterized by a low nuclearity [39]. Nevertheless, some deviations clearly appear between the weak precision on the nuclearity of FeO_x and the existence of well-defined di-nuclear species in Fe/SSZ-13. Interestingly, they also infer that the conclusion drawn for Fe-ZSM-11 would be no longer valid with Cu-ZSM-11 and Co-ZSM-11 where the active sites would be isolated Cu^{2+} in the framework and Co^{2+}, respectively.

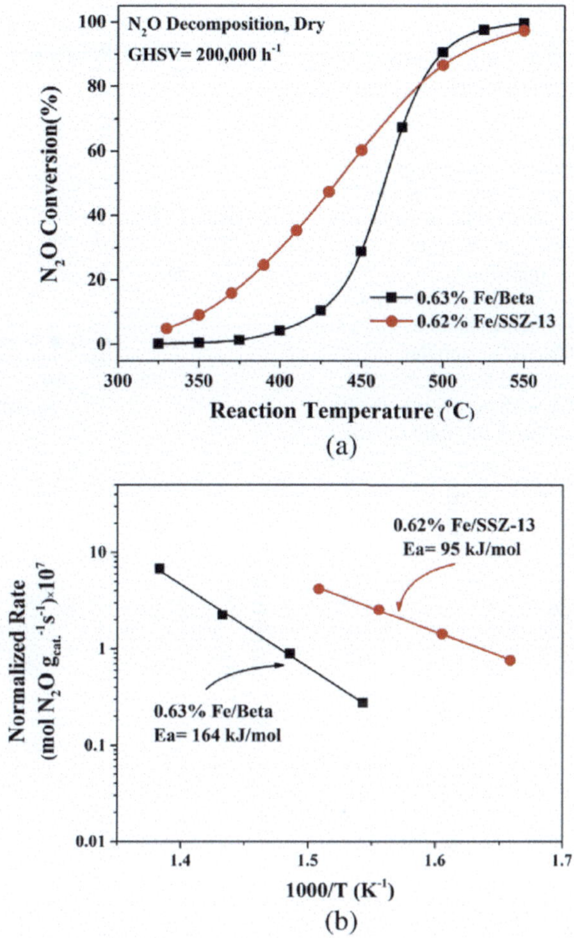

Fig. 3 **a** conversion vs. temperature curves for 'dry' N_2O decomposition on Fe/Beta (0.63 wt.%) and Fe/SSZ-13 (0.62 wt.%) samples. The reactant feed contains 540 pmm of N_2O balanced with N_2 at a GHSV of ~200,000h-1. **b** Arrhenius plots for 'Dry' N_2O decomposition decomposition on Fe/Beta (0.63 wt.%) and Fe/SSZ-13 (0.62 wt.%) samples. Reprinted from Wang et al. [72], with permission from Elsevier. Copyright (2018)

2.4 *Impact of Structural* and electronic properties of perovskites in their catalytic properties for N_2O decomposition

Perovskites (ABO_3) have been widely studied for many non-catalytic and catalytic processes [75]. In this latter case, exceptional structural and electronic features can be obtained through appropriate substitutions in A- and B-sites leading to remarkable redox properties. Both structural deformation, stabilization of B-site cations in unusual oxidation states and equilibration of the electronic charges through the creation of anionic vacancies or through overstoichiometry of oxygen can lead to remarkable catalytic features and hydrothermal stability [76]. The main disadvantage of perovskites lies in low specific surface area. However, as previously argued, this

Fig. 4 **a** conversion vs. temperature curves for 'wet' N_2O decomposition on Fe/Beta (0.63 wt.%) and Fe/SSZ-13 (0.62 wt.%) samples. The reactant feed contains 540 pmm of N_2O balanced with N_2 at a GHSV of ~200,000h-1. **b** Arrhenius plots for 'wet' N_2O decomposition decomposition on Fe/Beta (0.63 wt.%) and Fe/SSZ-13 (0.62 wt.%) samples. Reprinted from Wang et al. [72], with permission from Elsevier. Copyright (2018)

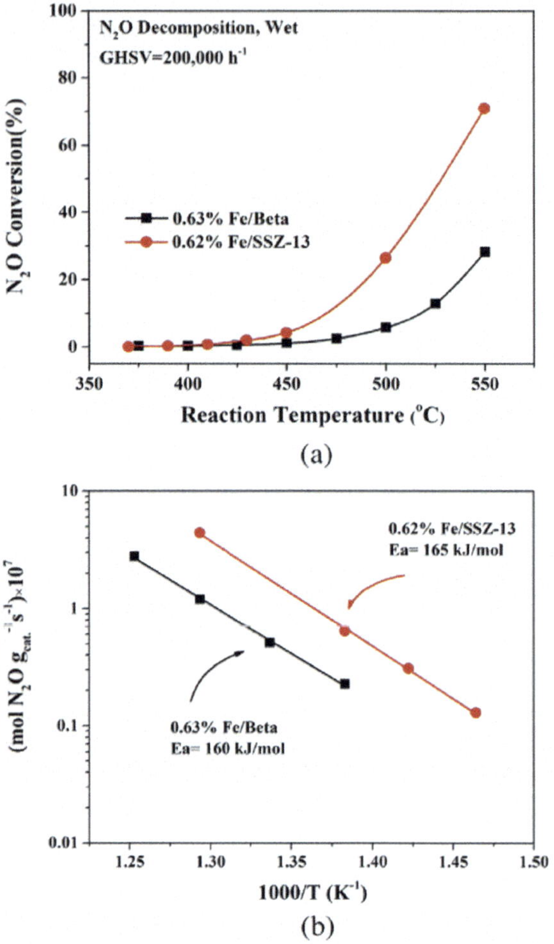

parameter is not critical when the catalyst runs at medium or high temperature. In practice, it is not easy to establish a reliable structure–activity relationship because in some extent partial exsolution process can lead to segregation of well-dispersed single oxide, which can also act as an active site in the selected reaction. Structural deformation and changes in the valence of B-cations can induce a lengthening of the B-O bond suggesting improved oxygen mobility, and an ease to form oxygen deficiencies. Based on this, a gain in catalytic activity should be expected thanks to a faster adsorption and dissociation which lead to the refilling of anionic vacancies. Previous study showed that partial substitution of Co^{3+} by Fe^{3+} in rhombohedral $LaCoO_3$ structure leads to orthorhombic $LaCo_{1-x}Fe_xO_3$ structure corresponding to an expansion of the unit cell volume and a lengthening of the B-O bond (with B = Co or Fe) [40]. As illustrated in Fig. 5, no clear correlation appears between the length of the B-O bond in $LaCo_{1-x}Fe_xO_3$ catalyst and the specific reaction rate with

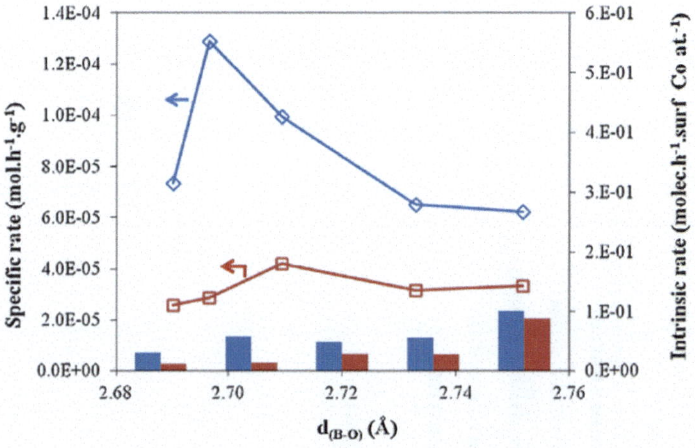

Fig. 5 Correlation between specific and intrinsic rate constants calculated at 525 °C on calcined (blue) and aged $LaCo_{1-x}Fe_xO_3$ catalyst (red) vs. the B-O bond length with B = Fe or Co. Reprinted from Wu et al. [40] with permission from Elsevier. Copyright (2013)

an optimum for $x = 0.2$ corresponding to the higher density of active cobalt sites at the surface. The evolution observed on the normalized reaction rate clearly shows a dependence of the B-O bond length and this tendency appears more distinctly on aged samples. This weakening of the B-O bond is accompanied with a lowering of the apparent activation energy. The deactivation in real gas mixture composition (0.1 vol.% N_2O, 5 vol.% NO, and 6 vol.% O_2, 15 vol.% H_2O) affects more significantly highy loaded Co samples first related to lower density of active cobalt species in octahedral coordination and secondly to partial exsolution leading to CoO_x oxidic species more sensitive to poisoning effect caused by stronger NO adsorption.

Improved hydrothermal stability was observed on La-deficient perovskite structures. $La_{0.8}CoO_3$ and $La_{0.9}Co_{0.8}Fe_{0.2}O_3$ catalysts were found to be the most resistant to deactivation and the most active systems. Based on this, the optimization of Co content and La-deficiency seems to be a good strategy for the development of a more active and stable perovskite catalyst for N_2O decomposition at high temperature [41]. Subsequent improvements can be obtained with appropriate preparation methods that could promote the specific surface area. Previous attempts comparing sol–gel (SG), reactive grinding (RG), and colloidal crystal templating method (T) lead to $LaCoO_3$ samples with specific surface area of 20, 50, and 12 m^2/g, respectively [77]. The best performances in terms of activity in N_2O decomposition was obtained on the sample prepared by reactive grinding. Figure 6 also emphasizes for this specific sample a narrower crystallite size distribution in comparison with the other preparation methods, which could correspond to more homogeneous oxygen mobility.

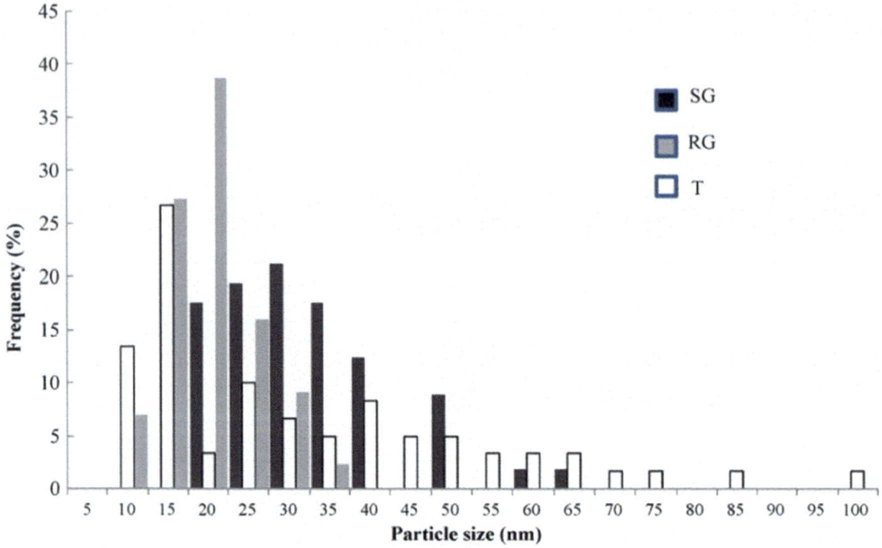

Fig. 6 Distribution of particle size obtained from TEM measurements on $LaCoO_3$(RG), $LaCoO_3$(T), and $LaCoO_3$(SG). Reprinted from Dacquin et al. [77] with permission from Elsevier. Copyright (2009)

2.5 *Kinetics of N_2O Decomposition–Reaction Mechanisms*

2.5.1 Steady-State Approaches

Kinetics and related reaction mechanisms have been widely studied. A first summary of the prominent kinetic features of this reaction was reported elsewhere [4]. The classical reaction schemes 1 and 2 are common to a broad panel of single metal oxides and mixed metal oxides catalysts. Several examples have been selected in this section, which agree with this mechanism proposal such as MgO nanocrystals [78], MoO_x/SiO_2 and ZSM5 exchanged with various transition metals [79], MgO-BaO [80], α-Cr_2O_3-Al_2O_3 [81], Cu-, Co- and FeSSZ13 [74], Fe-ZSM5 [82], Cu-containing zeolite [83], $Mn_xCo_{1-x}Co_2O_4$ [46]. As general trend, the slow step is usually related to O_2 desorption and structural requirements that could accelerate the recombination of two chemisorbed O atoms are often prerequisites. Hence, the evolution of activation energy can reflect the strength of oxygen bind to the surface as the function of the surface composition of the catalysts and the degree of isolation of the active centres, i.e. isolated or clusters. According to the geometry of the site, different reactions mechanisms can be proposed as reported elswhere on Cr-based catalysts [81] (see Fig. 7). As a general trend, the most active transition elements are characterized by a mixed valence. Among them, Co^{2+} appears as the most active species as electron donor. A charge transfer from the transition metal to the antibonding molecular orbitals of N_2O weakens the terminal N–O bond. The presence of vacancies with

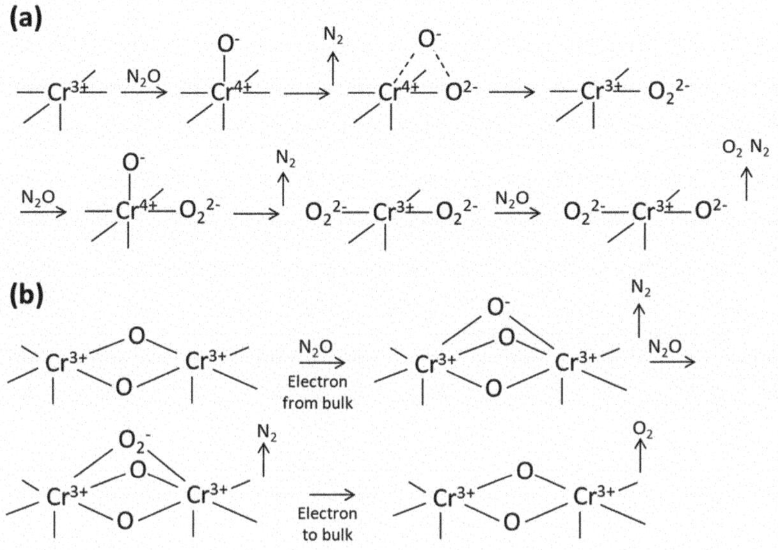

Fig. 7 Reaction scheme for N_2O decomposition over isolated Cr ions (**a**) and for bulk oxide (**b**). Reprinted from Egerton et al. [81] with permission from Elsevier. Copyright (1974)

a trapped electron can be also involved in this process. Indeed, an increase of the electron density on the transition metal can also weaken the Co–O bond and then enhancing O_2 desorption and related formation of anionic vacancies. In the final step of the catalytic cycle, the O_2 desorption would be accompanied by an electron back-donation to restore Co^{2+}. A cluster site combining Co^{3+} and Co^{2+} to ease these charge transfers is currently envisioned as already discussed for Mn^{4+}/Mn^{3+} [84] and Cu^{2+}/Cu^{+} [85].

As depicted in schemes 1 and 2, N_2O activation can go through adsorption, sequential terminal N–O bond breakage producing N_2 and adsorbed O atoms (O_a), surface oxygen diffusion and oxygen desorption. The rate-limiting step is usually assigned to oxygen desorption releasing active sites according to scheme 1. Recently Lin et al. [74] reported different kinetic features on Cu−, Co− and Fe-SSZ-13 zeolites which still meet the same reaction mechanistic insights illustrated in scheme 1. However, the authors assumed that N–O bond breakage was the slow step for this variety of catalysts. These authors also observed a good correspondence between the reaction rate and the activation energy associated with N–O bond scission.

Numerous investigations have combined theoretical and experimental approaches [7, 78, 79]. These investigations can complement the following reaction scheme 1 with cationic or anionic redox mechanisms. In the former one, the two-step sequence involves an electron transfer ($N_2O_{(g)} + e^- \rightarrow N_{2(g)} + O^-$). The transition metal, *i.e.* active Co site acts as the electron donor for the first step and then as electron acceptor in the course of oxygen desorption ($O^- \rightarrow 1/2O_{2(g)} + e^-$) [78]. In that case, a Langmuir–Hinshelwood mechanism can be envisaged involving the sequence in scheme 1. In the presence of NO in the inlet feed gas mixture alternative pathways can be envisioned mostly on Cu- and Fe-doped zeolites [82, 83] which led to a lowering of the apparent activation energy for the N_2O decomposition reaction due to the involvement of an additional step: $NO + O_{ads} \rightarrow NO_2 + S1$.

In case of an anionic redox mechanism, the following two reaction pathways would take place ($N_2O + O^{2-}{}_{(surf)} \rightarrow N_{2(g)} + O_2{}^{2-}{}_{(surf)}$) and then the restoration of active oxygen species ($2O_2{}^{2-}{}_{(surf)} \rightarrow O_{2(g)} + O^{2-}{}_{(surf)}$) [78]. Nevertheless, an alternative step can be proposed as follow: $N_2O + O_2{}^{2-}{}_{(surf)} \rightarrow N_{2(g)} + O_{2(g)} + O^{2-}{}_{(surf)}$ for the restoration of active oxygen species in agreement with scheme 2 [80]. Accordingly, adsorbed oxygen would no longer exert a strong inhibiting effect governed by a slow oxygen desorption.

Scheme 1

$$N_2O + S_1 \ominus N_2O_a \tag{1}$$

$$N_2O_a \rightarrow N_2 + O_a \tag{2}$$

$$2O_a \ominus O_2 + 2S_1 \tag{3}$$

S_1 stands for TM as adsorption site

$$r = k_2\theta_{N2O} = \frac{k_2 K_{N2O} P_{N2O}}{1 + K_{N2O} P_{N2O} + \sqrt{K_{O_2} P_{O_2}}} \tag{4}$$

$$r \cong \frac{k_2 K_{N2O} P_{N2O}}{\sqrt{K_{O_2} P_{O_2}}} \tag{5}$$

$$E_{app} = E_2 + \Delta H_{N2O} - \frac{1}{2}\Delta H_O \tag{6}$$

Scheme 2

$$N_2O + O_a \rightarrow N_2 + S_2 \tag{7}$$

$$N_2O + S_2 \rightarrow N_2 + O_a \tag{8}$$

S_2 stands for oxygen defective site

$$r = 2k_7 P_{N2O}\theta_O \tag{9}$$

$$r = \frac{2k_7 k_8 P_{N2O}}{(k_7 + k_8)} \tag{10}$$

$$E_{app} = f\,(E_7, E_8) \tag{11}$$

where r stands for the reaction rate, k_n the reaction rate constant for the elementary step (n), θ_{N2O} the surface coverage of adsorbed N_2O molecules, θ_O, the surface coverage of chemisorbed O atoms, K_{A_i} and P_{A_i} the adsorption equilibrium constant and the partial pressure of the reactant A_i with $A_i = N_2O$ or O_2, E_{app} the apparent energy of activation, E_n the activation energy for step (n), ΔH_{N2O} and ΔH_O the heat of adsorption for N_2O and O_2 respectively. E_{app} in Eq. (11) can be expressed as a function of E_7 and E_8, the activation energies for steps (7) and (8), respectively.

In the particular case of perovskite structures (ABO_3), their redox properties can be also related to the mixed valency of the transition metal cation in B-site. Substitution in A-site can also alter the valence of the B-site cations. As exemplified in Fig. 8, partial substitution of La^{3+} by Sr^{2+} leads to an optimum corresponding to the highest activity and the lowest activation energy at $x = 0.4$. This has been related to an optimal Mn^{4+}/Mn^{3+} ratio [84]. Most of these studies have been carried out under simulated operating conditions with, in most cases, inlet reaction mixtures containing only N_2O, which significantly differ from the real exhaust gas in the ammonia burner. Do we stand the fact that these results can be extrapolated in complex media in the presence of water, NO, O_2? Under typical industrial conditions, inhibiting effects must be taken into consideration and can definitively impact the value of the apparent activation energy and the relative rates of competitive reactions.

In order to answer the above question, a panel of perovskite catalysts have been compared where $LaCo_{0.8}Fe_{0.2}O_3$ served as benchmark. The effect of partial substitution of La^{3+} by Sr^{2+} has been studied on calcined samples from temperature-programmed reaction experiments (TPR) betweem 100 °C and 900 °C with inlet reaction mixture composed of 0.1 vol.% N_2O, 5 vol.% NO, 6 vol.% O_2, and 15 vol.% H_2O A second TPR experiment was performed on samples aged overnight at 900 °C. For $La_{1-y}Sr_yCo_{0.8}Fe_{0.2}O_3$, the substitution of trivalent La^{3+} by divalent

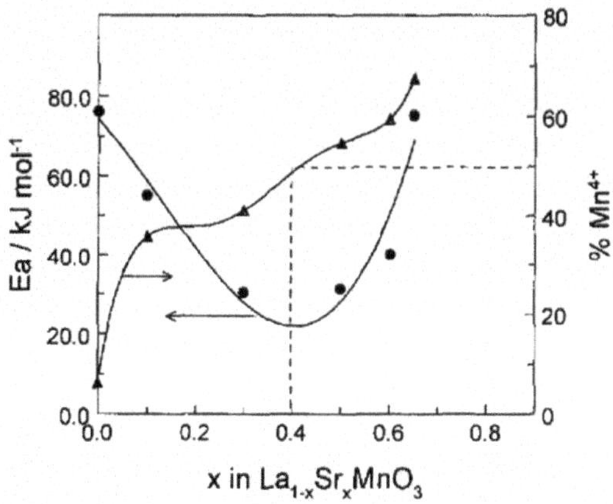

Fig. 8 Variation of the apparent activation energy and atomic % Mn^{4+} as a function of the fraction Mn in $La_{1-x}Sr_xMnO_3$. Reprinted from Raj et al. [84] with permission from Elsevier. Copyright (1982)

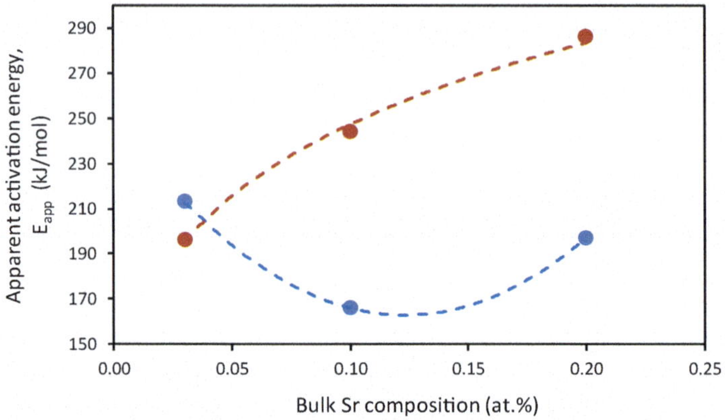

Fig. 9 Plots of apparent activation energy versus bulk Sr composition on calcined (blue symbol) and aged $La_{1-y}Sr_yCo_{0.8}Fe_{0.2}O_3$ catalysts (red symbol). After the first catalytic testing on calcined samples, the catalysts were aged overnight at 900 °C under reaction mixture composed of 0.1 vol.% N_2O, 5 vol.% NO, 6 vol.% O_2, and 15 vol.% H_2O

Sr^{2+} would in principle induce a greater stabilization of Co^{2+} sites to meet the electronic balance in case of pure structure. But, structural characterization also revealed the coexistence of additional CoO_x phase on calcined samples and the growth of $LaSrCoO_3$ on aged samples. A lower reducibility of Co^{3+} species has been assigned to the growth of $LaSrCoO_3$ phase which could explain the lower activity reflected by a continuous increase in the apparent activation energy on aged samples. In contrast, the evolution of E_{app} for calcined samples differs from that observed on aged samples with a minimum value recorded for $y = 0.1$. This could be rationalized by the presence of more reducible Co^{3+} species corresponding to an optimal Co^{3+}/Co^{2+} ratio as previously discussed for substituted manganite perovskite [84] (see Fig. 8).

2.5.2 Unsteady-State Kinetic Approaches

Oxygen isotopic oxygen exchange is helpful to investigate the competitive adsorption of O_2 and N_2O on same active sites corresponding to oxygen vacancies or metal cations as previously emphasized in Scheme 1. Indeed, this mechanism points out the influence of surface oxygen binding energy on the kinetics. Steady-state isotope transient kinetic analysis (SSITKA) is suited to relate the reactivity of surface species to the formation of reaction product. This method does not need assumptions for establishing rate expression and can be useful to investigate a reaction in a broad range of temperature especially when the identification of a single rate-limiting step is not evident. At high temperature, the catalytic process will involve not only the outermost layer, but also sub-surface oxygen species [76, 86] with improved bulk oxygen mobility. Ivanov et al. investigated N_2O decomposition of $LaMnO_{3-\delta}$ [86] and $La_{1-y}Sr_yMnO_3$ [76] at 900 °C with $y = 0$, 0.3, and 0.5 by steady-state isotope transient kinetic analysis. In practice, SSITKA consists of replacing $^{12}O_2$ by $^{18}O_2$ while maintaining the catalyst under steady-state conditions. The fraction of ^{18}O will depend both on adsorption and exchange at the surface and diffusion of labelled oxygen species in the bulk structure of the catalyst. $La_{1-y}Sr_yMnO_3$ perovskite materials can combine both the presence of metal oxide exhibiting mixed valence and anionic vacancies that can promote the bulk diffusion. According to the reaction temperature the stabilization of metal oxide cations or the formation of lattice oxygen species can be the driving force in determining the catalyst performances respectivity at low temperature [87] and high temperature [88]. The mechanism of oxygen transport in perovskite type oxide can reflect: the prevalence of faster oxygen diffusion along the grain boundaries than through the bulk [89]—the formation of high concentration of oxygen vacancies, which would ease oxygen diffusion [90]—the stabilization of highly dispersed metal oxide, with metal cation stabilized in its highest oxidation state, improving oxygen lability and related formation of surface oxygen vacancies. Starting from a simple mechanism scheme (Fig. 10b) assuming an immediate oxygen desorption coming from N_2O dissociation, Ivanov et al. found that this assumption does not match with the evolution of the isotopic distribution of oxygen in the gas phase, i.e. $^{16}O_2$ (f_{32}), $^{18}O_2$ (f_{36}), and $^{18}O^{16}O$ (f_{34}) versus time. Indeed, a sharp increase of $^{16}O_2$ (f_{32}) is expected according to this hypothesis while they observed a gradual increase suggesting a significant transfer to the bulk. The best agreement between experimental and predicted data from the model assumes that two oxygen ^{18}O lattices are involved in label oxygen transfer from N_2O dissociation to O_2 desorption according to Eq. (12)

$$2N_2^{16}O + 2^{18}O_{bulk} \xrightarrow{R_{N2O}} {}^{18}O_2 + 2^{16}O_{bulk} + 2N_2 \quad (12)$$

The lowest catalytic performances of $LaMnO_3$ in N_2O composition compared to Sr-substituted composition have been explained by the slowest oxygen exchange and the lowest bulk oxygen mobility. On the other, both processes are promoted on Sr-substituted samples with an optimum obtained for the composition $La_{0.5}Sr_{0.5}MnO_3$ [76]. As exemplified in Fig. 11 a direct correlation between the rate of N_2O conversion expressed per square metre, and the bulk oxygen diffusion coefficient D is observed. This comparison has been interpreted by the occurrence of fast diffusion of oxygen, thanks to the formation of oxygen vacancies to compensate the reduction of the oxidation state of cation.

N_2O decomposition has also been investigated in the temperature 550–700 °C from Temporal Analysis of Products (TAP) experiments over Fe-MFI and $BaFeAl_{11}O_{19}$ [23]. Transient responses of N_2O, N_2, and O_2 were modelled according to several micro-kinetic models. Results agree with different reactions' pathways on Fe-MFI and $BaFeAl_{11}O_{19}$. The simultaneous production of N_2 and O_2 from N_2O decomposition on $BaFeAl_{11}O_{19}$ can be explained by a reaction between gaseous N_2O and a bi-atomic ($*$–O_2) oxygen species according to step (13) releasing N_2 and O_2. On the contrary, the production of O_2 on Fe-MFI would involve a three steps sequence including a reaction between gaseous N_2O with an empty site leading to

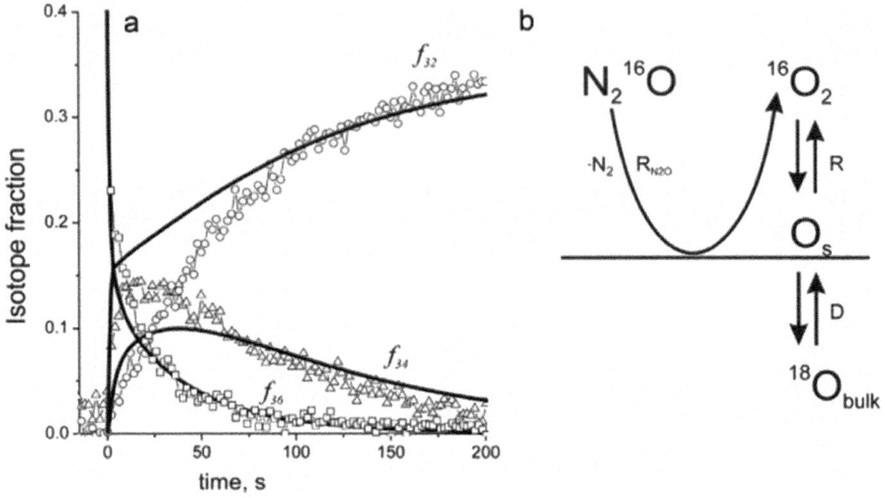

Fig. 10 **a** Experimental (symbols) and calculated (lines) isotopic responses of $^{16}O_2$ (f_{32}), $^{18}O_2$ (f_{36}) and $^{18}O^{16}O$ (f_{34}), obtained after switch $^{18}O_2/N_2{}^{16}O$ on $LaMnO_{3+\delta}$ (T = 900 °C) scheme of ^{16}O transfer during $^{18}O_2/N_2{}^{16}O$ decomposition. Reprinted from Ivanov et al. [86] with permission from Elsevier. Copyright (2016)

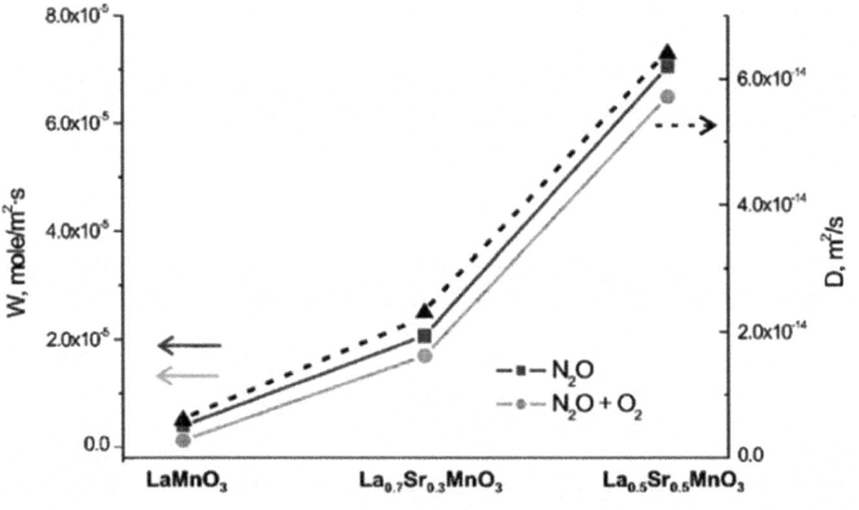

Fig. 11 A direct correlation between catalytic activity of the $La_{1-x}Sr_xMnO_3$ (x = 0, 0.3 and 0.5) samples in N_2O decomposition in the absence () or in the presence () of oxygen (900 °C, contact time 5 × 10^{-4} s) and diffusion coefficient (). Reprinted from Ivanov et al. [76] with permission from Elsevier. Copyright (2009)

adsorbed $*{-}O$ species (step (14)) on which N_2O would react to form $*{-}O_2$ species (step (15)). Finally, O_2 desorption according to step (16) would release $*$ active centres. The catalytic activity of Fe-MFI has been related to the collision frequency of N_2O on $*{-}O$ sites and the rearrangment of (O-$*$-O) adsorbed bi-atomic oxygen species to $*{-}O_2$ species in Fe-MFI depending on the degree of isolation of iron species [23].

$$N_2O +_* -O_2 \rightarrow N_2 + O_2 +_* -O \quad (13)$$

$$N_2O+_* \rightarrow N_2 +_* -O \quad (14)$$

$$N_2O +_* -O \rightarrow N_2 +_* -O_2 \quad (15)$$

$$_*-O_2 \rightarrow O_2+_* \quad (16)$$

2.6 *Theoretical Versus Experimental Approaches for the Identification of Active Sites*

2.6.1 Molecular Modelling

Density functional theory (DFT) calculation is a useful approach especially on single metal oxides as well as mixed metal oxides exhibiting mixed valence for the active element which can also correspond to different coordination geometries in the bulk and at the surface. The utilization of cobalt oxides in many reactions is generally explained by its capacity to form reactive oxygen species at the surface and release lattice oxygen to form anionic vacancies. DFT calculations can offer a quantitative approach through the prediction and the comparison of adsorption energies for the identification of the most stable adsorption configurations and energy barriers for subsequent elementary steps especially those corresponding to N_2O dissociation. Suo et al. have investigated the adsorption and dissociation of N_2O on CuO surfaces [91]. They found that the energy of N_2O dissociation is much lower on oxygen-deficient surface (oxygen vacancies) and precovered by oxygen on CuO(111) than on a perfect surface. DFT calculations can also predict optimal composition then demonstrating that a preliminary descrimination can precede an experimental approach. For instance, DFT calculations on (001) facet of $LaBO_3$ perovskite structure with B = Mn, Co or Ni can identify the most stable adsorption configuration through the terminal oxygen [92]. The computation of energy barrier for N–O dissociation of respectivity 0.99 eV, 1.65 eV, and 2.24 eV on $LaMnO_3$, $LaCoO_3$, and $LaNiO_3$ presumes a better catalytic activity for $LaMnO_3$. Another key point lies in the stability of intermediates from N_2O dissociation. Interestingly, desorption energy was found rather low in the range 0.2–0.5 eV, which suggests a fast release of oxygen from the surface. Such methodology is not restricted to specific facets, different local structures can be considered in iron-doped zeolite. For instance, isolated and dimer iron species are characterized in Fe/SSZ-13 by considering different structure. By calculating the energy of the system, it was found that isolated Fe species preferentially bind to a single Al site and dimer species bind to two Al sites in SSZ-13 whereas dimeric species in SAPO-34 would preferentially be coordinated to one Al and one Si [74]. The comparison of the overall barrier shows a lower value on the dimer (0.27 eV vs. 0.75 eV on monomer) showing that N_2O decomposition will occur more readily in the former case. In addition, O_2 desorption is more energy demanding on Fe–O monomer.

New materials have been recently developed for N_2O decomposition on Fe-embedded C_2N monolayer [93]. First considerations from DFT calculations provide interesting comparisons. Indeed, the authors computed the energy barriers of the elementary steps (17)-(19) in order to check if CO is needed to suppress a strong oxygen-inhibiting effect.

$$N_2O \rightarrow N_2 + O_* \tag{17}$$

$$CO + O_* \rightarrow CO_2 \tag{18}$$

$$N_2O + O_* \rightarrow N_2 + O_2 \tag{19}$$

In practice the comparison of the activation barriers can verify if the decomposition can be envisioned or if the use of CO as reducing agent is recommended to clean more readily the surface. The numerical values of respectivity 8.4 kcal/mol, 3.2 kcal/mol, and 18.8 kcal/mol emphasize the high value for step (19) but it remains rather low in comparison with a broad variety of catalysts. Thanks to the exothermicity of the overall reaction, the authors conclude that the energy barrier of step (19) is not enough high to slower the reaction. The heat released could help the crossing of energy barrier for step (19) then avoiding the oxygen inhibiting effect.

2.6.2 New Insights into Mechanistic Information from Operando Spectroscopic Investigations

The literature is scarce concerning in situ and *operando* studies for N_2O decomposition while such experiments could provide some information of capital importance on: (i) the nature of active sites in zeolite materials [55] and (ii) the impact of inhibiting effects of NO and H_2O [96].

The usefulness of this methodology has been proven in understanding the assistance of NO in N_2O decomposition. A comparative investigation on Co-MOR and Fe-MOR from in situ UV-visible and *operando* Fourier transform infrared (FTIR) investigation reveals opposite behaviour in the presence and in the absence of NO.

Indeed, Co-MOR is more active than Fe-MOR in the absence of NO, while the reverse tendency is observed in the presence of NO. Such deviation can be briefly summarized by examining Fig. 12. In the presence of NO, the growth of the 1876 cm^{-1} IR band ascribed to nitrosyl species at 300 °C progressively attenuates and disappears with a rise in temperature. Jointly additional IR bands assigned to nitrites/nitrates developed. Such evolutions do not appear on Co-MOR in connection with a weak effect of NO in N_2O decomposition. The beneficial effect of NO for Fe-MOR can be rationalized by an alternative reaction scheme which would be no longer involved Co^{3+}-O^- active site (Scheme 3) but dimeric Fe^{3+}-$O^{(1+\delta)+}$-$Fe^{(2+\delta)+}$ species (Scheme 4). This dimeric species exhibits high stability. Hence, a faster restoration of Fe^{2+}–Fe^{2+} can involve the addition of NO, which can act as an O-oxygen scavenger to desorb oxygen.

Scheme 3

$$N_2O + Co^{2+} \rightarrow N_2 + Co^{3+}O^-_{ads}. \tag{20}$$

$$Co^{3+}O^-_{ads} + N_2O \rightarrow N_2 + O_2 + Co^{2+} \tag{21}$$

Scheme 4

$$Fe^{2+}Fe^{2+} + N_2O \rightarrow Fe^{3+} - O^{(1+\delta)+} - Fe^{(2+\delta)+} + N_2 \tag{22}$$

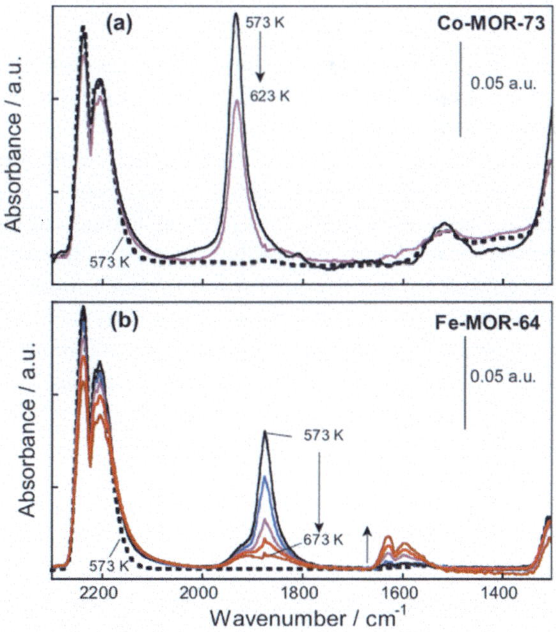

Fig. 12 Operando FTIR spectra of surface species on standard activated Co-MOR-73 (**a**), and Fe-MOR-64 (**b**) after saturation under N_2O flow at 573 K (dotted line spectra), under (N_2O + NO) flow at the same temperature (bold line spectra), and for increasing temperature. Reactant concentration: $[N_2O] = [NO] = 0.4\%$, total flow rate = 50 cm^3 STP/min, He as balance. Reprinted from Pietrogiacomi et al. [55] with permission from Elsevier. Copyright (2019)

$$Fe^{3+} - O^{(1+\delta)+} - Fe^{(2+\delta)+} + N_2O \rightarrow N_2 + O_2 + Fe^{2+}Fe^{2+} \quad (23)$$

The combination of operando UV–vis and DRIFT experiments also provides useful mechanistic insights involving $[Cu–O-Cu]^{2+}$ intermediate in CuO/CeO_2 [94]. The recombination of two oxygen ions is not favored from mono(oxo) dicopper sites. This recombination is promoted thanks to labile oxygen coming from CeO_2 then inducing a valence change for cerium from (+IV) to (+III). Hence, the redox couples Cu^{2+}/Cu^{+} and Ce^{4+}/Ce^{3+} cooperate according to reaction mechanims in Fig. 13a.

This reaction mechanism correctly explains the inhibiting effect associated with NO and water adsorption (see Fig. 13b) although these inhibiting effects differ in intensity. Indeed, NO has only a weak detrimental effect which is not related to competition for adsorption with N_2O on a single site, but would be more reasonably explained by a faster removal of active oxygen for N_2O decomposition through parallel NO oxidation. On the other hand, a strong inhibiting effect of water on the rate of N_2O decomposition is noticeable. In fact, hydroxyl groups, visualized by an intense and broad adsorption band at 3642 cm^{-1}, displace labile oxygen species, and prevent the regeneration of Cu^{+} dimeric species.

3 Challenges in Real Operating Conditions

3.1 Model of Predictions

The attempts to model heterogeneous reactors for N_2O decomposition are relatively scarce. Pseudo-homogeneous one-dimensional model of an ideal plug flow reactor has been earlier reported [95].

The mathematical models take essentially monolithic reactors into account. The level of description varies from 1 to 3D models, which account for the hydrodynamics, mass and heat transfers. Although numerous investigations have been already reported for modelling monolithic reactors such as catalytic converters, few of them have considered the decomposition of N_2O either for adipic acid plants or nitric acid plants [96, 97]. The level of description is closely related to the complexity of the model, and in general, simplifications are assumed to shorten the computation. Indeed, a laminar flow, an isothermal behaviour, *i.e.* strictly applicable in case of low N_2O concentration, small pressure drops, and ideal behaviour of gas inside the monolith are currently assumed. Let us note that computation can be also restricted to a single channel [97]. Bernauer et al. [96] obtained from a monolith washcoated by FeO_x/Al_2O_3 a complete conversion at high temperature and pressure corresponding to real operating conditions in the ammonia burner. Their predictions also conclude that the comparable size of a packed bed and a monolithic reactor leads to the same performance, then offering a cost-efficient solution to minimize the back pressure and reduce the amount of catalyst monolith.

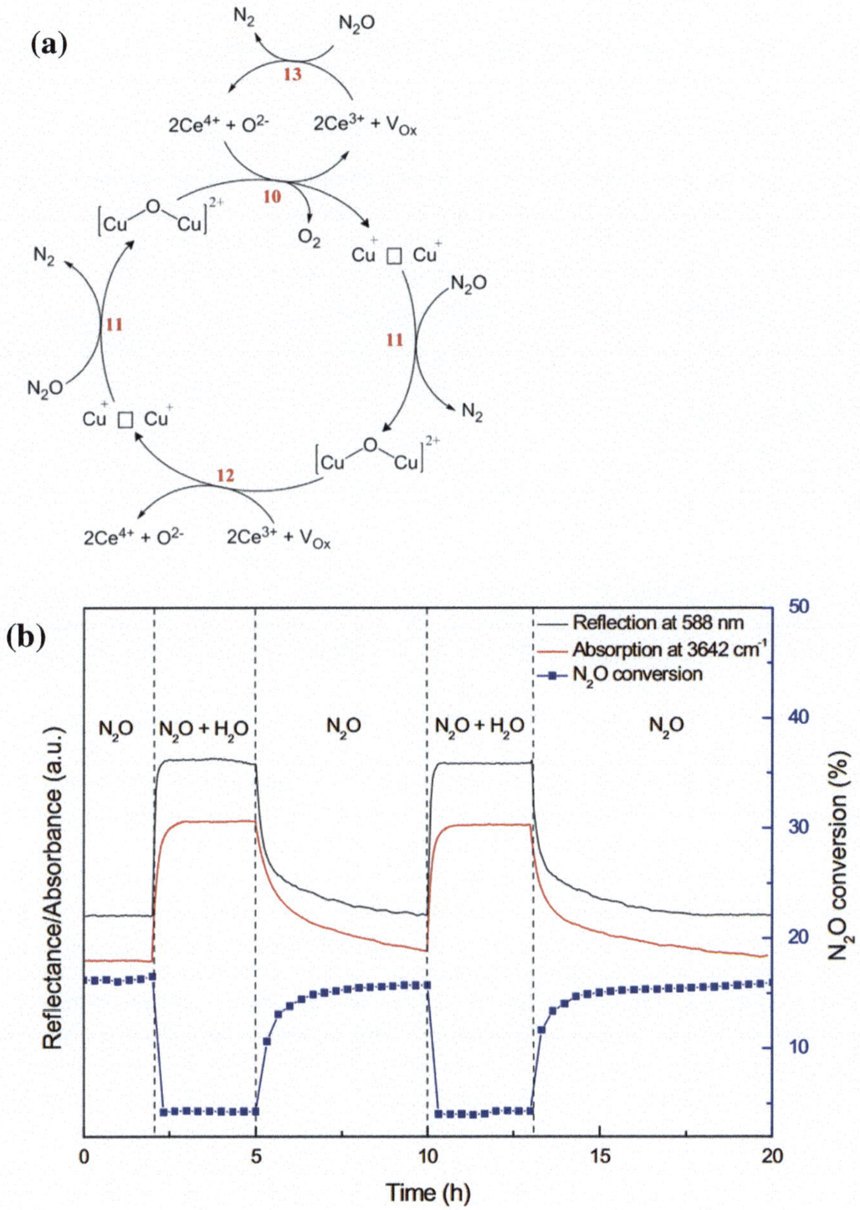

Fig. 13 Proposed reaction mechanism for N_2O decomposition on CuO/CeO_2 (**a**), and *operando* UV–vis reflection at 588 nm and DRIFT absorption at 3642 cm^{-1} with N_2O conversion during step-change catalytic experiments (**b**). Reprinted from Zabilskiy et al. [94] with permission from Elsevier. Copyright (2016)

The development of microreactor for the replacement of current industrial technologies has been recently explored [98]. Thanks to this type of reactor, the high surface-to-volume ratio improves heat transfer. Heshmatifar et al. checked the relevance of this methodology for low-temperature application in the range 280–360 °C on Pd/anodic γ-Al_2O_3/Al. The key point of this achievement lies in the uniformity of the porous support and its thickness. Such prerequisites can be fullfilled by the chemical vapour deposition [99], electrophoretic deposition, or anodization [100]. In the selected example, Pd deposition by anodization led to stronger interactions with the support and exhibited superior performances than a packed bed reactor operating in similar conditions, and gas-hourly-space-velocity.

3.2 Deactivation

The durability of a catalytic technology is a key point in the economic model for its subsequent development at the industrial scale. The topology will depend on the operating temperature conditions and gas composition. Hence, the catalyst features will strongly differ, *i.e.* supported vs. bulk catalysts, and the deactivation can be reversible or irreversible according to the catalyst design. For low-temperature application, the use of Platinum Metal Group highly dispersed on porous support materials, developing high specific surface area, is usually privileged. By way of illustration, Yentekakis et al. [101] found that the sensitivity of Ir nanoparticles to thermal sintering is closely related to the oxygen mobility of the support materials. For instance, negligible sintering appeared once Ir was supported on gadolinia-ceria support materials. In contrast, Pd/γ-Al_2O_3 was found highly sensitive because of the weak oxygen storage capacity (OSC) properties of alumina.

For medium and high-temperature application, the gas composition must be taken into account. Most investigations at the lab-scale were conducted under simulated conditions usually far from those characterizing industrial reactors. Typically, the exhaust gas of an ammonia burner is composed of a large amount of NO_x, O_2, steam, and only a trace amount of N_2O. Based on this, the competitive adsorptions must be considered at low and medium temperatures. This competition is also governed by the composition of the catalyst surface developing different affinities towards NO_x, O_2, and water adsorption. Langmuir models can be used for evaluating the inhibiting effect of those contaminants on the rate of N_2O decomposition. Grzybek et al. [102] found on K-$Zn_xCo_{3-x}O_4$ catalysts a stronger NO inhibition in the temperature range of 275–475 °C due to the stabilization of ad-NO_x species which decompose above 550° C. Strong NO adsorption also induces morphological changes with a stronger affinity of (111) facet than (100) surface. Such findings could be profitably used for engineering more resistant catalytic systems to poisoning effects. The sensitivity of Fe/ZSM5 to trace amount of steam has been observed with a loss of N_2O conversion. But the catalytic activity can be restored after calcination at 500 °C. This reversible deactivation has been assigned to the hydroxylation of dehydroxylated binuclear Fe sites [103]

As a matter of fact, the resistance to deactivation will also depend on the catalysts preactivation. The investigation of Fe/ZSM5 in the temperature range of 275–600 °C, with inlet gas mixture composed of 5000 ppm N_2O, 5% O_2, 500 ppm NO, and 5% H_2O diluted in Ar with a GHSV of 30,000 h^{-1}, revealed the beneficial effect of alkaline pretreatment on the hydrothermal stability of Fe-ZSM5. The resulting meso-microporous hybrid structure accompanied with higher density of iron active sites would originate the gain observed in the performances [104]. In the case of K-$Zn_xCo_{3-x}O_4$ catalysts, the calcination temperature is a crucial parameter to obtain an optimal K-promotional effect. Indeed, the migration of potassium enhanced at elevated calcination temperature leads to the formation of a K_xCoO_2 overlayer which induces a detrimental effect in the decomposition of N_2O below 500 °C [105].

3.3 Development of Structured Catalysts

In most cases, shaped catalysts are preferentially used in pilot reactors. Pelletized catalysts are commonly used [106]. However, they suffer from weak mechanical strength, inducing significant pressure drop and diffusion limitation affecting their efficiency. Remarkable results have been obtained on $Co_3O_4/\alpha\text{-}Al_2O_3$/cordierite and cobalt oxide electrochemically deposited on stainless steel sieves [32, 48]. Compared with cobalt oxide supported on porous TiO_2 substrate, higher conversion level is obtained on this non-porous catalyst due to negligible internal diffusion phenomena. Similar observation is also reported on $Co_3O_4/\alpha\text{-}Al_2O_3$/cordierite with a faster diffusion which facilitates obtaining a kinetic regime. The use of open-cell foams as support has been also explored. They exhibit the same advantages as monolithic reactors, but also exhibit high external surface area due to their porosity leading to superior performance than catalysts in powder form. The enhancement in catalytic activity lies in high dispersion of the active phase by using a simple wet impregnation, and also on the dopant to optimize the performances [25].

4 Conclusion and Outlooks

In the selected case study dedicated to the abatement of N_2O, several strategies can be envisioned. In terms of cost efficiency, the secondary option with catalysts inserted into the ammonia burner outstands. The development of catalyst for N_2O decomposition at medium temperature (200–400 °C) has received great attention. In this temperature range, mixed metal oxides and doped-zeolites have been extensively studied highlighting some differences among the nature of the transition metal, i.e. Cu, Co, or Fe. Some controversies also appear on the nature of active sites depending on the extent of dispersion and degree of isolation. The introduction of molecular modelling approach can a priori discriminate different scenarios in terms of the nature of active sites, structural requirement, and related reaction mechanism. Despite the scarcity of literature, *operando* spectroscopic investigation also contributes to elucidate the composition of active sites during the reaction. However, all these investigations have some limitations to describe precisely the catalyst functionalities and their stability under real operating conditions. This is even particularly true for high-temperature applications above 700 °C, where fewer investigations have been conducted.

The performances of perovskite type structures have been underlined in the temperature range 500–900 °C, exposed to gas composition similar to those encountered in the ammonia burner. Their bulk and surface properties can be modified through partial substitution in A- and B-site that change the oxidation state of active elements. It is found that calcined $LaCo_{0.8}Fe_{0.2}O_3$ outperforms all the other systems but strongly deactivates. Improved performances can be observed on Sr-substituted $La_{0.9}Sr_{0.1}Co_{0.8}Fe_{0.2}O_3$ catalyst for which an optimal Co^{3+}/Co^{2+} ratio would be reached inducing a lowering of the activation barrier. The gain obtained in catalytic performance can be rationalized based on SSITKA measurements, which demonstrated that catalytic performances can be related to improved oxygen mobility and related stabilization of anionic vacancies.

Acknowledgements The laboratory participates in the Institut de Recherche en ENvironnement Industriel (IRENI), which is financed by the Communauté Urbaine de Dunkerque, the Région Nord Pas-de-Calais, the Ministère de l'Enseignement Supérieur et de la Recherche, the CNRS, European Fund for Regional Development (FEDER), and the Agency for Ecological Transition (ADEME).

References

1. J. Wei, in *Advance in Catalysis*. ed. by D.D. Eley, H. Pines, P.B. Weisz (Academic Press, New York, 1975), p.57
2. M.M. Bettahar, G. Costentin, L. Savary, J.C. Lavalley, On the partial oxidation of propane and propylene on mixed metal oxide catalysts. Appl. Catal. A **145**, 1–48 (1996). https://doi.org/10.1016/0926-860X(96)00138-X
3. J. Pérez-Ramiréz, Prospects of N_2O emission regulations in the European fertilizer industry. Appl. Catal. B **70**, 31–35 (2007). https://doi.org/10.1016/j.apcatb.2005.11.019
4. J.A. Kapteijn, J.A. Rodriguez-Mirasol, Moulijn, Heterogeneous catalytic decomposition of nitrous oxide. Appl. Catal. B **9**, 25–64 (1996). https://doi.org/10.1016/0926-3373(96)90072-7
5. J. Pérez-Ramiréz, F. Kapteijn, K. Schöfel, J.A. Moulijn, Formation and control of N_2O in nitric acid production. Where do we stand today?. Appl. Catal. B **44**, 117–151 (2003). https://doi.org/10.1016/S0926-3373(03)00026-2
6. S. Alini, F. Basile, S. Blasioli, C. Rinaldi, A. Vaccari, Development of new catalysts for N_2O-decomposition from adipic acid plant. Appl. Catal. B **70**, 323–329 (2007). https://doi.org/10.1016/j.apcatb.2005.12.031
7. M. Konsolakis, Recent advance in nitrous oxide (N_2O) decomposition over non-noble-metal oxide catalysts: catalytic performance, mechanistic considerations and surface chemistry aspect. ACS Catal. **5**(11), 6397–6421 (2015). https://doi.org/10.1021/acscatal.5b01605
8. V. Schumacher, G. Bürger, T. Fetzer, M. Baier, M. Hesse (BASF), WO 9955621, 1999
9. P. Granger, P. Esteves, S. Kieger, L. Navascues, G. Leclercq, Effect of yttrium on the performances of zirconia based catalysts for the decomposition of N_2O at high temperature. Appl. Catal. B **62**, 236–243 (2006). https://doi.org/10.1016/j.apcatb.2005.07.015

10. F.J. Perez-Alonso, I. Melian-Cabrera, M. Lopez-Granados, F. Kapteijn, J.L.G. Fierro, Synergy of $Fe_xCe_{1-x}O_2$ mixed oxides for N_2O decomposition. J. Catal. **239**, 340–346 (2006). https://doi.org/10.1016/j.jcat.2006.02.008
11. P. Esteves, Y. Wu, C. Dujardin, M.K. Dongare, P. Granger, Ceria–zirconia mixed oxides as thermal resistant catalysts for the decomposition of nitrous oxide at high temperature. Catal. Today **176**, 453–457 (2011). https://doi.org/10.1016/j.cattod.2010.10.068
12. G. Giecko, T. Borowiecki, W. Gac, J. Kruk, Fe_2O_3/Al_2O_3 catalysts for the N_2O decomposition in the nitric acid industry. Catal. Today **137**, 403–409 (2008). https://doi.org/10.1016/j.cattod.2008.02.008
13. P. Stelmachowski, F. Zasada, G. Maniak, P. Granger, M. Inger, M. Wilk, A. Kotarba, Z. Sojka, Optimization of multicomponent cobalt spinel catalyst for N_2O abatement from nitric acid plant tail gases: laboratory and pilot plant studies. Catal. Lett. **130**, 637–641 (2009). https://doi.org/10.1007/s10562-009-0014-z
14. M. Inger, M. Wilk, M. Saramok, G. Grzybek, A. Grodzka, P. Stelmachowski, W. Makowski, A. Kotarba, Z. Sojka, Cobalt Spinel catalyst for N_2O abatement in the pilot plant operation–long-term activity and stability in tail gases. Ind. Eng. Chem. Res. **53**(25), 10335–10342 (2014). https://doi.org/10.1021/ie5014579
15. L. Obalová, K. Jirátová, F. Kovandac, M. Valášková, J. Balabánová, K. Pacultová, Structure–activity relationship in the N_2O decomposition over Ni-(Mg)-Al and Ni-(Mg)-Mn mixed oxides prepared from hydrotalcite-like precursors. J. Mol. Catal. A **248**, 210–219 (2006). https://doi.org/10.1016/j.molcata.2005.12.037
16. K. Pacultová, K. Karásková, F. Kovanda, K. Jirátová, J. Šrámek, P. Kustrowski, A. Kotarba, Ž Chromčáková, K. Kočí, L. Obalova, K-doped Co-Mn-Al mixed oxide catalyst for N_2O abatement from nitric acid plant waste gases: pilot plant studies. Ind. Eng. Chem. Res. **55**, 7076–7084 (2016). https://doi.org/10.1021/acsiecr.6b01206
17. Ž Chromčáková, L. Obalová, P. Kustrowski, M. Drozdek, K. Karásková, K. Jirátová, F. Kovanda, Optimization of Cs content in Co–Mn–Al mixed oxide as catalyst for N_2O decomposition. Res. Chem. Intermed. **41**, 9319–9332 (2015). https://doi.org/10.1007/s11164-015-2008-3
18. M. Jabłońska, M. Agote Arán , A.M. Beale, G. Delahay, C. Petitto, M. Nocuń, R. Palkovits, Understanding the origins of N_2O decomposition activity in $Mn(Fe)CoAlO_x$ hydrotalcite derived mixed metal oxides. Appl. Catal. B **243**, 66–75 (2019). https://doi.org/10.1016/j.apcatb.2018.10.010
19. H. Cheng, Y. Huang, A. Wang, L. Li, X. Wang, T. Zhang, N_2O decomposition over K-promoted Co-Al catalysts prepared from hydrotalcite-like precursors. Appl. Catal. B **89**, 391–397 (2009). https://doi.org/10.1016/j.apcatb.2008.12.018
20. J. Pérez-Ramiréz, M. Santiago, Metal-substituted hexaaluminates for high-temperature N_2O abatement. Chem. Commun. **6**, 619 (2007). https://doi.org/10.1039/b613602h
21. M. Tian, A. Wang, X. Wang, Y. Zhu, T. Zhang, Effect of large cations (La^{3+} and Ba^{2+}) on the catalytic performance of Mn-substituted hexaaluminates for N_2O decomposition. Appl. Catal. B **92**, 437–444 (2009). https://doi.org/10.1016/j.apcatb.2009.09.002
22. Y. Zhang, X. Wang, Y. Zhuc, T. Zhanga, Stabilization mechanism and crystallographic sites of Ru in Fe-promoted barium hexaaluminate under high-temperature condition for N_2O decomposition. Appl. Catal. B **129**, 382–393 (2013). https://doi.org/10.1016/j.apcatb.2012.10.001
23. E.V. Kondratenko, V.A. Kondratenko, M. Santiago, J. Pérez-Ramírez, Mechanism and micro-kinetics of direct N_2O decomposition over $BaFeAl_{11}O_{19}$ hexaaluminate and comparison with Fe-MFI zeolites. Appl. Catal. B **99**, 66–73 (2010). https://doi.org/10.1016/j.apcatb.2010.05.033
24. M. Santiago, J.C. Groen, J. Pérez-Ramírez, Carbon-templated hexaaluminates with enhanced surface area and catalytic performance. J. Catal. **257**, 152–162 (2008). https://doi.org/10.1016/j.jcat.2008.04.017
25. W. Burckardt, M. Voigt (Porzellanwerk Kloster Veilsdor), EP 1147813 A2, 2001

26. J. Neumann, L. Isopova, L. Pinaeva, N. Kulikovskaya, L. Zolotarskii (Umicore), WO2007104403 A1, 2007
27. Y. Wu, C. Dujardin, C. Lancelot, J.P. Dacquin, V.I. Parvulescu, M. Cabié, C.R. Henry, T. Neisius, P. Granger, Catalytic abatement of NO and N_2O from nitric acid plants: A novel approach using noble metal-modified perovskites. J. Catal. **328**, 236–247 (2015). https://doi.org/10.1016/j.jcat.2015.02.001
28. M. Zabilskiy, B. Erjavec, P. Djinović, A. Pintar, Ordered mesoporous CuO–CeO_2 mixed oxides as an effective catalyst for N_2O decomposition. Chem. Eng. J. **254**, 153–162 (2014). https://doi.org/10.1016/j.cej.2014.05.127
29. M. Lybaki, E. Papista, S.A.C. Carabineiro, P.B. Tavares, M. Konsolakis, Optimization of N_2O decomposition activity of CuO–CeO_2 mixed oxides by means of synthesis procedure and alkali (Cs) promotion. Catal. Sci. Technol. **8**, 2312–2322 (2018). https://doi.org/10.1039/c8cy00316e
30. P.J. Smeets, M.H. Grootheart, R.M. van Teeffelen, H. Leeman, E.J.M. Hensen, R.A. Schoonheydt, Direct NO and N_2O decomposition and NO-assisted N_2O decomposition over zeolites: Elucidating the influence of the Cu-Cu distance on oxygen migration. J. Catal. **245**, 358–368 (2007). https://doi.org/10.1016/j.jcat.2006.10.017
31. T. Franken, R. Palkovits, Investigation of potassium doped mixed spinels $Cu_xCo_{3-x}O_4$ as catalyst for an efficient N_2O decomposition in real conditions. Appl. Catal. B **176–177**, 298–305 (2015). https://doi.org/10.1016/j.apcatb.2015.04.002
32. A. Klyushina, K. Pacultová, S. Krejčová, G. Słowik, K. Jirátová, F. Kovanda, J. Ryczkowski, L. Obalova, Advantages of stainless steel sieves as support for catalytic N_2O decomposition over K-doped Co_3O_4. Catal. Today **257**, 2–10 (2015). https://doi.org/10.1016/j.cattod.2015.05.015
33. K. Pacultová, A. Klegova, T. Kiška, D. Fridichová, A. Martaus, A. Rokicińska, P. Kuśtrowski, L. Obalová, Effect of support on the catalytic activity of Co_3O_4-Cs deposited on open-cell ceramic foams for N_2O decomposition. Mater. Res. Bull. **129**, 110892 (2020). https://doi.org/10.1016/j.materresbull.2020.110892
34. J.B. Lim, S.H. Cha, S.B. Hong, Direct N_2O decomposition over iron-substituted small-pore zeolite with different pore topologies. Appl. Catal. B **243**, 750–759 (2019). https://doi.org/10.1016/j.apcatb.2018.10.068
35. P. Xie, Y. Luo, Z. Ma, L. Wang, C. Huang, Y. Yue, W. Hua, Z. Gao, CoZSM-11 catalysts for N_2O decomposition: Effect of preparation methods and nature of active sites. Appl. Catal. B **170–171**, 34–42 (2015). https://doi.org/10.1016/j.apcatb.2015.01.027
36. Y. You, H. Chang, L. Ma, L. Guo, X. Qin, J. Li, J. Li, Enhancement of N_2O decomposition performance by N_2O pretreatment over Ce-Co-O catalyst. Chem. Eng. J. **347**, 184–192 (2018). https://doi.org/10.1016/j.cej.2018.04.081
37. H.M. Choi, S.J. Lee, S.H. Moon, T.N. Phan, S.G. Jeon, C.H. Ko, Comparison between unsupported mesoporous Co_3O_4 and supported Co_3O_4 on mesoporous silicas as catalysts for N_2O decomposition. Catal. Comm. **82**, 50–54 (2016). https://doi.org/10.1016/j.catcom.2016.04.022
38. M. Zabilskiy, P. Djinovíc, B. Erjavec, G. Dražić, A. Pintar, Small CuO clusters on CeO_2 nanosphere as active species for catalytic N_2O decomposition. Appl. Catal. B **163**, 113–122 (2015). https://doi.org/10.1016/j.apcatb.2014.07.057
39. P. Xie, Y. Luo, Z. Ma, C. Huang, C. Miao, Y. Yue, W. Hua, Z. Gao, Catalytic decomposition of N_2O over Fe-ZSM11 catalysts prepared by different methods: Nature of active Fe species. J. Catal. **330**, 211–322 (2015). https://doi.org/10.1016/j.jcat.2015.07.010
40. Y. Wu, C. Cordier, E. Berrier, N. Nuns, C. Dujardin, P. Granger, Surface reconstructions of $LaCo_{1-x}Fe_xO_3$ at high temperature during N_2O decomposition in realistic exhaust gas composition: Impact on the catalytic properties. Appl. Catal. B **140–141**, 151–163 (2013). https://doi.org/10.1016/j.apcatb.2013.04.002
41. Y. Wu, X. Ni, A. Beaurain, C. Dujardin, P. Granger, Stoichiometric and non-stoichiometric perovskite-based catalysts: Consequences on surface properties and on catalytic performances in the decomposition of N_2O from nitric acid plants. Appl. Catal. B **125**, 149–157 (2012). https://doi.org/10.1016/j.apcatb.2012.05.033

42. H.J. Zhang, J. Wang, X.F. Xu, Catalytic decomposition of N_2O over $Ni_xCo_{1-x}CoAlO_4$ spinel oxides prepared by sol-gel method. J. Fuel Chem. Technol. **43**, 81–87 (2015). https://doi.org/10.1016/S1872-5813(15)60008-1
43. T.Q. Zhao, Q. Gao, H.J. Li, X.F. Xu, Catalytic decomposition of N2O over Y-Co3O4 composite oxides prepared by one-step hydrothermal method. J. Fuel Chem. Technol. **47**, 446–454 (2019). https://doi.org/10.1016/S1872-5813(19)30021-0
44. L. Zheng, H.J. Li, X.F. Xu, Catalytic decomposition of N_2O over Mg-Co composite oxides hydrothermally prepared by using carbon sphere as template. J. Fuel. Chem. Technol. **46**, 569–577 (2018). https://doi.org/10.1016/S1872-5813(18)30024-0
45. H. Yu, X. Wang, Y. Li, Strong impact of cobalt distribution on the activity for $Co_3O_4/CaCO_3$ catalyzing N_2O decomposition. Catal. Today **339**, 274–280 (2020). https://doi.org/10.1016/j.cattod.2018.10.036
46. B.M. Abu-Zied, L. Obalová, K. Pacultová, A. Klegova, A.M. Asiri, An investigation on the N_2O decomposition activity of $Mn_xCo_{1-x}Co_2O_4$ nanorods prepared by the thermal decomposition of their oxalate precursors. J. Ind. Eng. Chem. **93**, 279–289 (2021). https://doi.org/10.1016/j.jiec.2020.10.004
47. X. Hu, Y. Wang, R. Wu, Y. Zhao, Graphitic carbon nitride-supported cobalt oxides as a potential catalyst for decomposition of N_2O. Appl. Surf. Sci. **538**, 148157 (2021). https://doi.org/10.1016/j.apsusc.2020.148157
48. S. Wójcik, G. Ercolino, M. Gajewska, C.W.M. Quintero, S. Specchia, A. Kotarba, Robust $Co_3O_4|\alpha\text{-}Al_2O_3|$cordierite structured catalyst for N_2O abatement – Validation of the SCS method for active phase synthesis and deposition. Chem. Eng. J. **377**, 120088 (2019). https://doi.org/10.1016/j.cej.2018.10.025
49. S. Wójcik, G. Grzybek, J. Gryboś, A. Kotarba, Z. Sojka, Designing, optimization and performance evaluation of the $K\text{-}Zn_{0.4}Co_{2.6}O_4|\alpha\text{-}Al_2O_3|$cordierite catalyst for low-temperature N_2O decomposition. Catal. Comm. **110**, 64–67 (2018). https://doi.org/10.1016/j.catcom.2018.03.019
50. T. Zhao, Y. Li, Q. Gao, Z. Liu, X. Xu, Potassium promoted $Y_2O_3\text{-}Co_3O_4$ catalysts for N_2O decomposition Catal. Comm. **137**, 105948 (2020). https://doi.org/10.1016/j.catcom.2020.105948
51. Z. Xue, Y. Shen, S. Shen, C. Li, S. Zhu, Promotional effects of Ce^{4+}, La^{3+} and Nd^{3+} incorporations on catalytic performance of $Cu\text{-}Fe\text{-}O_x$ for decomposition of N_2O. J. Ind. Eng. Chem. **30**, 98–105 (2019). https://doi.org/10.1016/j.jiec.2015.05.008
52. M.J. Kim, S.J. Lee, I.S. Ryu, M.W. Jeon, S.H. Moon, H.S. Roh, S.G. Jeon, Catalytic decomposition of N_2O over cobalt based spinel oxides: The role of additive. Mol. Catal. **442**, 202–207 (2017). https://doi.org/10.1016/j.mcat.2017.05.029
53. F. Zasada, J. Grybós, E. Budiyanto, J. Janas, Z. Sojka, Oxygen species stabilized on the cobalt spinel nano-octahedra at various reaction conditions and their role in catalytic CO and CH_4 oxidation, N_2O decomposition and oxygen isotopic exchange. J. Catal. **371**, 224–235 (2019). https://doi.org/10.1016/j.jcat.2019.02.010
54. G.A. Zenkovets, R.A. Shutilov, V.I. Sobolev, V.Y. Gavrilov, Catalysts Cu/ZSM-5 for N_2O decomposition obtained with copper complexes of various structures. Catal. Comm. **144**, 106072 (2020). https://doi.org/10.1016/j.catcom.2020.106072
55. D. Pietrogiacomi, M.C. Campa, L.R. Carbone, M. Ochiuzzi, N_2O decomposition and reduction on Co-MOR, Fe-MOR and Ni-MOR catalysts: in situ UV-vis DRS and operando FTIR investigation. An insight on the reaction pathways. Appl. Catal. B **240**, 19–29 (2019). https://doi.org/10.1016/j.apcatb.2018.08.046
56. J. Pérez Ramirez, F. Kapteijn, A. Brückner, Active site structure sensitivity in N_2O conversion over FeMFI zeolites. J. Catal. **218**, 234–238 (2003). https://doi.org/10.1016/S0021-9517(03)00087-3
57. J. Pérez Ramirez, J.C. Groen, A. Brückner, M.S. Kumar, U. Bentrup, M.N. Debbagh, L.A. Villaescusa, Evolution of isomorphously substituted iron zeolites during activation: comparison of Fe-beta and Fe-ZSM-5. J. Catal. **232**, 318–334 (2005). https://doi.org/10.1016/j.jcat.2005.03.018

58. M. Santosh Kumar, J. Pérez Ramirez, M.N. Debbagh, B. Smarsly, U. Bentrup, A. Brückner. Evidence of the vital role of the pore network on various catalytic conversions of N_2O over Fe-silicalite and Fe-SBA-15 with the same iron constitution. Appl. Catal. B **62**, 244–254 (2006). https://doi.org/10.1016/j.apcatb.2005.07.012
59. B. Zhang, F. Liu, L. Xue, Role of agregated Fe oxo species in N_2O decomposition over Fe/ZSM5. Chin. J. Catal. **35**, 1972–1981 (2014). https://doi.org/10.1016/S1872-2067(14)60184-4
60. Q. Xiao, F.F. Yang, J. Zhuang, G.P. Qiu, Y.J. Zhong, W.D. Zhu, Facile synthesis of uniform FeZSM-5 crystals with controlled size and their application to N_2O decomposition. Microporous Mesoporous Mater. **167**, 38–43 (2013). https://doi.org/10.1016/j.micromeso.2012.05.029
61. T. Meng, Y. Lin, Z. Ma, Effect of the crystal size of Cu-ZSM5 on the catalytic performance in N_2O decomposition. Mater. Chem. Phys. **163**, 293–300 (2015). https://doi.org/10.1016/j.matchemphys.2015.07.043
62. G. Moretti, G. Fierro, G. Ferraris, G.B. Andreozzi, V. Naticchioni, N_2O decomposition over [Fe]-MFI catalysts: Influence of the Fe_xO_y nuclearity and the presence of framework aluminium on the catalytic activity. J. Catal. **318**, 1–13 (2014). https://doi.org/10.1016/j.jcat.2014.07.005
63. M. Rutkowska, L. Chmielarz, D. Macina, Z. Piwowarska, B. Dudek, A. Adamski, S. Witkowski, Z. Sojka, L. Obalová, C.J. Oers, P. Cool, Catalytic decomposition and reduction of N_2O over micro-mesoporous materials containing Beta zeolite nanoparticles. Appl. Catal. B **143**, 112–122 (2014). https://doi.org/10.1016/j.apcatb.2013.05.005
64. B.I. Palella, M. Cadoni, A. Frache, H.O. Pastore, R. Pirone, G. Russo, S. Coluccia, L. Marchese, On the hydrothermal stability of CuAPSO-34 microporous catalysts for N_2O decomposition: a comparison with Cu-ZSM5. J. Catal. **217**, 100–106 (2003). https://doi.org/10.1016/S0021-9517(03)00033-2
65. W.H. Eom, M. Ayoub, K.S. Yoo, Catalytic Decomposition of N_2O at Low Temperature by Reduced Cobalt Oxides. Nanosci. Nanotechnol. **16**, 4647–4654 (2016). https://doi.org/10.1166/jnn.2016.11026
66. P. Stelmachowski, G. Maniak, J. Kaczmarczyk, F. Zasada, W. Piskorz, A. Kotarba, Z. Sojka, Mg and Al substituted cobalt spinels as catalysts for low temperature deN_2O—Evidence for octahedral cobalt active sites. Appl. Catal. B **146**, 105–111 (2014). https://doi.org/10.1016/j.apcatb.2013.05.027
67. Y. Wang, X. Zhou, X. Wei, X. Li, R. Wu, X. Hu, Co/Hydroxyapatite catalysts for N_2O catalytic decomposition : Design of well-defined active sites with geometrical and spacing effect. Mol. Catal. **501**, 111370 (2021). https://doi.org/10.1016/j.mcat.2020.111370
68. P. Sazama, N.K. Sathu, E. Tabor, B. Wichterlová, S. Sklenák, Z. Sobalik, Structure and critical function of Fe and acid sites in Fe-ZSM-5 in propane oxidative dehydrogenation with N_2O and N_2O decomposition. J. Catal. **299**, 188–203 (2013). https://doi.org/10.1016/j.jcat.2012.12.010
69. A.J.J. Koekkoek, W. Kim, V. Degirmenci, H. Xin, R. Ryoo, E.J.M. Hensen, Catalytic performance of sheet-like Fe/ZSM-5 zeolites for the selective oxidation of benzene with nitrous oxide. J. Catal. **299**, 81–89 (2013). https://doi.org/10.1016/j.jcat.2012.12.002
70. K. Sun, H. Xia, E. Hensen, R. van Santen, C. Li, Chemistry of N_2O decomposition on active sites with different nature: Effect of high-temperature treatment of Fe/ZSM-5. J. Catal. **238**, 186–195 (2006). https://doi.org/10.1016/j.jcat.2005.12.013
71. G.D. Pirngruber, P.K. Roy, R. Prins, The role of autoreduction and of oxygen mobility in N_2O decomposition over Fe-ZSM-5. J. Catal. **246**, 340 (2007). https://doi.org/10.1016/j.jcat.2006.11.030
72. A. Wang, Y. Wang, E.D. Walter, R.K. Kukkadapu, Y. Guo, G. Lu, R.S. Weber, Y. Wang, C.H.F. Peden, F. Gao, Catalytic N_2O decomposition and reduction by NH_3 over Fe/Beta and Fe/SSZ-13 catalysts. J. Catal. **358**, 199–210 (2018). https://doi.org/10.1016/j.jcat.2017.12.011

73. T. Zhang, Y. Qiu, G. Liu, J. Chen, Y. Peng, B. Liu, J. Li, Nature of active Fe species and reaction mechanism over high-efficiency Fe/CHA catalysts in catalytic decomposition of N_2O. J. Catal. **392**, 322–335 (2020). https://doi.org/10.1016/j.jcat.2020.10.015
74. F. Lin, T. Andana, Y. Wu, J. Szanyi, Y. Wang, F. Gao, Catalytic site requirements for N_2O decomposition on Cu-, Co- and Fe-SSZ-13 zeolites. J. Catal. **401**, 70–80 (2021). https://doi.org/10.1016/j.jcat.2021.07.012
75. P. Granger, J.P. Dacquin, C. Dujardin, Catalytic abatement of N_2O from stationary sources, in *Perovskite materials and related mixed oxides,* ed. by P. Granger, S. Kaliaguine, V. Parvulescu, W. Prellier (Wiley-VCH Verlag GmbH, 2015), vol. 2, p. 611–630
76. D.V. Ivanov, E.M. Sadovskaya, L.G. Pinaeva, L.A. Isupova, Influence of oxygen mobility on catalytic activity of La–Sr–Mn–O composites in the reaction of high temperature N_2O decomposition. J. Catal. **267**, 5–13 (2009). https://doi.org/10.1016/j.jcat.2009.07.005
77. J.P. Dacquin, C. Lancelot, C. Dujardin, P. Da Costa, G. Djega-Mariadassou, P. Beaunier, S. Kaliaguine, S. Vaudreuil, S. Royer, P. Granger, Influence of preparation methods of $LaCoO_3$ on the catalytic performances in the decomposition of N_2O. Appl. Catal. B **91**, 596–604 (2009). https://doi.org/10.1016/j.apcatb.2009.06.032
78. W. Piskorz, F. Zasada, P. Stemachowski, O. Diwald, A. Kotarba, Z. Sojka, Computational and experimental investigation into N_2O decomposition over MgO nanocrystals from thorough molecular mechanism to ab initio microkinetics. J. Phys. Chem. C **115**(45), 22451–22460 (2011). https://doi.org/10.1021/jp2070826
79. P. Pietrzik, F. Zasada, W. Piskorz, A. Kotarba, Z. Sojka, Computational spectroscopy and DFT investigations into nitrogen and oxygen bond breaking and bond making processes in model $deNO_x$ and deN_2O reactions. Catal. Today **119**, 219–227 (2007). https://doi.org/10.1016/j.cattod.2006.08.054
80. E.J. Karlsen, M.A. Nygren, L.G.M. Pettersson, Theoretical study on the decomposition of N_2O over alkaline earth metal-oxides: MgO-BaO. J. Phys. Chem. A **106**, 7868–7875 (2002). https://doi.org/10.1021/jp025622g
81. T.A. Egerton, F.S. Stone, J.C. Vickerman, x-Cr_2O_3-Al_2O_3 Solid Solutions: II. The catalytic decomposition of nitrous oxide. J. Catal. **33**, 307–315 (1974). https://doi.org/10.1016/0021-9517(74)90275-9
82. H. Xia, K. Sun, Z. Liu, Z. Feng, P. Ying, C. Li, The promotional effect of NO on N_2O decomposition over the bi-nuclear Fe sites in Fe/ZSM-5. J. Catal. **270**, 103–109 (2010). https://doi.org/10.1016/j.jcat.2009.12.014
83. P.J. Smeets, B.F. Sels, R.M. van Teeffelen, H. Leeman, E.J.M. Hensen, R.A. Schoonheydt, The catalytic performance of Cu-containing zeolites in N_2O decomposition and the influence of O_2, NO and H_2O on recombination of oxygen. J. Catal. **256**, 183–191 (2008). https://doi.org/10.1016/j.jcat.2008.03.008
84. S.L. Raj, B. Viswanathan, V. Srinivasan, The activity of Mn^{3+} and Mn^{4+} in lanthanum strontium manganite for the decomposition of nitrous oxide. J. Catal. **75**, 185–187 (1982). https://doi.org/10.1016/0021-9517(82)90133-6
85. A. Dandekar, M.A. Vannice, Decomposition and reduction of N_2O over copper catalysts. Appl. Catal. B **22**, 179–200 (1999). https://doi.org/10.1016/S0926-3373(99)00049-1
86. D.V. Ivanov, L.G. Pinaeva, E.M. Sadovskaya, L.A. Isupova, Isotopic transient kinetic study of N_2O decomposition on $LaMnO_{3+\delta}$. J. Mol. Catal. A **412**, 34–38 (2016). https://doi.org/10.1016/j.molcata.2015.11.018
87. S. Ponce, M.A. Pena, J.L.G. Fierro, Surface properties and catalytic performance in methane combustion of Sr-substituted lanthanum manganites. Appl. Catal. B **24**, 193–205 (2000). https://doi.org/10.1016/S0926-3373(99)00111-3
88. I.K. Murwani, S. Scheurell, M. Feist, E. Kemnitz, ^{18}O-isotope exchange behavior and oxidation activity of $La_{1-x}Sr_xMnO_{3+\delta}$. J. Therm. Anal. Cal. **69**, 9–21 (2002). https://doi.org/10.1023/a:1019905819824
89. S. Royer, D. Duprez, S. Kaliaguine, Role of bulk and grain boundary oxygen mobility in the catalytic oxidation activity of $LaCo_{1-x}Fe_xO_3$. J. Catal. **234**, 364–375 (2005). https://doi.org/10.1016/j.jcat.2004.11.041

90. A.A. Taskin, A.N. Lavrov, Y. Ando, Achieving fast oxygen diffusion in perovskites by cation ordering. Appl. Phys. Lett. **86**, 0619110 (2005). https://doi.org/10.1063/1.1864244
91. W. Suo, S. Sun, N. Liu, X. Li, Y. Wang, The adsorption and dissociation of N_2O on CuO(111) surface: The effect of surface structures. Surf. Sci. **696**, 121596 (2020). https://doi.org/10.1016/j.susc.2020.121596
92. X. Gao, Y. Li, J. Chen, X. Zhuang, Z. Chang, Y. Li, First-principles study of N_2O decomposition on (001) facet of perovskite $LaBO_3$ (B = Mn Co, Ni). Mol. Catal. **510**, 111713 (2021). https://doi.org/10.1016/j.mcat.2021.111713
93. X. Liu, L. Sheng, Catalytic decomposition of N_2O on iron-embedded C_2N monolayer: A DFT study. Mater. Today Comm. **28**, 102585 (2021). https://doi.org/10.1016/j.mtcomm.2021.102585
94. M. Zabilskiy, P. Djinović, E. Tchernychova, A. Pintar, N_2O decomposition over CuO/CeO_2 catalyst: New insights into reaction mechanism and inhibiting action of H_2O and NO by operando techniques. Appl. Catal. B **197**, 146–158 (2016). https://doi.org/10.1016/j.apcatb.2016.02.024
95. L. Obalová, K. Jirátová, K. Karásková, Ž Chromčáková, N_2O catalytic decomposition – From laboratory experiment to industry reactor. Catal. Today **191**, 116–120 (2012). https://doi.org/10.1016/j.cattod.2012.03.045
96. M. Bernauer, B. Bernauer, G. Sádovská, Z. Sobalík, High-temperature decomposition of N_2O from HNO_3 production: Process feasibility using a structured catalyst. Chem. Eng. J. **220**, 115624 (2020). https://doi.org/10.1016/j.ces.2020.115624
97. R. Zhang, K. Hedjazi, B. Chen, Y. Li, Z. Lei, N. Liu, M(Fe, Co)-BEA washcoated honeycomb cordierite for N_2O catalytic decomposition. Catal. Today **273**, 273–285 (2016). https://doi.org/10.1016/j.cattod.2016.03.021
98. F. Heshmatifar, J. Karimi-Sabet, P. Khadiv-Parsi, The investigation of the efficient replaceable microreactor into the catalytic decomposition of N_2O over Pd/anodic N-Al_2O_3/Al. Chem. Eng. Proc. **168**, 108555 (2021). https://doi.org/10.1016/j.cep.2021.108555
99. G. Wang, Z. Li, C. Li, Recent progress in one-step synthesis of acrylic acid and methyl acrylatevia aldol reaction: Catalyst, mechanism, kinetics and separation. Chem. Eng. Sci. **247**, 117052 (2022). https://doi.org/10.1016/j.ces.2021.117052
100. A. Szydło, J.-D. Goossen, C. Linte, H. Uphoff, M. Bredol, Preparation of platinum-based electrocatalytic layers from catalyst dispersions with adjusted colloidal stability via a pulsed electrophoretic deposition method. Mater. Chem. Phys. **242**, 122532 (2020). https://doi.org/10.1016/j.matchemphys.2019.122532
101. I.V. Yentekakis, G. Goula, P. Panagiotopoulou, S. Kampouri, M.J. taylor, G. Kyriakou, R.M. Lambert, Stabilization of catalyst against sintering on oxide supports with high oxygen ion lability exemplified by-Ir-catalyzed decomposition of N_2O. Appl. Catal. B **192**, 357–364 (2016). https://doi.org/10.1016/j.apcatb.2016.04.011
102. G. Grzybek, J. Gryboś, P. Indyka, J. Janas, K. Ciura, B. Leszczyński, F. Zasada, A. Kotarba, Z. Sojka, Evaluation of the inhibiting effect of H_2O, O_2 and NO on the performance of laboratory and pilot K-$Zn_xCo_{3-x}O_4$ catalysts supported on α-Al_2O_3 for low-temperature N_2O decomposition. Appl. Catal. B **297**, 120435 (2021). https://doi.org/10.1016/j.apcatb.2021.120435
103. H. Xia, K. Sun, Z. Feng, C. Li, Effect of Water on Active Iron Sites for N_2O Decomposition over Fe/ZSM-5 Catalyst. J. Phys. Chem. C **115**, 542–548 (2011). https://doi.org/10.1021/jp1094917
104. Q. Shen, M. Wu, H. Wang, N. Sun, C. He, W. Wei, The influence of desilication on high-silica MFI and its catalytic performance for N_2O decomposition. Appl. Surf. Sci. **441**, 474–481 (2018). https://doi.org/10.1016/j.apsusc.2018.01.052
105. Crucial role of activation temperature on catalytic performances, G. Grzybek, S. Wójcik, P. Legutko, J. Gryboś, P. Indyka, B. Leszczyński, A. Kotarba, Z. Sojka, Thermal stability and repartition of potassium between the support and active phase in the K-$Zn_{0.4}Co_{2.6}O_4|\alpha$-$Al_2O_3$ catalyst for N_2O decomposition. Appl. Catal. B **205**, 597–604 (2017). https://doi.org/10.1016/j.apcatb.2017.01.005

106. M. Inger, P. Kowalik, M. Saramok, M. Wilk, P. Stelmachowski, G. Maniak, P. Granger, A. Kotarba, Z. Sojka, Laboratory and pilot scale synthesis, characterization and reactivity of multicomponent cobalt spinel catalyst for low temperature removal of N_2O from nitric acid plant tail gases. Catal. Today. Today **176**, 365–368 (2011). https://doi.org/10.1016/j.cattod.2010.11.044

Nitrite Removal from Water: New Support Materials for Pd-Based Catalysts Aiming for a Low Ammonium Production

F. M. Zoppas, N. Sacco, V. Aghemo, T. F. Beltrame, F. Battauz, A. Devard, E. Miró, and F. A. Marchesini

Abstract The concentration of nitrogen oxyanions in natural waters has steadily increased in recent decades, as a result of the intensification of agriculture and population growth. Reverse osmosis, ion exchange, adsorption and electrodialysis are currently used as separation technologies for water denitrification. Also, nitrite reduction technologies such as biological treatment and catalytic reduction, are able to convert nitrite into inert nitrogen gas. Several studies have proposed the catalytic reduction of nitrite to nitrogen in water over Pd supported on different materials as a promising alternative for water treatment. In this chapter, an overview of the current state of the art of catalytic nitrite reduction is presented. The use of catalysts supported on CeO_2, Nb_2O_5, ZrO_2, TiO_2, γ-Al_2O_3, SiO_2, and ZSM-5 for the removal of high concentrations of nitrite in water is reported in this chapter. All synthesized materials were evaluated and they showed catalytic activity in the reduction of nitrites in the water. The total conversion was achieved by catalysts supported in γAl_2O_3, ZSM-5 (Si: Al = 30), SiO_2, and TiO_2. Among this group, the most nitrogen-selective under the evaluated conditions were supported on ZSM-5 (Si: Al = 30) and TiO_2. These findings contribute to the existing data, providing insights into previously untested materials as supports for water removal nitrite catalysts.

Keywords Nitrite reduction · Support material · Pd catalyst · Water treatment

F. M. Zoppas · N. Sacco · V. Aghemo · F. Battauz · A. Devard · E. Miró · F. A. Marchesini (✉)
Instituto de Investigaciones en Catálisis y Petroquímica (FIQ, UNL-CONICET), Santiago del Estero, 2829, 3000 Santa Fe, Argentina
e-mail: albana.marchesini@gmail.com

T. F. Beltrame
Laboratório de Corrosão, Proteção e Reciclagem de Materiais (LACOR UFRGS), Av. Bento Gonçalves, Porto Alegre 9500, 91501-970, Brazil

S. J. Ikhmayies (ed.), *Advances in Catalysts Research*, Advances in Material Research and Technology, https://doi.org/10.1007/978-3-031-49108-5_8

1 Introduction

Water pollution is one of the most serious problems facing the scientific community worldwide. As water is an elementary need for human survival, it is one of the most valuable natural resources worldwide [1] and access to a safe supply of water is a human right. Despite this, over the last decade, the World Health Organization (WHO) reported that only 54% of the world's population can access water through a household connection to a water duct [2].

Groundwater is the main freshwater renewable supply for humans. In Argentina, although shallow aquifers and wells constitute the most accessible drinking water resource for human consumption, they are also susceptible to pollution from natural sources and anthropogenic activities [3]. Several studies concerning groundwater pollution in Argentina and Latin American countries have been reported in the last decades [4–7].

Nitrogen exists in the soil as nitrate (NO_3^-), nitrite (NO_2^-), and ammonium (NH_4^+) form and can easily transfer into the groundwater by leaching [8]. Nitrate pollution in surface and ground waters is one of this century's substantial engineering challenges because of the risks it represents to human health and the environment [9], and understanding the factors affecting groundwater nitrate pollution can help actions to guard water resources. Nitrate itself is regarded as being harmless [10, 11], but its toxicity is usually the result of the conversion of nitrate into nitrite. Nitrite is the primary reduction product of nitrate [12] and this ion is significantly more reactive than nitrate since it requires a less oxidizing medium for its formation. Nitrite (converted from ingested nitrates) can cause methemoglobinemia by the Fe^{2+} oxidation of the haem group to Fe^{3+} form with subsequent tissue hypoxia. In addition, nitrites can be combined with secondary amines or N-alkyl amides to form carcinogenic N-nitroso compounds [13].

1.1 Sources of Nitrate and Nitrite Water Pollution

Nitrates and nitrites are a hazard in the environment, mainly in groundwater and drinking water sources, as a result of different pollution fonts that habitually exceed maximum permissible concentrations [14].

Although nitrate reaches groundwater through anthropogenic or natural sources, the main contribution is due to the anthropogenic ones. Nitrate is highly soluble in water and can accumulate preferably in agricultural regions [15]. The inappropriate control of applying fertilizers and the consequent leaching is reported as the largest contributor to nitrate in the groundwaters. In Argentina, the problem of pollution with nitrates arises mainly from the severe fertilization required for the development of intensive crops [9, 16].

Several industries, such as textile, pharmaceuticals, agricultural and chemical ones, use or produce nitrogen-containing compounds such as urea, ammonia, and nitric acid, dyes, and pesticides [17]. A recent study assessed nitrate in polluted groundwater and soil and evaluated the impact on the inhabitant´s health of the Matanza-Riachuelo River basin [7]. The nitrate average values were 76 mg/L and 38 mg/L in the Upper and Puelche aquifers, respectively. Also, the researchers found that water ingestion and dermal contact were the main exposure ways in aquifers and soils, respectively, however, children and adults exposed to groundwater via ingestion and dermal contact faced an acceptable non-carcinogenic risk of nitrate. Nitrite and nitrate can also be originated from other treatment processes, such as concentrates of desalination and electrodialysis plants [18].

1.2 Water Quality Standards and Regulatory Limits

Human exposure to nitrates and nitrites is subject to regulatory limits from international health agencies. The alimentary exposition to nitrate and nitrite arises via three main fonts: occurrence in foods, food additives, and pollutants in drinking water. Moreover, nitrate is usually incorporated into the food chain through plant nutrients by the action of lightning and soil bacteria and this is the major dietary source of this compound while water sources are the minor contributor. Nitrite contribution is mainly from the reduction of nitrate via enterosalivary circulation [19]. Until the last decades, there were no dietary intake recommendations for nitrates and nitrites, except for regulatory limits in water and processed meat [20].

Several international organizations have recommended limits for human exposure to nitrate and its reduction products. Table 1 shows the limits set by regulatory organizations at the local and international levels [10, 21].

2 Nitrate and Nitrite Removal Technologies

To purify water with high amounts of nitrates and nitrites, researchers generally cite drinking water parameters as quality objectives to be achieved. Each of these technologies has certain advantages and disadvantages and the possibility of using them will depend on a large number of factors such as costs, quality and use of the water to be obtained, initial quality of the water to be treated, effluents generated

Table 1 Maximum contaminants limit for nitrates, nitrites, and ammonium in drinking water established by different worldwide and Argentina agencies

Regulatory agency	NO_3^-	NO_2^-	NH_4^+
World health organization	50 mg/L	3.0 mg/L	0.5 mg/L
European community	50 mg/L	0.1 mg/L	0.5 mg/L
Argentine food code	45 mg/L	0.1 mg/L	0.2 mg/L

during the purification process, and post-treatment requirements. It is possible to differentiate two groups of technologies among the variety of methods existing.

2.1 Separation Technologies

Nowadays, nitrate and nitrite removal from water is carried out mainly by several commercially offered physicochemical systems. Reverse osmosis (RO) and electrodialysis (ED), ion Exchange (IE) as physicochemical techniques have been used to reduce nitrites from water [22–24]. Conventional physicochemical techniques allow the effective removal of NO_3^-/NO_2^- ions concentrating them, but do not frequently reach their complete elimination.

2.1.1 Ion Exchange

The use of ion exchange (IE) is an alternative to the elimination of nitrites. Ion exchange is a relatively low-cost, harmless, fast, and adaptable technique in which ions can be separated from a solution by exchanging ions and separating compounds with great selectivity [25]. The substitution of an ion bound to an inert matrix with another ion by dissolving an ionic bond and forming a new bond occurs, without triggering significant structural changes.

Ion-exchange resins were also applied to the extraction of heavy metals like chromium (Cr), cobalt (Co), nickel (Ni), mercury (Hg), and zinc (Zn) from water and coolant water at the nuclear power plant.

Li et al. [26] have developed a new ion exchange polymer-modified capacitive deionization electrode (p-CDI). The results suggest that p-CDI has good application prospects for removing nitrite in dyeing wastewater treatment.

Metal regeneration, selectivity, too little sludge produced, strict discharge criteria, high treatment capacity, high removal efficiency, and fast kinetics are the key advantages of ion exchange over chemical precipitation. The major disadvantage is the management of the solution resulting from the regeneration of the resin, which must be disposed of or recovered by the addition of chemicals [27].

2.1.2 Reverse Osmosis

RO is a physical process that employs semipermeable membranes and can be used to remove several pollutants at the same time (ionic, particles, and organic compounds). Water is forced through the membrane under pressure, permitting water molecules to pass through while retaining most of the dissolved materials [27]. This technology works well with NO_3^- and NO_2^- and is one of the most expensive ways of centralized water treatment. It is only economically convenient if the water mandate is very low or if several pollutants need to be removed. RO requires cautious screening of the raw water and pretreatment to prevent membrane degradation, such as removing suspended solids from the water, which usually involves passing the water through a series of filters before the reverse osmosis stage.

The main advantages of RO are the production of high-quality water in comparison with raw water quality, the inclusion of multiple pollutants removal as well a high automation process easy to practice, and the suitability of the system for small stations. The main disadvantages of this technology include high energy costs and the water needs to be pre-treated.

2.1.3 Electrodialysis

ED is a flexible and efficient technology in terms of NO_3^- removal. The high cost involved in this technique makes it not extensively used in water treatment. This technology involves the use of electrodes that are introduced into a volume of water where a current is applied between them, leading to electrodialysis of the water [27]. Ions migrate through selective semi-permeable membranes as a consequence of the electrically charged membrane surfaces. Electric current makes anions and cations in solution move towards the anode and cathode, respectively.

As with all membrane processes, RO and ED require clean feed, vigilant operation, and periodic maintenance to prevent damage [28]. The main limitation of the membrane processes is the concentrated solution coming from the treatment process, which implies proper treatment or disposal. Albornoz et. al [29] applied ED to the treatment of university sewage for water recovery and results suggested the application of another type of treatment before or after ED.

2.1.4 Adsorption

In the adsorption method, a pollutant molecule is attached to the surface of an adsorbent material either through physisorption or chemisorption. Major advantages associated with adsorbents are high surface area, easy synthetic routes, high catalytic activity, and good electrical conductivity, and the adsorption ability of the nanomaterials [30]. Efficient denitrification removal rates have been achieved using these nanomaterials at differing pH, nanomaterial dosage, temperature, contact time, and initial nitrate concentration.

The adsorption method is cheap, reliable and its by-products could be recycled instead of disposing into expensive landfills [31, 32].

These nanomaterials applied as adsorbents have several limitations such as toxicity and particle agglomeration and may also become unstable at varying environmental conditions. Retrieval, separation, and recycling of nanomaterials that have been dispersed in a water source for nitrate removal may also be difficult.

2.2 *Transformation Techniques*

These techniques are intended to transform NO_2^-/NO_3^- into other harmless chemical compounds through biological or catalytic pathways. As regards transformation techniques, biological processes are commonly used in the treatment of wastewater and industrial water and incorporated in many treatment plants, giving very satisfactory results. However, although this technology is suitable for wastewater, it cannot be used for drinking water treatment due to the possible bacterial contamination of the treated water and the presence of organic residues after water treatment.

2.2.1 Biological Removal

Generally used in effluent treatment, biological reactors emerged in the past decades as a technique for denitrification of potable water with the potential to address multiple contaminants such as nitrate, perchlorate, chromate, and organic chemicals [33]. Unlike the separation methods mentioned above, nitrate is reduced and removed from the system instead of being displaced to a concentrated waste stream.

Denitrifying bacteria requires an electron donor (known as a substrate) for the reduction of nitrate to nitrogen gas. In conventional wastewater treatment, substrate addition is habitually not required, since wastewater contains enough carbon for denitrification to arise. However, substrate addition is necessary for drinking water biological denitrification [33]. One of the main goals is to minimize dissolved carbon in the water to minimize microbial growth.

He et al. [34] isolated a *Streptomyces mediolani strain*, designated EM-B2, in a cow manure fermentation biogas digester. *S. mediolani* is distributed in terrestrial and marine soils and freshwater ecosystems. They showed that hydroxylamine and nitrite could be effectively removed from wastewater; and can contribute to strategies to improve the rate of nitrogen removal from wastewater. Also, the simultaneous removal of nitrites and organic compounds has been performed by Li et al. [35], employing an improved biofilm for the treatment of high salt wastewater by mixotrophic bacteria.

The use of bioadsorbent has been one of the methods used to eliminate nitrites. Babu et al. [36] worked with a bio adsorbent derived from *Sesbania grandilora* sp/*Sesbania grandiflora activated carbon* (SGSP/SGSAC) and *Amorphophallus paeoniifolius sp/Amorphophallus paeoniifolius activated carbon* (APSP/APSAC) plants, investigated different extraction conditions, and studied the ability to remove nitrite from contaminated water. In their study, they prepared a powder of the stem of the plants on activated carbon. They observe adsorption capacities greater than those reported in the literature using bioadsorbents. These materials have the advantage of being simple (derived from plants) and effective for nitrite removal.

Innovative nitrite and ammonia removal is one of the key approaches to efficient aquaculture biotechnology. Tada et al. [37] proposed a technique for the rapid elimination of high levels of nitrite and ammonia by recirculating aquaculture waste at

30 ‰ salinity, using MBBR technology (moving bed biofilm reactors). The use of mature biofilms in MBBR was shown to increase bacterial diversity, improving nitrite and ammonia removal efficiency in 56 days or less.

It is important to take into account that biological treatments have a slow reaction rate and require a large-scale installation [24].

2.2.2 Catalytic Reduction

Several studies dealing with the development of heterogeneous catalysts including hydrogenating and oxidation processes for water pollutants removal have become of great interest in recent decades [38–40] since the use of homogeneous catalysis would require the adaptation separation processes that in most cases are economically and technically unfeasible [38]. In this sense, catalytic treatments have emerged as promising alternatives for nitrate/nitrite removal. In comparison to IE, the catalytic method produces less waste byproducts [41] and, in contrast to biological methods, catalytic reactors can be quickly applied [12]. Nevertheless, the implementation of an environmentally sustainable and financially viable catalytic technology requires the search for catalyst materials with improved activity and durability.

Similar to biological denitrification, nitrites can be reduced in the presence of a noble metal with reducing agents like hydrogen gas (H_2). Vorlop and Tacke [42] first performed a catalytic reduction using a single noble metal-supported catalyst and hydrogen gas as a reductant. The reactions are as follows (Eqs. (1) and (2)):

$$2NO_2^- + 3H_2 \rightarrow N_2 + 2H_2O + 2OH^- \tag{1}$$

$$NO_2^- + 3H_2 \rightarrow NH_4 + 2OH^- \tag{2}$$

The role of the noble metal is to activate the hydrogen, making the hydrogen available to act in the reduction of nitrite to ammonium or nitrogen gas. The main intermediate is $NO_{(ads)}$ in the generation of nitrogen and ammonium [43].

The nitrites are reduced in the monometallic sites of Pd, resulting in a mixture of two stable end products, nitrogen (N_2) and ammonium (NH_4^+). One of the key factors in the extent and selectivity of the reduction for any given noble metal is the support material. Among the most studied catalysts are those composed of a noble metal (Pd, Ir, Rh, or Pt) [44], with Pd having the best performance in conversion and selectivity towards nitrogen among them [20]. Langmuir–Hinshelwood kinetics has been proposed as a suitable model for describing nitrite reduction with hydrogen as a reducing agent [45].

The robustness of the catalyst can be measured in terms of stability over long operating times. In this sense, scientists are looking for materials that do not degrade, that are easily handled, that the metal of the active phase does not leach under working conditions, or even that the medium does not deactivate the active sites of the catalyst.

This chapter focuses on the use of catalytic reduction for the removal of nitrites from water using monometallic catalysts based on Pd supported on different supports. In addition, the influence of the latter on the catalytic results obtained will be evaluated.

Support Materials for Catalytic Treatment of Polluted Waters

It is widely reported the reaction rate, activity, and selectivity of catalysts are influenced by the preparation method, reaction conditions, the way noble metal is promoted, and the catalyst support [46–48]. Previous research on support materials for the reduction of aqueous oxyanions has focused on Al_2O_3 [45] and SiO_2 [8–12] and to a lesser extent also TiO_2 [52], CeO_2 [53], ZrO_2 [24] and zeolites [54].

The most-reported supports for catalysts employed in water denitrification are described below:

Al_2O_3

Al_2O_3 is the most frequent support used in the denitrification process, probably due it being a catalytic activity inducer of the high metal particles dispersion [55], or the positive net charge that alumina acquires in aqueous media at reaction pH which promotes the anions approximation to the catalytic surface [34, 37, 56, 57]. Monometallic Pd supported on alumina has high activity for nitrite reduction, but it is almost inactive for nitrate reduction, where it is necessary to use a second metal to promote the reduction of nitrate to nitrite [58].

SiO_2

SiO_2 is the second most used support in denitrification reactions [46, 59]. This solid has the disadvantage that at reaction pH it presents a net negative superficial charge, which reduces the anions adducting possibility to the active sites, thus reducing the total conversion reached. In addition, this former support presents narrow pores allowing the formation of microenvironments with OH gradients that diminish the activity and selectivity, leading the reaction products to the undesirable NH_3 [49].

These conventional supports allow the formation of an internal OH^- gradient that reduces their activity and selectivity to N_2, increasing the undesirable NH_4^+ production due to the presence of narrow pores. The development of catalysts with a higher pore width decreases the effect of alkalinity inside the pore system, increasing the selectivity to N_2 compared with catalysts with narrow pores [48].

Zaki et al. studied Pt/SiO_2 catalyst for water denitrification and suggested that the redox process during the reaction could explain changes observed in the particle size distribution partially linked to the deterioration of the mesoporous silica support under these operating conditions [60].

TiO_2

The use of TiO_2 as support in denitrification reactions is associated with its photocatalytic properties and the high electron disponibility [61], which induce an over-reduction of reaction products [19, 27–29]. Titanium dioxide has favorable characteristics such as low costs and non-toxicity. In addition to its important photocatalytic activity, it has a strong oxidizing power being able to oxidize by itself ammonia to NO_2 or NO_3 [62]. Pd/TiO_2 catalyst is reported to exhibit high activity for nitrate removal, but low selectivities to nitrogen in comparison with bimetallic catalysts [58]. Kominami et al. have studied the performance of bare TiO_2 and Pd-loaded TiO_2 systems in the photocatalytic reaction of NO_2 in aqueous suspension [63, 64]. They found the bare TiO_2 system was active in the disproportionation of nitrite ions to dinitrogen and nitrate, although the reaction was slow. Pd/TiO_2 system exhibited a high rate of disproportionation of nitrite and this ion was almost completely converted to N_2 and NO_3 [63].

CeO_2

Another frequent support used in the interest reaction is CeO_2 [23, 24, 65]. CeO_2-based materials have been widely explored for several applications due to their ability to easily alternate their oxidative state between Ce^{3+} and Ce^{4+} to store and release oxygen [66]. Pd/CeO_2 has proven to be much more active than Pd/Al_2O_3 catalyst due to the difference in the metallic surface. The authors reported the Pd dispersion was much higher on CeO_2 than on Al_2O_3 [67]. Despite the important promoting effect of the ceria support on the catalytic activity of monometallic Pd for nitrite reduction, and the fact that it allows nitrate reduction without the addition of a second metal, the reaction is too selective towards ammonium, which is an undesired product.

ZrO_2

Zirconia as catalytic support has received considerable attention due to the combination of its surface properties. The acidity or basicity as well as its oxidizing and reducing ability could be of great importance in the performance of catalytic systems [68]. Pd-loaded ZrO_2 has been applied in several oxidation processes [69–71]. For reduction reactions such as the denitrification process, the use of ZrO_2 results in great interest since it could improve the performance of supported transition metals in the hydrogenation process [68, 72]. Despite the beneficial effects of this support on catalytic performance, ZrO_2 has a low surface area and is frequently deposited on oxides with higher surface areas, such as SiO_2 or Al_2O_3 supports. Sakamoto et al. [73] reported the monometallic Pd and bimetallic Cu-Pd catalysts loaded different supports including ZrO_2 performance on the catalytic reduction of NO_3 and NO_2 in water and found these catalysts were inactive in nitrate hydrogenation and Pd/ZrO_2 produces large amounts of NH_3 from NO_2.

ZrO_2-loaded CeO_2 support is also widely reported in catalytic applications since the addition of ZrO_2 could increase the oxygen vacancies and the redox properties of cerium [74, 75]. Peroni et al. [76] reported the use of PdCu/CeO-ZrO_2 catalysts in the elimination of NO_3 and NO_2 from synthetic and real water.

Zeolites

Recent studies reported an excellent catalytic activity of Pd-based catalysts supported by different types of zeolites in nitrate reduction and attributed this associated catalytic performance to the physicochemical properties of the support, such as high external surface area, shape selectivity, uniform channel size, and high hydrothermal stability among others [77]. Knowing that zeolites are natural materials, a group of aluminosilicate minerals, that have been widely used as an efficient and inexpensive adsorbent for removing water contaminants. They can be synthetically produced to modify the properties for specific applications [28, 78].

In addition to described supports, several materials such as glass fiber [79], pumice [80], hydrotalcites [81–83], resins [84–86], polymers [87, 88], carbon nanotubes [89], and active carbon [89–93] have been tested in catalytic denitrification processes. All of them had different physicochemical properties which induce their possible use.

3 Pd-Based Catalysts

In this section, the preparation, characterization, and catalytic performance of Pd loaded onto the supports described above are described. Moreover, although there are not many papers reporting the use of Nb_2O_5 as a support for catalytic denitrification, the Pd/Nb_2O_5 catalytic performance on nitrite reduction is studied, since this oxide yielded similar results to alumina-supported catalysts in terms of nitrate reduction and nitrogen selectivity [43, 88].

3.1 Synthesis and Characterization

One of the most commonly used methods for the synthesis of catalysts is wet impregnation. This method uses the following methodology: the support, deionized water (about 100 mL), and the solutions of the precursor salts are placed in a container. In the case of this study, a volume of palladium chloride ($PdCl_2$) dissolved in 0.1 M HCl is added to obtain 1% by weight of Pd. The suspension is then heated (about 80 °C) with stirring until total evaporation of the liquid. The next step consists of calcination at 500 °C, which serves to better anchor the metal of the active phase to the support and to remove synthesis precursors. The reduction of the active phase is generally carried out with a solution of a reducing agent, e.g. hydrazine hydrate. In this study, 0.2 M hydrazine hydrate was used for 1 h at the temperature of 40 °C. Finally, the catalyst is rinsed many times and dried in an oven at 80 °C overnight.

Catalysts with 1 wt.% Pd were prepared as described elsewhere [94, 95]. Pd based catalysts were prepared by wet impregnation method, as described in Fig. 1, using the following materials as support: α-Al_2O_3 (Sigma Aldrich, pa, 300 m^2/g), γ-Al_2O_3 (KEDJEN CK300, 180 m^2/g, vol. Pore 0.5 $cm^3/$ g), SiO_2 (Alfa Aesar 300 m^2/g,

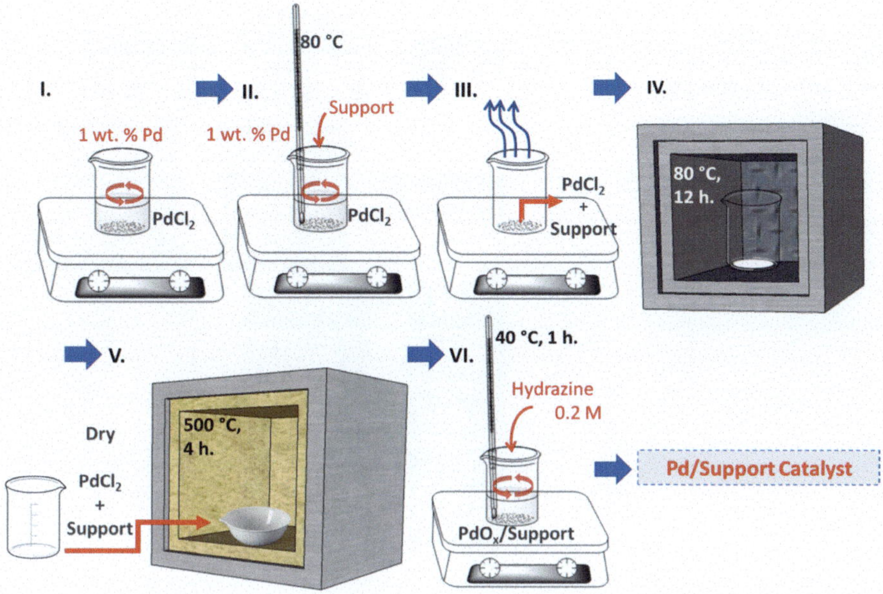

Fig. 1 Schematic of the wet impregnation method for the preparation of monometallic Pd catalysts on different supports

vol. Pore 1 cm^3/g), CeO_2 (Rieden de Haen, pa), ZrO_2 (JMC, purathronic, 99.99%), Nb_2O_5 (65 180 m^2/g, CBMM, HY340), TiO_2 (Degussa, p25, 48 m^2/g), and zeolite ZMS-5 (Shanghai, 400 m^2/g) with different Si/Al proportions (Si/Al = 30, 80 and 300). Figure 2 summarizes by a scheme the catalysts assessed in this chapter.

Isoelectric point (IEP) of the supports employed for the conformation of Pd-based catalysts shown in Fig. 2 are summarized in Table 1. These IEP values are compared with those reported in the literature.

3.2 *Catalytic Performance of Pd-Based Catalysts in Nitrite Removal from Water*

In the following section, the catalytic performance of the studied Pd-based catalysts is described.

The catalytic reduction of nitrite usually presents a curve of nitrite concentration versus reaction time characteristic of first-order reactions. In the reaction medium all the ammonium that is generated, remains in solution, as long as the pH is not higher than 9. Conversions are expressed as % and represent the amount of nitrite removed during the evaluation time as expressed in Eq. (3), where C^0 is N ppm of nitrite at the beginning of the reduction process and C is N ppm of NO_2 at time t. They are calculated as the quotient of the initial nitrite concentration (C^0) minus the

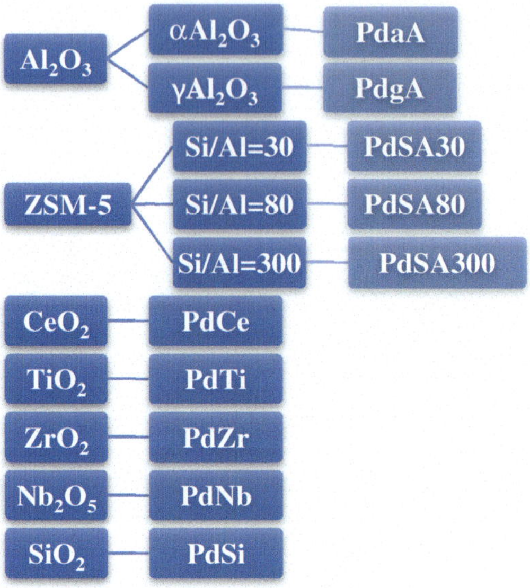

Fig. 2 Scheme of the monometallic catalysts supported on metal oxides with 1 wt.% Pd studied in this chapter

residual nitrite concentration (C) over the initial nitrite concentration multiplied by 100%. With both nitrite and ammonium concentrations, it is possible to calculate the selectivity to N_2 by a mass balance between the formed product (C^A) and the converted nitrite concentration (C^0–C), exposed in Eq. (4).

$$X(\%) = \left[1 - \left(C/C^0\right)\right] \times 100\% \tag{3}$$

$$S(\%) = [C^A/(C^0 - C)] \times 100\% \tag{4}$$

The most employed techniques for nitrate and their byproducts reduction monitoring are ionic chromatography, Griess reaction, and analytical standard methods [104]. In this study, samples were taken at fixed times and analyzed through colorimetric methods. The residual content of NO_2^- was measured using Griess reaction (absorbance at wavelength $\lambda = 543$ nm) [105], while the production of NH_4^+ was quantified by Berthelots's [25] modified reaction (absorbance at $\lambda = 633$ nm).

The percentage of nitrite elimination was calculated with Eq. (3), and the values are expressed as the N concentration stemming from nitrite (ppm $N - NO_2^-$).

Among the catalysts described in Sect. 3.1, it was possible to differentiate three categories of catalytic performance taking into account the yields in the nitrite conversion. Group I, II, and III represent catalysts with nitrite conversions between (30–79)%, (80–97)%, and (98–100)%, respectively. Figure 3 shows the reaction profiles for nitrite consumption (Fig. 3A, C, and E) and ammonium production (Fig. 3B, D, and F). The lowest nitrite conversion in Group I was observed for the PdCe catalyst (Fig. 3.A), with 32% conversion and the highest ammonium production. The ammonia selectivity could be related to Ce(III) species that are formed after the catalyst reduction, due to the redox ability of CeO_2. Moreover, the generation of oxygen vacancies on the Ce–O interface would favor the over-reduction of reaction intermediates.

Lee et al. [24] used a 3 wt.% Pd/CeO_2 catalyst yielding 100% conversion of 28 ppm $N - NO_2^-$ i n120 minutes. As can be seen in Fig. 3.A, PdCe catalyst yielded 32/100 ppm $N - NO_2^-$ which was accomplished by employing 1 wt.% Pd, within the same reaction time. Similar data for the other catalysts from group I has not been found for aqueous nitrite removal.

Within Group I, the PdNb catalyst showed an intermediate performance, with 38% of nitrite conversion and low ammonium production (Fig. 3B), which is desirable. This could be related to the strong surface acidity and stability of Nb_2O_5 in an aqueous medium [107], which would promote a very good reduction capacity and low selectivity towards ammonium [108, 109]. Besides, the higher mesoporosity and the drastic surface area decrease after the calcination at 550 °C [88] could contribute to efficient local pH control which plays an important role in the balance between the desired N_2 and NH_3 production [110].

The best nitrite conversion within this group (57%) was obtained for the catalyst supported on zeolite with a Si/Al ratio of 300. This better performance compared to the other catalysts in the group would be associated with the greater surface area (see Table 2), and consequent increase in the number of active sites compared to the other supports. On the other hand, the high ammonium production by this catalyst could be related to the negative surface charge this zeolite provides, which hinders the approach of anions (NO_2^-) to the catalytic surface sites [111] and promotes a high H:N species surface favoring the ammonium production over N_2 [45, 112].

Within group II, the zeolite-supported catalyst PdSA80 attained the highest nitrite conversion (97%) in 120 min (Fig. 3C), but a high proportion of ammonium was also observed (Fig. 3D). As the reaction occurred rapidly, the pH control system could be unable to neutralize the hydroxyl anions produced, thus favoring the selectivity towards ammonium [48]. It is also important to notice that the PdZr catalyst performed very well, with 84% nitrite conversion and low ammonium production.

Among Group III, all catalysts achieved complete nitrite removal (Fig. 3E). PdSi, PdTi, and PdSA30 reached 100% conversion in 80 min, while PdgA catalyst reached the same result at 120 min. According to the literature [113], alumina catalysts are initially more active than those supports that have an IEP closer to the reaction pH.

Devard et al. [114] evaluated nitrite reduction using monometallic Pd catalysts supported on TiO_2. The monometallic Pd/TiO_2 showed 100% conversion and a yield lower than 10 ppm of ammonium, according to the PdTi catalyst in this work. At the same nitrite conversion (32%), the catalysts that produced the smallest ammonium ratio were PdTi and PdSA30 (Fig. 3F).

Among the ZMS-5 catalysts, the fastest total conversion of nitrites was observed with PdSA30, while the lowest conversion was achieved with PdSA300. This could be explained in terms of the chemical properties of zeolite. The negatively charged Al atoms are the source of acid sites and their strength is determined by the extent of their isolation from neighboring acid sites. Dealumination (an increase of the Si /Al

Fig. 3 Catalytic performance corresponding to *Group I* *Group II* *Group III* *Group I (X% = 30%–79%)*, *Group II (X% = 80%–97%)* Group *III (X% = 98%–100%)*. ***(A)***, ***(C)***, ***(E)***: *Nitrites Consumption* ***(B)***, ***(D)***, ***(F)***: *ammonium Production*

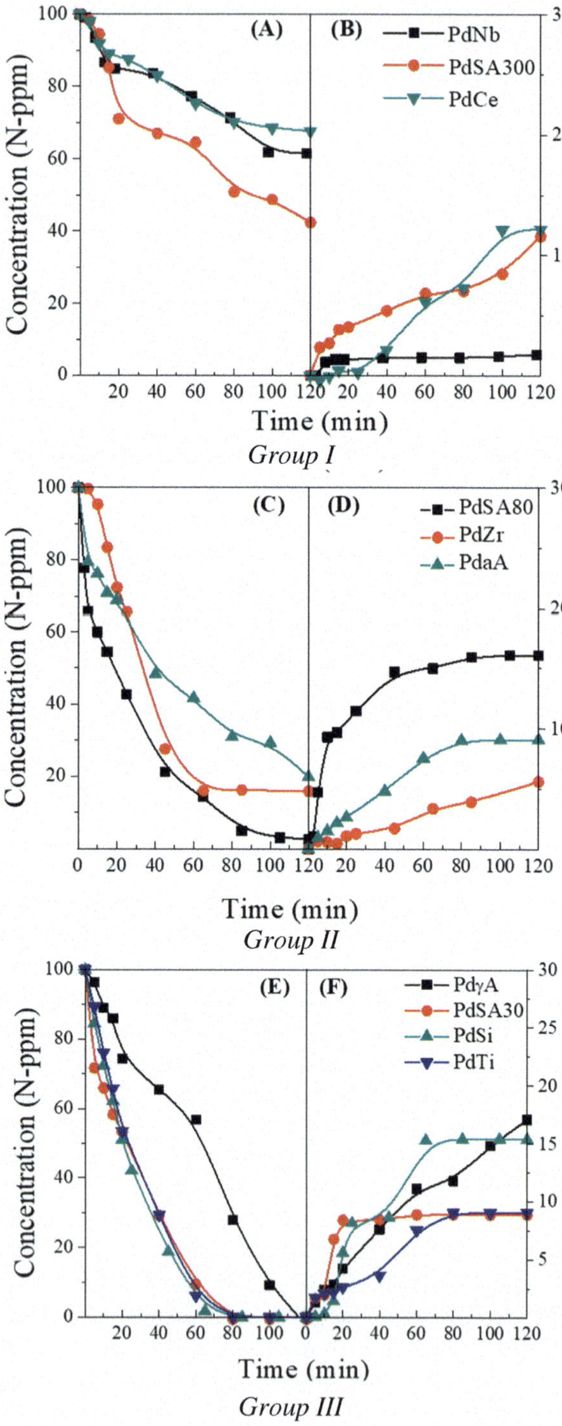

Table 2 IEP and supports Brunauer–Emmett–Teller (BET)-specific surface area obtained from the characterization of the supports and reported ones

Catalyst	(IEP)	Reported (IEP)	(BET)-specific Surface area (m^2/g)
PdaA	9.1	7.0–9.0 [96]	300
PdgA	8.3	7.0–9.0 [49, 97]	180
PdCe	6.7	6.7–6.8 [96]	200
PdZr	6.5	5.8 [98]; 6.0–6.7 [96]	68
PdTi	6.1	6.3 [99]; 7.0–8.0*	48
PdSA30	0.5	3.6** [100]	400
PdSA80	0.8	< 1.0 [101]	400
PdSA300	0.7	< 1.0 [101]	400
PdSi	2.0	< 2.0 [99]	300
PdNb	3.1	< 1.0 [102]; 2.7 [99], 4 [103]	180

*TiO_2 Degussa P25. **The mass rate of Si:Al = 28.5:1

ratio) decreases the number of acid centers, while it enhances their strength [115]. Experimental data suggests that an intermediate number of acid sites is essential to the catalytic activity, whereas their strength controls the selectivity [116]. Liu et al. [78] also reached this conclusion in their work. They tested different treatments to determine which favored nitrite adsorption in aqueous media. The acidity modification of the zeolite changed the surface area and strongly protonated the material favoring the adsorption of nitrite onto the resulting zeolite.

Nitrite conversion of the catalysts and selectivity values calculated at the same extent of reaction are shown in Table 3. Values are expressed as nitrite conversion (X%), selectivity to ammonium (S_{NH4}), and selectivity to nitrogen (S_{N2}).

Table 3 Nitrite conversion in water and selectivity attained

Catalyst	(X = 32%)		(X% 120 min)	% SN_2 (X = 100%)	N-NO_2^- Conversion
	%SNH_4^+*	%SN_2*			
PdCe	12.2	87.8	32.0	–	Group I (X% = 30%–60%)
PdNb	1.6	98.4	38.0		
PdSA300	5.3	94.7	57.0		
PdaA	2.6	97.4	80.0	–	Group II (X% = 80%–97%)
PdZr	1.2	98.8	84.0		
PdSA80	3.5	96.5	97.0		
PdgA	6.7	93.3	100.0	83.0	Group III (X% = 100%)
PdSA30	2.3	97.7	100.0	91.2	
PdSi	0.9	99.1	100.0	84.6	
PdTi	2.2	97.8	100.0	92.0	

* *selectivityat the same nitrite conversion (32%)*

The most selective to N_2 were catalysts PdNb of group I (98.4%), PdZr of group II (98.8%), and PdSi of group III (99.1%) being comparable to each other.

Regarding Group III, it was observed that at 100% nitrite conversion, selectivities were different from those at 32%. PdTi (92.0%), and PdSA30 (91.2%) showed higher SN_2 values, also comparable to each other.

Figures 4 and 5 summarize the conversion percent values and N_2 selectivity (at 32% of conversion) for the catalysts analyzed.

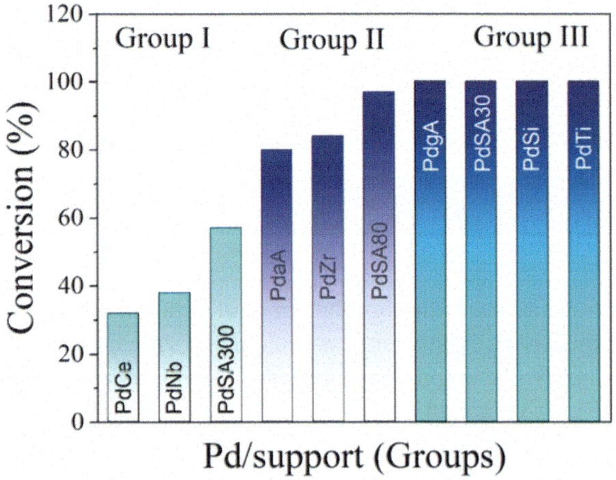

Fig. 4 Conversion % of the different labeled groups

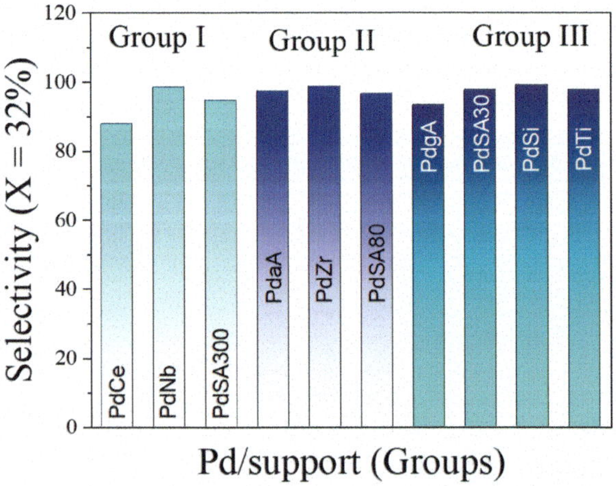

Fig. 5 Selectivity at 32% of conversion

4 Conclusions

This chapter provides insights into the performance of support materials for the catalytic removal of nitrites from water. Pd-based catalysts for nitrite removal are reported employing the following supports: TiO_2, CeO_2, Nb_2O_5, ZrO_2, Al_2O_3, SiO_2, and ZSM-5.

All catalysts showed catalytic activity towards nitrites reduction in water. It was possible to differentiate three groups according to their catalytic performance. Groups I, II, and III belong to increasing conversions, from 30 to 100%. In group III, the most active catalysts were PdgA, PdSA30, PdSi, and PdTi, reaching total conversion of nitrite.

At the same nitrite conversion (32%), PdNb, PdZr, and PdSi were the most selective catalysts for N_2 formation (98.4%, 98.8%, and 99.1%, respectively). The synthesis of catalysts allowing for a low ammonium formation could enhance the efficacy and competitiveness of this method.

Nevertheless, the lack of information about other factors affecting the full-scale applicability, such as catalyst stability in long operating times requires attention to solve the current drawbacks of this technology.

Acknowledgements Thanks are given to UNL (CAI+D 50420150100037LI and 50620190100148LI), ANPCyT (PICT 2019 02970), and CONICET.

References

1. K. Eman, G. Mesko, in *Emerald Handb. Crime, Justice Sustain. Dev.*, ed. by J. Blaustein, K. Fitz-Gibbon, N.W. Pino, R. White (Emerald Publishing Limited, 2020), pp. 465–484
2. M.G. Rivas, Latin Am Res Review **49**, 129 (2020)
3. D. Machiwal, M.K. Jha, V.P. Singh, C. Mohan, Earth-Science Rev. **185**, 901 (2018)
4. C. Bustingorri, R.S. Lavado, Agric. Water Manag.Manag. **144**, 134 (2014)
5. L. Borgnino, M.G. Garcia, G. Bia, Y.V. Stupar, P. Le, P.J. Depetris, Sci. Total. Environ. **443**, 245 (2013)
6. S. Kim, K. Kim, K. Ko, Y. Kim, K. Lee, Chemosphere **87**, 851 (2012)
7. E. Ceballos, S. Dubny, N. Othax, M.E. Zabala, F. Peluso, Expo. Heal. **13**, 323 (2021)
8. Q. Zhang, H. Qian, P. Xu, W. Li, W. Feng, R. Liu, J. Clean. Prod. **298**, 126783 (2021)
9. G. Tokazhanov, E. Ramazanova, S. Hamid, S. Bae, W. Lee, Chem. Eng. J. **384**, 123252 (2019)
10. World Health Organization, Backgr. Doc. Dev. WHO Guidel. Drink. Water Qual. 31 (2007)
11. S. Jung, S. Bae, W. Lee, Water Res. **36**(13), 3330 (2014)
12. J.P. Troutman, H. Li, A.M. Haddix, B.A. Kienzle, G. Henkelman, S.M. Humphrey, C.J. Werth, ACS Catal.Catal. **10**, 7979 (2020)
13. T.T. Mensinga, G.J.A. Speijers, J. Meulenbelt, Toxicol. Rev.. Rev. **22**, 41 (2003)
14. L. Balejčíková, A. Tall, B Kandra, D Pavelková, Acta Hydrol Slovaca, **21**, 74 (2020)
15. S.G. Gardner, J. Levison, B.L. Parker, R.C. Martin, Hydrogeology J. **28**, 1891 (2020)
16. J.G. Albo, Ingeniería **19**, 24 (2015)
17. D. Bamba, M. Coulibaly, D. Robert, Sci. Total. Environ. **580**, 1489 (2017)
18. C. Wisniewski, F. Persin, T. Cherif, R. Sandeaux, A. Grasmick, C. Gavach, F. Lutin, Desalination **149**, 331 (2002)

19. N.G. Hord, M.N. Conley, Nitrite Nitrate Hum. Heal. Dis. 153 (2017)
20. N.G. Hord, Curr. Atheroscler. Rep.. Atheroscler. Rep. **13**, 484 (2011)
21. Codigo Alimentario Argentino, Instituto Nacional de Normalización, R. del Ecuador, and Instituto Ecuatoriano de Normalización, **003**, 1 (1989)
22. V. Höller, K. Rådevik, I. Yuranov, A. Renken, Appl. Catal. BCatal. B **32**, 143 (2001)
23. N. Barrabes, A. Dafinov, F. Medina, J.E. Sueiras, Catal. Today. Today **149**, 341 (2010)
24. J. Lee, Y.G. Hur, M.S. Kim, K.Y. Lee, J. Mol. Catal. A Chem. **399**, 48 (2015)
25. B.S. Rathi, P.S. Kumar, Environ. Chem. Lett. **18**, 1209 (2020)
26. D. Li, X. An Ning, Y. Yuan, Y. Hong, J. Zhang, J. Environ. Sci. (China) **91**, 62 (2020)
27. A. Matei, G. Racoviteanu, IOP Conf. Ser. Earth Environ. Sci. **664**, (2021)
28. R. Soni, S. Bhardwaj, D.P. Shukla, in *Inorganic pollutants in water*, ed. by P. Devi, P. Singh, and S. K. Kansal (Elsevier, 2020), pp. 273–295
29. L.L. Albornoz, L. Marder, T. Benvenuti, A.M. Bernardes, J. Environ. Chem. Eng. **7**, 102982 (2019)
30. S. Tyagi, D. Rawtani, N. Khatri, M. Tharmavaram, J. Water Process Eng. **21**, 84 (2018)
31. M. Abdulredha, R. Al Khaddar, D. Jordan, P. Kot, A. Abdulridha, K. Hashim, Waste Manag. **77**, 388 (2018)
32. I.A. Idowu, W. Atherton, K. Hashim, P. Kot, R. Alkhaddar, B.I. Alo, A. Shaw, Waste Manag.Manag. **87**, 761 (2019)
33. V.B. Jensen, J.L. Darby, C. Seidel, C. Gorman, Crit. Rev. Environ. Sci. Technol. **44**, 2203 (2014)
34. T. He, Q. Wu, C. Ding, M. Chen, M. Zhang, Ecotoxicol. Environ. Saf.. Environ. Saf. **224**, 112693 (2021)
35. W. Li, J. Liu, Y. Zhen, M. Lin, X. Sui, W. Zhao, X. Bing, J. Lin, L. Zhai, J. Water Process Eng. **40**, 101976 (2021)
36. D. K. Babu, K. Ravindhranath, and S. Mekala, Biomass Convers. Biorefinery (2021).
37. M.A. Tadda, C. Li, M. Gouda, A.E.F. Abomohra, A. Shitu, A. Ahsan, S. Zhu, D. Liu, J. Environ. Chem. Eng. **9**, 105947 (2021)
38. K. Pirkanniemi, M. Sillanpää, Chemosphere **48**, 1047 (2002)
39. Q. Zhao, Q. Mao, Y. Zhou, J. Wei, X. Liu, J. Yang, L. Luo, J. Zhang, H. Chen, H. Chen, L. Tang, Chemosphere **189**, 224 (2017)
40. W. Oh, T. Lim, Chem. Eng. J. **358**, 110 (2019)
41. S. Seraj, P. Kunal, H. Li, G. Henkelman, S.M. Humphrey, C.J. Werth, ACS Catal.Catal. **7**, 3268 (2017)
42. K. Vorlop, T. Tacke, Chemie Ing. Tech. **61**, 836 (1989)
43. J. Martínez, A. Ortiz, I. Ortiz, Appl. Catal. B Environ.Catal. B Environ. **207**, 42 (2017)
44. O.S.G.P. Soares, J.J.M. Órfão, M.F.R. Pereira, Catal. Letters **126**, 253 (2008)
45. P. Xu, S. Agarwal, L. Lefferts, J. Catal.Catal. **383**, 124 (2020)
46. A. S. G. G. Santos, J. Restivo, C. A. Orge, M. F. R. Pereira, and O. S. G. P. Soares, C **6**, 78 (2020).
47. S. Hamid, Y. Niaz, S. Bae, W. Lee, J. Environ. Chem. Eng. **8**, 103754 (2020)
48. A.H. Pizarro, I. Torija, V.M. Monsalvo, J. Water Supply Res. Technol.—AQUA **67**, 615 (2018).
49. F.A. Marchesini, N. Picard, E.E. Miró, Catal. Commun.. Commun. **21**, 9 (2012)
50. F.A. Marchesini, S. Irusta, C. Querini, E. Miró, Appl. Catal. A Gen. **348**, 60 (2008)
51. A. Devard, M.A. Ulla, F.A. Marchesini, Catal. Commun.. Commun. **34**, 26 (2013)
52. X. Zhao, G. Zhang, Z. Zhang, Environ. Int. **136**, 105453 (2020)
53. M.B. Navas, H.P. Bideberrripe, C.I. Cabello, D. Gazzoli, M.L. Casella, M.A. Jaworski, Catal. Today. Today **372**, 154 (2021)
54. M. Ying, M. Zhang, Y. Liu, Z. Wu, Sci. Rep. **9**, 1 (2019)
55. M. Kobune, D. Takizawa, J. Nojima, R. Otomo, Y. Kamiya, Catal. Today. Today **352**, 204 (2020)
56. C. Song, Y. Wei, J. Sun, Y. Song, S. Li, Y. Kitamura, Bioresour. Technol.. Technol. **296**, 122320 (2020)

57. S. Li, X. Zheng, Y. Chen, C. Song, Z. Lei, Z. Zhang, Bioresour. Technol.. Technol. **313**, 123743 (2020)
58. J. Sá, J.A. Anderson, Appl. Catal. B Environ.Catal. B Environ. **77**, 409 (2008)
59. A. Zaki, M. Wastiaux, S. Casale, A. Mussi, J.F. Dhenin, C. Lancelot, J.P. Dacquin, P. Granger, Catal. Sci. Technol. **8**, 4604 (2018)
60. A. Zaki, C. Lancelot, J.P. Dacquin, P. Granger, Catal. Commun.. Commun. **148**, 106168 (2021)
61. K. Sathishkumar, Y. Li, M.S. Alsalhi, B. Muthukumar, G.K. Gaurav, S. Devanesan, A. Rajasekar, R. Manikandan, Environ. Res. **207**, 112158 (2022)
62. S. Haga, H. Nagakawa, T. Ochiai, M. Nagata, Chem. Lett. **47**, 1542 (2018)
63. H. Kominami, H. Gekko, K. Hashimoto, Phys. Chem. Chem. Phys. **12**, 15423 (2010)
64. H. Gekko, K. Hashimoto, H. Kominami, Phys. Chem. Chem. Phys. **14**, 7965 (2012)
65. L. Liu, J. Cai, L. Qi, Q. Yu, K. Sun, B. Liu, F. Gao, L. Dong, Y. Chen, J. Mol. Catal. A Chem. **327**, 1 (2010)
66. Y.S. Ryou, J. Lee, H. Lee, C.H. Kim, D.H. Kim, Catal. Today. Today **307**, 93 (2018)
67. F. Epron, F. Gauthard, J. Barbier, J. Catal.Catal. **206**, 363 (2002)
68. M.A. Jaworski, I.D. Lick, G.J. Siri, M.L. Casella, Appl. Catal. B Environ.Catal. B Environ. **156–157**, 53 (2014)
69. J. Xiong, P. Zhang, Y. Li, Y. Wei, Y. Zhang, J. Liu, Z. Zhao, Chem. Eng. Sci. **239**, 116635 (2021)
70. C.A. Akinnawo, D.J. Maheso, N. Bingwa, R. Meijboom, Appl. Catal. A Gen. **613**, 118022 (2021)
71. W. Xue, B. Gu, H. Wu, M. Liu, S. He, J. Li, X. Rong, C. Sun, Appl. Catal. A Gen. **616**, 118107 (2021)
72. C. García-Sancho, C.P. Jiménez-Gómez, N. Viar-Antuñano, J.A. Cecilia, R. Moreno-Tost, J.M. Mérida-Robles, J. Requies, P. Maireles-Torres, Appl. Catal. A Gen. **609**, 117905 (2021)
73. Y. Sakamoto, K. Nakamura, R. Kushibiki, Y. Kamiya, T. Okuhara, Chem. Lett. **34**, 1510 (2005)
74. B. Liu, Y. Li, S. Qing, K. Wang, J. Xie, Y. Cao, Cryst. Eng. Comm. **22**, 4005 (2020)
75. J.P. Dacquin, S. Troncéa, V.I. Parvulescu, P. Granger, Catal. Today. Today **383**, 330 (2022)
76. B. Peroni, M. Navas, H.P. Bideberripe, B. Barbero, M.L. Casella, M.A. Jaworski, Ind. Eng. Chem. Res. **60**, 12767 (2021)
77. W.S. Hamid, M.A. Kumar, J.I. Han, H. Kim, Lee, Green Chem. **19**, 1 (2017)
78. J. Liu, X. Cheng, Y. Zhang, X. Wang, Q. Zou, L. Fu, Microporous Mesoporous Mater. **252**, 179 (2017)
79. Y. Matatov-Meytal, V. Barelko, I. Yuranov, M. Sheintuch, Appl. Catal. B Environ.Catal. B Environ. **27**, 127 (2000)
80. F. Deganello, L.F. Liotta, A. Macaluso, A.M. Venezia, G. Deganello, Appl. Catal. B Environ.Catal. B Environ. **24**, 265 (2000)
81. A.E. Palomares, J.G. Prato, F. Márquez, A. Corma, Appl. Catal. B Environ.Catal. B Environ. **41**, 3 (2003)
82. A.E. Palomares, J.G. Prato, F. Rey, A. Corma, J. Catal.Catal. **221**, 62 (2004)
83. Y. Wang, J. Qu, H. Liu, J. Mol. Catal. A Chem. **272**, 31 (2007)
84. A. Pintar, J. Batista, J. Levec, Chem. Eng. Sci. **56**, 1551 (2001)
85. D. Gašparovičová, M. Králik, M. Hronec, A. Biffis, M. Zecca, B. Corain, J. Mol. Catal. A Chem. **244**, 258 (2006)
86. D. Gašparovičová, M. Králik, M. Hronec, Z. Vallušová, H. Vinek, B. Corain, J. Mol. Catal. A Chem. **264**, 93 (2007)
87. D. Gašparovičová, M. Králik, M. Hronec, Collect. Czechoslov. Chem. Commun.Czechoslov. Chem. Commun. **64**, 502 (1999)
88. M.P. Maia, M.A. Rodrigues, F.B. Passos, Catal. Today. Today **123**, 171 (2007)
89. G.P. Soares, M.F.R. Pereira, Ind. Eng. Chem. Res. **49**, 7183 (2010)
90. L. Lemaignen, C. Tong, V. Begon, R. Burch, D. Chadwick, Catal. Today. Today **75**, 43 (2002)
91. N. Barrabes, J. Just, A. Dafinov, F. Medina, J.L.G. Fierro, J.E. Sueiras, P. Salagre, Y. Cesteros, Appl. Catal. B Environ.Catal. B Environ. **62**, 77 (2006)

92. U. Matatov-Meytal, M. Sheintuch, Catal. Commun.. Commun. **10**, 1137 (2009)
93. Y. Yoshinaga, T. Akita, I. Mikami, T. Okuhara, J. Catal.Catal. **207**, 37 (2002)
94. F.M. Zoppas, F. Beltrame, F.A. Sosa, A.M. Bernardes, E. Miró, F.A. Marchesini, Environ. Sci. Pollut. Res.Pollut. Res. **27**, 40405 (2020)
95. F.M. Zoppas, A.M. Bernardes, E.E. Miró, F.A. Marchesini, Catal. Letters **148**, 2572 (2018)
96. M. Rahmani, K. Badii, M. Faghihi, M. Sanati, N. Cruise, O. Augustsson, J.J. Spivey, in *Catalysis*, vol. 17, ed. by J.J. Spivey, G.W. Roberts(Royal Society of Chemistry, Cambridge, 2004), p. 210-257
97. S.W. Vierck, C.A. Leclerc, Adv. Powder Technol. **28**, 1792 (2017)
98. S. Muhammad, S.T. Hussain, M. Waseem, A. Naeem, J. Hussain, M.T. Jan, Iran. J. Sci. Technol. (Sci.) **36**(4), 481 (2012)
99. M. Kosmulski, J. Colloid Interface Sci. **337**, 439 (2009)
100. D. Shao, X. Wang, Q. Fan, Microporous Mesoporous Mater. **117**, 243 (2009)
101. S. Suárez, I. Jansson, B. Ohtani, B. Sánchez, Catal. Today. Today **326**, 2 (2018)
102. L. C. A. Oliveira, M. F. Portilho, A. C. Silva, H. A. Taroco, and P. P. Souza, Appl. Catal. B, Environ. **117–118**, 29 (2012)
103. M. Kosmulski, Adv. Colloid Interface Sci. **238**, 1 (2016)
104. APHA/AWWA/WEF, *Standard Methods for the Examination of Water and Wastewater*, 23rd edn, American Public Health Association. (Washington DC, USA, 2017).
105. APHA/AWWA/WEF, *Standard Methods for the Examination of Water and Wastewater*, 22nd edn. American Public Health Association, American Water Works Association, Water Environment Federation (2012)
106. P.L. Searle, Analyst **109**, 549 (1984)
107. Y. Zhao, X. Zhou, L. Ye, and S. Chi Edman Tsang, Nano Rev. **3**, 17631 (2012)
108. M. Al Bahri, L. Calvo, M.A. Gilarranz, J.J. Rodriguez, F. Epron, Appl. Catal. B Environ. **138–139**, 141 (2013)
109. D.P. Barbosa, P. Tchiéta, M.D.C. Rangel, F. Epron, J. Mol. Catal. A Chem. **366**, 294 (2013)
110. M. D'Arino, F. Pinna, G. Strukul, Appl. Catal. B Environ.Catal. B Environ. **53**, 161 (2004)
111. T. Ma, S.P. Mu, B. Kraushaar-czarnetzki, Industrial **40**, 2573 (2001)
112. B.P. Chaplin, M. Reinhard, W.F. Schneider, C. Schüth, J.R. Shapley, T.J. Strathmann, C.J. Werth, Environ. Sci. Technol. **46**, 11469 (2012)
113. M. S. Heise, J. Colloid. Interf. Sci. **135**, (1990)
114. A. Devard, V. S. Aghemo, C. A. Caballero Dorantes, M. Gutierrez Arzaluz, F. A. Marchesini, and M. A. Ulla, React. Kinet. Mech. Catal. **120**, 39 (2017).
115. S. Mohan, P. Dinesha, S. Kumar, Chem. Eng. J. **384**, 123253 (2019)
116. N. Rahimi, R. Karimzadeh, Appl. Catal. A Gen. **398**, 1 (2011)

Noble-Metal-Free Bifunctional Electrocatalysts for Overall Water Splitting in Alkaline Medium

Subhasis Shit, Tapas Kuila, and Suneel Kumar Srivastava

Abstract The ever-growing energy crisis and environmental pollution motivated scientific researchers to develop a green energy economy. Hydrogen generation through electrochemical water splitting has shown immense potentiality to be an important energy conversion technology towards the said goal. The electrochemical water splitting consists of two half-cell reactions: hydrogen evolution reaction (HER) and oxygen evolution reaction (OER). The overall water splitting is sluggish as both of these half-cell reactions involve multiple proton and electron transfer steps. Theoretical as well as experimental results have revealed that the noble-metal-based electrocatalysts (such as Pt and RuO_2) possess superior electrocatalytic activity toward water splitting. However, these materials lack commercial-scale applicability due to their high price and lower abundance. Recently, numerous investigations have been conducted to develop cheap and efficient noble-metal-free electrocatalysts. Pure water is a poor conductor of electricity and thus an electrolyte is added to facilitate the overall water-splitting process. The efficiency and cost-effectiveness of water electrolysis are higher in an alkaline medium if all the technical aspects related to it are considered together. An electrocatalyst with sufficient bifunctional activity towards both HER and OER in the same operational condition simplifies the electrolyzer optimization process and consequently, higher efficiency in full water splitting is achieved. So, a cheap and efficient bifunctional noble-metal-free electrocatalyst for water splitting in an alkaline medium is highly desirable. Borides, carbides, pnictides, and chalcogenides of transition metals like Fe, Co, Ni, etc. show significant promise in that context. Few transition metal-based heterostructures and engineered electrocatalysts also exhibit desirable bifunctional activity towards overall water splitting.

S. Shit · T. Kuila
Surface Engineering & Tribology Division, Council of Scientific and Industrial Research-Central Mechanical Engineering Research Institute, Mahatma Gandhi Avenue, Durgapur, West Bengal, Tapas Kuila 713209, India

CSIR-Human Resource Development Centre, (CSIR-HRDC) Campus, Academy of Scientific and Innovative Research (AcSIR), Postal Staff College Area, Sector 19, Kamla Nehru Nagar, Ghaziabad, Uttar Pradesh 201002, India

S. K. Srivastava (✉)
Department of Chemistry, Indian Institute of Technology Kharagpur, Kharagpur 721302, India
e-mail: falsesuneelchemkgp@gmail.com

S. J. Ikhmayies (ed.), *Advances in Catalysts Research*, Advances in Material Research and Technology, https://doi.org/10.1007/978-3-031-49108-5_9

This chapter provides a brief idea about electrocatalytic water splitting and also summarizes some recent advancements in noble-metal-free bifunctional electrocatalyst development. Finally, the future prospects in the development of efficient and cheap bifunctional electrocatalysts for water splitting have been discussed.

Keywords Electrocatalyst · Overall water splitting · Bifunctional electrocatalyst · Electrolysis · Transition-metal-based materials

1 Introduction

The ever-growing technological enhancement augmented the energy demand throughout the globe, however, the common sources of energy, i.e., fossil fuels are going to end soon. In addition, the usage of fossil fuels generates ample amount of greenhouse gases such as CO_2 which causes environmental pollution and global warming. A recent report about the global energy outlook has shown that the energy demand is going to increase by (20–30)% in 2040 [1]. An analysis has shown that global fossil fuel extraction will reach a peak (11.42 gigatonnes of oil equivalent per year (Gtoe/yr)) and will start to decline at a rapid rate [2]. Intergovernmental Panel on Climate Change (IPCC) has predicted that the median CO_2 concentration can reach a value of more than 1000 ppm by 2100. Wang et al. have shown that due to the increment in the usage of non-conventional fossil fuels, the median CO_2 concentration may not reach 1000 ppm rather it would be 610 ppm by 2100, which is also quite high [2]. In the present scenario, the development of a greener energy economy is of high importance to cope with the energy crisis and environmental pollution. Huge research efforts are therefore being deployed to develop technologies for the utilization of renewable energy sources like sun light, wind power, flywheel, biomass, etc. However, the lack of suitable energy conversion and storage technology hinders the large-scale applicability of these intermittent energy sources [3]. The hydrogen having very high specific energy (142 MJ kg^{-1}) shows great potentiality to be the greener and cleaner energy carrier of the future. Besides that, the combustion of H_2 produces environmentally benign water along with energy [4]. Although H is the most abundant element in the universe, it is not freely available for utilization. A naturally abundant source of hydrogen is water and hydrogen can be generated after passing electrical charges through it [5, 6]. The electrolysis of water utilizing the electricity generated from renewable sources is therefore a viable technique to produce large amounts of pure hydrogen gas with zero emission of carbonaceous by-product [6].

The electrolysis of water comprises two simultaneously occurring half-reactions: hydrogen evolution reaction (HER) and oxygen evolution reaction (OER). The involvement of multiple proton-coupled charge transfer steps in HER and OER makes these reactions sluggish in nature, and also reduces the water splitting efficiency [7, 8]. An electrocatalyst is therefore employed to enhance the efficiency of these reactions. Theoretical combined with experimental results have revealed that noble-metal-based materials are superior electrocatalysts for water-splitting applications. Additionally, an electrolyte is added to enhance the conductivity of the water, which in turn facilitates the water-splitting process [9]. The platinum on activated carbon (Pt/C) is used for HER catalysis and the oxide of ruthenium or iridium (RuO_2, IrO_2) is used for OER catalysis in an acidic medium [10, 11]. The high cost and lower natural abundance hinder large-scale applicability of these noble-metal-based electrocatalysts [12, 13]. Extensive research to develop noble-metal-free electrocatalysts with comparable or superior activity and stability is going on. The operational stability of the OER electrocatalysts is lower in the acidic medium as compared to the alkaline one. Additionally, the kinetics of the generally-sluggish OER is faster in the alkaline medium in comparison to an acidic one [14]. Solid polymeric electrolyte membrane electrolysis shows superiority to alkaline water electrolysis however, the former faces serious drawbacks like lower stability and cost-effectiveness. Low-cost and stable Ni-based alloys are utilized as electrodes for alkaline liquid electrolyte water electrolysis resulting in the generation of lower current density. If all the aspects of water electrolysis are compared, the one containing alkaline electrolyte shows superiority in terms of overall performance and cost-effectiveness [15]. A disparity between the conditions favorable for HER and OER electrocatalysts results in a mediocre full water splitting performance and hydrogen generation [16]. In that context, a bifunctional electrocatalyst that can simultaneously catalyze both HER and OER in an alkaline medium is beneficial. The development of a bifunctional electrocatalyst also simplifies the electrolyzer fabrication procedure which further enhances the efficiency of the hydrogen generation process [17]. A bifunctional electrocatalyst that can catalyze both HER and OER in an alkaline medium is, therefore, highly desirable to achieve better efficiency in overall water splitting [18].

2 Electrocatalytic Water Splitting and Electrocatalyst

In the late eighteenth century, scientists observed that the water decomposed to produce gaseous hydrogen and oxygen when electric charges passed through it [19]. However, two English scientists William Nicholson and Anthony Carlisle successfully demonstrated the water electrolysis process using Voltaic piles in 1800 [20]. The technologies for large-scale production of hydrogen started to grow between 1920 and 1970, with increasing demand on ammonia and petroleum refining industries. Rapid innovation in water electrolysis was initiated after 1970 due to the awareness regarding the energy crisis and environmental pollution [21]. The hydrogen was considered as a potential green fuel of the future. Modern innovations like the integration of the electrolyzer with renewable energy source harvesters (like PV cells) started to develop in the twenty-first century [21]. In recent times, a significant part of the scientific community has been focusing on water electrolysis technology to achieve better efficiency in the process.

The electrolysis occurs according to the following reaction:

$$H_2H_2O_{(l)} \rightarrow H_{2(g)} + 1/2O_{2(g)} \quad (1)$$

The thermodynamics suggests that the reversible voltage required for successful water electrolysis is 1.23 V [22]. The reaction is highly endothermic and the Gibb's free energy change (ΔG) is positive causing the reaction to be non-spontaneous. The ΔG becomes negative at temperature > 2500 K, and therefore direct thermal decomposition of water is not practically viable [13]. The thermoneutral voltage for water splitting calculated from the enthalpy is 1.48 V. At least 1.48 V potential should be externally applied to perform the electrolysis of water without supplying any thermal energy [9, 22].

The electrolysis of water is comprised of two half-cell reactions:

$$HER: 4H_2O + 4e^- \rightarrow 2H_2 + 4OH^- \quad (2)$$

$$OER: 4OH^- \rightarrow 2H_2O + O_2 + 4e^- \quad (3)$$

The standard reduction potential for HER and OER is –0.829 and 0.401 V (at pH = 14), respectively [23]. The thermodynamics provide the minimal potential required to drive the electrochemical reactions however, due to the kinetic barriers some extra potentials have to be applied to drive the reaction. That extra potential is termed as "Overpotential" (η) [21, 24].

$$\eta = E_{applied} - E^0_{rev} - iR_u \quad (4)$$

The iR_u signifies the potential gain due to various unavoidable aspects such as wire resistance, bubble generation at the interface, and electrolyte diffusion blockage. [21].

The Butler–Volmer equation is the fundamental equation for the kinetics of any electrochemical reaction. It describes the dependency of the electrical current on the applied overpotential and it considers both the forward and backward steps of a reaction. The equation is as follows:

$$i = i_0[exp(\alpha_f \eta F_n/RT) - exp(-\alpha_b \eta F_n/RT)] \quad (5)$$

where *i, F, R, n,* and *T* stand for current density, Faraday constant, universal gas constant, number of electrons involved in the reaction and absolute temperature. The α_f and α_b are the charge transfer coefficients of the forward and backward reactions, respectively. The i_0 is the exchange current density. It is the current density observed at zero overpotential i.e. at the equilibrium condition [24, 25]. High overpotential has to be applied to drive the electrochemical reaction either in the forward way or backward way. At higher overpotential, the current density related to the forward reaction is higher and the backward is negligible [24]. The Butler-Volmer reaction at that condition reduces to the Tafel equation [26].

$$i \approx i_0 exp(\alpha\eta F_n/RT) \tag{6}$$

The following equation is obtained by converting Eq. (6) into logarithmic form [26].

$$log(i) = log(i_0) + \eta/b \tag{7}$$

The b is termed as "Tafel slope" and

$$b = \delta\eta/\delta log(i) = 2.303RT/\alpha F_n \tag{8}$$

The b value signifies how faster the current response for an electrochemical reaction increases with applied overpotential. The smaller the b faster the electrochemical reaction. In addition, the b value can provide insightful information about the reaction mechanism and helps in identifying the probable rate-determining step (r.d.s.) [24, 26].

Two and four elementary steps are involved in HER and OER catalysis, respectively, as evident from the schematic energy profile diagrams shown in Fig. 1 [27, 28]. In addition to a thermodynamic barrier, a kinetic one will also be associated with each and every elementary step, and these barriers cumulatively contribute towards the overall η of the reaction. However; the corresponding kinetic barriers are not shown in these diagrams for simplification. The free energy change associated with each, and every elemental step involved in OER is positive and thus is thermodynamically unfavorable. A very large potential (~1.8 V) has to be applied to make the OER thermodynamically feasible/downhill (Fig. 1b). On the other hand, one elementary step is thermodynamically downhill and the other is uphill for HER (Fig. 1a). So, the thermodynamic aspects suggest that the OER is more sluggish as compared to HER.

The constituent steps for HER are different in acidic and alkaline medium. The H gets adsorbed on the electrode surface in the initial step which is termed as "Volmer

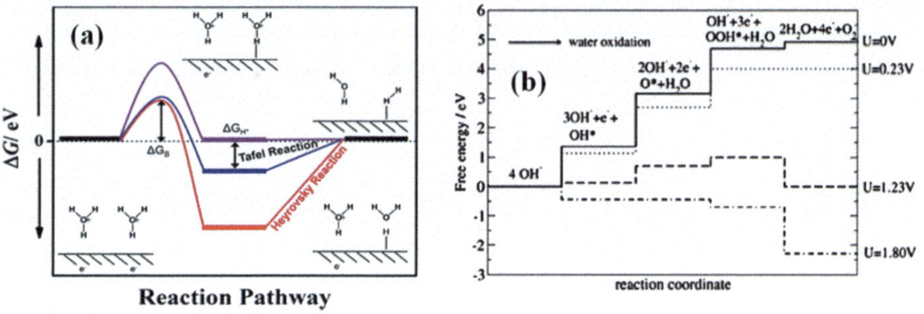

Fig. 1 Energy profile diagram for **a** HER. Reprinted with permission from Li et al. [27], copyright (2018). The Royal Society of Chemistry. **b** OER. Reprinted from Serov et al. [28], under the Creative Commons Attribution 4.0 International License (CC BY 4.0). http://creativecommons.org/licenses/by/4.0/

step". The adsorbed H gets desorbed to form H_2 via a chemical or electrochemical pathway. The chemical and electrochemical pathways are termed as the "Tafel step" and "Heyrovsky step", respectively [26, 29, 30].

The elemental steps of HER in an acidic medium:

Volmer step:

$$H_3O + +e^- \rightarrow H_{ads} \tag{9}$$

Tafel step:

$$H_{ads} + H_{ads} \rightarrow H_2 \tag{10}$$

Heyrovsky step:

$$H_{ads} + H_3O + +e^- \rightarrow H_2 + H_2O \tag{11}$$

And in alkaline medium:

Volmer step:

$$H_2O + e^- \rightarrow H_{ads} + OH^- \tag{12}$$

Tafel step:

$$H_{ads} + H_{ads} \rightarrow H_2 \tag{13}$$

Heyrovsky step:

$$H_{ads} + H_2O + e^- \rightarrow H_2 + OH^- \tag{14}$$

The Tafel slope value will be 118, 39, and 29.5 mV dec^{-1} if the r.d.s. is Volmer, Heyrovsky, and Tafel step, respectively [26].

Similarly, the constituent steps for OER are also different in acidic and alkaline mediums [24]. The proposed mechanism of OER in an acidic medium is as follows:

$$H_2O \rightarrow OH_{ads} + H^+ + e^- \tag{15}$$

$$OH_{ads} + OH_- \rightarrow O_{ads} + H_2O + e^- \tag{16}$$

$$2O_{ads} \rightarrow O_2 \tag{17}$$

$$O_{ads} + H_2O \rightarrow OOH_{ads} + H^+ + e^- \tag{18}$$

$$OOH_{ads} + H_2O \rightarrow O_2 + H^+ + e^- \tag{19}$$

The probable mechanism of OER in alkaline medium is as follows:

$$OH^- \rightarrow OH_{ads} + e^- \tag{20}$$

$$OH_{ads} + OH^- \rightarrow O_{ads} + H_2O + e^- \tag{16}$$

$$2O_{ads} \rightarrow O_2 \tag{17}$$

$$O_{ads} + OH^- \rightarrow OOH_{ads} + e^- \tag{18}$$

$$OOH_{ads} + OH^- \rightarrow O_2 + H_2O + e^- \tag{19}$$

Both the oxo group coupling path and terminal peroxide group formation path are considered for the generation of gaseous oxygen. Predicting the Tafel slope value for any step being the r.d.s. is difficult due to the complexity of the OER process. The Tafel slope value will be 120, 40, 15 mV dec^{-1} when the r.d.s. is 1st, 2nd, and 3rd step, respectively [31, 32]. The Tafel slope value can range between 21and 120 mV dec^{-1} when the r.d.s is the 4th or the 5th step [33].

A catalytic pathway should be undertaken to reduce the hindrances appearing due to the kinetic barriers in the reaction steps and to minimize the overpotential. A catalyst that facilitates the electrochemical reactions and stays unreacted after it is termed as "electrocatalyst". Two important aspects that affect the electrocatalytic activity of a material are active species adsorption-desorption and the charge transfer between the electrocatalyst and that species. The hydrogen adsorption-desorption and the charge transfer are the two crucial factors influencing the HER [34]. The ΔG_{H*} can give an idea about the hydrogen adsorption-desorption proficiency of the electrocatalyst and the i_0 can provide an idea about the charge transfer efficiency [24, 34, 35]. According to the Sabatier principle, the catalytically active site should hold the substrate neither too strong nor too weak. A volcano-type curve is obtained when the experimentally obtained i_0 is plotted against ΔG_{H*} (Fig. 2a) [36]. The plot suggests that the noble metals are superiorly active towards HER. Pt is situated at the top of the volcano suggesting it to be the superior electrocatalyst for HER. The noble-metal-based electrocatalysts follow the Tafel pathway as understood from the lower ΔG_{H*} value [27]. The distance between two adjacent active sites in these electrocatalysts is smaller as compared to the van der Waals radius of two adsorbed hydrogen atoms [37]. Most of the non-precious metal-based electrocatalysts follow the Volmer-Heyrovsky pathway due to the unavailability of the catalytically active sites in close proximity [37]. The electronic configuration of the electrocatalyst plays a significant role in its activity. The Ni, Pd, and Pt show superior catalytic activity due to their favorable electronic configuration, whereas the Zn, Cd, and Hg show inferiority [21]. The physical properties of the electrocatalyst like morphology and surface area also influence its catalytic activity [21].

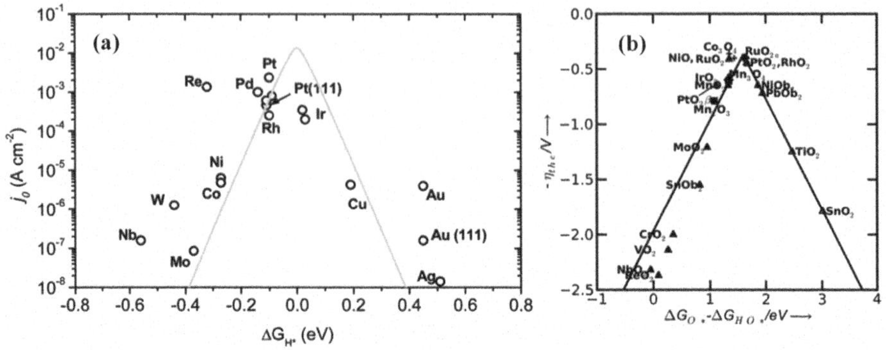

Fig. 2 **a** HER volcano plot for metals. Reprinted with permission from Morales-Guio et al. [36]. Copyright (2014). The Royal Society of Chemistry. **b** OER volcano plots for metal oxides. Reprinted with permission from Man et al. [38]. Copyright (2011). Wiley–VCH Verlag GmbH & Co. KGaA

The theoretical calculations for OER suggest that the difference between ΔG_{OH^*} and ΔG_{OOH^*} is constant for all the oxide-based electrocatalysts. The catalytic activity therefore varies with varying O adsorption efficiency [38]. Either the OH_{ads} or the O_{ads} formation step acts as r.d.s. for most of the OER electrocatalysts. The ΔG_{O^*} − ΔG_{OH^*} has been considered as the descriptor for OER activity, and a volcano-type plot is obtained when these values are plotted against the overpotential values (Fig. 2b) [35, 38]. The volcano plots suggested that the noble-metal-containing oxides especially the oxides of ruthenium and iridium possess superior electrocatalytic activity, and the experimental results also support the same [39, 40].

The state-of-the-art noble-metal-based electrocatalysts have few drawbacks that impede their large-scale applicability. In recent times a huge research effort has been made to design noble-metal-free electrocatalysts with efficient catalytic performance [14, 24, 37, 41–46].

3 Classification of Water Electrolysis

The electrical conductivity of pure water is significantly low [47]. The auto-ionization of water may occur if a huge potential is applied, and then water splitting will also occur, however; the rate will be very low. A water-soluble electrolyte is, therefore, added to increase the electrical conductivity of water and enhance the water-splitting efficiency. Either strong acid or strong base is used as the electrolyte in the commercially available electrolyzers [48]. The electrolysis of water is operable at ambient conditions and can be classified into four categories: **a** acid liquid electrolyte water electrolysis (ACIWE), **b** alkaline liquid electrolyte water electrolysis (ALKWE), **c** Polymeric electrolyte membrane water electrolysis (PEMWE) and **d** anion exchange membrane water electrolysis (AEMWE) [16, 49–51]. Though significant research has been conducted in the past few years, the AEMWE technology is still in the R&D stage [51].

ACIWE and ALKWE are the electrolysis processes where the electrolytes used are acidic and alkaline in nature, respectively. In general, the acid used is H_2SO_4 and the alkali is KOH or NaOH [16, 21, 49]. The anode and the cathode components are separated with a dielectric membrane (also called separator) or diaphragm (Fig. 3a, b). The separator lets the ions pass through it and prevents the mixing of produced H_2 and O_2 [48, 49]. The ALKWE is more advantageous as compared to ACIWE as only the noble-metal-based electrocatalysts can sustain against electrochemical corrosion in an acidic medium. The non-noble-metal-based materials are fairly stable in an alkaline medium, which makes the ALKWE more cost-effective [49]. Among the two reactions, the OER is more complex in nature, and thus it is more sluggish as compared to HER. The HER electrocatalysts are efficiently active and stable in both acidic and alkaline media however; the OER electrocatalysts show efficiency and stability only in alkaline medium [14, 49, 52, 53]. From these perspectives, the ALKWE is the favorable electrolysis process for achieving greater water-splitting efficiency. PEMWE is the electrolysis process where the solid polymer electrolyte membrane (PEM) is used to conduct the ions, separate the produced gases, and provide electrical insulations to the electrodes (Fig. 3c) [15]. The advantages of PEMWE is that pure water can be fed into the cell, and the only components that are in direct contact with the membrane evidences the electrochemical corrosion [49]. A recent study has suggested that the ALKWE can show not only comparable but also superior efficiency as compared to the PEMWE [49]. The main drawbacks of the PEMWE process are (a) lower cost-effectiveness due to the usage of noble metal electrocatalysts, and (b) lower stability due to the corrosive environment and high pressure [15]. Polymeric anion exchange membrane (AEM) is used instead of PEM in AEMWE technology. Similar to the ALKWE, non-noble-metal-based electrocatalysts can be used for AEMWE. Distilled water or slightly alkaline water is fed to the electrolyzer, which reduces the chance of electrocatalyst corrosion in AEMWE [48]. The major drawbacks of AEMWE are (a) lower durability owing to the rapid degradation of AEM, (b) low current density generation, and (c) excessive catalyst loading [51]. A significant exploration is still required to enhance the commercial viability of AEMWE. Although having a few drawbacks like lower current density generation and gas crossover, the ALKWE shows greater promise in terms of cost-effectiveness when all the technical aspects are considered together [15, 49].

4 Overall Water Splitting and Bifunctional Electrocatalyst

The efficiency of the overall water splitting depends on the HER and OER. The HER proceeds at a faster rate in an acidic medium; whereas the alkaline medium is favorable for OER [16, 52]. Thus, most of the research efforts are made in developing the HER electrocatalyst for an acidic medium and the OER electrocatalyst in an alkaline medium. However, both the electrocatalysts for HER and OER should perform

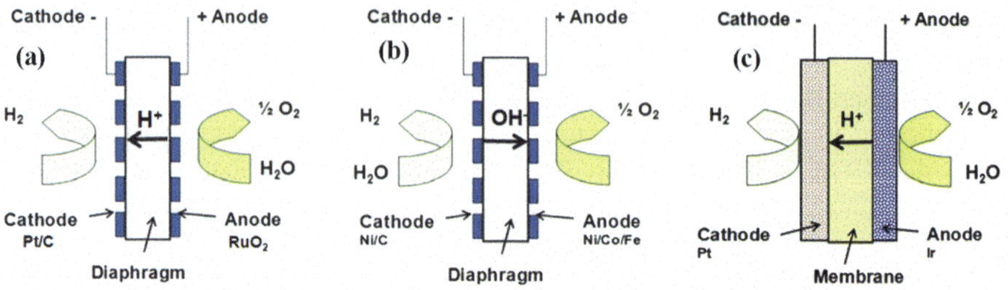

Fig. 3 Schematic of the acid liquid electrolyte water electrolysis (ACIWE) (**a**), alkaline liquid electrolyte water electrolysis (ALKWE) (**b**), and polymer electrolyte membrane water electrolysis (PEMWE) process (**c**). Reprinted with permission from Carmo et al. [15]. Copyright (2013). Elsevier

efficiently in the same electrolyte if the desired overall water-splitting efficiency has to be achieved. The disparity in the operational condition for the HER and OER electrocatalysts complicates the integration process and results in mediocre overall water-splitting performance [16]. Therefore, an electrocatalyst that can concurrently catalyze both HER and OER in the same electrolyte is highly desirable. This type of electrocatalyst is termed as a "bifunctional electrocatalyst". The bifunctional electrocatalysts are developed by combining the merits of the HER and OER electrocatalysts and these electrocatalysts have the potential to bind both the H-containing and O-containing species [17]. Moreover, the fabrication and integration of the electrolyzer will be simplified on the successful development of the efficient bifunctional electrocatalyst, which will make the overall water-splitting process more cost-effective [17, 52]. As discussed earlier the ALKWE is more cost effective than ACIWE and therefore, the efficient bifunctional electrocatalysts in alkaline medium are highly advantageous to achieve a greater overall water splitting efficiency [52]. In recent times, an extensive investigation has been employed to develop bifunctional electrocatalysts consisting of earth-abundant metals, mainly transition metals. One important aspect of bifunctional electrocatalysts is that the as-prepared catalyst goes through a transformation during the electrochemical reactions. The transformation is not significant for HER catalysis; however, a complete conversion of it into corresponding oxyhydroxide is evidenced for OER catalysis [53].

5 Recently Reported Noble-Metal-Free Bifunctional Electrocatalysts

Most of the recently reported noble-metal-free bifunctional electrocatalysts comprise transition metals such as Fe, Co, and Ni. A common strategy to develop a bifunctional electrocatalyst is to form a heterostructure by combining two transition metal-based components, one of which is catalytically active towards HER and the other towards OER. In addition to the heterostructures, the overall water-splitting efficiency of a few bimetallic compounds has been investigated in recent times. There are also few reports of bifunctional electrocatalysts comprising a single transition metal. The recently reported bifunctional electrocatalysts are hereby categorized according to their constituting metal elements. Their properties and electrocatalytic activity are discussed in the subsequent sections. The researchers are also adopting physicochemical modification strategies such as doping, and vacancy creation to engineer the electrocatalysts and achieve desirable bifunctional activity. The activity of these engineered bifunctional electrocatalysts is discussed in the dedicated section.

5.1 Monometallic Electrocatalysts

The crustal abundance of the metals that are utilized for electrocatalyst preparation follows the trend W $\approx$ Mo < Co < Cu < Ni <<< Fe [54]. The activity of the Co and Ni-containing monometallic bifunctional electrocatalysts has been investigated extensively in recent times. In addition, few investigations have been conducted to establish the catalytic activity of Fe and other metal-containing electrocatalysts too.

5.1.1 Co-Based Electrocatalysts

The bifunctional electrocatalytic activity of the chalcogen-containing cobalt-based electrocatalysts has been investigated extensively. An electrochemical deposition technique is employed to grow nanostructured α-$Co(OH)_2$ on carbon fiber paper. The introduction of glycine in the solution helps in the nanostructuring of the electrocatalyst and stays coordinated with it after the reaction. The presence of mixed valent Co helps the α-$Co(OH)_2$ film to achieve bifunctional electrocatalytic activity [55]. The cobalt chalcogenides ranging from oxide to telluride show satisfactory bifunctional electrocatalytic activity [56–61]. Ultrathin film of highly crystalline CoO_x nanoparticles (NPs) is fabricated on an electrode following a spray-coating technique. The fabricated thin film possesses a high electrochemically active surface (ECSA) of 301 cm^2 and superior charge transfer efficiency. The CoO_x thin film achieves an overall water splitting current density > 10 mA cm^{-2} at 1.64 V and is able to retain its activity for at least 12 h [56]. The activity of the crystalline cobalt oxide can be enhanced by incorporating defects in it. The highly ordered crystalline structure of cobalt oxide (CoO_x) is disrupted by incorporating electronegative S via a facile room-temperature ion exchange method [57]. The disorderedness in the CoO_x enhances low oxygen coordination and amount of catalytically active defect sites. These physicochemical changes optimize the H and O-containing species adsorption efficiency (Fig. 4a, b), and as a result, the S-CoO_x show superior bifunctional activity. The S-CoO_x requires 1.63 V to achieve 10 mA cm^{-2} current density in a two-electrode setup [57]. Co_3S_4 and Ni-doped Co_3S_4 (Ni:Co_3S_4) nanowires are grown on the conducting nickel foam (NF) in a hydrothermal method followed by a real-annealing treatment [58]. The Co_3S_4 shows superior OER catalytic activity whereas the Ni:Co_3S_4 shows superiority towards HER. Theoretical calculations show that the H-adsorption energy of Ni:Co_3S_4 and the calculated OER overpotential for Co_3S_4 are the lowest. The asymmetric electrolyzer constructed with Ni:Co_3S_4 as cathode and Co_3S_4 as anode achieves 10 mA cm^{-2} current density at an applied voltage of 1.70 V [58]. The thiospinel-type Co_3S_4 shows inferior activity which is enhanced through the incorporation of hydrophilic nature to it. Hydrophilic Co_3S_4 (Co_3S_4-L) is formed when triethylene glycol is added to the reaction mixture whereas the hydrophobic one (Co_3S_4-B) is formed when Oleyl amine is added [59]. The dynamic contact angle analysis confirms the hydrophobic and hydrophilic nature of the electrocatalyst-coated electrodes. The Co_3S_4-L shows superior bifunctional electrocatalytic activity as compared to Co_3S_4-B, which confirms that the hydrophilicity of the surface is an important parameter for electrocatalytic activity. The Co_3S_4-L in a two-electrode setup achieves a current density of 10 mA cm^{-2} at 1.63 V [59]. The as-grown $Co(OH)_2$ nanosheet arrays (NSAs) on carbon cloth (CC) are converted into porous $CoSe_2$ (p-$CoSe_2$/CC) following a vapor selenization process [60]. The highly porous nature of the architecture in p-$CoSe_2$/CC is confirmed 'by its scanning electron microscopic images. The p-$CoSe_2$/CC requires 183 and 243 mV overpotential to achieve the benchmarking current density of 10 mA cm^{-2} for HER and OER, respectively. The binder-free growth of the $CoSe_2$ on the conducting substrate and the large surface area are the reasons behind the augmented catalytic activity of p-$CoSe_2$/CC. The symmetric two-electrode setup constructed with p-$CoSe_2$/CC achieves a current density of 10 mA cm^{-2} at a cell voltage of 1.62 [60]. Cobalt telluride (CoTe) nanoflakes are directly grown on the NF following a hydrothermal reaction [61]. The porous three-dimensional (3D) morphology of the synthesized CoTe helps it to achieve higher ECSA, and it also possesses lower charge transfer resistance. The fabricated electrode shows superior bifunctional electrocatalytic activity and achieves 10 mA cm^{-2} current density for full water splitting at 1.45 V. CoTe is able to retain a current density of 30 mA cm^{-2} in the two-electrode setup for at least 18 h [61]. The electrocatalytic activity of the cobalt chalcogenides is enhanced as the electronegativity of the counter anion decreases moving from oxide to telluride. The lower electronegativity facilitates the substrate adsorption-desorption processes related to the electrochemical water splitting.

The overall water-splitting activity of CoP has been investigated greatly among all the pnictides. Co_4N shows inferior bifunctional electrocatalytic activity despite being metallic in nature [62]. It requires a voltage of 1.78 V to achieve a current density of 20 mA cm^{-2} for overall water splitting. The activity of Co_4N is enhanced through doping Nb which lowers the d-band center and facilitates H-desorption (Fig. 4c, d). On the other hand, abundant Co^{3+} is formed due to doping, which facilitates the formation of OER catalytically active CoOOH [62]. Amorphous cobalt-phosphorous-derived film (Co-P) with a thickness ranging between 1 and 3 μm uniformly grows over copper foil on performing cyclic voltammetry in the solution containing $CoSO_4$ and NaH_2PO_2. A few Co content in that film is present in its elemental state and few form linkages with the phosphide ions [63]. The as-grown film achieves a current density of 10 mA cm^{-2} at an overpotential of 94 and 345 mV for HER and OER, respectively. The film is able to retain its chemical state and morphology in HER catalysis conditions. The film oxidizes to form Co_3O_4 and cobalt

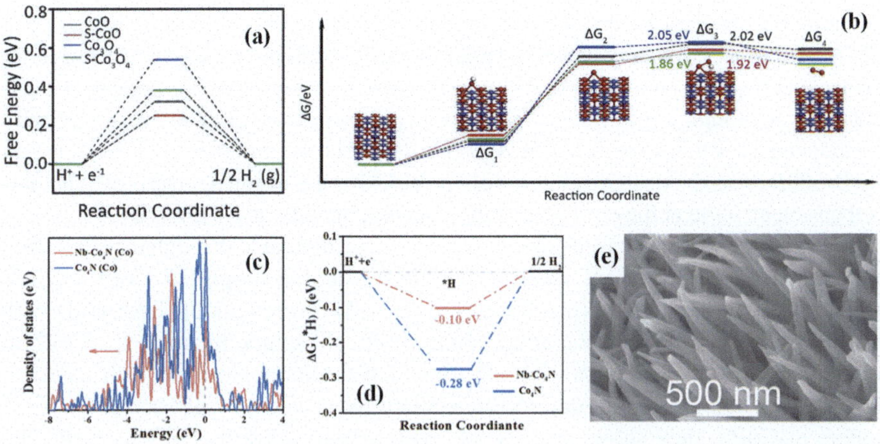

Fig. 4 **a** HER and **b** OER free energy diagram for (100) surface of Co_3O_4 (blue line), S-incorporated Co_3O_4 (S-Co_3O_4: green line), CoO (black line) and S-incorporated CoO (S-CoO: red line) (Reprinted with permission from Yu et al. [57]. Copyright (2020). Elsevier Ltd. **c** Calculated density of states (DOS) of Co_4N and Nb-Co_4N; **d** free energy diagram for HER on Co_4N and Nb-Co_4N. Reprinted with permission from Chen et al. [62]. Copyright (2020). Elsevier Ltd.). **e** Scanning electron microscopic (SEM) image of CoP/TM. Reprinted with permission from [64]. Copyright (2016). IOP Publishing Ltd

phosphate, and its morphology also changes significantly when the OER catalysis is performed. The Co-P film requires $\approx$1.744 V to achieve a current density of 100 mA cm^{-2} in an electrolyzer full cell [63]. Co_3O_4 nanowires grown on Ti mesh get converted into CoP nanowires (CoP/TM) on reaction with NaH_2PO_2 at 300 °C inside a furnace [64]. The electrocatalytic activity of CoP is enhanced through nanostructuring and forming a well-organized architecture (Fig. 4e). CoP/TM shows bifunctional catalytic activity, and achieves 10 mA cm^{-2} current density at 1.64 V for overall water splitting. The reasons behind the superior electrocatalytic activity of CoP/TM are; (a) good electrical conductivity owing to the direct fabrication of CoP on metallic mesh, and elimination of polymeric binder, (2) facilitated "vectorial electron transport" owing to its wire-like morphology, and **c** improved ion diffusion towards the electrocatalytically active sites due to the porous nature [64]. The carbides and borides of cobalt show decent electrocatalytic activity. The activity of the inferior Co_3C electrocatalyst is enhanced through B and N doping [65]. Individual B and N doping enhances the OER catalytic activity to a similar extent whereas an enormous enhancement is observed in simultaneous doping (Fig. 5a). The B doping reduces the HER catalytic activity of Co_3C, however, N doping enhances it (Fig. 5b). The B, N co-doped Co_3C achieves a current density of 10 mA cm^{-2} at 154 and 358 mV overpotentials for HER and OER, respectively [65]. Amorphous cobalt boride is synthesized following a soft chemical route which gets converted into Co_2B after annealing at 500 and 1000 °C temperatures. These electrocatalysts get activated and the OER catalytically active CoOOH moieties are formed on performing cyclic voltammetry between 1.0 and 1.5 V (vs. reversible hydrogen electrode, Fig. 5c) [66]. The activated electrocatalysts show superior OER catalytic activity and its activity further increases on the incorporation of 30 wt.% nitrogen-doped graphene (Co_2B-500/NG). Although Co_2B-500 shows inferiority towards HER catalysis, its activity is enhanced after the formation of Co_2B-500/NG. The asymmetric electrolyzer with Co_2B-500/NG as the positive electrode, Co_2B-500 as the negative electrode, and 3 M KOH as electrolyte achieves 10 mA cm^{-2} current density at 1.81 V [66].

The activity of cobalt-based electrocatalyst can be regulated through incorporating two anions having different electronegativity. Alumina-supported CoPS (CoPS/Al_2O_3) is synthesized by thermal annealing of CoAl layered double hydroxide with P and S [67]. The electron-donating feature of the phosphide ligand helps CoPS to achieve the favorable H-adsorption ability which results in superior HER catalytic activity. On the other hand, due to the availability of catalytically active O_h Co^{3+} centers in CoPS, they show superior catalytic activity towards OER as well. The alumina support helps the electrocatalysts to sustain their activity for a longer time in its operational condition. CoPS/Al_2O_3 requires 1.75 V to achieve a current density of 10 mA cm^{-2} for overall water splitting [67]. Amorphous cobalt phosphosulfide (Co-P-S) can be fabricated on NF substrate following an electrochemical deposition method [68]. The Co-P-S shows superior bifunctional activity as compared to the

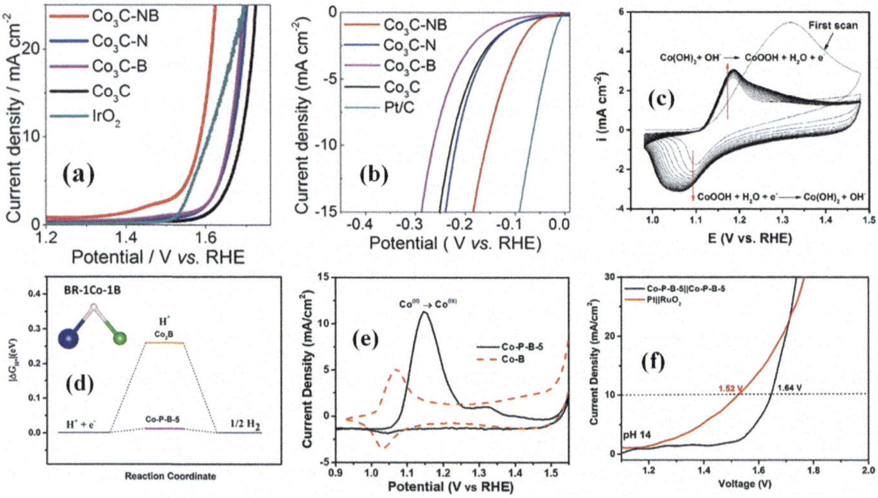

Fig. 5 **a** OER and **b** HER polarization curves for Co_3C, N-doped Co_3C (Co_3C-N), B-doped Co_3C (Co_3C-B), B, N-codoped Co_3C (Co_3C-NB), Pt/C (HER only), and RuO_2 (OER only). Reprinted with permission from Ma et al. [65]. Copyright (2019). The Royal Society of Chemistry. **c** Cyclic voltammetric (CV) curves obtained during electrochemical activation of Co_2B. Reprinted with permission from Masa et al. [66]. Copyright (2016). John Wiley and Sons. **d** HER free energy diagram for bridge site connecting Co (blue sphere) and B (green sphere) in Co_2B and Co-P-B-5. **e** CV curves for Co-B and Co-P-B-5. **f** Overall water splitting polarization curves for RuO_2||Pt, and Co-P-B-5||Co-P-B-5. Reprinted with permission from Chunduri et al. [69]. Copyright (2019), Elsevier

Co-S and Co-P deposits. The Co-P-S achieves 10 mA cm^{-2} current density for HER and OER at 58, and 283 mV overpotentials, respectively. The amorphous nature of Co-P-S helps it to achieve optimized electrochemically active surface area, and local electronic structure [68]. Spherical particles of cobalt-phosphide-boride (Co-P-B) are synthesized by a facile chemical reduction method [69]. The Co-P-B with a B/P molar ratio of 5 (Co-P-B-5) shows superior electrocatalytic activity towards both HER and OER. The B in Co-P-B loses an electron to Co, which in turn transfers it to the P, and as a result, an optimized H-adsorption efficiency is achieved. Theoretical calculations suggest that the bridge site connecting the Co and B acts as the active site for hydrogen adsorption in Co-P-B-5 (Fig. 5d). The formation of OER catalytically active CoOOH species is facile for Co-P-B-5 among all the synthesized materials (Fig. 5e) and thus it shows superior activity towards OER. The Co-P-B-5 gains a current density of 10 mA cm^{-2} at an applied voltage of 1.64 V (Fig. 5f) [69]. The transformation of two anions containing cobalt compound into CoOOH during OER is facile as compared to a single anion containing one. On the other hand, the anion with lower electronegativity transfers electrons towards the higher electronegative anion and consequently, the H-adsorption energy reduces.

5.1.2 Ni-Based Electrocatalysts

The bifunctional electrocatalytic activity of Ni-based electrocatalysts is found to be superior as compared to Co-containing one and thus investigated extensively. The electrocatalysts are categorized and discussed on the basis of their counter anions in the following sections. $Ni(OH)_2$ nanosheets over the NF ($Ni(OH)_2$/NF) are fabricated following a hydrothermal route, where NF itself acts as the Ni-source. The fabricated electrode shows efficiency towards both HER and OER in 1.0 M KOH solution [70]. The unique nanoporous architecture of $Ni(OH)_2$ helps in gas bubble generation and release during the catalysis process. The presence of Ni^{2+} in the electrocatalyst facilitates the water dissociation step due to its affinity towards the OH^- ions. $Ni(OH)_2$/NF requires 127 and 170 mV overpotentials to achieve 10 mA cm^{-2} current density for HER and OER catalysis, respectively. The electrolyzer fabricated for overall water splitting with $Ni(OH)_2$/NF electrode attains the benchmarking current density of 10 mA cm^{-2} at a cell voltage of 1.68 V [70]. The Ni of the foam helps $Ni(OH)_2$ to achieve superior electrocatalytic activity which is revealed from another investigation. $Ni(OH)_2$ fabricated on CC gets enhanced when metallic Ni is incorporated with it. Theoretical calculations reveal that the substrate adsorption energy for $Ni(OH)_2$ reduces after interacting with Ni [71]. Doping of another transition metal like Fe can enhance the bifunctional activity of $Ni(OH)_2$. Fe-doped $Ni(OH)_2$ nanopyramid arrays are directly fabricated on NF following a two-step process. The Fe doping in $Ni(OH)_2$ regulates the electronic environment of the O and alters the OER pathway (Fig. 6a). The Fe doping facilitates the water activation and the H-adsorption process in HER. The Fe-doped $Ni(OH)_2$ containing NF shows bifunctional electrocatalytic activity and achieves 10 mA cm^{-2} at 1.58 V in a two-electrode setup [72]. The electrocatalytic activity of NiO is inferior as compared to $Ni(OH)_2$ as the later can easily adsorb proton and also get transformed into a catalytically active phase. Nickel ammonium carbonate synthesized via a soft chemical pathway is calcined at different temperatures to obtain the NiO hollow microspheres [72]. The NiO obtained after calcination at 300 °C (NiO-300) achieves 10 mA cm^{-2} current density for HER and OER at 424 and 370 mV overpotentials, respectively. An electrolyzer cell for overall water splitting was constructed using the indigenously developed plastic chip electrode modified with NiO-300. The generation of H_2 and O_2 bubbles on the surfaces of the electrodes was clearly visible on the application of 1.6 V (Fig. 6b, d) [73].

The sulfides with different stoichiometric ratios show varied electrocatalytic activity. NiS_2 nanotube and nanorod arrays are grown on NF through hydrothermal reactions (Fig. 6e, f) [74]. The nanotube arrays show superior bifunctional electrocatalytic activity as compared to the nanorod one. The high specific surface area, fast ion diffusion process, and abundant catalytically active sites help the NiS_2 nanotube

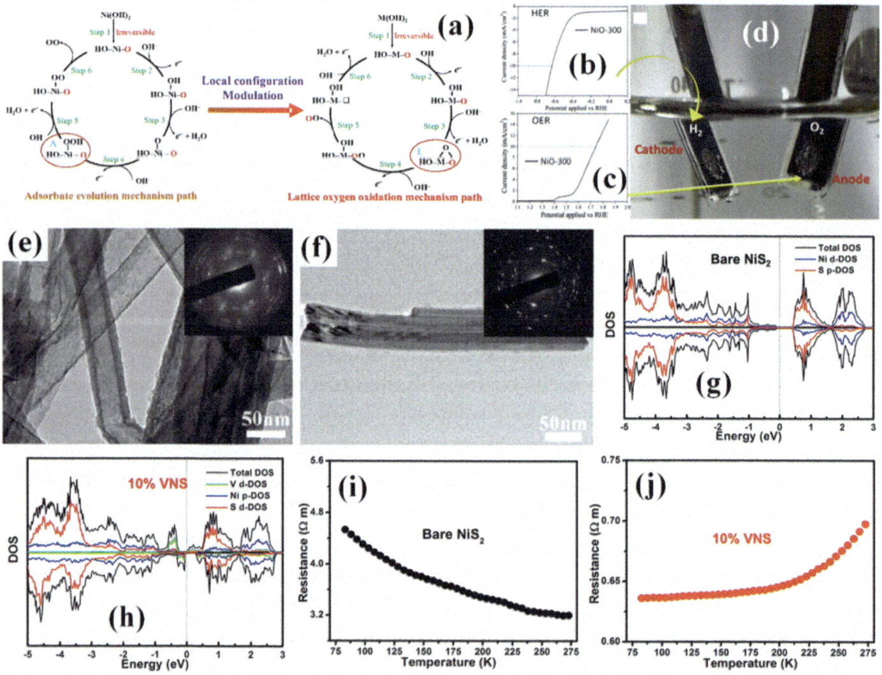

Fig. 6 **a** Pathways of OER catalyzed by $Ni(OH)_2$ and Fe-doped $Ni(OH)_2$ (M denotes Fe and Ni). Reprinted with permission from Qiao et al. [72]. Copyright (2020). American Chemical Society. **b** HER, and **c** OER polarization curves for plastic chip electrode modified with NiO-300; **d** photograph of electrolyzer constructed with NiO-300 modified electrodes. **b**, **d** reprinted with permission from Mondal et al. [73]. Copyright (2018). Elsevier. Transmission electron microscopic (TEM) images of NiS_2 **e** nanotube, and **f** nanorod arrays. **e** and **f** reprinted with permission from Yang et al. [74]. Copyright (2021) Elsevier. Calculated DOS of **g** NiS_2, and (h) 10% V-doped NiS_2 (10% VNS). Temperature-dependent resistance curves for **i** NiS_2, and **j** 10% VNS. **g–j** reprinted with permission from Liu et al. [75]. Copyright (2017). American Chemical Society

to achieve superior bifunctional activity. The NF-containing NiS_2 nanotubes achieve a catalytic current density of 10 mA cm^{-2} at 1.58 V for full water splitting [74]. V doping can further enhance the activity of the pyritic NiS_2. V-doped NiS_2 is synthesized by gas-phase sulfidation of the nickel hydroxide/NiV layered double hydroxide (LDH). Experimental combined with theoretical calculations confirm that the semiconducting nature of NiS_2 changes into metallic on V-doping (Fig. 6g-j) [75]. This change augments the charge transfer kinetics which in turn facilitates the electrocatalytic activity of NiS_2. The NiS_2 shows superior HER and OER catalytic activity on 10% molar doping of V. An overall cell voltage of 1.56 V is required to achieve the 10 mA cm^{-2} current density for overall water splitting when the 10% V-doped NiS_2 acts as bifunctional electrocatalyst [75]. Ni_3S_2 NSAs on NF (Ni_3S_2/NF) with exposed (−210) facets are synthesized by a one-pot hydrothermal route [76]. Ni_3S_2/NF achieves 10 mA cm^{-2} current density at 223 and 260 mV overpotentials for HER and OER catalysis, respectively. The theoretical calculations suggest that the Gibbs free energy for hydrogen adsorption as well as the overpotential for OER are lowest for the high-index (−210) facets of Ni_3S_2. The (−210) facets undergo a catalytic process due to the favorable architecture of the Ni_3S_2/NF which in turn helps them achieve superior catalytic activity [76]. Ni_3S_4 nanosheets over NF (Ni_3S_4/NF) are obtained when Na_2S is used as the S-source instead of thiourea in the hydrothermal reaction [76, 77]. The nanosheets grown on the NF skeleton have 20–30 nm thickness and some of them form nanospheres as a result of mutual stacking. The Ni_3S_4/NF requires 122 mV overpotential to afford a current density of 10 mA cm^{-2} for HER catalysis and 320 mV overpotential to attain 20 mA cm^{-2} current density in OER catalysis. The higher active site accessibility owing to the 3D porous architecture is the reason behind the superior catalytic activity of Ni_3S_4/NF. The fabricated electrode requires 1.61 and 1.67 V in the overall water-splitting process to achieve current densities of 10 and 20 mA cm^{-2}, respectively [77]. Nickel oxalate (NiC_2O_4) on the skeleton of the NF is synthesized by treating the NF with a mixture of oxalic acid, ethanol, and water. The as-synthesized NiC_2O_4 is converted into $NiSe_2$ by means of selenization inside a tube furnace using Se powder [78]. The skeleton of the NF is found to be covered with the nanowrinkles of the porous $NiSe_2$. The fabricated electrode requires 1.64 V to attain the 10 mA cm^{-2} current density in the full water splitting process. The large surface area, improved electrical conductivity, and faster charge transfer kinetics are the reasons behind the superior electrocatalytic activity of the fabricated electrode [78]. The electrocatalytic OER activity of nickel chalcogenide is enhanced with the decrement in electronegativity of the counter anion [79]. The covalent nature of the bonds present in the compound is enhanced with changing the anion from oxide to telluride. This change modulates the coordinate environment of the Ni, and OER catalytic activity gradually increases. The Ni_3Te_2 shows bifunctional electrocatalytic activity and achieves a current density of 10 mA cm^{-2} on application of 1.62 V. Electrocatalytic activity of Ni_3Te_2 obtained through different reaction and deposited on different substrates shows similar electrocatalytic activity which proves its intrinsic activity [79].

Black Ni nanorod arrays are obtained through electrochemical deposition followed by template removal technique. Ni_3N is coated over the nanorod in a nitrogen plasma treatment method (Fig. 7a). The high specific surface area owing to the architecture helps the electrocatalyst to achieve higher ECSA which further enhances after Ni_3N coating. The Ni_3N-coated Ni nanorod arrays achieve 5 mA cm^{-2} current density at 1.64 V [80]. Porous Ni_8P_3 and Ni_9S_8 are fabricated on NF (Ni/NiP and Ni/NiS) following a two-step process. Initially, the NiO is grown on the NF by acid activation which converts into sulfide and phosphide on reaction with the S and P vapors, respectively [81]. The Ni/NiP exhibits superior HER and OER catalyzing activity as compared to the Ni/NiS which confirms that the decrement in electronegativity of the anion enhances the activity of the compound. The Ni_8P_3 has the capability to efficiently dissociate the H_2O molecule which helps in achieving superior HER catalyzing activity. The Ni with a higher valence state is formed during OER due to the oxidation of Ni-sites of the Ni/NiP which augments the catalysis process. The two-electrode setup requires 1.61 V to achieve a current density of 10 mA cm^{-2} [81]. The porous $Ni(OH)_2$ grown on the CC can be converted into NiP_2 (NiP_2/CC) by calcination with NaH_2PO_2 inside the furnace without disrupting the morphology [82]. The porous architecture of NiP_2/CC helps it to attain a large number of catalytically active sites and to enhance the charge transfer efficiency. The symmetrical overall water-splitting electrolyzer fabricated with NiP_2/CC attains a 10 mA cm^{-2} current density at 1.65 V [82]. Ni_2P NSAs with cationic vacancies (V-Ni_2P) can be synthesized by etching out the dopants of the Mn-doped Ni_2P using HNO_3 [83]. The incorporation of defects in the Ni_2P enhances its electrical conductivity and the amount of abundant surface active sites. The defect reduces the energy barriers for water adsorption and optimizes the substrate adsorption process in the HER and OER (Fig. 7b, c). V-Ni_2P directly grown on the NF achieves an overall water-splitting current density of 10 mA cm^{-2} at 1.59 V and is able to maintain it for at least 50 h [83]. Ni_2P nanosheets are fabricated on the iron foam by phosphidizing the as-grown $Ni(OH)_2$ inside a furnace [84]. The fabricated electrode achieves 10 mA cm^{-2} current density at 334 mV overpotential for overall water splitting. The fast charge transport efficiency, high ECSA, and formation of phosphate layer on the surface of Ni_2P are the reasons behind its excellent water-splitting activity [84]. The electrocatalytic activity of nickel phosphide enhances with the increment in stoichiometric ratio between Ni and P present in it.

Superior bifunctional electrocatalytic activity is achieved through the incorporation of more than one counter anion in nickel-based electrocatalyst. Surface oxidized nickel borides are obtained via a soft chemical pathway followed by aerial oxidation. Nickel borides with different compositions (Ni–B–O@Ni_3B, Ni–B–O@Ni_2B, and Ni–B–O@NiB_2) are obtained by changing the Ni and B precursors [85]. The surface oxidation of the borides degrades the HER catalytic activity, but enhances the OER catalyzing efficiency. The B plays a significant role in the formation of OER catalytically active nickel oxide/hydroxide moieties. The Ni–B–O@Ni_3B among all the above-mentioned nickel borides is superiorly active towards both HER and OER. The two-electrode setup comprised with Ni–B–O@Ni_3B as an electrocatalyst requires 1.58 V to achieve 10 mA cm^{-2} current density [85]. Thin $NiPS_3$ flakes are exfoliated from its bulk sample following a sonication-free electrochemical cathodic exfoliation technique [86]. The tetra-n-butylammonium salt gets intercalated during

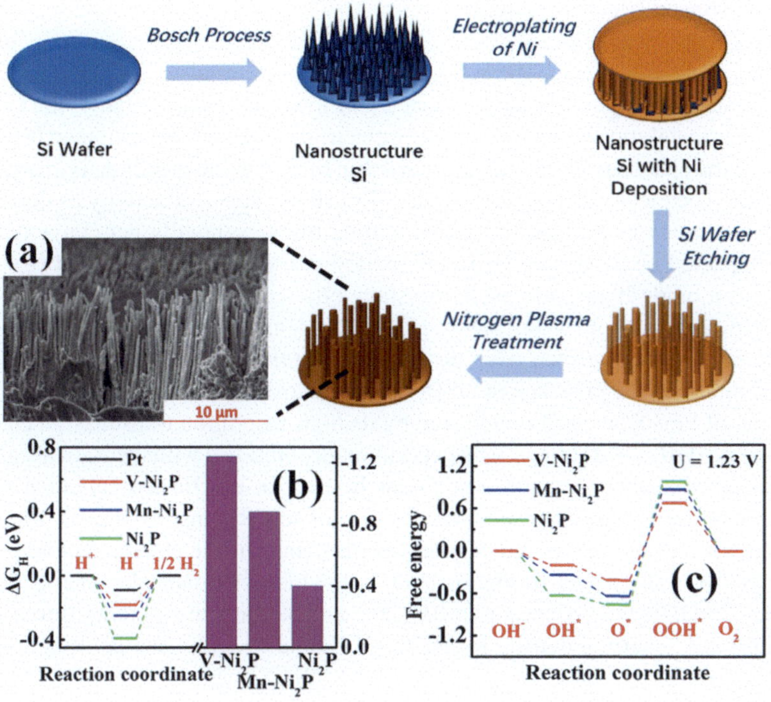

Fig. 7 **a** Schematic diagram showing the preparation process of Ni_3N-coated Ni nanorod arrays and SEM image of the final product. Reprinted with permission from Wang et al. [80]. Copyright (2020). American Chemical Society. **b** Free energy diagram for proton and H_2O adsorption, and **c** OER on Ni_2P, Mn-doped Ni_2P (Mn-Ni_2P) and cationic vacancy containing Ni_2P (V-Ni_2P). **b**, **c** reprinted with permission from Zhang et al. [83]. Copyright (2019). Elsevier

the electrochemical process which decomposes into gaseous species and expands the $NiPS_3$ layers in the bulk crystals. Large-sized $NiPS_3$ ultrathin flakes with minimum structural irregularity can be obtained by slight manual shaking. The oxygen groups are absent in the exfoliated samples which helps them to achieve superior catalytic activity. The P of $NiPS_3$ acts as the OER catalytic active site whereas the S as the HER active site. A voltage of ~1.56 V is required to achieve 10 mA cm^{-2} current density when $NiPS_3$ flakes act as the bifunctional catalyst [86]. Conducting Ni-P-B shows excellent overall water-splitting efficiency even if it is deposited over non-conducting substrates like cloth, paper, etc. The film fabricated over paper can achieve a current density of 50 mA cm^{-2} in a two-electrode setup at 1.661 V [87]. So, the electrocatalytic activity of nickel-based materials with more than one anion needs further exploration.

5.1.3 Fe-Based Electrocatalysts

The bifunctional electrocatalytic activity of Fe-based materials is not explored significantly despite its high natural abundance and lower toxicity. Initially, the bifunctional electrocatalytic activity of iron-only electrodes has been investigated [88]. The activity of the electrode is enhanced when a thin film of oxide or sulfide is grown over its surface. The electrode containing FeS_x film shows superior activity as compared to FeO_x-containing one. The oxide and sulfide films eventually get converted into oxide supported Fe (0) during HER and oxide-hydroxide during the OER [88]. The pristine iron oxide-hydroxide itself can show efficient bifunctional electrocatalytic activity towards overall water splitting. A high-density networked structure of δ-FeOOH nanosheet is grown over the NF (δ-FeOOH NSs/NF) by oxidizing iron-boron complex in the presence of air [89]. The nanosheets are defect-rich due to the formation of Fe-vacancies as evidenced by its high-resolution transmission electron microscopic (HR-TEM) image. The experimental combined with theoretical calculations suggest that the Fe-vacancies augment the electrocatalytic activity of FeOOH in two ways: (i) one Fe-vacancy activates the second neighboring Fe (Fig. 8a, b), which enhances the amount of catalytically active sites, and (ii) the electrical conductivity increases due to the defect formation (Fig. 8c), which facilitates the charge transfer process. δ-FeOOH NSs/NF requires 108, 265, and 390 mV overpotentials to achieve the 10 mA cm^{-2} current density for HER, OER, and overall water splitting, respectively. In addition to the efficiency, δ-FeOOH NSs/NF shows superior stability in operational conditions [89].

Few investigations have been conducted to understand the electrocatalytic bifunctional activity of the Fe-S-based materials. Fe_3S_4 with varied morphology and crystal parameters are synthesized in a modified polyol method by changing the ratio between H_2O and ethylene glycol (EG). The change in these physicochemical features of the Fe_3S_4 regulates its charge transfer, substrate adsorption-desorption efficiency, and the amount of exposed active sites. The Fe_3S_4 obtained from the solvent containing EG: H_2O = 3: 1 shows both efficiency and stability in overall water splitting [90]. Mackinawite FeS nanosheet arrays which act as precatalyst for HER and OER are formed when iron foam is sulphidized in a hydrothermal reaction [91]. Experimental results suggest that the FeS gets converted into iron@iron oxysulfide ($Fe@FeO_xS_y$) NPs and amorphous iron oxide (FeO_x) during the HER and OER catalysis process, respectively. $Fe@FeO_xS_y$ and FeO_x act as the real catalysts in these reactions. Theoretical calculations suggest that the Fe surface covered with O atoms along with the S atom is beneficial for the H-adsorption (Fig. 8d) which in turn facilitates the HER catalysis process. The fabricated electrode achieves 10 mA cm^{-2} current density for full water splitting operation at 1.65 V [91]. The pyritic iron chalcogenides also show reasonable efficiency towards overall water splitting. FeS_2/C nanoparticles synthesized through hydrothermal reaction achieve 10 mA cm^{-2} at 1.72 V in a two-electrode setup. The H and O of water bind the Fe and S terminals of FeS_2 during the catalysis [92]. The electrocatalytic activity of FeS_2 can be enhanced through doping and incorporation of oxygen. Ni-doped O-containing FeS_2 is directly fabricated on the NF via an electrodeposition technique where NF itself

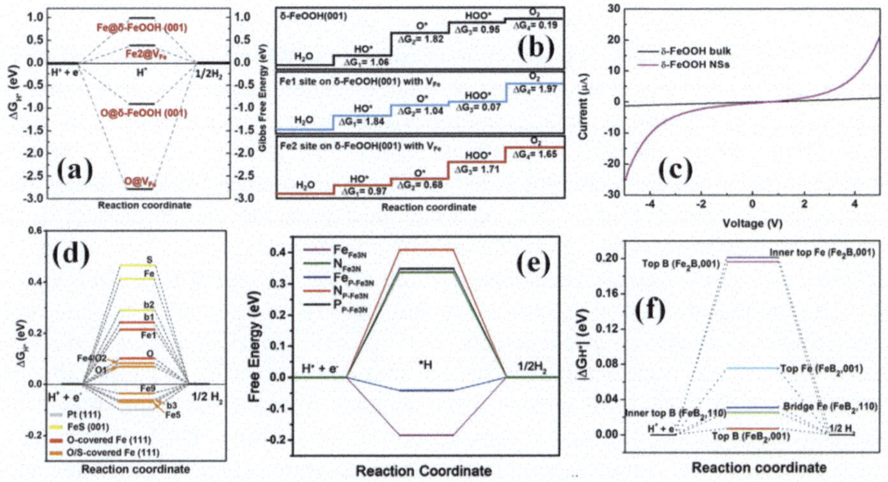

Fig. 8 **a** Free energy diagram for HER on Fe and O in δ-FeOOH NSs with no defect [labeled as δ-FeOOH (001)] and O neighboring to the defect, and second neighboring Fe in iron-vacancy rich δ-FeOOH NSs [labeled as V_{Fe}]; **b** OER-free energy diagram for δ-FeOOH (001), first (Fe1) and second (Fe2) sites on δ-FeOOH (001) with V_{Fe}; **c** current-voltage characteristics of bulk δ-FeOOH and δ-FeOOH NSs. **a–c** Reprinted with permission from Liu et al. [89]. Copyright (2018). John Wiley and Sons. **d** Free energy diagram for HER on Pt (111), FeS (001), O-covered Fe (111) and O/S-covered Fe (111) surfaces. Reprinted with permission from Zou et al. [91]. Copyright (2018). Elsevier. **e** Free energy diagram for HER on different models of P-doped and undoped Fe_3N surfaces. Reprinted with permission from Li et al. [95]. Copyright (2020). John Wiley and Sons. **f** Free energy diagram for HER on low-index (001) and high-index (110) surfaces of FeB_2 and Fe_2B. Reprinted with permission from Li et al. [97]. Copyright (2017). John Wiley and Sons

acts as Ni source and conducting substrate [93]. The morphology, electronic structure and charge carrier density of the electrodeposit alters with the change in duration of deposition. The Ni-doped FeS_2 obtained after 30 and 45 min electrodeposition (deposition potential = -0.9 V vs. Ag/AgCl) shows superior activity towards OER and HER, respectively. The asymmetric electrolyzer constructed using these two electrodes achieves 10 mA cm^{-2} at 1.588 V [93]. $FeSe_2$ electrocatalysts are synthesized by thermal degradation of the β-diketiminato supported 2Fe-2Se clusters [94]. $FeSe_2$ supported over NF ($FeSe_2$/NF) achieves 10 mA cm^{-2} current density for OER at an overpotential of 245 mV. The surface of the electrocatalyst gets oxidized during the OER process to form $Fe(OH)_2$/FeOOH, which acts as the true electrocatalyst for OER. The vacancies created due to the loss of Se-atoms at the surface during the OER process facilitate the adsorption of OH^- which in turn augments the OER process. The HER catalytic activity of $FeSe_2$/NF is not up to the mark as compared to its activity towards OER. The symmetrical two-electrode setup with $FeSe_2$/NF electrodes achieves 10 mA cm^{-2} current density at a voltage of 1.73 V [94].

P-doped Fe_3N is directly fabricated on iron foam through high-temperature reaction where *Saccharomycetes cerevisiae* acts as the source of both N and P. The way of N and P diffusion in the microorganism regulates the phase and composition of the final product. The Fe in the electrocatalyst acts as the active site, whereas the N and P regulate the substrate adsorption-desorption efficiency. The doping of P enhances the proton adsorption efficiency of the Fe center and reduces that of the N center (Fig. 8e) [95]. Iron phosphide nanotubes (IPNTs) are grown over CC following a three-step process. Initially, ZnO nanowire arrays are grown over the CC which acts as the sacrificial template [96]. The $Fe(OH)_2$ is grown over the ZnO nanowires which then is converted into IPNT via a chemical vapor deposition process. Iron oxide/phosphate layer is present on the rough surfaces of the IPNTs which plays a significant role in the bifunctional catalytic activity of the synthesized electrocatalyst. The fabricated electrode requires 120 and 288 mV overpotentials to achieve a geometrically normalized current density of 10 mA cm^{-2} for HER and OER, respectively. The two-electrode setup achieves the said current density at an overall cell voltage of 1.69 V [96]. FeB_2 nanoparticles obtained from the reduction of Fe^{2+} in tetrahydrofuran show efficiency towards overall water splitting. The B-rich surface of the electrocatalyst achieves optimized proton adsorption (Fig. 8f) efficiency and therefore, shows excellent HER catalytic activity. The facile conversion of FeB_2 into catalytically active FeOOH during OER catalysis helps it to achieve superior efficiency. FeB_2 achieves a catalytic current density of 10 mA cm^{-2} at 1.57 V for full water splitting [97].

5.1.4 Other Transition Metal-Based Electrocatalysts

The bifunctional activity of the electrocatalysts comprising transition metals other than Fe, Co, and Ni has also been investigated. Among them, the materials containing Mn, Mo, and Cu show sufficient promise. Lepidocrocite VOOH hollow nanospheres with different morphological features are synthesized by varying reaction temperature and time [98]. The roughness of the nanosphere surface alters with changing reaction temperatures, which in turn modulates the electrocatalytic activity of the VOOH. The nanospheres obtained after the reaction at 160 °C for 8 h show superior electrocatalytic activity towards both HER and OER. The said electrocatalyst is able to achieve 10 mA cm^{-2} current density for HER and OER at 164 and 270 mV overpotentials, respectively [98]. A two-step metal–organic framework (MOF)-mediated synthetic technique is adapted to directly fabricate Fe-doped vanadium phosphide functionalized porous carbon on NF [99]. Successful interaction between V and Fe facilitates the formation of a higher valence V. The increment in valence of V facilitates the binding of the HER and OER intermediates. The Fe doping optimizes the electrical conductivity of the vanadium phosphide. The developed electrocatalysts with optimum activity achieve 10 mA cm^{-2} at 1.56 V [99].

Bifunctional activity can be induced in MnO_2 through doping or defect generation. MnO_2 nanosheets (NS-MnO_2) with two-atomic layer thickness are synthesized following a hydrothermal route [100]. O-vacancies in the nanosheets are formed during the synthesis and as a result, coordinately unsaturated Mn^{3+} active sites are generated. Theoretical calculations suggest that the Mn^{3+} induces half-metallicity into the MnO_2 and the O-vacancies enhance the water adsorption efficiency of MnO_2.

The NS-MnO_2 achieves 10 mA cm^{-2} current density at 197 and 320 mV overpotentials for HER and OER, respectively [100]. A controllable Prussian-blue-analogue (PBA)-mediated sacrificial strategy (Fig. 9a) is utilized to synthesize defect-rich Ni and Fe-doped $K_{0.23}MnO_2$ cubic nanoflowers (Ni–Fe-$K_{0.23}MnO_2$ CNFs-300) [101]. Highly abundant defect, doping of transition metals, and favourable morphology help Ni–Fe-$K_{0.23}MnO_2$ CNFs-300 to achieve superior bifunctional activity. The incorporation of defects and dopants enhances the electrical conductivity and facilitates the elemental steps of the water-splitting reaction. Ni–Fe-$K_{0.23}MnO_2$ CNFs-300 || Ni–Fe-$K_{0.23}MnO_2$ CNFs-300 electrolyzer achieves 10 and 100 mA cm^{-2} at 1.62 and 1.81 V, respectively (Fig. 9b) [101].

Pristine Mo-based electrocatalysts show inferiority towards overall water splitting, and doping foreign elements is necessary to achieve the desired activity. Fe-doped MoO_2 is epitaxially grown over etched NF through a hydrothermal route followed by an annealing process. Fe-doping enhances the substrate adsorption efficiency and electrical conductivity of MoO_2 and exposes catalytically active sites. The Fe-doped MoO_2 achieves 10 mA cm^{-2} for HER and 100 mA cm^{-2} for OER at 36 and 310 mV overpotentials, respectively [102]. Co-doped MoS_2 nanosheets are synthesized following a hydrothermal reaction, which forms an aggregate with flowerlike morphology [103]. Theoretical calculation suggests that the band gap

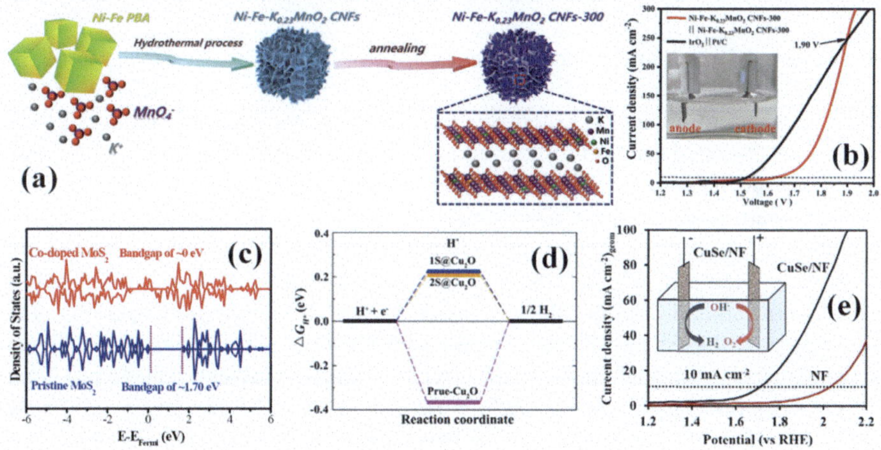

Fig. 9 **a** Schematic diagram of the Ni–Fe-$K_{0.23}MnO_2$ CNFs-300 synthesis procedure; **b** overall water splitting polarization curves for RuO_2||Pt and Ni–Fe-$K_{0.23}MnO_2$ CNFs-300|| Ni–Fe-$K_{0.23}MnO_2$ CNFs-300 with the photograph of the two-electrode setup at inset. (a and b) reprinted with permission from Liao et al. [101]. Copyright (2020) John Wiley and Sons. **c** Calculated DOS of Co-doped and undoped MoS_2. Reprinted with permission from Xiong et al. [103]. Copyright (2018). The Royal Society of Chemistry. **d** Free energy diagram for HER on pure-Cu_2O, 1S doped Cu_2O and 2S doped Cu_2O. Reprinted with permission from Zhang et al. [106]. Copyright (2017). American Chemical Society. **e** Overall water splitting polarization curves for CuSe/NF||CuSe/NF and NF||NF. Reprinted with permission from Chakraborty et al. [107]. Copyright (2020). John Wiley and Sons

value of MoS_2 can be reduced up to ≈0 eV by doping the desired amount of Co (Fig. 9c). The hydrogen adsorption-free energy of MoS_2 also reduces significantly upon Co-doping. These two factors help MoS_2 to achieve superior HER catalytic activity. The Co with higher valence states ($3^+/4^+$) is found to be present in the post-OER electrocatalysts which are responsible for its OER catalytic activity. The Co-doped MoS_2 achieves 10 mA cm^{-2} current density at 1.58 V for the overall water-splitting process [103]. Molybdenum carbide-supported carbon sheets (Mo_2C@CS) are synthesized following a two-step thermal degradation mechanism where ammonium molybdate and glucose act as the Mo and C sources, respectively [104]. Mo_2C NPs get anchored over the carbon sheet in this process. Higher amounts of NPs are formed on the nanosheets and the specific surface area of the material decreases by increasing the amount of Mo-precursor in the reaction mixture. The electrocatalyst with optimized properties requires 60 and 320 mV overpotentials to achieve the current density of 10 mA cm^{-2} for HER and OER, respectively. The lamellar structure and the abundant active sites at the basal plane are the reasons behind its superior electrocatalytic activity. The two-electrode electrolyzer yields 10 mA cm^{-2} current density at a cell voltage of 1.73 V. In addition to efficiency, the Mo_2C@CS is able to retain its electrocatalytic activity in 100 h long operation [104].

The Cu-based materials are inferiorly active towards electrocatalytic water splitting and thus different strategies are adopted to achieve reasonable activity. La-doped CuO which shows bifunctional electrocatalytic activity towards overall water splitting is synthesized following a coprecipitation method [105]. The 1% La-doped CuO achieves 10 mA cm^{-2} for HER and OER at 173 and 322 mV overpotentials, respectively. The surface-to-volume ratio of the particles decreases when higher than 1% La is doped into CuO and therefore the electrocatalytic activity reduces [105]. Nanoporous S-doped C_2O is synthesized on the Cu foam ($Cu_2O_xS_{1-x}$/Cu) following a three-step mechanism. Initially, $Cu(OH)_2$ nanowires are synthesized on the Cu foam which are then converted into Cu_2S using Na_2S in a soft chemical method [106]. The obtained Cu_2S is then converted into a nanoporous film of $Cu_2O_xS_{1-x}$ by electrochemical activation in 0.5 M H_2SO_4. The theoretical calculation and experimental results confirm that S-doping enhances the HER catalytic activity of Cu_2O (Fig. 9d). The symmetrical electrolyzer constructed with $Cu_2O_xS_{1-x}$/Cu electrodes achieves the benchmarking current density of 10 mA cm^{-2} at an overall cell voltage of 1.56 V [106]. Crystalline klockmannite CuSe is deposited on NF following a solid-state reaction at elevated temperatures [107]. This CuSe acts as a pre-catalyst which gets converted into crystalline $Cu(OH)_2$ and amorphous $Cu(OH)_2/CuO_x$ during OER and HER, respectively. The $Cu(OH)_2$ layer acts as the active catalyst for OER and facilitates the O–O bond formation. The low valent Cu present in the $Cu(OH)_2/CuO_x$ layer acts as the proton-adsorption center and facilitates the HER. The excellent electrical conductivity of CuSe reduces the charge transfer resistance of the fabricated electrode. The defects generated during the electrochemical activation process enhance the ECSA of the electrocatalyst. The two-electrode setup constructed using the CuSe electrode achieves 10 mA cm^{-2} at 1.68 V (Fig. 9e) [107]. Doping of heteroatoms like Ru and N regulates the electronic structure of the active sites in Sb_2S_3 and therefore, its electrocatalytic activity is enhanced [108]. The activity of the Sb_2S_3 enhances in the pseudo-steady-state region after the N-doping, however; it matches with the undoped one in the steady-state-region. The Ru doping further enhances the activity of the N-doped Sb_2S_3. The dual-doped Sb_2S_3 forms pellet-drum type aggregates which helps it to achieve higher ECSA [108]. Table 1 summarizes the activity of the recently reported monometallic bifunctional electrocatalysts for overall water splitting.

The Ni and Co-based chalcogenides and pnictides have shown significant promise and the incorporation of more than one anion is found beneficial. No significant research attention is being shown towards the development of Fe-based bifunctional electrocatalysts despite its higher natural abundance [54, 93]. The researchers are mostly focusing on the activity of the S, O, and P-based compounds of transition metals. The incorporation of more than one anion in a material is beneficial in terms of its electrocatalytic activity. The carbides and borides of transition metals mostly show inferior bifunctional activity towards overall water splitting [45, 46]. Further investigation and modification are still required to establish the bifunctional activity of the electrocatalysts containing transition metals like V, Mn, Mo, or Cu. The effect of incorporating more than one metal towards the bifunctional electrocatalytic activity is also investigated.

5.2 *Bimetallic Electrocatalysts*

The incorporation of more than one anion in an electrocatalyst has been found beneficial and a similar phenomenon is observed when an electrocatalyst comprising of two-metal ions is developed. The incorporation of two metals optimizes their individual oxidation state and consequently, superior activity is achieved. NiFe-oxide porous nanocubes (NiFe-NCs) are synthesized by calcining the NiFe PBA metal–organic framework (MOF) in the presence of air. NiFe-oxide spherical NPs (NiFe-NPs) are synthesized via a co-precipitation method followed by calcination [109]. The average length of the NiFe-NCs side is $\approx$200 nm and their surfaces are rough in nature. The NiFe-NCs show better catalytic efficiency as compared to the NiFe-NPs. The employment of MOF precursor for synthesis helps NiFe-NCs to achieve superior electrocatalytic activity in two ways: (i) the agglomeration of NiFe-NCs can be impeded, and (ii) the NiFe-NCs formed in this procedure are mesoporous in nature which results in larger catalytically active surface area. The NiFe-NCs attain 10 mA cm^{-2} current density for full water splitting at 1.67 V [109]. $Co_{0.75}Ni_{0.25}(OH)_2$ nanosheets with abundant pores are synthesized via laser ablation in liquid methodology (Fig. 10a) [110]. Theoretical combined with experimental results suggest that trivalent Co and Ni ions are formed on the wall of the pores in these nanosheets (Fig. 10b). The Co^{3+} ions simultaneously facilitate the electrocatalytic process (Fig. 10c), and enhance the electrical conductivity of the $Co_{0.75}Ni_{0.25}(OH)_2$. The synthesized porous nanosheets achieve 10 mA cm^{-2} at 95 and 235 mV overpotentials for HER and OER, respectively [110]. $Co_{1.9}Ni_{0.1}(CO_3)(OH)_2$ (NiCCH), and $Co_{0.95}Mn_{0.05}CO_3$ (MnCCH) are formed when a few Cos of cobalt carbonate hydroxide (CCH) are substituted with Ni

Table 1 Recently reported monometallic bifunctional electrocatalysts

Electrocatalyst	Overpotential @ 10 mA cm^{-2} for HER (mV)	Overpotential @ 10 mA cm^{-2} for OER (mV)	Cell voltage @ 10 mA cm^{-2} for overall water splitting (V)	Refs.
Co-based electrocatalysts				
α-$Co(OH)_2$	145	320	1.60	[55]
CoO_x thin film	220	250	1.64	[56]
S-…CoO_x	136	370	1.63	[57]
Co_3S_4	–	283	1.70	[58]
Ni:Co_3S_4	199			
Hydrophilic Co_3S_4	270	360	1.63	[59]
$CoSe_2$ nanosheet	138	243	1.62	[60]
CoTe nanoflakes	90	241	1.45	[61]
Nb-Co_4N	90 (20 mA cm^{-2})	273 (20 mA cm^{-2})	1.60 (20 mA cm^{-2})	[62]
CoP film	94	345	1.744 (100 mA cm^{-2})	[63]
CoP nanowire	72	310	1.64	[64]
Co_3C-NB	154	358	–	[65]
Amorphous Co_2B	~420	380	–	[66]
Al_2O_3-supported CoPS	130	–	1.75	[67]
Co-P-S	58	283	1.60	[68]
Co-P-B	145	290	1.64	[69]
Ni-based electrocatalysts				
$Ni(OH)_2$ nanosheets	172 (20 mA cm^{-2})	330 (50 mA cm^{-2})	1.68	[70]
$Ni(OH)_2$@Ni film	68	458 (100 mA cm^{-2})	1.58	[71]
Fe-doped-$Ni(OH)_2$	121	218	1.58	[72]
NiO hollow spheres	424	370	–	[73]
NiS_2 nanotubes	209 (50 mA cm^{-2})	367 (50 mA cm^{-2})	1.58	[74]
V-doped NiS_2	110	290	1.56	[75]
Ni_3S_2 nanosheet	223	260	–	[76]
Ni_3S_4 nanosheets	122	320 (20 mA cm^{-2})	1.61	[77]
$NiSe_2$ nanowrinkles	166	235	1.64	[78]
Ni_3Te_2	167	220	1.66	[79]
Ni_3N coating	165	330	1.64 (5 mA cm^{-2})	[80]

(continued)

Table 1 (continued)

Electrocatalyst	Overpotential @ 10 mA cm^{-2} for HER (mV)	Overpotential @ 10 mA cm^{-2} for OER (mV)	Cell voltage @ 10 mA cm^{-2} for overall water splitting (V)	Refs.
Ni_8P_3	130	270 (30 mA cm^{-2})	1.61	[81]
NiP_2 nanosheet	–	310 (20 mA cm^{-2})	1.65	[82]
Ni_2P NSAs	55	250	1.59	[83]
Ni_2P nanosheets	–	198	1.564	[84]
Ni–B–O@Ni_3B	64	264	1.58	[85]
$NiPS_3$ flakes	158	300	1.56	[86]
Ni–P-B	76 (50 mA cm^{-2})	263 (50 mA cm^{-2})	1.661 (50 mA cm^{-2})	[87]
Fe-based electrocatalysts				
δ-FeOOH	108	265	1.62	[89]
Fe_3S_4	279 (50 mA cm^{-2})	255 (50 mA cm^{-2})	1.68	[90]
FeS nanosheets	300 (100 mA cm^{-2})	238	1.65	[91]
FeS_2/C nanoparticle	202	240	1.72	[92]
Ni-doped FeS_2	157	255 (100 mA cm^{-2})	1.588	[93]
$FeSe_2$	-	245	1.73	[94]
P-Fe_3N	102	270	1.61	[95]
FeP nanotubes	120	288	1.69	[96]
FeB_2	61	296	1.57	[97]
Other transition metal-based electrocatalysts				
VOOH nanospheres	164	270	1.62	[98]
Fe-doped V_2P	66	201	1.56	[99]
δ-MnO_2 nanosheet	197	320	–	[100]
Ni–Fe-doped $K_{0.23}MnO_2$	116	270	1.62	[101]
Fe-doped MoO_2	36	310 (100 mA cm^{-2})	1.57	[102]
Co-doped MoS_2	90	190	1.58	[103]
Mo_2C@C	60	320	1.73	[104]
La-doped CuO	173	322	1.6	[105]
$Cu_2O_xS_{1-x}$	40	–	1.56	[106]
CuSe	162	297	1.68	[107]

and Mn via hydrothermal reaction [111]. The morphology and the electronic structure of the CCH vary due to this type of transition-metal-substitution. Therefore, the electrocatalytic activity of CCH enhances and the NiCCH shows superiority among them. The occurrence of constituent metal atoms with varied oxidation states and easy-accessibility of active sites can be attributed to this enhancement in catalytic activity [111].

The bifunctional electrocatalytic activity of layered double hydroxide (LDH) comprising of two different metal ions is investigated extensively. NiCo-LDH is directly grown on the NF following two processes: (i) solvothermal and (ii) coprecipitation method [112]. The ultrathin nanosheets grow uniformly and vertically on

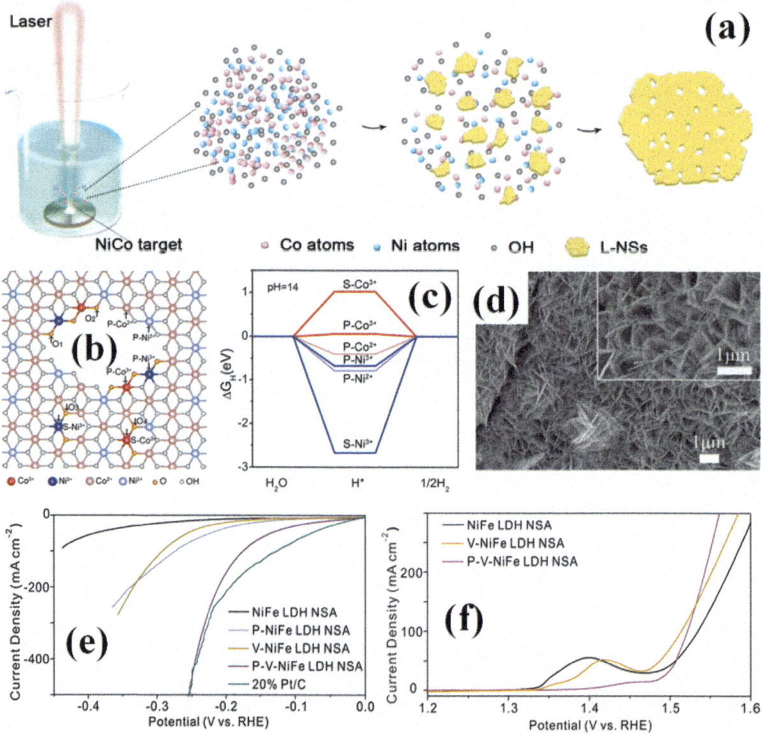

Fig. 10 **a** Schematic diagram showing the formation of $Co_{0.75}Ni_{0.25}(OH)_2$ nanosheets via laser ablation in liquid methodology (L-NSs); **b** Structural model of L-NSs with a pore for theoretical calculations (red, blue, pink, light blue, yellow and orange spheres represent Co^{3+}, Ni^{3+}, Co^{2+}, Ni^{2+}, OH, and O ions, respectively); **c** HER free energy diagrams for the metallic ions on the surface (prefixed S-), and pore walls (prefixed P-). (a-c) reprinted with permission from Wang et al. [110]. Copyright (2019 John Wiley and Sons. **d** SEM image of NiCo-LDH NSAs. Reprinted with permission from Liu et al. [112]. Copyright (2017). The Royal Society of Chemistry. **e** HER and **f** OER polarization curves for NiFe LDH NSAs, plasma-reduced NiFe LDH NSAs (P-NiFe LDH NSA; HER only), V-doped NiFe LDH NSAs (V-NiFe LDH NSA), P–V-NiFe LDH NSAs and Pt/C (HER only). (e and f) reprinted with permission from Tang et al. [114]. Copyright (2020). Elsevier

the skeleton of the NF (Fig. 10d) when the solvothermal procedure is utilized for electrode fabrication. This morphology of LDH facilitates ion diffusion, gas release and helps in exposing a greater number of active sites for catalysis. The solvothermally fabricated electrode (NiCo-LDH/NF) shows superiority towards electrocatalysis as compared to that obtained from the coprecipitation method. NiCo-LDH/NF achieves 20 mA cm^{-2} current density at 1.73 V in a symmetrical electrolyzer constructed for overall water splitting [112]. CoMo-LDH, CoMn-LDH, and $Co(OH)_2$ ultrathin nanosheets are synthesized following a facile coprecipitation method [113]. CoMo-LDH shows superior electrocatalytic activity as compared to the other two electrocatalysts. CoMo-LDH achieves 10 mA cm^{-2} current density for HER and OER at 115 and 290 mV overpotentials, respectively. The ultralow thickness of the electrocatalyst and the presence of Co^{6+} synergistically adjust the number and the reactivity of the active centers. The presence of Mo^{6+} in the electrocatalyst is beneficial in two aspects: (i) its electron drawing power helps Co to maintain its high valence state which acts as the active center, and (ii) Mo-ion itself has HER catalyzing efficiency. The CoMo-LDH requires 1.63 V to achieve 10 mA cm^{-2} for overall water splitting [113]. Incorporation of metal atoms having intrinsic electrocatalytic activity can enhance the bifunctional electrocatalytic activity of an LDH. NiFe-LDH is electrocatalytically inert towards HER- is activated through simultaneous doping of V and vacancy generation (Fig. 10e) [114]. The simultaneous doping and vacancy generation strategy alters its OER catalytic activity to some extent (Fig. 10f). The vacancy generation reduces the energy barrier for the Volmer step and adsorption-free energy. The doping coupled with vacancy generation enhances the electrical conductivity of NiFe-LDH. The Ni^{2+} of the LDH acts as the active site for OER and the doping coupled with vacancy generation reduces the energy barrier for the r.d.s. The V-doped vacancy containing NiFe-LDH NSAs achieves a current density of 10 mA cm^{-2} for overall water splitting at 1.43 V [114].

The spinel-type bimetallic materials have shown promising bifunctional electrocatalytic activity towards overall water splitting. The higher valent metal atoms in these electrocatalysts act as the OER catalytically active sites and the anion terminals as the HER catalytically active ones. Different crystallographic planes of an electrocatalyst show varied activity and thus morphology modulation is an effective strategy to achieve superior activity. The incorporation of second metal in a single-metal-containing spinel enhances the formal oxidation state of the initially present metal atom and consequently, its electrocatalytic OER activity. Iron-incorporated Co_3O_4 spinel (FeCoO) is directly fabricated on the NF via a chemical bath deposition followed by an aerial oxidation technique [115]. The 50 nm thick FeCoO sheets with hierarchical white-fungus-like assemblies are formed in this process which facilitates the mass transfer process. The formal oxidation state of Co in Co_3O_4 enhances after incorporation of Fe which accelerates the nucleophilic surface reactions. The electrolyzer constructed with the FeCoO containing NF shows 76% electrical fuel efficiency [115]. Copper cobaltite ($Cu_{0.3}Co_{2.7}O_4$, CCO) with varied particle size and morphology is synthesized in a solvothermal followed by a calcination procedure [116]. Interconnected 1D nanochains of CCO are formed when the solvothermal reaction is performed at 180 °C temperature. Monoclinic CuO and polycrystalline Co_3O_4 are formed instead of the desired CCO when the solvothermal reaction is performed at 210 °C. The interconnected CCO nanochains show superior electrocatalytic activity due to their high specific surface area [116]. Hydrothermally synthesized $CoMoO_4$ NSAs are annealed in reducing gases like H_2 or NH_3 to create oxygen vacancies in it [117]. The superior electrocatalytic activity in spinel-type materials is achieved through vacancy generation, morphology modulation, etc. The incorporation of oxygen vacancy facilitates the HER and OER catalytic activity of $CoMoO_4$ by reducing its band gap. The defect-rich $CoMoO_4$ obtained from the heat treatment in H_2 shows superior activity as compared to that obtained after a similar treatment in the NH_3 environment. The defect-rich $CoMoO_4$ achieves 10 mA cm^{-2} current density on application of 1.56 V in a two-electrode setup [117]. $NiCo_2O_4$ nanosheet, octahedron, and truncated octahedron (Fig. 11a–c) are synthesized in a hydrothermal route after varying the amount of reactants [118]. (110) and (111) crystal planes of $NiCo_2O_4$ are exposed in the nanosheet and octahedron, respectively. On the other hand, both (111) and (100) crystal planes of $NiCo_2O_4$ are exposed in the truncated octahedron. The catalytic activity towards both HER and OER follows the order: nanosheet > octahedron > truncated octahedron. Theoretical calculations suggest that the (110) crystal plane has higher surface energy, smaller H-adsorption free energy (Fig. 11d), and smaller overpotential related to OER as compared to the other two planes. An electrolyzer can be constructed after directly growing the $NiCo_2O_4$ NSAs with an exposed (110) plane on the NF. The fabricated electrolyzer achieves 10 mA cm^{-2} at 1.59 V (Fig. 11e) [118]. Hierarchical $NiCo_2S_4$ nanowire arrays on the NF ($NiCo_2S_4$ NW/NF) which show efficiency towards OER are synthesized following a two-step process. $NiCo_2S_4$ NW/NF achieves 10 mA cm^{-2} current density at 260 mV overpotential for OER [119]. The Ni and Co with various oxidation states are present in the octahedral and tetrahedral sites of the thiospinel-type $NiCo_2S_4$. The higher abundance of octahedrally coordinated Co^{3+} in the structure helps it to attain better OER efficiency at lower activation potential. The high surface area and highly hydrophilic nature of the $NiCo_2S_4$ NW/NF also enhance its electrocatalytic activity. The $NiCo_2S_4$ NW/NF achieves 10 mA cm^{-2} current density for HER at an overpotential of 210 mV. An electrolyzer prototype constructed using $NiCo_2S_4$ NW/NF electrodes achieves 10 mA cm^{-2} current density at 1.63 V and its efficiency degrades only 10% after 50 h performance (Fig. 11f). The electrolyzer generates H_2 and O_2 gases on illuminating an integrated solar cell with an 18 W light emitting diode (LED) desk night light (Fig. 11g) [119]. Cobalt nickel selenide with the chemical structure of $CoNi_2Se_4$ can be electrodeposited on conducting substrates like carbon fiber paper [120]. A rough film with granular morphology is formed as the result of electrodeposition. The $CoNi_2Se_4$ is one of the rare materials that contain abundant Ni^{3+} ions in its as-prepared state. The presence of Ni^{3+} in the electrocatalysts significantly augments the OER catalytic activity of the material. The electrodeposits are able to achieve 10 mA cm^{-2} current density for OER, HER, and overall water splitting at 160, 220, and 380 mV overpotentials, respectively [120].

Initially, NiFe-LDH is grown on the NF following a hydrothermal route which is then converted into nickel iron sulfide by reacting with S-vapor inside a furnace [121]. $Ni_{0.7}Fe_{0.3}S_2$ nanoplatelets with flowerlike aggregates are formed in this synthetic

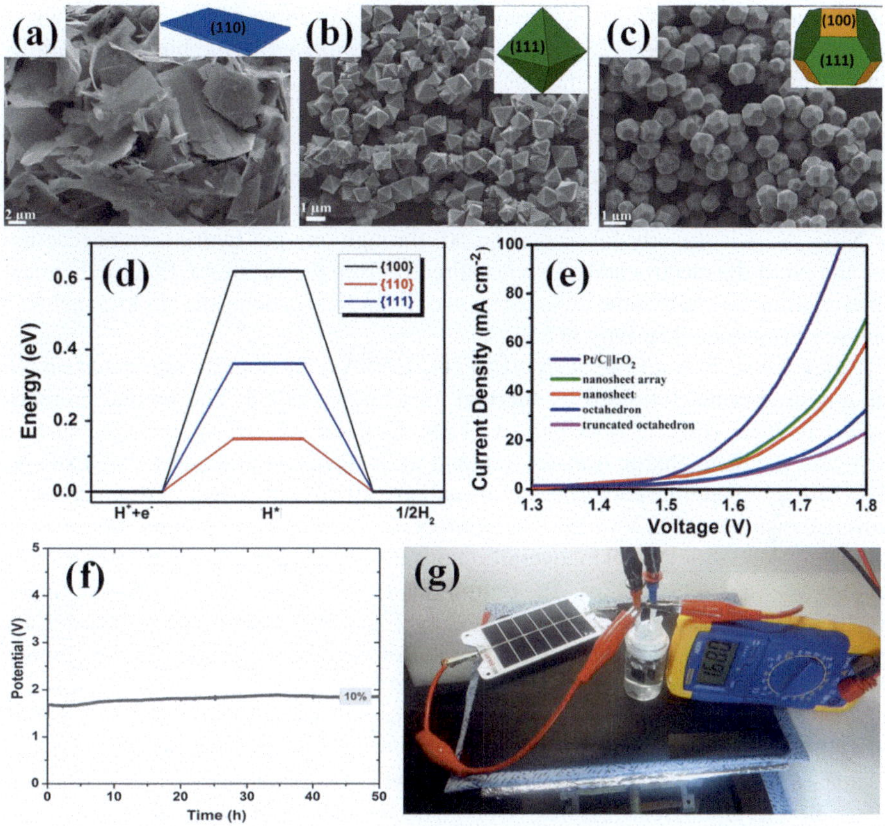

Fig. 11 SEM images of $NiCo_2O_4$ **a** nanosheet, **b** octahedron, and **c** truncated octahedron [in inset showing the orientation of the exposed crystal planes]; **d** HER free energy diagram for the (100), (110) and (111) planes of $NiCo_2O_4$; **e** overall water splitting polarization curves for NSAs, nanosheet, octahedron and truncated octahedron of $NiCo_2O_4$ and RuO_2||Pt/C. **a–e** reprinted with permission from Fang et al. [118]. Copyright (2017). Elsevier. **f** The chronopotentiometric curve for $NiCo_2S_4$ NW/NF||$NiCo_2S_4$ NW/NF at 10 mA cm^{-2} current density; **g** photograph of the $NiCo_2S_4$ NW/NF||$NiCo_2S_4$ NW/NF two-electrode setup integrated with the solar cell and multi-meter. **f** and **g** reprinted with permission from Sivanantham et al. [119]. Copyright (2016). John Wiley and Sons

procedure. HR-TEM image reveals that the Fe atoms consistently replace Ni atoms in the NiS_2 to form $Ni_{0.7}Fe_{0.3}S_2$ and therefore, the electronic structure of NiS_2 changes. The $Ni_{0.7}Fe_{0.3}S_2$ attains 10 mA cm^{-2} current density for overall water splitting at 1.625 V [121]. Fe-Co PBAs are synthesized via a self-assembly method which is selenized to obtain $FeCoSe_2$ NPs. The porous nature of the NPs, higher ECSA, and suitable electron transfer efficiency help the NPs to achieve superior bifunctional catalytic activity. The synthesized $FeCoSe_2$ NPs require 1.59 V to achieve full water-splitting current density of 10 mA cm^{-2} [122]. The metal-interlinking chalcogenide atoms in this type of electrocatalysts act as active sites for water splitting. Cobalt molybdenum sulfide ($CoMoS_x$) containing Co-S-Mo moiety is synthesized via a hydrothermal technique [123]. The amount of this moiety in the bimetallic sulfide can be regulated by varying the reaction time. The variation in reaction time also regulates the crystal and electronic structure of $CoMoS_x$. These physiochemical changes influence the catalysis pathway and the efficiency of the bimetallic sulfide. The $CoMoS_x$ obtained after 12 h reaction shows superior overall water splitting efficiency and surpasses the activity of the RuO_2-Pt/C couple [123]. Ultra-thin NSAs of $NiFe_{10}Se_{10}$ is directly fabricated on NF through one-pot hydrothermal reaction and the Ni of the electrocatalyst comes directly from the substrate. Electrochemical investigations reveal that the incorporation of Fe in $NiSe_{10}$ helps the NSAs to achieve superior bifunctional activity [124].

Holey cobalt-iron nitride NSAs ($CoFeN_x$ HNSAs) are directly fabricated on NF via a hydrothermal followed by thermal nitridation method. The unique morphological features (Fig. 12a) of these $CoFeN_x$ HNSAs enhance their active surface area; and facilitate charge transfer, ion and mass diffusion processes. The $CoFeN_x$ HNSAs having optimal catalytic activity achieve 10 mA cm^{-2} overall water splitting current density at 1.592 V [125]. The NiMoN nanowires with exposed (100) crystallographic plane show excellent overall water splitting and achieve a current density of 20 mA cm^{-2} at 1.498 V [126]. The proton adsorption efficiency of the N-terminated NiMoN (100) plane is superior to that of Pt (Fig. 12b). The charge carrier density of NiMoN near the Fermi level is higher as compared to Pt (Fig. 12c), and consequently, it shows superior HER catalytic activity [126]. Ni-Co carbonate hydroxide nanowires are phosphorized to obtain cone-shaped nanowire arrays of NiCoP on NF (NiCoP-NWAs/NF) [127]. The NiCoP-NWAs/NF shows bifunctional activity as compared to Ni-Co carbonate hydroxide nanowire, Co_2P nanowire, Ni_2P NSAs. The favorable morphological features, synergistic interaction between metals, and fast electron transfer help NiCoP-NWAs/NF to achieve superior activity. NiCoP-NWAs/NF requires 1.64 V to achieve 20 mA cm^{-2} in a fabricated electrolyzer prototype [127]. Vertically oriented nanoarrays of Fe-Co-P are fabricated on NF following an electroless plating process [128]. The physicochemical characterizations suggest that body-centered-cubic Co_7Fe_3 is present in the deposit. The incorporation of Fe in the deposit significantly increases the catalytic activity of Co-P. The Co_7Fe_3 gets partially oxidized to form FeOOH and CoOOH which act as the OER active sites, and the P atoms surrounding Co_7Fe_3 act as active sites for HER. The Fe-Co-P deposit achieves 10 mA cm^{-2} current density for full water splitting at an applied voltage of 1.68 V [128]. Fe-rich macroporous CoFeP triangular plate arrays (CoFeP TPAs) are synthesized from CoFe MOF precursor. The porosity of the CoFeP is regulated by selectively and controllably etching organic ligands from the MOF precursor. The interconnected pores formed in the CoFeP plates facilitate the ion diffusion and mass transport process. The CoFeP TPAs show long-term (for ~100 h) operational stability in addition to their superior catalytic performance [129]. $Co_{0.6}Fe_{0.4}P$ nanocages with open 3D architectures are synthesized through a PBA-mediated strategy [130]. Partially oxidized $Co_{0.6}Fe_{0.4}P$ and CoFe oxide/hydroxide heterostructure act as the true HER and OER catalysts, respectively. The structural openness of the electrocatalyst facilitates mass transfer, enhances charge transfer efficiency, and exposes a greater amount of catalytically active sites. $Co_{0.6}Fe_{0.4}P$ with optimized physicochemical properties achieves a current density of 10 mA cm^{-2} on the application of 1.57 V [130].

Synergistic interaction between metals in bimetallic borides helps in achieving optimized proton adsorption energy. The bimetallic borides easily get converted into oxyhydroxide on the application of positive bias and consequently show superior OER catalytic activity. Cobalt nickel boride on NF (Co–Ni-B@NF) is fabricated by an electroless plating process followed by calcination [7]. The surface of the electrode becomes rough and porous after the fabrication. The average thickness of the deposited layer is found to be around 40 nm. Metallic Ni and Co is predominantly present in the as-deposited layer which gets converted into NiOOH and CoOOH on electrochemical activation (Fig. 12d, e). These oxyhydroxide phases act as the catalytically active sites for OER. Annealed Co–Ni–B@NF requires 313 mV overpotential to achieve 10 mA cm^{-2} current density in OER catalysis. The formation of Co–Ni–B (core)/Co(Ni)O_xH_y (shell) facilitates the OER catalytic activity as the core enhances the electron transport efficiency and the shell provides catalytically active sites. Annealed Co–Ni-B@NF shows decent HER catalytic activity and achieves 10 mA cm^{-2} current density at 205 mV overpotential. The fabricated electrode achieves 10 mA cm^{-2} current density in overall water splitting at an applied voltage of 1.72 V [7]. FeNiB is synthesized over FeNi foam through a solid-state boronization process [132]. The porous bimetallic borides are formed on the surface

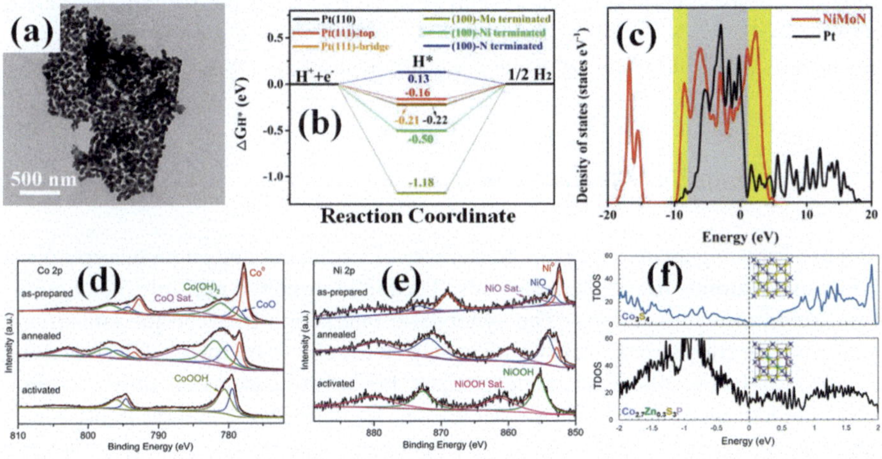

Fig. 12 **a** High-resolution TEM (HR-TEM) image of $CoFeN_x$ HNSAs. Reprinted with permission from Li et al. [125]. Copyright ()0.2020 American Chemical Society. **b** Free energy diagram for HER on (100) facet of NiMoN, and (110) and (111) facets of Pt; **c** calculated DOS of NiMoN and Pt. Reprinted with permission from Chang et al. [126]. Copyright (2018) John Wiley and Sons. **d** Co 2p, and **e** Ni 2p X-ray photoelectron spectroscopic (XPS) profiles of as-prepared, annealed, and electrochemically activated Co–Ni–B@NFs. **d, e** reprinted with permission from Xu et al. [7]. Copyright (2017). The Royal Society of Chemistry. **f** Calculated DOS of Co_3S_4 and $Co_{2.7}Zn_{0.3}S_3P$. Reprinted with permission from Liang et al. [134]. Copyright (2019). John Wiley and Sons

of the foam which provides abundant catalytically active sites. Additionally, the hydroxylation of the bimetallic boride is facile and consequently, it provides abundant OER catalytically active OOH intermediates. FeNiB in a two-electrode setup achieves 10 mA cm^{-2} on application of 1.65 V [132]. $Ni_xFe_{1-x}B$ NPs assemblies are synthesized at 800 °C temperature via a borothermal reduction reaction [133]. The as-prepared electrocatalyst is able to achieve 10 mA cm^{-2} current density for HER and OER at 63.5 and 282 mV overpotentials, respectively. Theoretical calculations suggest that the $Ni_xFe_{1-x}B$ has lower H-adsorption energy as compared to NiB and FeB, which is advantageous for HER. On the other hand, the synergistic interaction between Ni and Fe helps $Ni_xFe_{1-x}B$ to achieve superior activity towards OER [133].

The electrocatalytically active bimetallic dual anion-containing electrocatalyst is also not rare. The incorporation of different metals and anions in it affects its electrocatalytic activity differently. A solvothermal reaction is carried out to convert ZnCo zeolitic imidazolate frameworks into hollow ZnCo sulfide particles [134]. These particles are then converted into ZnCo phosphosulfide ($Zn_{0.3}Co_{2.7}S_3P$) following a solid/gas-phase reaction. The bifunctional activity of these phosphosulfide particles alters with the variation in their size. Theoretical calculation shows that the band gap of Co_3S_4 reduces after the simultaneous incorporation of Zn and P (Fig. 12f) which enhances its electrocatalytic activity. The $Zn_{0.3}Co_{2.7}S_3P$ with a particle size of 50 nm achieves an overall water-splitting current density of 10 mA cm^{-2} at 1.7 V [134]. Amorphous Ni–Fe-P-B NPs are synthesized following a co-reduction method using $NaBH_4$ and NaH_2PO_2 [135]. The incorporation of P in the Ni–B enhances its HER catalytic activity, whereas the incorporation of Fe facilitates the formation of OER catalytically active NiFe-oxyhydroxides. The Ni-Fe-P-B NPs achieve 10 mA cm^{-2} current density for full water splitting on the application of 1.58 V cell voltage [135].

5.3 *Multimetallic Electrocatalysts*

After experiencing an increment in the activity of an electrocatalyst on incorporation of dual metal atoms, researchers have also tried to incorporate multiple metal atoms to further improve its activity. The synergistic interaction between the constituent metal atoms in alloy based electrocatalyst helps it to achieve optimized HER catalytic activity. N-doped graphene layer encapsulated FeCoNi ternary metal alloy is synthesized by calcining the hybrid of two MOFs i.e., $Ni_3[Co(CN)_6]_2$ and $Fe_3[Co(CN)_6]_2$. Binary alloys such as CoNi, FeNi, and FeCo can also be synthesized following a similar procedure [136]. Spherical particles with rough surfaces are obtained in this procedure and a higher amount of tubular structures formed at the surface on increasing the Ni content in the precursor. The metals retain their zero valence state in the alloy however; some metals with higher oxidation states are also present in it. The binary FeCo alloy shows superior HER catalytic activity whereas the ternary metal alloy shows superiority towards the OER (Fig. 13a, b). Theoretical calculations suggest that the change in composition of alloy modulates its electronic properties which in turn regulate the electrocatalytic activity. The electrolyzer fabricated with

FeCo as cathode material and FeCoNi as anode material achieves 10 mA cm^{-2} current density at a cell voltage of 1.687 V. The incorporation of Ni in the bimetallic electrocatalyst is found to adversely affect its HER catalytic activity [136]. NiMoCo hydride nanowire coated on the NF is synthesized following a hydrothermal route. The presence of $NiMo_2Co$ and NiMo in synthesized hybrid nanowires is confirmed by its X-ray diffraction (XRD) pattern and HR-TEM image analysis [137]. The ternary alloy shows the superior electrocatalytic activity as compared to the binary alloys, i.e., NiMo, NiCo, and MoCo. The mixed valence state of the constituent elements facilitates the incorporation of Co into the binary alloy NiMo, which in turn tunes the surface electronic structure of the alloy and optimizes Gibb's free energy of the active species adsorption. Theoretical calculations suggest that the Co atoms on (027) plane of $NiMo_2Co$ are the potential electrocatalytically active sites for both HER and OER. The NiMoCo electrocatalyst shows both efficiency and stability towards the overall water splitting [137]. A different phenomenon in HER catalytic activity is observed for the trimetallic NiFeMo electrocatalyst. The Mo having intrinsic HER catalytic activity helps bimetallic NiMo to achieve superior activity as compared to the trimetallic one. On the other hand, the trimetallic alloy easily gets converted into an oxyhydroxide phase and provides a higher number of OER catalytically active sites [138]. Bimetallic alloy-based electrocatalysts show superior HER catalytic activity as compared to the corresponding trimetallic one however; the latter one shows superior activity towards OER.

Bimetallic oxides/hydroxides show superior OER catalytic activity however their HER catalytic activity is enhanced through incorporation of another metal atom. Zirconium (Zr) is incorporated in $CoFeO_4$ spinel following a hydrothermal reaction to obtain the CoFeZr oxide [139]. The incorporation of Zr changes the morphology of $CoFeO_4$ from nanoparticle to nanosheet. The Zr incorporation also optimizes the substrate adsorption efficiency of the spinel (Fig. 13c, e) by regulating the electronic environment around Co and Fe. The as-prepared CoFeZr oxide NSAs on NF require 1.63 V to achieve 10 mA cm^{-2} current density in a two-electrode setup [139]. 3D hierarchical nanochannels of CoFeNi-O are synthesized through a one-step dealloying method. The porous architecture and synergistic interaction between constituent elements help the electrocatalyst to achieve superior bifunctional activity. NF loaded with CoFeNi-O electrocatalyst achieves a current density of 10 mA cm^{-2} at 1.558 V in a two-electrode setup [140]. Multi-metallic phosphide shows excellent bifunctional electrocatalytic activity owing to successful interaction between constituent elements. Nanoporous NiFeMoP ribbons are synthesized by selectively removing constituting elements (dealloying process) from the $Ni_{45}Fe_{40}Mo_{10}P_5$ precursor using 1 M HNO_3 [141]. A synergistic interaction between the multi-metallic phosphide and $Ni_xFe_{1-x}OOH$ (formed during electrochemical experiments) enhances the specific surface area of the nanoribbons and facilitates electron transfer. The Mo in the NiFeMoP enhances the structural robustness and operational stability of the nanoribbons. The porous metal phosphide nanoribbons require 1.41 V to reach 10 mA cm^{-2} while acting as a bifunctional electrocatalyst [141]. Ternary cobalt-molybdenum-vanadium LDH NSAs which show bifunctional activity towards full water splitting are directly grown on NF through a hydrothermal reaction. The incorporation of Mo

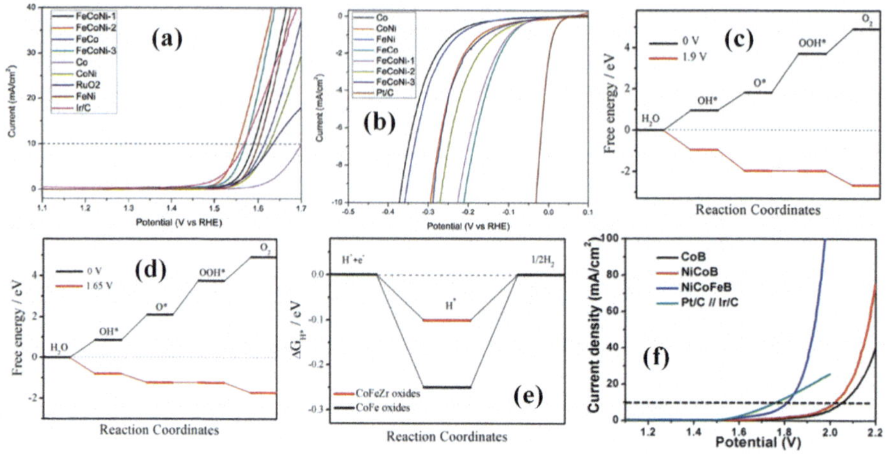

Fig. 13 **a** OER and **b** HER polarization curves for Co, CoNi, FeCo, FeNi, FeCoNi, Pt/C (HER only), RuO_2 (OER only) and Ir/C (OER only). (a and b) reprinted with permission from Yang et al. [136]. Copyright (2016). American Chemical Society. OER free energy diagram for **c** $CoFeO_4$, and **d** Zr-incorporated $CoFeO_4$ at 0 V and corresponding theoretical limiting potentials; **e** HER free energy diagram for bare $CoFeO_4$ (CoFe oxides) and Zr-incorporated $CoFeO_4$ (CoFeZr oxides). **c–e** reprinted with permission from Huang et al. [139]. Copyright (2019). John Wiley and Sons. **f** Overall water splitting polarization curves for CoB||CoB, NiCoB||NiCoB, NiCoFeB||NiCoFeB and Pt/C||Ir/C. Reprinted with permission from Huang et al. [143]. Copyright (2018). John Wiley and Sons

and V with a high valance state optimizes the electronic environment and catalytic activity of the LDH. The CoMoV LDH attains 10 mA cm^{-2} current density at 1.61 V for overall water splitting [142]. Nanochains of cobalt-based borides having other transition metals are synthesized via soft chemical routes [143]. A strong electronic coupling between the constituting elements is evidenced for NiCoFeB and the Co is found to be present in its metallic state. The HER and OER catalytic activities of CoB increase on the incorporation of Ni and Fe within the structure (Fig. 13f) however, the increment is not significant enough for HER catalysis. The theoretical calculations suggest that the increment in water-splitting activity is due to the presence of Ni and Fe atom-modulated metallic Co surface states. The calculation also suggests that the Fe 3d spin majority states can modulate the 3d band of Co and so the electrocatalytic activity of the material. The electrolyzer achieves 10 mA cm^{-2} current density at 1.81 V when NiCoFeB acts as a bifunctional electrocatalyst [143]. Table 2 summarizes the activity of the recently reported bifunctional electrocatalysts for overall water splitting comprising more than one metal atom.

The incorporation of more than one metal atom regulates their individual oxidation state and synergistically enhances the substrate adsorption efficiency of an electrocatalyst. An electrocatalyst that can easily get converted into oxyhydroxide form is beneficial for OER catalytic activity. The incorporation of metal atoms having intrinsic HER catalytic activity (like V and Mo) helps the electrocatalyst to achieve superior

Table 2 Recently reported bifunctional electrocatalysts comprising more than one metal atom

Electrocatalyst	Overpotential @ 10 mA cm^{-2} for HER (mV)	Overpotential @ 10 mA cm^{-2} for OER (mV)	Cell voltage @ 10 mA cm^{-2} for overall water splitting (V)	Refs.
Bimetallic electrocatalysts				
NiFe-oxide nanocubes	197	271	1.67	[109]
$Co_{0.75}Ni_{0.25}(OH)_2$	95	235	1.56	[110]
NiCo-LDH	162	271	1.66	[112]
CoMo-LDH	115	290	1.63	[113]
V-doped NiFe-LDH with vacancy	19	–	1.43	[114]
Iron cobalt oxide	205	244	1.62	[115]
$CoMoO_4$ nanosheet	42	295	1.56	[117]
$NiCo_2O_4$ nanosheet	157 (5 mA cm^{-2})	330 (5 mA cm^{-2})	1.59	[118]
$NiCo_2S_4$ nanowire	210	260	1.63	[119]
$CoNi_2Se_4$	220	160	1.61	[120]
$Ni_{0.7}Fe_{0.3}S_2$	155	198	1.625	[121]
$FeCoSe_2$	90	251	1.59	[122]
$CoMoS_x$	190	280	1.574	[123]
$NiFe_{10}Se_{10}$ nanosheet	154	199	161	[124]
Holey $CoFeN_x$ nanosheets	200	260 (50 mA cm^{-2})	1.592	[125]
NiMoN nanowires	38	290 (50 mA cm^{-2})	1.498 (20 mA cm^{-2})	[126]
NiCoP NWAs	197 (100 mA cm^{-2})	370 (100 mA cm^{-2})	1.64 (20 mA cm^{-2})	[127]
Fe-Co-P	163	250	1.68	[128]
CoFeP triangular plate arrays	43	198	1.47	[129]
$Co_{0.6}Fe_{0.4}P$ nanocages	133	298	1.57	[130]
Co–Ni-B	205	313	1.72	[7]
FeNiB	–	272	1.65	[132]
$Ni_xFe_{1-x}B$	63.5	282	1.57	[133]
$Zn_{0.3}Co_{2.7}S_3P$	188	261	1.70	[134]
Ni–Fe-P-B	220	269	1.58	[135]
Multimetallic electrocatalysts				
FeCo	149	–	1.687	[136]
FeCoNi	–	288		
NiMoCo NWAs	23	277	1.56	[137]
NiFeMo	45	238	1.45	[138]

(continued)

Table 2 (continued)

Electrocatalyst	Overpotential @ 10 mA cm^{-2} for HER (mV)	Overpotential @ 10 mA cm^{-2} for OER (mV)	Cell voltage @ 10 mA cm^{-2} for overall water splitting (V)	Refs.
CoFeZr oxides	104	264 (20 mA cm^{-2})	1.63	[139]
CoFeNi-O	57.9	200	1.558	[140]
NiFeMoP	223	197 (20 mA cm^{-2})	1.41	[141]
CoMoV-LDH NSAs	150	270	1.61	[142]
NiCoFeB nanochains	345	284	1.81	[143]

bifunctional activity. The spinel-type materials show satisfactory bifunctional electrocatalytic activity due to the presence of metal atoms in the higher valence state. The morphology which exposes the superiorly active crystal plane of these electrocatalysts is highly desirable. The incorporation of the higher number of elements in an electrocatalyst may adversely affect its activity, and thus the elements should be chosen cautiously. Heterostructure or composite is another form of multimetallic material, where two or more different compounds stay integrated.

5.4 *Heterostructure-Based Electrocatalysts*

A metal with intrinsic activity towards a reaction is incorporated in an electrocatalyst to form bimetallic and multimetallic electrocatalysts and to achieve bifunctional activity. This observation motivated the researchers to investigate the electrocatalytic activity of hetero-structure-based electrocatalysts. A well-established strategy to develop superior bifunctional electrocatalytic activity is to integrate two different compounds active towards different electrocatalytic processes. This strategy provides a wide scope for alteration in the bifunctional activity of an electrocatalyst. The following section categorically discusses that strategy.

Previous thorough discussion revealed that the oxide/hydroxide/oxyhydroxides show efficiency towards OER and thus those are integrated with their metallic counterpart to achieve bifunctionality. Hexagonal Ni/$Ni(OH)_2$ nanoplates are synthesized in a hydrothermal reaction using Ni-precursor and hydrazine, a strong reducing agent. The hybrid electrocatalyst with optimum efficiency achieves 10 mA cm^{-2} for HER and OER at 168 and 310 mV overpotentials, respectively [144]. $NiMoO_4$ nanowire arrays grown on NF are treated with hydrogen for constructing Ni_4Mo nanoalloys on their surfaces to obtain the $Ni_4Mo/NiMoO_x$ with oxygen vacancies [145]. N-doped carbon sheath is fabricated over the heterostructure via the carbonization process (Fig. 14a) to enhance its electrical conductivity. A higher abundance of active sites and electrical conductivity of the heterostructure helps it to achieve superior bifunctional activity. The NF containing carbon coated $Ni_4Mo/NiMoO_x$ achieves 10 mA cm^{-2} overall water-splitting current density at a cell voltage of 1.57 V and is able to retain its activity for at least 60 h [145]. Chalcogenides and pnictides of transition metal-based bifunctional electrocatalysts are integrated with OER catalytically active hydroxides or LDH to enhance their activity [146–150]. Ni_3S_2 is fabricated on the NF following a hydrothermal route and then $Co(OH)_2$ is electrodeposited over it to obtain the Ni_3S_2@$Co(OH)_2$/NF heterostructure [146]. The XRD pattern of Ni_3S_2@$Co(OH)_2$/NF confirms the crystalline and amorphous nature of the Ni_3S_2 and electrodeposited $Co(OH)_2$, respectively (Fig. 14b). Ultrathin Ni_3S_2 nanosheets are vertically formed over the NF skeleton which is covered with amorphous $Co(OH)_2$ NPs (Fig. 14c). The charge redistribution between Ni_3S_2 and $Co(OH)_2$ and the formation of heterointerface augments the charge transferability and surface reactivity of the electrocatalyst, which in turn enhances the electrocatalytic activity. The Ni_3S_2@$Co(OH)_2$/NF affords 10 mA cm^{-2} current density on application of 1.61 V in a two-electrode setup [146]. $Co_{0.85}Se$ NSAs are grown on the electrochemically exfoliated graphite (ExGr) foil which is then coated with NiFe-LDH to obtain the ExGr/$Co_{0.85}Se$/NiFe-LDH heterostructure [147]. The specific surface area of the heterostructure is calculated to be 156 m^2g^{-1} from the Brunauer–Emmett–Teller (BET) surface area analysis. The hybrid achieves 150 mA cm^{-2} current density for OER and 20 mA cm^{-2} current density for HER at 270 and 260 mV overpotentials, respectively. The successful coupling between the components of the hybrid and large specific surface area are the main reasons behind its superior electrocatalytic activity. The two-electrode setup comprising ExGr/$Co_{0.85}Se$/NiFe-LDH as catalyst achieves an overall water-splitting current density of 20 mA cm^{-2} at an applied voltage of 1.71 V [147]. NiFe-LDH is electrodeposited over previously constructed Ni_3N microsheet arrays on NF to obtain the NiFe-LDH@Ni_3N hybrid [148]. The intrinsic metallic character of Ni_3N enhances the electrical conductivity of the hybrid and facilitates the charge transfer process. The 3D highly porous hierarchical architecture of the hybrid exposes a higher amount of catalytically active sites. All these factors help the NiFe-LDH@Ni_3N hybrid to achieve a desirable bifunctional electrocatalytic activity and operational stability [148]. Hierarchical NiFe LDH coupled NiCoP heterostructure (NiFe LDH@NiCoP/NF) is fabricated on NF substrate following a three-step hydrothermal-phosphorization-hydrothermal process [149]. A successful coupling between the ultrathin NiFe LDH nanosheets and NiCoP nanowires is confirmed from its HR-TEM image (Fig. 14d). NiFe LDH@NiCoP/NF shows superior stability in addition to its efficient HER and OER catalyzing activity. The heterostructure is able to retain its initial architecture even after long-term catalytic performance (Fig. 14e–f). NiFe LDH@NiCoP/NF requires ~1.57 V cell voltage to achieve 10 mA cm^{-2} current density in a two-electrode setup [149]. Amorphous NiFe hydroxide is electrodeposited on $MnCo_2O_4$ nanoflowers previously fabricated on NiFe foam. The synergistic interaction between the NiFe hydroxide and $MnCo_2O_4$ facilitates the electron transfer ability of the hybrid and augments the reaction kinetics. The hierarchical morphology of the hybrid can be modulated by changing the deposition potential which in turn regulates its electrocatalytic activity [150].

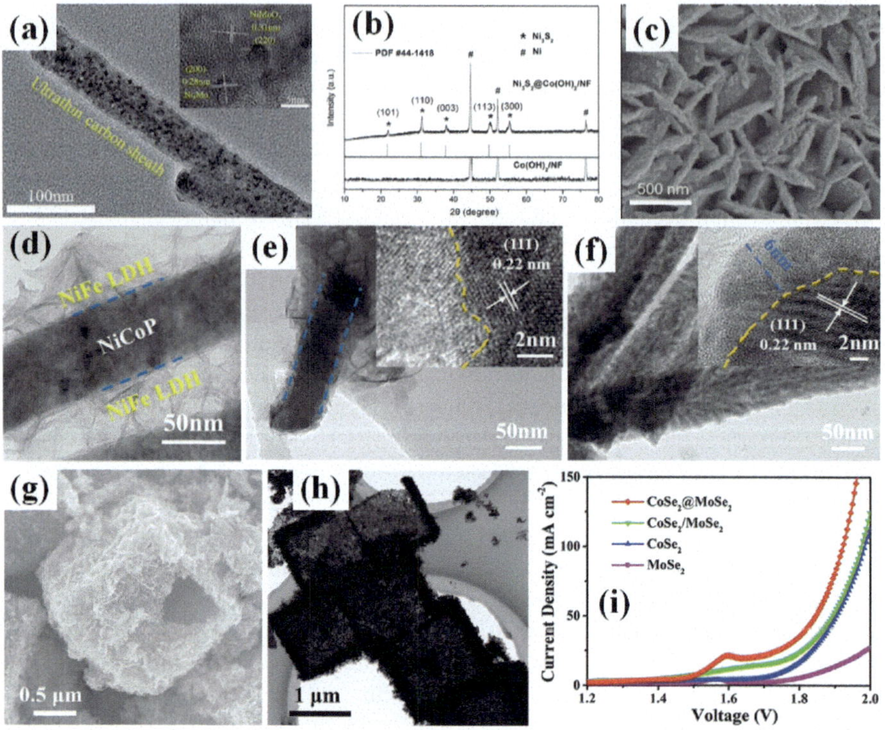

Fig. 14 **a** HR-TEM image of carbon sheath coated $Ni_4Mo/NiMoO_x$. Reprinted with permission from Hou et al. [145]. Copyright (2017). John Wiley and Sons. **b** X-ray diffraction pattern of $Ni_3S_2@Co(OH)_2$/NF and $Co(OH)_2$/NF; **c** SEM image of $Ni_3S_2@Co(OH)_2$/NF. (b and c) reprinted with permission from Wang et al. [146]. Copyright (2018). Elsevier. HR-TEM images of **d** as-synthesized, **e** post-HER, and **f** post-OER NiFe LDH@NiCoP/NF (d-f) reprinted with permission from Lin et al. [149]. Copyright (2018). John Wiley and Sons. **g** SEM, and (h) HR-TEM images of $CoSe_2@MoSe_2$; **i** overall water splitting polarization curves for $CoSe_2@MoSe_2 \| CoSe_2@MoSe_2$, $CoSe_2/MoSe_2 \| CoSe_2/MoSe_2$, $CoSe_2 \| CoSe_2$ and $MoSe_2 \| MoSe_2$ ($CoSe_2/MoSe_2$ obtained after physically mixing $CoSe_2$ and $MoSe_2$). **g–i** reprinted with permission from [155]. Copyright (2020). The Royal Society of Chemistry

The MoS_2 and WS_2 are superiorly active towards HER in an acidic medium; however, their activity in an alkaline medium is relatively inferior [151]. The integration of transition metals with these transition metal dichalcogenides (TMDs) enhances the activity in an alkaline medium. Additionally, the integration of TMDs with other transition metal-based electrocatalyst synergistically induces bifunctional electrocatalytic activity to the later one [151]. This is a widely adopted strategy for the development of bifunctional electrocatalysts. The heterostructure interface of the $Co_3O_4@MoS_2$ hybrid shows synergism towards water dissociation and active species adsorption [152]. XRD analysis confirms the presence of both Co_3O_4 and MoS_2 in the hybrid fabricated on CC ($Co_3O_4@MoS_2$/CC) following a two-step hydrothermal process. The charge redistribution occurs between Co_3O_4 and MoS_2

due to this hybridization. The Co_3O_4@MoS_2/CC affords 10 mA cm^{-2} current density in a two-electrode setup on application of 1.59 V [152]. Hierarchical CoS_x@MoS_2 heterostructure is constructed via a one-pot solvothermal process where EG helps to achieve the desired architecture [153]. The formation of Co-S-Mo at the interfacial region of the heterostructure facilitates the charge transfer process. On the other hand, integration between CoS_x and MoS_2 modifies the valence band structure of the CoS_x. These factors help the heterostructure to achieve superior bifunctional activity towards overall water splitting [153]. Co_9S_8/WS_2 heterostructure is synthesized on Ti foil via a three-step hydrothermal-sulfurization-hybridization process [154]. Heterostructure with three different morphologies (i.e., nanobelts, nanoneedles, and nanorhombuses) can be obtained by altering the different parameters of the hydrothermal process. The Co_9S_8/WS_2 nanobelts show superior HER and OER catalytic activity as compared to the heterostructure with other morphologies. The superior activity of the nanobelts is attributed to their higher surface owing to the presence of macropores in them. The Co_9S_8/WS_2 nanobelts attain 10 mA cm^{-2} current density for overall water splitting at 1.65 V cell voltage [154]. Hollow $CoSe_2$ nanocubes are initially synthesized by selenizing the Co-PBA precursor and then $MoSe_2$ nanosheets are decorated over them to obtain $CoSe_2$@$MoSe_2$ heterostructure [155]. The unique morphological features of the heterostructure (Fig. 14g–h) provide abundant active sites and facilitate ion/electron transfer. The formation of heterointerface in the hybrid provides additional active species adsorption sites and facilitates the water dissociation process. 1.524 V is required to achieve 10 mA cm^{-2} current density when $CoSe_2$@$MoSe_2$ heterostructure acts as a bifunctional electrocatalyst (Fig. 14i) [155]. MoS_2 is grown over the FeNi foam/$Fe_5Ni_4S_8$ hybrid following a chemical vapor deposition process [156]. The vertically oriented MoS_2 nanosheets grow uniformly over the substrate and form a heterointerface with the $Fe_5Ni_4S_8$. The MoS_2/$Fe_5Ni_4S_8$ achieves 10 mA cm^{-2} current density at 120 and 240 mV overpotentials for HER and OER, respectively. The high specific surface area, heterointerface generation, and MoS_2 with exposed edge formation synergistically augment the electrocatalytic water-splitting activity of the heterostructure [156]. MoS_2/Ni_3S_2 nanorod arrays are directly grown on the NF following a hydrothermal reaction which shows superior overall water-splitting activity. In situ generation of the heterostructure on the 3D NF substrate exposes its catalytically active sites and enhances its electrical conductivity and stability. The electrolyzer constructed with MoS_2/Ni_3S_2 containing NF electrode achieves 10, 100, 200, and 300 mA cm^{-2} at 1.467, 1.593, 1.640, and 1.661 V, respectively [157].

Coupling two different compounds with the same counter anion is another strategy for developing a bifunctional electrocatalyst. The desired heterostructure can be synthesized in a one-pot procedure however; special care should be taken so that instead multimetallic compound does not form. The heterostructures synthesized through a one-step process are directly grown on hard metallic foam where the substrate sometimes acts as the source of metal. Otherwise, the heterostructure can be synthesized following more than one reaction step. CuO_x@Co_3O_4 nanorods over the Cu foam (CuO_x@Co_3O_4 NRs/CF) are fabricated following a four-step process [158]. The hierarchical shell of Co_3O_4 having a thickness of 120 nm wraps the

CuO_x NRs formed with a diameter of 200 nm. The Co_3O_4 in the heterostructure gets oxidized to form Co^{3+} O_6 octahedra and Co^{3+} OH containing β-CoOH and the CuO_x forms Cu-oxo active species during the OER catalysis. These Co^{3+} containing species and Cu-oxo species act as the active centers for the OER catalysis. The CuO_x@Co_3O_4 NRs/CF attains 50 mA cm^{-2} for HER and OER at an overpotential of 242 and 240 mV, respectively. In addition to the activity, the heterostructure shows fair stability in catalyzing conditions [158]. $NiCo_2O_4$@$CoMoO_4$ hierarchical heterostructure is fabricated over the NF ($NiCo_2O_4$@$CoMoO_4$/NF) using a two-step hydrothermal process followed by annealing [159]. The $NiCo_2O_4$ nanowires are uniformly distributed over the NF and the $CoMoO_4$ nanosheets form flower-like morphologies over those nanowires. (1) High electrochemically active surface area due to the formation of hierarchical structure and (2) synergistic effect between $CoMoO_4$ and $NiCo_2O_4$ results in enhanced water splitting activity. The heterostructure attains 10 mA cm^{-2} current density for HER at 121 mV overpotential and 20 mA cm^{-2} for OER at 256 mV overpotential. The fabricated electrolyzer with $NiCo_2O_4$@$CoMoO_4$/NF electrode requires 1.55 V to attain 10 mA cm^{-2} current density [159]. $CoMoO_4$ nanocolumns are hydrothermally grown over the NF and then are converted into $CoMoS_4$ via an ion exchange mechanism (Fig. 15a) [160]. Additional Ni_3S_2 is formed and the nanocolumn morphology of $CoMoO_4$ changes into porous nanosheets during this sulfurization process. Ni_3S_2 synergistically enhances the OER catalytic activity of $CoMoS_4$ by suppressing the charge recombination process and enhancing the ECSA. The $CoMoS_4$/Ni_3S_2 hybrid requires 1.568 V cell voltage to afford the overall water-splitting current density of 10 mA cm^{-2} (Fig. 15b) [160]. $Ni_xCo_{3-x}S_4$-decorated Ni_3S_2 is fabricated on the NF following a partial anion exchange process between Ni_3S_2 and Co^{2+} ions [18]. The fabricated electrode attains 10 mA cm^{-2} current density at 1.53 V in an electrolyzer prototype. Theoretical combined with experimental results suggest that $Ni_xCo_{3-x}S_4$ is a superior electrocatalyst than Ni_3S_2 and their interfacial region in the heterostructure provides additional catalytically active sites [18]. The CoS_x/Ni_3S_2 heterostructure is constructed on NF via a one-pot hydrothermal reaction where NF itself acts as the Ni source [131]. The elimination of catalyst binder, formation of heterointerface, and increment in charge transfer efficiency help the heterostructure to achieve superior overall water splitting efficiency. The CoS_x/Ni_3S_2 is able to achieve an overall water-splitting current density of 10 and 50 mA cm^{-2} at 1.572 and 1.684 V, respectively [131]. MoP/Ni_2P heterostructure is fabricated on the NF (MoP/Ni_2P/NF) following a two-step process [163]. The XRD pattern shows that the Ni_2P is formed with good crystallinity, however; the MoP is poorly crystalline in nature. The Ni_2P nanoparticles form aggregates over the skeleton of the NF and the MoP nanosheets assemble together to form nanoflowers. MoP/Ni_2P/NF composite requires 309 mV overpotential to achieve 20 mA cm^{-2} current density for OER and 75 mV to attain 10 mA cm^{-2} current density for HER. The two-electrode setup comprised of MoP/Ni_2P/NF electrode achieves 10 mA cm^{-2} current density on application of 1.55 V [163].

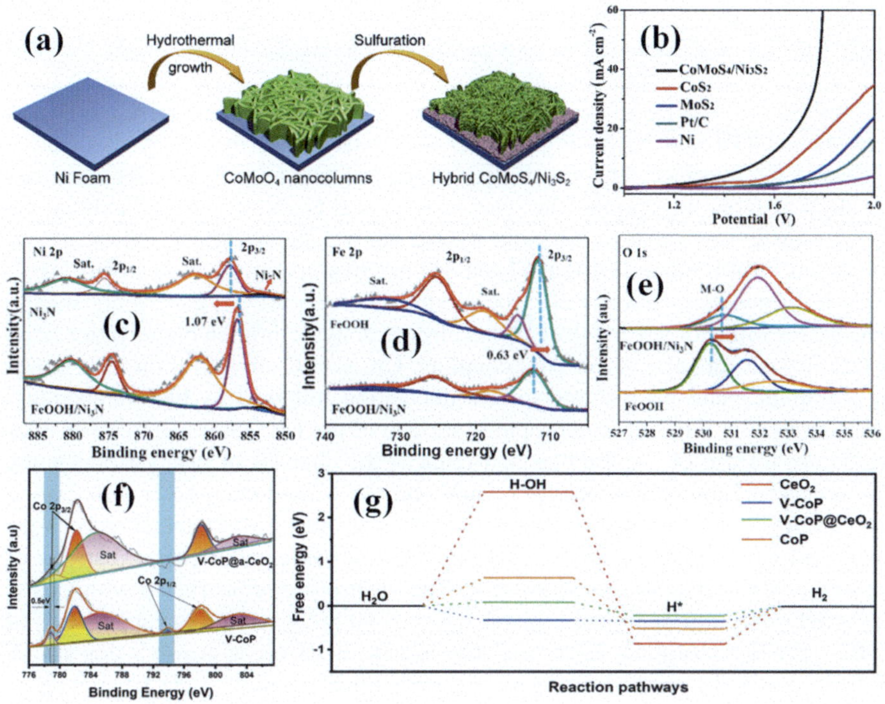

Fig. 15 **a** Schematic diagram showing $CoMoS_4/Ni_3S_2$ hybrid fabrication procedure; **b** overall water splitting polarization curves for $CoMoS_4/Ni_3S_2$‖$CoMoS_4/Ni_3S_2$, CoS_2‖CoS_2, MoS_2‖MoS_2, Pt/C‖Pt/C and NF‖NF. (a and b) reprinted with permission from Hu et al. [160]. Copyright (2019). Elsevier. **c** Ni 2p, **d** Fe 2p and **e** O 1 s XPS profiles for Ni_3N, FeOOH and FeOOH/Ni_3N hybrid. (c-e) reprinted with permission from Guan et al. [166]. Copyright (2020) Elsevier. **f** Co 2p XPS profile for V-CoP and V-CoP@a-CeO_2 hybrid; **g** Free energy diagram for CeO_2, V-CoP, V-CoP@a-CeO_2, and CoP catalyzed H-adsorption from alkaline H_2O. **f** and **g** reprinted with permission from Yang et al. [168]. Copyright (2020). John Wiley and Sons

More than one-step reaction processes are adopted to synthesize heterostructure containing metal compounds with different counter anions. A two-step electrodeposition method is adapted to fabricate heterostructured Ni-S-P film on NF [164]. Physicochemical analyses reveal that the ternary film is actually comprised of Ni_2P and NiS nanoparticles embedded $Ni(OH)_2$ nanosheets. In addition to superior bifunctional catalytic activity, the fabricated electrode shows excellent activity retention for at least 160 h. 12.5% solar-to-hydrogen conversion efficiency is achieved when the fabricated electrolyzer is integrated with a silicon planar solar cell [164]. Cobalt nitride-vanadium oxynitride nanohybrid is fabricated on CC via a polyaniline meditated synthetic process [165]. The nanohybrid shows bifunctional electrocatalytic activity as compared to its counterparts and achieves 10 mA cm^{-2} for HER and OER at 118 and 263 mV overpotentials, respectively. Post-catalytic sample analysis reveals that the defect-rich amorphous CoO_x acts as a true OER catalyst whereas $Co(OH)_2$-$Co^{\delta+}$-N ($0 < \delta < 2$) centers are responsible for the HER catalysis [165]. FeOOH nanosheets on Ni_3N nanotubes arrays are fabricated on CC via solid-state reaction followed by an electrodeposition method [166]. Charge reorganization occurs between FeOOH and Ni_3N (Fig. 15c–e) which facilitates the water dissociation process at the interface. The positively charged FeOOH adsorbs the OH^- whereas the Ni_3N adsorbs the proton. The relatively vertical Ni_3N nanotubes facilitate the charge transfer and ion diffusion in the hybrid catalyst. FeOOH/Ni_3N hybrid affords 10 mA cm^{-2} for HER and OER at 67 and 244 mV overpotentials, respectively [166]. CeO_2 itself possesses inferior electrocatalytic activity however; can synergistically enhance the bifunctional activity of other electrocatalysts through optimization of its electronic structure after successful coupling. Co_4N-CeO_2 hybrid is directly grown on a graphite plate via electrochemical deposition followed by high-temperature nitridation [167]. The super-hydrophilic surface of the fabricated electrode enables the timely release of gas bubbles and reduces the chances of catalyst poisoning. The synergistic interaction between Co_4N and CeO_2 facilitates water dissociation and optimizes the H-adsorption energy of the hybrid. Successful interaction between CeO_2 and Co_3O_4 formed on the surface of Co_4N reduces the energy barriers of the intermediate steps involved in OER. The electrolyzer constructed with Co_4N-CeO_2 hybrid as a bifunctional catalyst achieves 10 mA cm^{-2} on the application of 1.507 V [167]. Amorphous CeO_2 (a-CeO_2) is deposited over as-grown V-doped CoP (V-CoP) nanorod arrays following an electrochemical reaction [168]. Charge redistribution occurs between two counterparts of the V-CoP@a-CeO_2 hybrid and the bifunctional electrocatalytic activity enhances synergistically. A successful interaction between the V-dopant and CeO_2 enhances the electron density at Co-active sites (Fig. 15f) which in turn optimizes the H-adsorption energy of the hybrid (Fig. 15g). The V-CoP@a-CeO_2 hybrid in a two-electrode setup reaches 10 and 100 mA cm^{-2} on application of 1.56 and 1.71 V, respectively [168]. Similarly, $CoCr_2O_4$ possesses inferior activity towards HER which increases drastically after successful integration with bifunctional electrocatalyst CoP. Mesoporous CoP/$CoCr_2O_4$ heterostructure is synthesized using a KIT-6 hard silica template via a nanocasting-phosphorization process [169]. The phosphide-spinel oxide hybrid shows superior electrocatalytic activity owing to its 3D mesoporous architecture and large surface area. The CoP/$CoCr_2O_4$ heterostructure achieves 10 mA cm^{-2} for full water splitting at 1.68 V [169].

The electrocatalytic activity and operational stability of transition metal-based materials are enhanced after heterostructure formation with highly conducting carbonaceous materials like graphene, carbon nanotube (CNT), etc. [104, 170–175]. Semi-spherical nickel vanadate ($Na_3V_2O_8$) particles anchored N-doped reduced graphene oxide (NRGO) heterostructure is fabricated via a reflux method [171]. The presence of the optimum amount NRGO (~5.6 wt.%) in the hybrid helps it to achieve higher ECSA and lower charge transfer resistance. The NRGO in the heterostructure acts as its conducting backbone, and also enhances its operational stability [171]. D Ni_2P nanocrystals encapsulated in 2D N,-S codoped graphene are synthesized through solution phase reaction followed by ex situ sonication method

(Fig. 16a–c) [172]. The formation of 0D@2D heterostructure provides a high specific surface area and exposes abundant catalytically active sites. The successful interaction between the N, S-dopant of graphene and Ni_2P optimizes the water dissociation and proton adsorption step of the electrocatalysis process. The heterostructure achieves 10 mA cm^{-2} in a two-electrode setup on the application of 1.572 V and is able to sustain the activity for at least 40 h (Fig. 16d) [172]. Formation of heterostructure with reduced graphene oxide (rGO) and successful electronic interaction enhances the specific surface area and charge transfer efficiency of NiVB. The heterostructure achieves 10 mA cm^{-2} on application of 1.46 V for full water splitting [173]. A ternary composite of graphitic carbon nitride (g-C_3N_4), CeO_2 and Fe_3O_4 is synthesized following a three-step process. A synergistic interaction between these counterparts enhances the charge transfer efficiency of the composite and helps it to achieve superior bifunctional activity [174]. The interfacial engineering between Co and Co_2P lowers the proton adsorption energy of the heterostructure. The bifunctional electrocatalytic activity of the Co/Co_2P heterostructure further increases after coupling with N, P dual-doped CNT. The doped CNT not only protects the electrocatalyst from electrochemical degradation (Fig. 16e) but also assures facile charge transfer in it [175]. Table 3 summarizes the activity of the recently reported heterostructure-based bifunctional electrocatalysts for overall water splitting.

An electrocatalyst with hierarchical architecture shows the superior activity as compared to its bimetallic or multimetallic analog [120, 123, 153, 176]. Multi-step

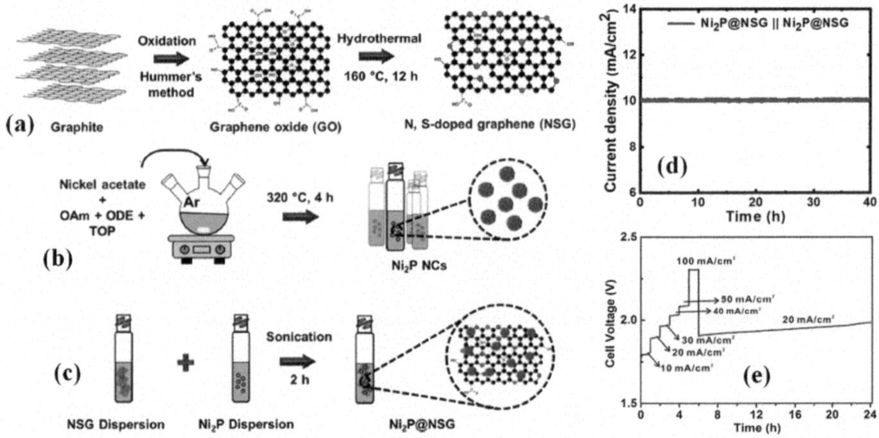

Fig. 16 Schematic diagram showing the synthesis of **a** defective N,S-doped graphene (NSG) nanosheets, **b** colloidal 0D Ni_2P nanocrystals, and **c** 0D@2D Ni_2P@NSG heterostructure; **d** chronoamperometric curve for Ni_2P@NSG|| Ni_2P@NSG obtained at 1.572 V (a-d) reprinted with permission from Suryawanshi et al. [172]. Copyright (2021). American Chemical Society. **e** Chronoamperometric curve for Part-Ph Co@Co–P@NPCNTs||Part-Ph Co@Co–P@NPCNTs at different current densities. Reprinted with permission from Jiao et al. [175]. Copyright (2020). John Wiley and Sons

Table 3 Recently reported heterostructure-based bifunctional electrocatalysts

Electrocatalyst	Overpotential @ 10 mA cm^{-2} for HER (mV)	Overpotential @ 10 mA cm^{-2} for OER (mV)	Cell voltage @ 10 mA cm^{-2} for overall water splitting (V)	Refs.
$Ni/Ni(OH)_2$ nanoplates	168	310	1.68	[144]
$Ni_4Mo/NiMoO_x/NF$	29	284	1.57	[145]
$Ni_3S_2@Co(OH)_2/NF$	110	257	1.61	[146]
$ExGr/Co_{0.85}Se/NiFe$-LDH	260	270 (150 mA cm^{-2})	1.71 (20 mA cm^{-2})	[147]
NiFe LDH@Ni_3N/NF	142 (100 mA cm^{-2})	238 (100 mA cm^{-2})	1.63 (100 mA cm^{-2})	[148]
NiFe LDH@NiCoP/NF	120	220	1.57	[149]
NiFe–$MnCo_2O_4$	98	272 (100 mA cm^{-2})	1.49	[150]
$Co_3O_4@MoS_2/CC$	90	269	1.59	[152]
$CoS_x@MoS_2$	146	276	1.668	[153]
Co_9S_8/WS_2	138	–	1.65	[154]
$CoSe_2@MoSe_2$	183	309	1.524	[155]
$MoS_2/Fe_5Ni_4S_8$	120	204	–	[156]
$MoS_2/Ni_3S_2/NF$	187	217	1.467	[157]
$CuO_x@Co_3O_4$	242 (50 mA cm^{-2})	240 (50 mA cm^{-2})	–	[158]
$NiCo_2O_4@CoMoO_4/NF$	121	265 (20 mA cm^{-2})	1.55	[159]
$CoMoS_4/Ni_3S_2$	158	200	1.568	[160]
$Ni_xCo_{3-x}S_4/Ni_3S_2/NF$	136	160	1.53	[18]
$CoS_x/Ni_3S_2@NF$	204	280 (20 mA cm^{-2})	1.572	[131]
$MoP/Ni_2P/NF$	75	309 (20 mA cm^{-2})	1.55	[163]
$NiS/Ni_2P@Ni(OH)_2$	120	219	1.58	[164]
Co_4N-$VN_{1-x}O_x$	118	263	1.64	[165]
$FeOOH/Ni_3N$	67	244	1.58	[166]
Co_4N-CeO_2	24	239	1.507	[167]
V-CoP@a-CeO_2	68	225	1.56	[168]
$CoP/CoCr_2O_4$	212	290	1.68	[169]
Ni_2P@NSG	110 Stability ≈60 h	240 Stability ≈60 h	1.572 Stability ≈60	[172]
NiVB/rGO	151	267	1.46	[173]
Part-Ph Co@Co–P@NPCNTs	160 Stability ~24 h	290 Stability ~24 h	1.63 Stability ~24 h	[175]

synthetic procedure instead of one-step is more suitable for this kind of heterostructure development. One-pot synthesis may be applicable for designing heterostructure containing compounds with similar counter anions. However, there lies a possibility of multimetallic compound formation instead of the desired heterostructure in that case. The heterostructure containing two or more compounds with different counter anions shows superior activity as a bifunctional electrocatalyst. The formation of heterostructure with carbonaceous materials not only enhances its activity but also improves long-term operational stability.

5.5 *Engineered Electrocatalysts*

The performance of a transition metal-based electrocatalyst is enhanced after heteroatom doping or defect formation as discussed earlier. To take few examples, Nb-doping enhances the activity of Co_4N and V-doping serves the same purpose for NiS_2 [62, 75]. Fe-doping augments the bifunctional activity of $Ni(OH)_2$ by facilitating an alternate electrocatalytic pathway [72]. On the other hand, the S-incorporation in CoO_x enhances its bifunctional activity through the generation of defects and vacancies created during the synthesis process augments the performance of MnO_2 [57, 100]. Recently, researchers have adopting these particular strategies to design the electrocatalysts and achieve desirable bifunctional activity. These new sets of materials may be termed as "engineered electrocatalysts" and deserve special mention.

Engineering defect sites or vacancies in a material, mainly the transition metal oxides, is an efficient strategy to develop efficient bifunctional electrocatalysts. A vacancy in a material is generated due to the removal of either the metal atom or the counter anion however; the latter one is commonly observed [89, 100, 101]. As mentioned earlier, the defect-engineered δ-MnO_2 can be synthesized through hydrothermal reaction or annealing which shows electrocatalytic activity towards overall water splitting. The defect engineering enhances the electrical conductivity of MnO_2 through the formation of abundant metal atoms with higher valency [100, 101]. The electrocatalytic activity of $CoFe_2O_4$ tremendously increases after incorporation of surface defects [177]. The surface oxygen vacancy in $CoFe_2O_4$ is introduced by treating it with $NaBH_4$ solution. The incorporation of oxygen vacancy enhances its electron density near the Fermi level which reduces the band gap and facilitates the adsorption of proton and O-containing species. The defect-engineered $CoFe_2O_4$ achieves the benchmarking 10 mA cm^{-2} for overall water splitting on application of 1.53 V [177]. Similar to MnO_2, annealing is able to create O-vacancies in other metal oxides, e.g., Co_3O_4 and $CoMoO_4$ [117, 178]. This high-temperature strategy is able to enhance the specific surface area of Co_3O_4 up to 57.4 m^2 g^{-1}. The removal of O from Co_3O_4 delocalizes its electron on the adjacent metal atoms and creates defect states in between the band gap. These changes optimize the substrate adsorption and charge transfer efficiency of Co_3O_4 and it consequently achieves superior bifunctional activity [178]. Theoretical calculations conducted taking into account the $CoMoO_4$ suggest an interesting phenomenon. The electronic level that generates within the band gap due to the defect formation, serves as springboards and thereby facilitates the electronic transition [117]. Electrochemical corrosion and deposition techniques are also adopted to develop defect-engineered bifunctional electrocatalysts [179]. O-vacancy-rich NiFe hydroxide is directly fabricated on NF by electrochemically treating it with $FeCl_3$. Two ions present in the said salt play different roles in achieving the suitable bifunctional electrocatalytic activity. Cl^- ions enhance the amount of active sites by etching the NF and creating O-vacancies in the hydroxide. On the other hand, Fe^{3+} enhances its overall water-splitting activity by facilitating the OER. The O-vacancy incorporated NiFe hydroxide achieves 10 mA cm^{-2} for overall water splitting on the application of 1.7 V [179]. The metal vacancy is also able to enhance the electrical conductivity of a material and the amount of catalytically active site. As mentioned earlier, the Fe defect-engineered δ-FeOOH is able to achieve a current density of 10 mA cm^{-2} on application of 1.62 V. A Fe-B precursor is oxidized in the presence of air to create abundant metal vacancies [89]. Another strategy for creating a metal vacancy in a material is the alkali-etching of the precursor. Al is one of such metals which can be easily etched out from a compound with the help of a strong alkaline solution [180–182]. Al from solvothermally synthesized NiAl-LDH is etched and subjected to phosphorization to obtain defect-incorporated $NiAl_{\delta}P$ (Fig. 17a) [180]. The incorporation of defect regulates the surface electronic structure of NiAlP by increasing the electron density around the catalytically active Ni and P sites (Fig. 17b). These changes improve the proton as well as water adsorption efficacy of NiAlP (Fig. 17c). The $NiAl_{\delta}P$ achieves the benchmarking current density of 10 mA cm^{-2} at a cell voltage of 1.55 V in an alkaline medium [180].

The bifunctional electrocatalytic activity of doped transition metal compounds has been investigated in recent times. Various dopants like S, P, N, B, C, etc. are doped into the materials to tailor their activity and obtain this kind of engineered electrocatalysts [183–192]. N is doped into phosphides of Fe and Co through a high-temperature reaction inside a tube furnace [183, 184]. A stark difference in electrocatalytic activity is observed between P-doped Co foam and CoP nanorod [183]. Doping N into the CoP enhances the polarity of the Co-P bond and a δ- charge is created on the P which acts as the proton binding site. A d-p mixing between the N 2p orbital and Co 3d orbital occurs which enhances the electron density at the Fermi region and facilitates the conversion of CoP into the corresponding oxyhydroxide. Consequently, the N-doped CoP shows efficiency towards overall water splitting in an alkaline medium and is evident to achieve 50 mA cm^{-2} on the application of 1.61 V [183]. A similar improvement in O-containing substrate adsorption efficiency is observed also for FeP resulting in enhanced OER catalytic activity [184]. The N dopants improve the HER catalytic activity of FeP by facilitating the water dissociation and proton adsorption process. The N-doped FeP requires 256 mV overpotential to attain the 10 mA cm^{-2} for HER and 440 mV to attain 100 mA cm^{-2} for OER [184]. Another pnictogen P is doped into chalcogenides of various transition metals, e.g., Ni_3S_2, Co_3O_4, etc. [185, 186]. The free energy change for the potential determining step in OER reduces from 1.87 to 1.63 eV on doping the Co_3O_4 with P (Fig. 17d, e) and consequently, its activity is enhanced. A similar improvement in HER catalytic

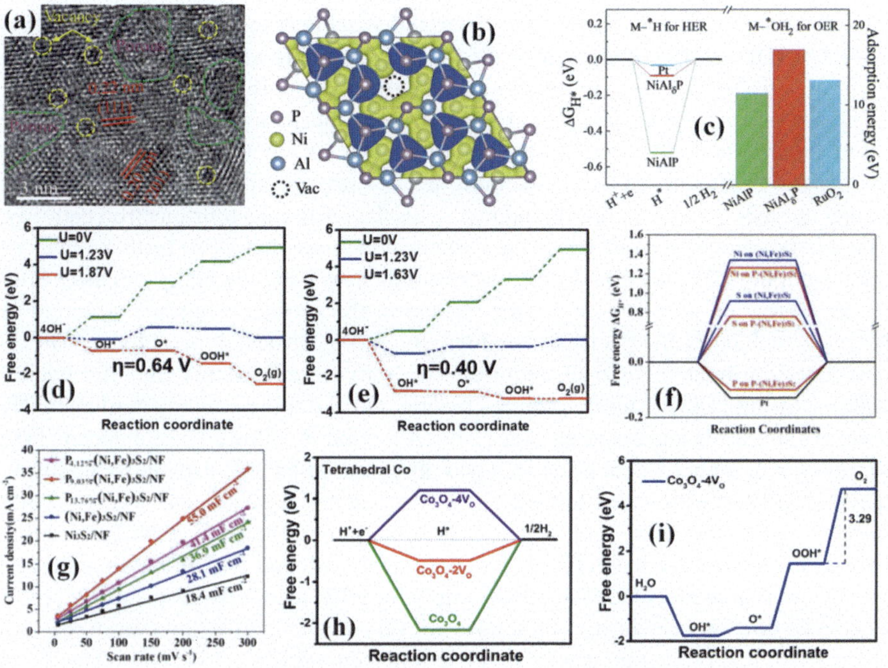

Fig. 17 **a** HR-TEM image of $NiAl_8P$ nanowall arrays; **b** theoretically calculated local charge density on different elements of $NiAl_8P$; **c** free energy diagram for HER and water adsorption on $NiAl_8P$, NiAlP, Pt, and RuO_2. **a–c** reprinted with permission from Cheng et al. [180]. Copyright (2018). The Royal Society of Chemistry. OER free energy diagram for **d** Co_3O_4, and **e** P-doped Co_3O_4. **d, e** reprinted with permission from Wang et al. [185]. Copyright (2018). American Chemical Society. **f** HER free energy diagram for different sites of P-doped and undoped (Ni, Fe)$_3S_2$; **g** representative plot showing the double layer capacitance of (Ni, Fe)$_3S_2$ with varied levels of P doping. (f and g) reprinted with permission from Liu et al. [187]. Copyright (2019). American Chemical Society. Free energy diagram for **h** HER on tetrahedral Co^{2+} sites, and (**i**) OER on octahedral Co^{3+} sites of Co_3O_4 with different levels of O-vacancies. **h, i** reprinted with permission from Yuan et al. [191]. Copyright (2020) Elsevier

activity of Co_3O_4 is also observed on P doping and the activity enhances with the increment in dopant concentration. The P-doped Co_3O_4 with optimum bifunctional activity achieves 10 mA cm^{-2} at an applied cell voltage of 1.63 V [185]. The P-doping enhances the proton adsorption efficacy of a material as evidenced from the theoretical calculations related to Ni_3S_2. In addition to that, the P doping also enhances the electrical conductivity and ECSA of Ni_3S_2. On optimum P-doping, Ni_3S_2 requires 101 and 256 mV overpotentials to achieve 10 mA cm^{-2} for HER and OER, respectively [186]. Some of the Ni sites in Ni_3S_2 are replaced with Fe to obtain (Ni, Fe)$_3S_2$ and thus a similar improvement in bifunctional activity on P doping is observed for the later one also. Similar to the Ni_3S_2, Gibb's adsorption-free energy change of (Ni, Fe)$_3S_2$ reduces, and ECSA enhances on P doping (Fig. 17f, g) [186, 187]. Similar to the N dopant, P is also evidenced to facilitate the water dissociation process and enhance the electron density near the Fermi level [183, 184, 187]. Overall water splitting can be driven by a commercial 1.5 V battery utilizing the P-doped (Ni, Fe)$_3$S$_2$ as a bifunctional electrocatalyst [187].

S is doped into $NiCo_2O_4$ via an ion exchange mechanism which enhances the bifunctional activity of the oxide [188]. The O vacancies generated in $NiCo_2O_4$ during the plasma treatment reduce the amount of Co^{3+} present in it. The Co^{3+} exposes a higher number of vacant orbitals and facilitates the OER via a four-electron pathway. The relative abundance of Co^{3+} in the vacancy containing $NiCo_2O_4$ further enhances after S-doping. The electron density in the conduction band gradually enhances after vacancy creation and S-doping. This change in the surface electronic structure of $NiCo_2O_4$ enhances its electrical conductivity and electrocatalytic activity. The S-doped plasma-treated $NiCo_2O_4$ requires 1.63 V to achieve 10 mA cm^{-2} for overall water splitting [188]. Anion vacancy creation and S doping can be simultaneously done in a single plasma treatment step [189]. A reduction followed by enhancement in the amount of higher valent Ni and Co was observed for NiCoP also due to the subsequent vacancy creation and doping. The P vacancy and the S dopant synergistically improve the conductivity and the substrate adsorption efficiency of NiCoP leading to a superior bifunctional activity. The optimized S-doped NiCoP requires 88 and 264 mV overpotentials to reach 10 mA cm^{-2} for HER and OER, respectively [189]. The metal phosphide synthesis process is not very cost-effective however; a scalable electrochemical procedure can be a viable replacement [183, 184, 189]. The S can also be doped during the formation of the phosphide film in the electrodeposition process [190]. The pulse electrodeposition technique is applied to fabricate S-doped Ni–P nanospheres over Cu substrate. The frequency of the deposition process plays a crucial role in the electrocatalytic activity of the fabricated electrode. The increment in frequency enhances its active surface area and wettability. The P present in the Ni–P and the S-dopant synergistically enhances its bifunctional electrocatalytic activity. The S-doped Ni–P requires 1.51 V to achieve an overall water-splitting current density of 10 mA cm^{-2} [190]. Among the previously mentioned dopants, the affectivity of B and C doping towards the bifunctional activity enhancement in an electrocatalyst is rarely investigated. The doping and defect engineering synergistically improves the electrocatalytic activity of a material and thus, surface defects are also engineered in addition to the doping of B or C [188, 189, 191, 192]. The Co_3O_4 is treated with $NaBH_4$ to engineer the O-vacancies and dope B into it [191]. The electronegative B redistributes the electron cloud in Co_3O_4 and facilitates the water dissociation followed by the proton adsorption step. On the other hand, the O vacancy significantly improves the charge transfer efficiency of Co_3O_4 by regulating its surface electronic structure. The tetrahedral Co^{2+} sites and octahedral Co^{3+} sites adjacent to the O vacancies act as HER and OER catalytically active sites, respectively (Fig. 17h, i). The B-doped and O vacancy-engineered Co_3O_4 require 184 and 315 mV overpotentials to achieve 50 mA cm^{-2} current density for HER and OER, respectively [191]. Thermal phosphorization of a PBA precursor not only leads to the formation of CoNiFe phosphide nanocubes but also engineers its surface defects and dope C into it [192]. The synergistic interactions among the elements present in the ternary phosphide result in superior bifunctional activity. The C-doping optimizes the electronic structure of the electrocatalyst and the defect engineering exposes a higher amount of catalytically active sites. The doped ternary phosphide with optimized physicochemical features achieves a current density of 10 mA cm^{-2} at 1.52 V [192]. Table 4 summarizes the activity of the recently reported bifunctional-engineered electrocatalysts for overall water splitting.

The defect or vacancy creation in an electrocatalyst enhances its electrical conductivity through the formation of defect states within its band gap. The vacancy generation also regulates its surface electronic structure and exposes a higher amount of catalytically active sites [177–179]. Doping of heteroatoms regulates the surface electronic structure of an electrocatalyst depending on the electronegativity of the dopant [183, 188, 191]. The heteroatom doping is also able to enhance the electrical conductivity and ECSA of a material [186, 187]. These changes in an electrocatalyst optimize the substrate adsorption and charge transfer efficiency of an electrocatalyst and consequently improve its bifunctional activity [186, 191]. The simultaneous doping and defect engineering synergistically help an electrocatalyst to achieve superior bifunctional activity [188, 189, 191, 192]. The strategies generally adopted to dope heteroatom into a material or create defects in it are hardly controllable.

Table 4 Recently reported bifunctional engineered electrocatalysts

Electrocatalyst	Overpotential @ 10 mA cm^{-2} for HER (mV)	Overpotential @ 10 mA cm^{-2} for OER (mV)	Cell voltage @ 10 mA cm^{-2} for overall water splitting (V)	Refs.
O-vacancy enriched $CoFe_2O_4$	121	280	1.53	[177]
Defect-rich Co_3O_4	-	256 (20 mA cm^{-2})	1.49 (20 mA cm^{-2})	[178]
$NiAl_\delta P$	80	240	1.55	[180]
N-doped CoP	100 (50 mA cm^{-2})	260 (50 mA cm^{-2})	1.61 (50 mA cm^{-2})	[183]
N-doped FeP	226	440 (100 mA cm^{-2})	1.72 (100 mA cm^{-2})	[184]
P-doped Co_3O_4	97	260 (20 mA cm^{-2})	1.63	[185]
P-doped Ni_3S_2	101	256	1.63	[186]
P-doped $(Ni, Fe)_3S_2$	98	196	1.54	[187]
S-doped Ni–P	55	229	1.51	[190]
S-doped defect-rich $NiCo_2O_4$	137	256	1.63	[188]
S-doped P-vacancy enriched NiCoP	88	264	1.60	[189]
O-vacancy engineered B-doped Co_3O_4	111	315 (50 mA cm^{-2})	1.67	[191]
Defect enriched P-doped CoNiFe phosphide	200.7	273.1	1.52	[192]

More sophisticated techniques should be adopted to develop bifunctional engineered electrocatalysts to achieve desirable activity.

6 Summary and Future Prospects

The chapter provides a basic idea about the bifunctional electrocatalyst for overall water splitting and also summarizes some recent advancements in that field. Monometallic electrocatalysts show sufficient activity; however, the incorporation of other metals and more than one anion further enhances it. The inclusion of metals with inherent electrocatalytic activity is more beneficial for that purpose. Incorporation of more than one metal in an electrocatalyst enhances the activity, however, it may sometimes affect adversely if a higher number of elements is incorporated. The heterostructure-based electrocatalysts show superior activity as compared to their bimetallic or multimetallic analogues. The heterointerface in any heterostructure plays a significant role in enhancing its electrocatalytic activity. The integration of one HER electrocatalyst with another OER catalyst is the most efficient technique for achieving superior bifunctional activity. The recently reported electrocatalysts have shown immense promise; however, still several investigations have to be conducted to come up with a viable replacement for noble-metal-based electrocatalysts. The following important points should be kept in mind before conducting future investigations in this field:

1. The choice of metal is an important aspect of electrocatalyst development. A balance between activity and natural abundance should be maintained to minimize the price-performance ratio of the electrocatalyst. The crustal abundance of Fe is the highest among all the transition metals utilized for developing electrocatalysts. However, the very least research efforts have been employed to develop a bifunctional monometallic electrocatalyst comprising Fe.
2. Compounds of transition metals like Mo, Cu, V, etc. have shown promise to be bifunctional electrocatalysts for overall water splitting in an alkaline medium. Those electrocatalysts need further modification to achieve sufficient activity and thus further exploration is required in that regard.
3. Most of the recently developed bifunctional electrocatalysts contain O, S, and P as anions. The electrocatalytic activity of an electrocatalyst is enhanced with the reduction in electronegativity of the counter anion. In that context, the activity of electrocatalysts containing anion other than the previously mentioned ones should also be explored greatly. Borides and carbides of transition metals show decent bifunctional activity and thus require further modification.
4. The doping and defect creation techniques are evidenced to improve the electrocatalytic activity of a material. The activity of the electrocatalysts engineered through these strategies should be investigated vastly to achieve the desirable performance. Most importantly, the synthetic techniques adopted in these

processes should be optimized in such a way that the level of doping and defect formation can be controlled up to a desirable limit.
5. The developed electrocatalysts eventually get converted into oxyhydroxide during OER catalysis and the electrocatalyst containing metal in its higher valence shows higher activity in that regard. The bifunctional activity of transition metal oxyhydroxides should be investigated vastly and required modifications should also be performed.
6. The integration of transition metal compounds having different anions helps the resulting heterostructure to achieve higher bifunctional activity. Moreover, the interfacial region of a heterostructure plays a significant role in achieving bifunctional activity. Precise but cost-effective synthetic strategies should be developed to tailor the activity of heterostructure-based electrocatalysts and achieve superior efficiency.
7. The development of bifunctional electrocatalysts with excellent efficiency totally depends on the understanding of the structural-activity relationship. More and more fundamental research should be conducted to understand that relationship.
8. The operational stability is an important feature of an electrocatalyst, which has been overlooked in most of the recent investigations. Heterostructure formation between transition metal compounds and carbonaceous materials has shown promise in that regard. More innovative strategies should be taken during the electrocatalyst development so that its stability can be enhanced significantly.
9. Finally, the synthetic techniques should be optimized in such a way that a large amount of cheap but efficient bifunctional electrocatalysts can be developed. The price-performance ratio of water electrolysis can only be reduced if a higher amount of superiorly active electrocatalysts is developed through a cost-effective synthetic procedure.

In conclusion, enormous investigations should be conducted in the near future keeping these points in mind to finally come up with the desired noble-metal-free bifunctional electrocatalyst for alkaline medium with excellent efficiency. It can be definitely said that these investigations would lead human civilization one step forward towards a cleaner, greener, and more sustainable future.

References

1. R. Newell, D. Raimi, G. Aldana, Global energy outlook 2019: the next generation of energy. Resources for the Future **1**, 8 (2019)
2. J. Wang, L. Feng, X. Tang, Y. Bentley, M. Höök, Futures **86**, 58 (2017)
3. L. Zhang, J. Xiao, H. Wang, M. Shao, ACS Catal. **7**, 7855 (2017)
4. J. Luo, J.H. Im, M.T. Mayer, M. Schreier, M.K. Nazeeruddin, N.G. Park, S.D. Tilley, H.J. Fan, M. Grätzel, Science **345**, 1593 (2014)
5. S.Z. Baykara, Int. J. Hydrogen Energy **43**, 10605 (2018)
6. M.T.M. Koper, Nat. Chem. **5**, 255 (2013)
7. N. Xu, G. Cao, Z. Chen, Q. Kang, H. Dai, P. Wang, J. Mater. Chem. A **5**, 12379 (2017)
8. S. Shit, S. Ghosh, S. Bolar, N.C. Murmu, T. Kuila, J. Electrochem. Soc. **167**, 116514 (2020)

9. P. Millet, in *Hydrogen Production: by Electrolysis*, Fundamentals of Water Electrolysis, ed. by A. Godula-Jopek (Wiley-VCH Verlag GmbH & Co. KGaA, 2015), pp. 33–62
10. J. Tian, Q. Liu, A.M. Asiri, X. Sun, J. Am. Chem. Soc. **136**, 7587 (2014)
11. Z. Ma, Y. Zhang, S. Liu, W. Xu, L. Wu, Y. C. Hsieh, P. Liu, Y. Zhu, K. Sasaki, J. N. Renner, K. E. Ayers, Radoslav R. Adzic, J. X. Wang, J. Electroanal. Chem. **819**, 296 (2018)
12. H.A. Bandal, A.R. Jadhav, A.H. Tamboli, H. Kim, Electrochim. Acta **249**, 253 (2017)
13. X. Zhao, X. Ma, Q. Lu, Q. Li, C. Han, Z. Xing, X. Yang, Electrochim. Acta **249**, 72 (2017)
14. J. Mohammed-Ibrahim, X. Sun, J. Energy Chem. **34**, 111 (2019)
15. M. Carmo, D.L. Fritz, J. Mergel, D. Stolten, Int. J. Hydrogen Energy **38**, 4901 (2013)
16. Q. Lei, B. Wang, P. Wang, S. Liu, J. Energy Chem. **38**, 162 (2019)
17. Y. Yan, B.Y. Xia, B. Zhao, X. Wang, J. Mater. Chem. A **4**, 17587 (2016)
18. Y. Wu, Y. Liu, G.D. Li, X. Zou, X. Lian, D. Wang, L. Sun, T. Asefa, X. Zou, Nano Energy **35**, 161–170 (2017)
19. S. Trasatti, J. Electroanal. Chem. **476**, 90 (1999)
20. E. Fabbri, T.J. Schmidt, ACS Catal. **8**, 9765 (2018)
21. K. Zeng, D. Zhang, Prog. Energy Combust. Sci. **36**, 307 (2010)
22. G. Chisholm, L. Cronin, in *Storing Energy*, ed. by T. M. Letcher (Elsevier Inc., 2016), pp. 315–343.
23. Y. Cheng, S.P. Jiang, Prog. Nat. Sci.: Mater. Int. **25**, 545 (2015)
24. N.T. Suen, S.F. Hung, Q. Quan, N. Zhang, Y.J. Xu, H.M. Chen, Chem. Soc. Rev. **46**, 337 (2017)
25. S.B. Adler, in *High-temperature Solid Oxide Fuel Cells for the 21st Century*, ed. by K. Kendall, M. Kendall (Elsevier Ltd., 2015) pp. 357–381.
26. M. Zeng, Y. Li, J. Mater. Chem. A **3**, 14942 (2015)
27. K. Li, Y. Li, Y. Wang, J. Ge, C. Liu, W. Xing, Energy Environ. Sci. **11**, 1232 (2018)
28. A. Serov, N.I. Andersen, A.J. Roy, I. Matanovic, K. Artyushkova, P. Atanassov, J. Electrochem. Soc. **162**, F449 (2015)
29. Y. Wang, B. Kong, D. Zhao, H. Wanga, C. Selomulya, Nano Today **15**, 26 (2017)
30. S. Jayabal, G. Saranya, J. Wu, Y. Liu, D. Geng, X. Meng, J. Mater. Chem. A **5**, 24540 (2017)
31. B.E. Conway, P.L. Bourgault, Can. J. Chem. **40**, 1690 (1962)
32. J.O.M. Bockris, J. Chem. Phys. **24**, 817 (1956)
33. T. Shinagawa, A.T. Garcia-Esparza, K. Takanabe, Sci. Rep. **5**, 13801 (2015)
34. C. Hu, L. Zhang, J. Gong, Energy Environ. Sci. **12**, 2620 (2019)
35. Z.W. Seh, J. Kibsgaard, C.F. Dickens, I. Chorkendorff, J.K. Nørskov, T.F. Jaramillo, Science **355**, 146 (2017)
36. C.G. Morales-Guio, L.A. Stern, X. Hu, Chem. Soc. Rev. **43**, 6555 (2014)
37. S. Anantharaj, S.R. Ede, K. Karthick, S.S. Sankar, K. Sangeetha, P.E. Karthik, S. Kundu, Energy Environ. Sci. **11**, 744 (2018)
38. I.C. Man, H.Y. Su, F. Calle-Vallejo, H.A. Hansen, J.I. Martínez, N.G. Inoglu, J. Kitchin, T.F. Jaramillo, J.K. Nørskov, J. Rossmeisl, ChemCatChem **3**, 1159 (2011)
39. Y. Lee, J. Suntivich, K.J. May, E.E. Perry, Y. Shao-Horn, J. Phys. Chem. Lett. **3**, 399 (2012)
40. S. Moon, Y.B. Cho, A. Yu, M.H. Kim, C. Lee, Y. Lee, A.C.S. Appl, Mater. Interfaces **11**, 1979 (2019)
41. Z. Chen, X. Duan, W. Wei, S. Wang, B.J. Ni, J. Mater. Chem. A **7**, 14971 (2019)
42. M.I. Jamesh, X. Sun, J. Power. Sources **40**, 31 (2018)
43. F. Lyu, Q. Wang, S.M. Choi, Y. Yin, Small **15**, 1804201 (2019)
44. F. Wang, T.A. Shifa, X. Zhan, Y. Huang, K. Liu, Z. Cheng, C. Jiang, J. He, Nanoscale **7**, 19764 (2015)
45. S. Gupta, M.K. Patel, A. Miotello, N. Patel, Adv. Funct. Mater. **30**, 1906481 (2020)
46. K.N. Dinh, Q. Liang, C.-F. Du, J. Zhao, A.I.Y. Tok, H. Mao, Q. Yan, Nano Today **25**, 99 (2019)
47. T.S. Light, S. Licht, A.C. Bevilacqua, K.R. Morash, Electrochem. Solid-State Lett. **8**, E16 (2005)
48. D.V. Esposito, Joule **1**, 651–658 (2017)

49. M. Schalenbach, G. Tjarks, M. Carmo, W. Lueke, M. Mueller, D. Stolten, J. Electrochem. Soc. **163**, F3197 (2016)
50. D. Pletcher, X. Li, Int. J. Hydrogen Energy **36**, 15089 (2011)
51. I. Vincent, D. Bessarabov, Renewable Sustainable Energy Rev. **81**, 1690 (2018)
52. B. You, Y. Sun, Acc. Chem. Res. **51**, 1571 (2018)
53. B.R. Wygant, K. Kawashima, C.B. Mullins, ACS Energy Lett. **3**, 2956 (2018)
54. X. Zou, Y. Zhang, Chem. Soc. Rev. **44**, 5148 (2015)
55. N. Shilpa, A. Nadeema, S. Kurungot, Chemsuschem **12**, 5300 (2019)
56. N.-U.-A. Babar, M.N. Asghar, F. Hussain, K.S. Joya, Mater. Today. Energy **17**, 100434 (2020)
57. X. Yu, Z.Y. Yu, X.L. Zhang, P. Li, B. Sun, X. Gao, K. Yan, H. Liu, Y. Duan, M.R. Gao, G. Wang, S.H. Yu, Nano Energy **71**, 104652 (2020)
58. S. Tang, X. Wang, Y. Zhang, M. Courté, H.J. Fan, D. Fichou, Nanoscale **11**, 2202 (2019)
59. M. Zhu, Z. Zhang, H. Zhang, H. Zhang, X. Zhang, L. Zhang, S. Wang, J. Colloid Interface Sci. **509**, 522 (2018)
60. S. Wan, W. Jin, X. Guo, J. Mao, L. Zheng, J. Zhao, J. Zhang, H. Liu, C. Tang, ACS Sustainable Chem. Eng. **6**, 15374 (2018)
61. L. Nisar, M. Sadaqat, A. Hassan, N.-U.-A. Babar, A. Shah, M. Najam-Ul-Haq, M.N. Ashiq, M.F. Ehsan, K.S. Joya, Fuel **280**, 118666 (2020)
62. X.H. Chen, Q. Zhang, L.L. Wu, L. Shen, H.C. Fu, J. Luo, X.L. Li, J.L. Lei, H.Q. Luo, N.B. Li, Mater. Today Phys. **15**, 100268 (2020)
63. N. Jiang, B. You, M. Sheng, Y. Sun, Angew. Chem. Int. Ed. **54**, 6251 (2015)
64. L. Yang, H. Qi, C. Zhang, X. Sun, Nanotechnology **27**, 23LT01 (2016)
65. X. Ma, K. Li, X. Zhang, B. Wei, H. Yang, L. Liu, M. Zhang, X. Zhang, Y. Chen, J. Mater. Chem. A **7**, 14904 (2019)
66. J. Masa, P. Weide, D. Peeters, I. Sinev, W. Xia, Z. Sun, C. Somsen, M. Muhler, W. Schuhmann, Adv. Energy Mater. **6**, 1502313 (2016)
67. T. Wang, Y. Zhang, Y. Wang, J. Zhou, L. Wu, Y. Sun, X. Xu, W. Hou, X. Zhou, Y. Du, W. Zhong, ACS Sustainable Chem. Eng. **6**, 10087 (2018)
68. F. Du, Y. Zhang, H. He, T. Li, G. Wen, Y. Zhou, Z. Zou, J. Power. Sources **431**, 182 (2019)
69. A. Chunduri, S. Gupta, O. Bapat, A. Bhide, R. Fernandes, M.K. Patel, V. Bambole, A. Miotello, N. Patel, Appl. Catal., B **259**, 118051 (2019)
70. Y. Rao, Y. Wang, H. Ning, P. Li, M. Wu, A.C.S. Appl, Mater. Interfaces **8**, 33601 (2016)
71. Z. Xing, L. Gan, J. Wang, X. Yang, J. Mater. Chem. A **5**, 7744 (2017)
72. X. Qiao, H. Kang, Y. Li, K. Cui, X. Jia, H. Liu, W. Qin, M. Pupucevski, G. Wu, A.C.S. Appl, Mater. Interfaces **12**, 36208 (2020)
73. A. Mondal, A. Paul, D.N. Srivastava, A.B. Panda, Int. J. Hydrogen Energy **43**, 21665 (2018)
74. Y. Yang, Y. Liang, M. Guo, T. Yu, K. Xu, L. Lu, C. Yuan, Int. J. Hydrogen Energy **46**, 50 (2021)
75. H. Liu, Q. He, H. Jiang, Y. Lin, Y. Zhang, M. Habib, S. Chen, L. Song, ACS Nano **11**, 11574 (2017)
76. L.L. Feng, G. Yu, Y. Wu, G.D. Li, H. Li, Y. Sun, T. Asefa, W. Chen, X. Zou, J. Am. Chem. Soc. **137**, 14023 (2015)
77. J.T. Ren, Z.Y. Yuan, ACS Sustainable Chem. Eng. **5**, 7203 (2017)
78. J. Zhang, Y. Wang, C. Zhang, H. Gao, L. Lv, L. Han, Z. Zhang, ACS Sustainable Chem. Eng. **6**, 2231 (2018)
79. U.D. Silva, J. Masud, N. Zhang, Y. Hong, W.P.R. Liyanage, M.A. Zaeem, M. Nath, J. Mater. Chem. A **6**, 7608 (2018)
80. H. Wang, J. Xiong, X. Cheng, M. Fritz, A. Ispas, A. Bund, G. Chen, D. Wang, P. Schaaf, A.C.S. Appl, Nano Mater. **3**, 10986 (2020)
81. G.F. Chen, T.Y. Ma, Z.Q. Liu, N. Li, Y.Z. Su, K. Davey, S.Z. Qiao, Adv. Funct. Mater. **26**, 3314 (2016)
82. Z. Pu, Y. Xue, W. Li, I.S. Amiinu, S. Mu, New J. Chem. **41**, 2154 (2017)
83. W.Z. Zhang, G.Y. Chen, J. Zhao, J.C. Liang, L.F. Sun, G.F. Liu, B.W. Ji, X.Y. Yan, J.R. Zhang, J. Colloid Interface Sci. **561**, 638 (2020)

84. M. Zhang, T. Wang, H. Cao, S. Cui, P. Du, J. Energy Chem. **42**, 71 (2020)
85. W. Yuan, X. Zhao, W. Hao, J. Li, L. Wang, X. Ma, Y. Guo, ChemElectroChem **6**, 764 (2019)
86. X. Li, Y. Fang, J. Wang, B. Wei, K. Qi, H.Y. Hoh, Q. Hao, T. Sun, Z. Wang, Z. Yin, Y. Zhang, J. Lu, Q. Bao, C. Su, Small **15**, 1902427 (2019)
87. W. Hao, R. Wu, H. Huang, X. Ou, L. Wang, D. Sun, X. Ma, Y. Guo, Energy Environ. Sci. **13**, 102 (2020)
88. B.C.M. Martindale, E. Reisner, Adv. Energy Mater. **6**, 1502095 (2016)
89. B. Liu, Y. Wang, H.Q. Peng, R. Yang, Z. Jiang, X. Zhou, C.S. Lee, H. Zhao, W. Zhang, Adv. Mater. **30**, 1803144 (2018)
90. S. Shit, W. Jang, S. Bolar, N.C. Murmu, H. Koo, T. Kuila, ChemElectroChem **6**, 3199 (2019)
91. X. Zou, Y. Wu, Y. Liu, D. Liu, W. Li, L. Gu, H. Liu, P. Wang, L. Sun, Y. Zhang, Chem **4**, 1139 (2018)
92. Z. Li, M. Xiao, Y. Zhou, D. Zhang, H. Wang, X. Liu, D. Wang, W. Wang, Dalton Trans. **47**, 14917 (2018)
93. S. Shit, S. Bolar, N.C. Murmu, T. Kuila, ACS Sustainable Chem. Eng. **7**, 18015 (2019)
94. C. Panda, P.W. Menezes, C. Walter, S. Yao, M.E. Miehlich, V. Gutkin, K. Meyer, M. Driess, Angew. Chem. Int. Ed. **56**, 10506 (2017)
95. G. Li, J. Yu, W. Yu, L. Yang, X. Zhang, X. Liu, H. Liu, W. Zhou, Small **16**, 2001980 (2020)
96. Y. Yan, B. Yu Xia, X. Ge, Z. Liu, A. Fisher, X. Wang, Chem. Eur. J. **21**, 18062 (2015)
97. H. Li, P. Wen, Q. Li, C. Dun, J. Xing, C. Lu, S. Adhikari, L. Jiang, D.L. Carroll, S.M. Geyer, Adv. Energy Mater. **7**, 1700513 (2017)
98. H. Shi, H. Liang, F. Ming, Z. Wang, Angew. Chem. Int. Ed. **56**, 573 (2017)
99. N. Suo, X. Han, C. Chen, X. He, Z. Dou, Z. Lin, L. Cui, J. Xiang, Electrochim. Acta **333**, 135531 (2020)
100. Y. Zhao, C. Chang, F. Teng, Y. Zhao, G. Chen, R. Shi, G.I.N. Waterhouse, W. Huang, T. Zhang, Adv. Energy Mater. **7**, 1700005 (2017)
101. H. Liao, X. Guo, Y. Hou, H. Liang, Z. Zhou, H. Yang, Small **16**, 1905223 (2020)
102. J. Chen, Q. Zeng, X. Qi, B. Peng, L. Xu, C. Liu, T. Liang, Int. J. Hydrogen Energy **45**, 24828 (2020)
103. Q. Xiong, X. Zhang, H. Wang, G. Liu, G. Wang, H. Zhang, H. Zhao, Chem. Commun. **54**, 3859 (2018)
104. H. Wang, Y. Cao, C. Sun, G. Zou, J. Huang, X. Kuai, J. Zhao, Lijun Gao. Chemsuschem **10**, 3540 (2017)
105. J.D. Rodney, S. Deepapriya, M.C. Robinson, C.J. Raj, S. Perumal, B.C. Kim, S.J. Das, Int. J. Hydrogen Energy **45**, 24684 (2020)
106. X. Zhang, X. Cui, Y. Sun, K. Qi, Z. Jin, S. Wei, W. Li, L. Zhang, W. Zheng, A.C.S. Appl, Mater. Interfaces **10**, 745 (2018)
107. B. Chakraborty, R. Beltrán-Suito, V. Hlukhyy, J. Schmidt, P.W. Menezes, M. Driess, Chemsuschem **13**, 3222 (2020)
108. A. Maiti, S.K. Srivastava, A.C.S. Appl, Mater. Interfaces **12**, 7057 (2020)
109. A. Kumar, S. Bhattacharyya, A.C.S. Appl, Mater. Interfaces **9**, 41906 (2017)
110. X. Wang, Z. Li, D.Y. Wu, G.R. Shen, C. Zou, Y. Feng, H. Liu, C.K. Dong, X.W. Du, Small **15**, 1804832 (2019)
111. A. Karmakar, S.K. Srivastava, A.C.S. Appl, Energy Mater. **3**, 7335 (2020)
112. W. Liu, J. Bao, M. Guan, Y. Zhao, J. Lian, J. Qiu, L. Xu, Y. Huang, J. Qian, H. Li, Dalton Trans. **46**, 8372 (2017)
113. J. Bao, Z. Wang, J. Xie, L. Xu, F. Lei, M. Guan, Y. Huang, Y. Zhao, J. Xia, H. Li, Inorg. Chem. Front. **5**, 2964 (2018)
114. Y. Tang, Q. Liu, L. Dong, H. B. Wu, X.-Y. Yu, Appl. Catal., B **266**, 118627 (2020)
115. H.A. Bandal, A.R. Jadhav, A.H. Tamboli, J.M.C. Puguan, H. Kim, Electrochim. Acta **249**, 253 (2017)
116. A. Karmakar, S.K. Srivastava, A.C.S. Appl, Mater. Interfaces **9**, 22378 (2017)
117. K. Chi, X. Tian, Q. Wang, Z. Zhang, X. Zhang, Y. Zhang, F. Jing, Q. Lv, W. Yao, F. Xiao, S. Wang, J. Catal. **381**, 44 (2020)

118. L. Fang, Z. Jiang, H. Xu, L. Liu, Y. Guan, X. Gu, Y. Wang, J. Catal. **357**, 238 (2018)
119. A. Sivanantham, P. Ganesan, S. Shanmugam, Adv. Funct. Mater. **26**, 4661 (2016)
120. B.G. Amin, A.T. Swesi, J. Masud, M. Nath, Chem. Commun. **53**, 5412 (2017)
121. J. Yu, G. Cheng, W. Luo, J. Mater. Chem. A **5**, 15838 (2017)
122. W. Zhang, H. Zhang, R. Luo, M. Zhang, X. Yan, X. Sun, J. Shen, W. Han, L. Wang, J. Li, J. Colloid Interface Sci. **548**, 48 (2019)
123. S. Shit, W. Jang, S. Bolar, N.C. Murmu, H. Koo, T. Kuila, A.C.S. Appl, Mater. Interfaces **11**, 21634 (2019)
124. Y.-Y. Sun, M.-Y. Jiang, L.-K. Wu, G.-Y. Hou, Y.-P. Tang, M. Liu, Sustainable Energy Fuels **4**, 582 (2020)
125. D. Li, Y. Xing, R. Yang, T. Wen, D. Jiang, W. Shi, S. Yuan, A.C.S. Appl, Mater. Interfaces **12**, 29253 (2020)
126. B. Chang, J. Yang, Y. Shao, L. Zhang, W. Fan, B. Huang, Y. Wu, X. Hao, Chemsuschem **11**, 3198 (2018)
127. J. Li, G. Wei, Y. Zhu, Y. Xi, X. Pan, Y. Ji, I.V. Zatovsky, W. Han, J. Mater. Chem. A **5**, 14828 (2017)
128. K. Sun, K. Wang, T. Yu, X. Liu, G. Wang, L. Jiang, Y. Bu, G. Xie, Int. J. Hydrogen Energy **44**, 1328 (2019)
129. L. Zhang, X. Wang, A. Li, X. Zheng, L. Peng, J. Huang, Z. Deng, H. Chen, Z. Wei, J. Mater. Chem. A **7**, 17529 (2019)
130. Y. Lian, H. Sun, X. Wang, P. Qi, Q. Mu, Y. Chen, J. Ye, X. Zhao, Z. Deng, Y. Peng, Chem. Sci. **10**, 464 (2019)
131. N. Xu, G. Cao, Z. Chen, Q. Kang, H. Dai, P. Wang, J. Mater. Chem. A 5, 12379 (2017)
132. H. Yuan, S. Wang, X. Gu, B. Tang, J. Li, X. Wang, J. Mater. Chem. A **7**, 19554 (2019)
133. W. Hong, S. Sun, Y. Kong, Y. Hu, G. Chen, J. Mater. Chem. A **8**, 7360 (2020)
134. Z. Liang, Z. Yang, J. Dang, J. Qi, H. Yuan, J. Gao, W. Zhang, H. Zheng, R. Cao, Chem. Eur. J. **25**, 621 (2019)
135. W. Tang, X. Liu, Y. Li, Y. Pu, Y. Lu, Z. Song, Q. Wang, R. Yu, J. Shui, Nano Res. **13**, 447 (2020)
136. Y. Yang, Z. Lin, S. Gao, J. Su, Z.Y. Lun, G. Xia, J. Chen, R. Zhang, Q. Chen, ACS Catal. **7**, 469 (2017)
137. K. Hu, M. Wu, S. Hinokuma, T. Ohto, M. Wakisaka, J.I. Fujita, Y. Ito, J. Mater. Chem. A **7**, 2156 (2019)
138. F. Qin, Z. Zhao, M.K. Alam, Y. Ni, F. Robles-Hernandez, L. Yu, S. Chen, Z. Ren, Z. Wang, J. Bao, ACS Energy Lett. **3**, 546 (2018)
139. L. Huang, D. Chen, G. Luo, Y.R. Lu, C. Chen, Y. Zou, C.L. Dong, Y. Li, S. Wang, Adv. Mater. **31**, 1901439 (2019)
140. L. Han, L. Guo, C. Dong, C. Zhang, H. Gao, J. Niu, Z. Peng, Z. Zhang, Nano Res. **12**, 2281 (2019)
141. H. Qian, K. Li, X. Mu, J. Zou, S. Xie, X. Xiong, X. Zeng, Int. J. Hydrogen Energy **45**, 16447 (2020)
142. J. Bao, Z. Wang, J. Xie, L. Xu, F. Lei, M. Guan, Y. Zhao, Y. Huang, H. Li, Chem. Commun. **55**, 3521 (2019)
143. Y. Li, B. Huang, Y. Sun, M. Luo, Y. Yang, Y. Qin, L. Wang, C. Li, F. Lv, W. Zhang, S. Guo, Small **15**, 1804212 (2019)
144. D. Lim, S. Kim, N. Kim, E. Oh, S.E. Shim, S.-H. Baeck, ACS Sustainable Chem. Eng. **8**, 4431 (2020)
145. J. Hou, Y. Wu, S. Cao, Y. Sun, L. Sun, Small **13**, 1702018 (2017)
146. S. Wang, L. Xu, W. Lu, Appl. Surf. Sci. **457**, 156 (2018)
147. Y. Hou, M.R. Lohe, J. Zhang, S. Liu, X. Zhuang, X. Feng, Energy Environ. Sci. **9**, 478 (2016)
148. B. Wang, S. Jiao, Z. Wang, M. Lu, D. Chen, Y. Kang, G. Pang, S. Feng, J. Mater. Chem. A **8**, 17202 (2020)
149. H. Zhang, X. Li, A. Hähnel, V. Naumann, C. Lin, S. Azimi, S.L. Schweizer, A.W. Maijenburg, R.B. Wehrspohn, Adv. Funct. Mater. **28**, 1706847 (2018)

150. Y. Lin, Z. Yang, D. Cao, Y. Gong, CrystEngComm **22**, 1425 (2020)
151. F. Zhou, X. Zhang, R. Sa, S. Zhang, Z. Wen, R. Wang, Chem. Eng. J. **397**, 125454 (2020)
152. J. Liu, J. Wang, B. Zhang, Y. Ruan, H. Wan, X. Ji, K. Xu, D. Zha, L. Miao, J. Jiang, J. Mater. Chem. A **6**, 2067 (2018)
153. S. Shit, S. Chhetri, S. Bolar, N.C. Murmu, W. Jang, H. Koo, T. Kuila, ChemElectroChem **6**, 430 (2019)
154. S. Peng, L. Li, J. Zhang, T.L. Tan, T. Zhang, D. Ji, X. Han, F. Cheng, S. Ramakrishna, J. Mater. Chem. A **5**, 23361 (2017)
155. Z. Chen, W. Wang, S. Huang, P. Ning, Y. Wu, C. Gao, T.-T. Le, J. Zai, Y. Jiang, Z. Hua, X. Qian, Nanoscale **12**, 326 (2020)
156. Y. Wu, F. Li, W. Chen, Q. Xiang, Y. Ma, H. Zhu, P. Tao, C. Song, W. Shang, T. Deng, J. Wu, Adv. Mater. **30**, 1803151 (2018)
157. N. Zhang, J. Lei, J. Xie, H. Huang, Y. Yu, RSC Adv. **7**, 46286 (2017)
158. Q. Zhou, T.T. Li, J. Qian, Y. Hu, F. Guo, Y.Q. Zheng, J. Mater. Chem. A **6**, 14431 (2018)
159. Y. Gong, Z. Yang, Y. Lin, J.L. Wang, H. Pan, Z. Xu, J. Mater. Chem. A **6**, 16950 (2018)
160. P. Hu, Z. Jia, H. Che, W. Zhou, N. Liu, F. Li, J. Wang, J. Power. Sources **416**, 95 (2019)
161. Y. Wu, Y. Liu, G. D. Li, X. Zou, X. Lian, D. Wang, L. Sun, T. Asefa, X. Zou, Nano Energy 35, 161 (2017)
162. S. Shit, S. Chhetri, W. Jang, N. C. Murmu, Hyeyoung Koo, P. Samanta, T. Kuila, ACS Appl. Mater. Interfaces **10**, 27712 (2018)
163. C. Du, M. Shang, J. Mao, W. Song, J. Mater. Chem. A **5**, 15940 (2017)
164. Q. Xu, W. Gao, M. Wang, G. Yuan, X. Ren, R. Zhao, S. Zhao, Q. Wang, Int. J. Hydrogen Energy **45**, 2546 (2020)
165. S. Dutta, A. Indra, Y. Feng, H. Han, T. Song, Appl. Catal., B **241**, 521 (2019)
166. J. Guan, C. Li, J. Zhao, Y. Yang, W. Zhou, Y. Wang, G.-R. Li, Appl. Catal., B **269**, 118600 (2020)
167. H. Sun, C. Tian, G. Fan, J. Qi, Z. Liu, Z. Yan, F. Cheng, J. Chen, C.-P. Li, M. Du, Adv. Funct. Mater. **30**, 1910596 (2020)
168. L. Yang, R. Liu, L. Jiao, Adv. Funct. Mater. **30**, 1909618 (2020)
169. A. Saad, H. Shen, Z. Cheng, Q. Ju, H. Guo, M. Munir, A. Turak, J. Wang, M. Yang, A.C.S. Appl, Energy Mater. **3**, 1684 (2020)
170. S. Shit, S. Bolar, N.C. Murmu, T. Kuila, J. Energy Chem. **59**, 160 (2021)
171. A. Karmakar, S.K. Srivastava, J. Mater. Chem. A **7**, 15054 (2019)
172. U.P. Suryawanshi, U.V. Ghorpade, D.M. Lee, M. He, S.W. Shin, P.V. Kumar, J.S. Jang, H.R. Jung, M.P. Suryawanshi, J.H. Kim, Chem. Mater. **33**, 234 (2021)
173. M. Arif, G. Yasin, M. Shakeel, M.A. Mushtaq, W. Ye, X. Fang, S. Ji, D. Yan, J. Energy Chem. **58**, 237 (2021)
174. J. Rashid, N. Parveen, T. ul Haq, A. Iqbal, S. H. Talib, S. U. Awan, N. Hussain, M. Zaheer, ChemCatChem **10**, 5587 (2018)
175. J. Jiao, W. Yang, Y. Pan, C. Zhang, S. Liu, C. Chen, D. Wang, Small **16**, 2002124 (2020)
176. X. Zhang, Y. Ding, G. Wu, X. Du, Int. J. Hydrogen Energy **55**, 30611 (2020)
177. C. Guo, X. Liu, L. Gao, X. Ma, M. Zhao, J. Zhou, X. Kuang, W. Deng, X. Sun, Q. Wei, J. Mater. Chem. A **7**, 21704 (2019)
178. P. Yan, M. Huang, B. Wang, Z. Wan, M. Qian, H. Yan, T.T. Isimjan, J. Tian, X. Yang, J. Energy Chem. **47**, 299 (2020)
179. W.H. Lee, M.H. Han, U. Lee, K.H. Chae, H. Kim, Y.J. Hwang, B.K. Min, C.H. Choi, H.-S. Oh, ACS Sustainable Chem. Eng. **8**, 14071 (2020)
180. W. Cheng, H. Zhang, X. Zhao, H. Su, F. Tang, J. Tian, Q. Liu, J. Mater. Chem. A **6**, 9420 (2018)
181. Y. Wang, M. Qiao, Y. Li, S. Wang, Small **14**, 1800136 (2018)
182. Y. Sun, Y. Xia, L. Kuai, H. Sun, W. Cao, M. Huttula, A.-P. Honkanen, M. Viljanen, S. Huotari, B. Geng, Chemsuschem **12**, 2564 (2019)
183. Z. Liu, X. Yu, H. Xue, L. Feng, J. Mater. Chem. A **7**, 13242 (2019)

184. M. Yang, J.-Y. Xie, Z.-Y. Lin, B. Dong, Y. Chen, X. Ma, M.-L. Wen, Y.-N. Zhou, L. Wang, Y.-M. Chai, Appl. Surf. Sci. **507**, 145096 (2020)
185. Z. Wang, H. Liu, R. Ge, X. Ren, J. Ren, D. Yang, L. Zhang, X. Sun, ACS Catal. **8**(3), 2236 (2018)
186. W. He, D. Jia, J. Cheng, F. Wang, L. Zhang, Y. Li, C. Liu, Q. Hao, J. Zhao, Catal. Sci. Technol. **10**, 7581 (2020)
187. C. Liu, D. Jia, Q. Hao, X. Zheng, Y. Li, C. Tang, H. Liu, J. Zhang, X. Zheng, A.C.S. Appl, Mater. Interfaces **11**, 27667 (2019)
188. J.H. Lin, Y.T. Yan, T.X. Xu, C.Q. Qu, J. Li, J. Cao, J.C. Feng, J.L. Qi, J. Colloid Interface Sci. **560**, 34 (2020)
189. J. Lin, Y. Yan, T. Liu, J. Cao, X. Zhou, J. Feng, J. Qi, Int. J. Hydrogen Energy **45**, 16161 (2020)
190. M.A. Ashraf, Y. Yang, D. Zhang, B.T. Pham, J. Colloid Interface Sci. **577**, 265 (2020)
191. H. Yuan, S. Wang, Z. Ma, M. Kundu, B. Tang, J. Li, X. Wang, Chem. Eng. J. **404**, 126474 (2021)
192. W.-K. Gao, M. Yang, J.-Q. Chi, X.-Y. Zhang, J.-Y. Xie, B.-Y. Guo, L. Wang, Y.-M. Chai, B. Dong, Sci. China Mater. **62**, 1285 (2019)

Electrochemical Approach for Hydrogen Technology: Fundamental Concepts and Materials

Victor Márquez, Eva Ng, Daniel Torres, Carlos Borrás, Benjamín R. Scharifker, Franco M. Cabrerizo, Lorean Madriz, and Ronald Vargas

Abstract This chapter presents aspects related to the electrochemical approach to hydrogen technologies, considering key concepts that drive both the thermodynamic and kinetic phenomena of the redox processes involved. Strategies to improve surface processes on various electrode materials are considered. The fundamental approach to the development of applied technologies illustrates the impact on the environment and energy, as well as the role of related physicochemical processes.

Keywords Electrocatalysis · Hydrogen evolution reaction · Storage · Hydrogen transfer reaction · Fuel cells

1 Introduction

Molecular hydrogen (H_2) as an energy vector represents an option with zero greenhouse gas emissions. The H_2 molecule is also considered a key in the new technologies that society needs to materialize the aim of sustainable global development through alternative energies [1, 2]. However, advances in novel, and more efficient technological strategies to either efficiently generate, store, and convert this molecule are still a real need [2]. Even though this concept has become increasingly attractive lately, electrochemists have been studying this quintessential process for decades [1, 2]. Much of the acquired knowledge can be directed to the production,

V. Márquez · E. Ng · D. Torres · C. Borrás · B. R. Scharifker · L. Madriz · R. Vargas (✉)
Departamento de Química, Universidad Simón Bolívar (USB), Caracas, Venezuela, US
e-mail: falseronaldevb@gmail.com

F. M. Cabrerizo · L. Madriz · R. Vargas
Instituto Tecnológico de Chascomús (INTECH) - Consejo Nacional de Investigaciones Científicas y Técnicas (CONICET) / Universidad Nacional de San Martín (UNSAM), Provincia de Buenos Aires, Chascomús Argentina,, US

L. Madriz
Instituto de Investigaciones Fisicoquímicas Teóricas y Aplicadas (INIFTA), CCT-La Plata-CONICET, Universidad Nacional de La Plata (UNLP), La Plata Argentina, US

S. J. Ikhmayies (ed.), *Advances in Catalysts Research*, Advances in Material Research and Technology, https://doi.org/10.1007/978-3-031-49108-5_10

safe storage, and applications of H_2, mainly through oxidation–reduction (redox) reactions. Thus, the assertive linking of redox reactions to the most suitable materials defines the framework of fundamentals required to achieve the necessary developments involving the functionality of the related technology.

Redox reactions as electrochemical processes have a predominant place in energy conversion, both in living beings and in human-developed technologies. Electrochemical energy conversion is a cornerstone of today's technologies and is also essential to the sustainability of life on planet Earth. The oxygen reduction reaction (ORR) to water (Eq. 1) is a key for energy conversion [3]. It is crucial in cellular respiration, photocatalysis, and especially for the operation of fuel cells, which are battery-like devices capable of supplying energy, highlighting that these devices work with a constant supply of reagents.

$$O_2 + 4H^+ + 4e^- \rightarrow 2H_2O \tag{1}$$

On the other hand, an appropriate fuel is an energy carrier molecule such as hydrogen, which can be oxidized to produce water [2],

$$2H_2 + O_2 \rightarrow 2H_2O \tag{2}$$

This reaction, widely studied, occurs spontaneously and releases a considerable amount of energy although with slow kinetics, but in the presence of suitable catalysts can yield substantial power. If it is separated into two electrodes, one in which the oxidation of hydrogen to protons occurs on the anode,

$$2H_2 \rightarrow 4H^+ + 4e^- \tag{3}$$

and the reduction of oxygen to water taking place in the cathode (Eq. 1), then the overall reaction can be carried out with electron flow through an external circuit, which means that electrical work can be harvested as electricity. Figure 1 shows a diagram of a fuel cell based on a proton-conducting electrolyte, emphasizing that, the electron flow through the external circuit is possible due to the redox reactions on the electrodes and the ionic flux in the electrolyte. Further details about fuel cells will be commented on in Sect. 6.

The standard potential of this cell is 1.23 V, and therefore the oxidation of one mol (2 g) of H_2 in a fuel cell can produce, in principle, 2 × 96,500 Coulombs × 1.23 V = 237 kJ of energy. Thus, the specific energy of the hydrogen fuel cell is around 120 MJ/kg. In fact, H_2 as fuel provides nearly three times the energy content of gasoline, 44 MJ/kg [4].

Similar energy output possibilities exist with other fuels, e.g. methanol [5]. In the case of methanol, the following oxidation reaction can take place on the anode,

$$2CH_3OH + 2H_2O \rightarrow 2CO_2 + 12H^+ + 12e^- \tag{4}$$

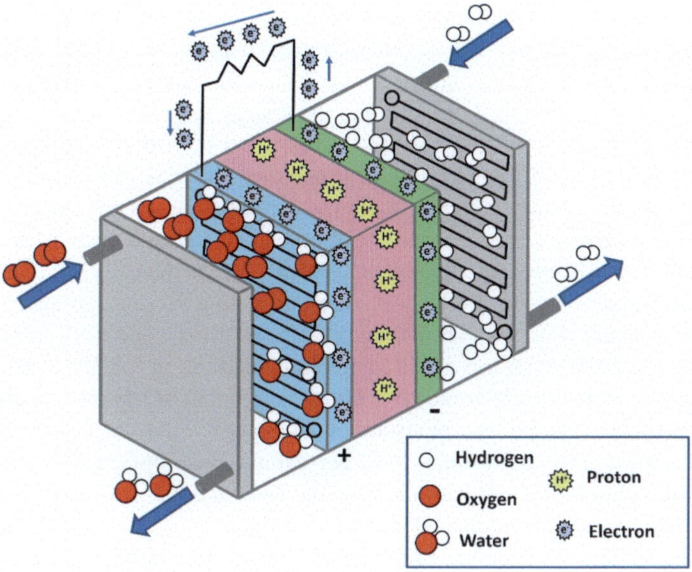

Fig. 1 Scheme of an H_2 fuel cell based on a proton-conducting electrolyte

while on the cathode, likewise in the hydrogen fuel cell, oxygen is reduced (Eq. 1). The standard potential for methanol reduction is -0.39 V, which, in addition to that of oxygen (1.23 V), accounts for the standard potential, 1.62 V, of the overall reaction of the methanol fuel cell,

$$2CH_3OH + 3O_2 \rightarrow 2CO_2 + 4H_2O \tag{5}$$

The oxidation of one mol of methanol (32 g) in a fuel cell could then, in theory, produce 6 × 96,500 Coulombs × 1.62 V = 938 kJ, for a specific methanol energy (energy density) of 29.3 MJ/kg.

The processes described open many possible pathways for electrochemistry to become a major tiebreaker when it comes to the development of sustainable energy-conversion processes, currently engaged by many researchers on a worldwide scale. Hydrogen can be used in fuel cells to power cars, trucks, ships, airplanes, electronic devices, and on a larger scale powering industries and cities. In this context, a key question arises: where will we get the H_2 from? Electrochemistry also helps us to build a proper response. As a start, H_2 can be produced electrochemically by water electrolysis, using electrical energy generated by hydroelectric, wind, or solar photovoltaic conversion [6]. Or it can be obtained directly through energy-conversion photoelectrochemical processes, as first described by Fujishima and Honda in the 1970s, who managed to separate oxygen and hydrogen from water by irradiating a TiO_2 photo-anode with sunlight [7]. After its suitable generation, the safe storage and use of hydrogen pose technological challenges to springboard its applications.

Modern society has intensified its activities and significantly increased its energy demands, thus the traditional way of generating and using energy has been negatively impacting the environment for some time due to the consumption of natural resources and the emissions generated. For example, the energy requirements of the World's population increased significantly in past years, from 93 million barrels of oil per day (92.96 mb/d) in 2015 to nearly 99 million barrels per day (98.85 mb/d) in 2018 [8]. This represents a yearly increase of almost 2 million barrels per day (1.96 mb/d), equivalent to an increase of more than 2 billion barrels in three years (2×10^9 mb). Thanks to combined efforts and agreements and also due to the COVID-19 pandemic, these numbers experienced a slight improvement, with a decrease of approximately 8 million barrels per day, with an average of almost ninety-one million barrels per day for the year 2020 (90.6 mb/d) but keeping an estimation for 2021 of ninety-six million barrels per day (96.58 mb/d) [9]. The development of advanced technologies capable of further reducing the environmental impact while satisfying the energy demand of a growing population, whose per capita consumption also increases as a function of technological development, is the fundamental challenge. The role of science in achieving these goals is decisive.

In all cases, through a combination of H_2 generation, safe storage and distribution, and the current approach to localized generation technology, the goal of meeting society's energy demands with sources other than fossil fuels is possible. Multidisciplinary research in both basic and applied sciences, especially chemistry, electrochemistry, materials science, physics, and engineering, will be able to continue bringing us closer to the realization of a sustainable world. Now, to meet this technological challenge, it is necessary to recognize that we must delve into the mechanisms and understanding of the related physicochemical processes. This is the purpose in this chapter, in which the basic aspects and concepts relevant to H_2-based electrochemical processes will be discussed.

2 Hydrogen-Based Electrochemical Technologies: An Overview

Figure 2 depicts a scheme of hydrogen-related technological processes. This can be divided into three main areas: ***generation***, ***safe storage***, and ***utilization***. To date, generation is based on hydrogen production, which can be obtained by water (ammonia/urea/others) electrolysis, with current efficiencies ranging near around 60–80% [2, 10]. Safe storage of H_2 can be done using materials that allow hydrogen to accumulate in the form of metal hydride. An advantage is that this process is reversible, so that accumulated hydrogen can be released when convenient. These processes have a current efficiency of around 90% [11]. Finally, the use of hydrogen to generate electricity in fuel cells has an efficiency of over 50% [12]. It should be noticed that the combustion of hydrogen produces water steam, so it is considered a

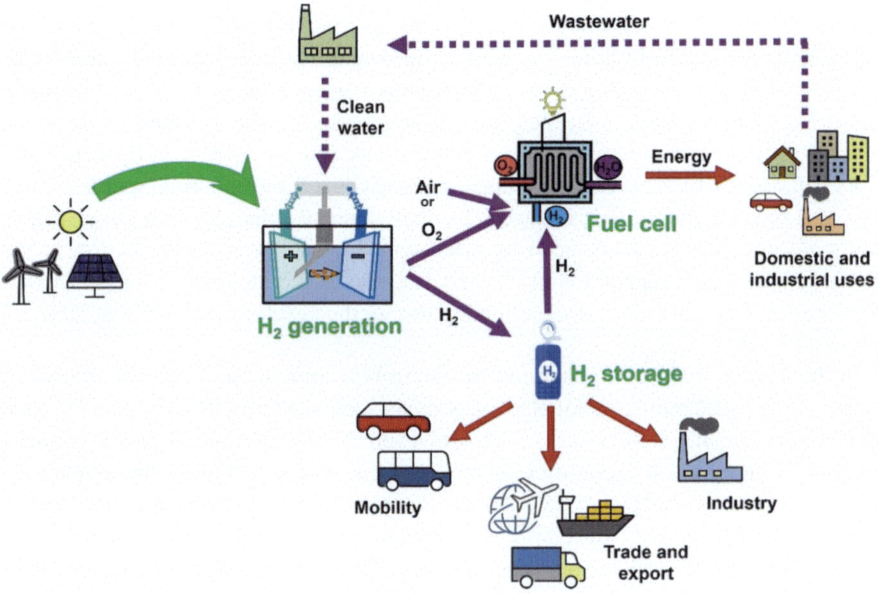

Fig. 2 Scheme of hydrogen-related technological processes

clean energy. In turn, the steam can be reused in the hydrolysis process, thus closing the cycle of a technologically interesting and convenient process.

Generation. Hydrogen generation by various pathways has been widely studied and its development has accelerated with the implementation of new international policies. In this regard, hydrogen is essential to support the World's commitment to reach carbon neutrality by 2050 and for the global efforts to implement the 2015 Paris Agreement [13]. Among the technologies that have seen the greatest growth are the electrolysis of water, photocatalytic generation, and the use of biological organisms [10]. Promising results have been reported on these different options, but they have in turn various complications.

In the case of the use of microorganisms, contaminated effluents can be exploited and during the purification process, electricity and hydrogen are produced as a by-product [14]. Also, these devices require relatively small installations, which favor localized energy production, thus allowing the independence of a centralized distribution network. However, a relatively stable and controlled environment is required to prevent loss of conditions, which influences the increase in maintenance and operation costs.

In the case of photocatalytic generation, the processes are much more robust and stable, so the maintenance conditions are not so strict, and can be used in conjunction with different cleaning applications of wastewater [15], a double benefit that makes it more attractive. Adding to the fact that its operation depends solely on solar energy, it has been a striking option in terms of research and development of new materials that are becoming more efficient. However, running in sunlight also involves disadvantages, *e.g.*, it does not allow constant energy or hydrogen generation due to the change in daylight hours, as well as the annual variability of the intensity of the solar radiation depending on the geographical area of the planet and/or climatic conditions. Additionally, most materials showing good results in terms of degradation rate or hydrogen generation capacity generally require activation by ultraviolet (UV) radiation, which accounts for approximately just 5% of the total incident radiation that reaches the Earth´s surface. The challenges of solar research are plentiful, new materials have recently been reported with the ability to use radiation in the visible range, which represents the largest fraction of the incident radiation (50%). The latter clearly represents an excellent option for the development of photocatalytic processes [16].

In the case of hydrogen generated by electrolysis in reactors with polymer electrolyte membranes, noble metal electrocatalysts such as Pt and Rh consistently have the best performance benchmarks [17]. However, the scarcity and high cost of these materials limit their use in commercial applications. As the cost of the anode/cathode can account for nearly 50% of the total stack cost [17, 18], the development of earth-abundant electrodes that can retain excellent catalytic properties is indispensable to make the overall process economically viable. The search to develop cheap, earth-abundant, high-performance materials for water splitting has become an essential topic of research over the last decades. Further complications related to the difficulty of producing efficient and stable catalysts over time, as well as the fact that the electrical energy used for this process usually comes from burning fossil fuels or using non-renewable energies altogether, are key challenges that are currently being addressed [2, 18]. Nonetheless, according to the economic and technical projections reported, these disadvantages would be overcome by the ability of constant gaseous hydrogen production, totally independent of the environmental conditions as well as the very low operational costs [18].

In addition, the use of electrolysis in reactors with solid oxide membranes that operate at high temperatures allows for defining higher efficiencies and lower losses due to the fast kinetics of the reactions. In fact, these electrolyzers work with electrodes made of non-noble materials, normally metal and ceramics containing nickel and zirconia are used as cathodes, and transition metal oxides are used as anodes, for example, perovskites. To review the details of some high-temperature electrolysis systems, the reader is encouraged to review the following literature [19, 20].

Storage. Once hydrogen is generated, its storage is a serious challenge. Traditional methods such as high-pressure cylinders or cryogenic tanks are not very convenient, especially for use in mobile devices [21] and due to hydrogen embrittlement of ferrous materials. Thus, this problem has been approached in different ways. The use of alloys or organic compounds that can store hydrogen and then release it in a controlled manner has been extensively studied in recent years [2, 11, 21, 22]. All this research has led to the development of different devices based on this technology. The advantage of using organic compounds in storage alloys lies mainly in the difference in weight, with alloys being much heavier than solutions based on organic compounds [22]. In both cases it is very important to have a catalytic material that can interact with hydrogen in ways that allow its absorption, in the case of alloys, or their transfer, in the case of organic compounds, and inhibit the evolution of hydrogen bubbles that are not stored and signify a loss of process efficiency. One of the difficulties usually encountered in the case of hydrogen alloys is the energy needed to extract it, as it is often necessary to heat the alloy materials to temperatures between 100 °C and 300 °C to obtain adequate flows, which is not the case for organic composite-based storage devices.

The storage through organic compounds can also be used for the generation of compounds whose energy content is higher, representing another form of energy use in chemical form or the synthesis of compounds of interest for different applications [22]. In this sense, efforts have been directed towards the area of biomass conversion. Compounds such as lignin, obtained as waste in the stationery industry, are a desired raw material to be used as biofuels, or as a source of aromatic compounds used in the chemical industry [23]. To this aim, homogeneous or heterogeneous hydrogenation processes have been used with quite promising results.

Utilization. As discussed in the introduction, H_2 in fuel cells is particularly interesting due to its energy density as a fuel. The history of fuel cells began almost two centuries ago with the first Schonbein's studies, which tested fundamental concepts, and in parallel, with the gaseous battery study conducted by Grove, whose results were published in 1843 [2, 18]. Different types of fuel cells are currently known, and some will be discussed in a later section of this chapter (6.- Fuel cells). Now, going back to the processes involved in Fig. 1, the electrochemical reaction between H_2 and O_2 produces electricity and water, making it a particularly clean process. The catalysis involved in the oxygen reduction reaction (ORR) and engineering designs for efficient devices were limited for decades [12]. Proton (H^+) reduction on Pt is particularly favored and there have been many reports on the use of functional electrocatalysts with representative efficiencies, however, still far from those obtained from using Pt [24]. The lower costs of these new materials, relative to Pt, make them very interesting alternatives for the design of possible massive-production strategies development. Today, there is a promising projection for the use of fuel cells, with some of them being already available on the market [25].

Fine chemistry-based processes such as the formation of C–C bonds through couplings between different compounds have been one of the main objectives of chemistry over the past 100 years [26, 27]. Currently, using reaction optimization based on Trost atomic efficiency [28], Wender step number [29], and Baran redox efficiency [30], green chemistry processes have been implemented. In general, these methodologies have caused a change in the traditional organic synthetic methods, being hydrogen transfer as a fundamental step throughout this process.

Hydrogenation or hydrogen transfer reaction (HTR) has previously been used to obtain hydrogenated compounds or to degrade polluting compounds [31]. Yépez and Scharifker [32, 33] have described hydrogen transfer as an effective method for strongly adsorbed CO desorption on palladium leading to ethanol, formaldehyde, and CO_2 formation, also demonstrating that the oxidation of formic acid (formate in basic medium) can be sustained indefinitely by displacement of adsorbed CO, via reaction with hydrogen. In any case, the micro-kinetic process defines the performance of the reactions [34]. Then, the development of theories and methodologies to determine the effect of different parameters such as the applied potential, concentration of the organic compound, or the type of catalyst used, is crucial to move more accurately in this field.

To further contribute to the understanding and implementation of processes based on hydrogen technology, it is necessary to define the conceptual basis of the physicochemical processes that take place during the reactions described above. The energetics or thermodynamics of each reaction allow us to understand why the processes occur, as well as to define the conditions that maximize the energy efficiency of each stage. Unraveling the kinetics of the chemical processes allows us to understand the interrelationship between physical variables involved in chemical processes, as well as to determine the appropriate time frames for each reaction. Thus, understanding the effect of the chemical structure of the electrocatalysts used in relation to their activity to promote chemical processes is essential for the design of new and adequate materials to achieve the conversion goals, bringing with it the correct generation of renewable energy. The following sections will address the basic physicochemical aspects related to these processes: hydrogen ***generation***, ***storage***, and ***utilization***. Key chemical reactions will be collected in each case and results of particular interest will be discussed.

3 Hydrogen Evolution Reaction

The hydrogen evolution reaction (HER) has a special place in the history of electrochemistry; is considered one of the simplest and most studied reactions in electrochemistry and has remained an active topic of research for more than a century. R. de Levie summarized the contribution of various scientists in the early days of such an interesting reaction [35]. However, the first reports with a modern description of a reaction where hydrogen evolves from an aqueous solution were presented by Tafel in 1905 [36]. Since then, important contributions have been made from both a fundamental and an applied perspective, bringing HER to the archetypal state for the study of electrochemical reactions. Its technological importance, being the cathodic reaction in the electrolysis of water or a competitive reaction in the electrocrystallization of many metals, is only comparable to its fundamental interest since the reaction has also been a cornerstone in developing the fundamental concepts of electrocatalysis, such as the Sabatier principle [37, 38].

The "simplest" electrocatalytic reaction is found to be unexpectedly complicated. As in any heterogeneous process, a relationship between the properties of the catalyst and the kinetics of the reaction is expected. To establish such a link, a traditional approach is to build a theoretical–experimental framework to predict activity based on descriptors that help to estimate the electrocatalytic performance of the electrode material. Usually, the descriptor should be related to the rate-limiting step (RLS) of the reaction. The ability to choose/identify the appropriate descriptors is based on the mechanistic understanding of the reaction, which has evolved in the last decades. Several frameworks have emerged to try to establish design guidelines to address the question: *what physicochemical principles should be taken into account to improve the performance of HER electrocatalysts?* Now, to answer this question, the physicochemical aspects that take place during the occurrence of electrochemical phenomena related to HER will be addressed.

3.1 Kinetics of the Hydrogen Evolution Reaction

Since the pioneering works of Tafel [36], Heyrovsky [39], and Volmer [40], HER can today be considered a well-studied reaction. In fact, its mechanism and behavior on many materials has been widely characterized in the annals of electrochemistry [2, 17, 18, 34, 41]. In 1928, Bowden and Rideal [42] recognized the formation of an adsorbed layer of hydrogen (H_{ads}) during the HER, with the H_{ads} coverage being dependent in some way on the nature of the surface and the electrode potential. In any case, the most accepted description today was developed by Volmer [40, 41], who proposed electroadsorption as the first step of the HER, where the solvated protons in an acidic medium (or water in a basic medium) react with electrons to form a neutral species H_{ads} adsorbed on a free site on the surface:

$$H_3O^+ + e^- = H_{ads} + H_2O \quad \text{(low pH)} \tag{6a}$$

$$H_2O + e^- = H_{ads} + HO^- \quad \text{(high pH)} \tag{6b}$$

Upon adsorption, hydrogen must react to form H_2. By 1930, Kobosew and Nekrassow [43] proposed for the first time an electrochemical desorption step, where the neutral species formed in the first step (H_{ads}) interact with proton (low pH) or water (high pH) from solution, to form molecular hydrogen. This step was proved and extensively elucidated by Heryovsky [39, 41] and is now known by his name:

$$H_3O^+ + e^- + H_{ads} = H_2 + H_2O \quad \text{(low pH)} \tag{7a}$$

$$H_2O + e^- + H_{ads} = H_2 + HO^- \quad \text{(high pH)} \tag{7b}$$

On the other hand, Tafel suggested the possibility of a different mechanism [36, 41], where desorption does not involve a charge transfer process and is independent of the pH. The latter chemical step is known as the Tafel recombination mechanism, where two molecules adsorbed on the surface interact and combine to form H_2:

$$2H_{ads} = H_2 \tag{8}$$

The existence of the adsorbed state on the surface opens the possibility of alternative reaction pathways [44]. The mechanisms widely reported in the literature for HER in aqueous environments are Volmer-Tafel and Volmer-Heryovsky [18, 34, 41,

44]. In any case, the reaction described in Eqs. (6) to (8) can have a decisive impact on the overall kinetics.

In the stationary state, the overall reaction proceeds at the rate-limiting step (RLS). For a process whose determining step is Volmer's electroadsorption at low pH, it is possible to write the rate law as [45]:

$$r_v = k_v C_{H^+}(1-\theta) \tag{9}$$

where r_v is the reaction rate, k_v the rate constant (Volmer reaction), C_{H+} the proton concentration and θ is the surface coverage with species H_{ads}. According to Faraday's laws, the reaction rate (r) can be related to the current density that crosses the interphase (j), and according to Butler and Volmer's theory, the rate constant (k) depends exponentially on the potential (E) [34, 45],

$$j = zFr \tag{10}$$

$$k = k^0 exp\left\{\frac{\left[zF\alpha\left(E - E^0\right)\right]}{RT}\right\} \tag{11}$$

where z is the number of electrons transferred, F the Faraday constant, α the electronic transfer coefficient, E^0 the reversible potential, k^0 the rate constant at $E = E^0$, R the universal constant of gases, and T the absolute temperature. From Eqs. (10) and (11), it is then possible to express the rate laws during HER. If electroadsorption is RLS and the steps that occur after are very fast, then it follows that the adsorbed species quickly disappear from the surface and $(1-\theta) \approx 1$. Therefore,

$$j = Fk_v C_{H^+} \tag{12}$$

and

$$\ln(j) = \ln\left(Fk_v^0\right) + \ln(C_{H^+}) - (\alpha F/RT)E \tag{13}$$

It is then observed that the resulting current is of the first order with respect to the concentration of protons C_{H+}, and for $\alpha = 0.5$ and T = 298 °C, the Tafel slope $\partial(E)/\partial[\log(j)]$ yields $\approx$ 120 mV.dec^{-1}.

Now, for a mechanism where the RLS is Heyrovsky's electrochemical desorption, involving the transfer of 2 electrons, acidic pH, and a high coverage θ, the rate law is elucidated under the stationary state approximation $d\theta/dt = 0$ and the current density is expressed as [45]

$$j = 2Fk_h\theta^2 \tag{14}$$

and

$$\ln(j) = ln\left[2Fk_h^0(k_v'/k_v)^2\right] + 2\ln(C_{H^+}) - (2F/RT)E \quad (15)$$

where k_h the rate constant (Heyrovsky reaction) and k_v is the rate constant of the reverse reaction to Volmer in equilibrium condition. For this mechanism, after using $\alpha = 0.5$ and T = 298 ºC, a Tafel slope of 30 mV dec^{-1} is achieved. In addition, the current density varies with the square of the proton concentration.

On the other hand, when the consumption of H_{ads} is limited by a chemical reaction step, Tafel's reaction (8) is decisive. Thus, the rate of the reaction depends on both the adsorbed hydrogen coverage and the proton concentration in the solution,

$$r_t = k_t C_{H^+}\theta \quad (16)$$

where k_t the rate constant (Tafel reaction) and with similar arguments, the logarithm of the current density is expressed as:

$$\ln(j) = ln\left[2Fk_h^0(k_v'/k_v)^2\right] + 2\ln(C_{H^+}) - [(1+\alpha)F/RT]E \quad (17)$$

Assuming again that $\alpha = 0.5$ and T = 298 ºC, the Tafel slope is 40 mV dec^{-1}.

The above development shows that different mechanisms can lead to different values on the Tafel slope. However, this experimental observation alone is not a complete diagnosis of the reaction mechanism since different routes can proceed with the same RLS. For example, the Volmer electroadsorption step is common for both mechanisms and leads to the same values on the slope. The development of the rate laws at basic pH is analogous. A more rigorous discussion of the kinetics should be based on a broader body of consistent evidence [34]. Figure 3 illustrates the reaction pathways described for the HER.

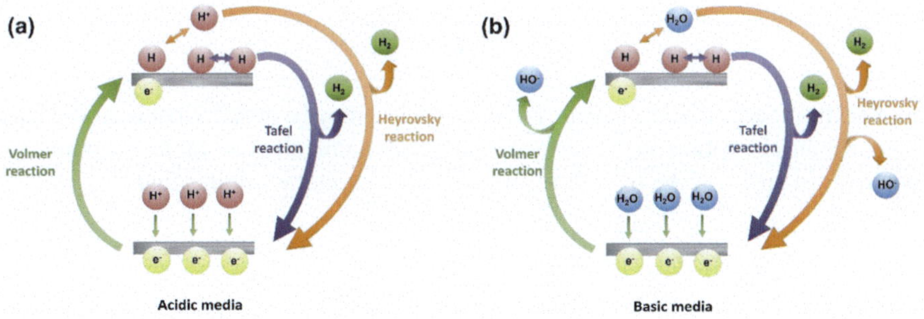

Fig. 3 HER reaction pathways in **a** Acidic media and **b** Basic media

3.2 Thermodynamic Considerations of the Hydrogen Evolution Reaction

The global reaction of hydrogen evolution at low pH is a classic example of a multi-step reaction involving the transfer of multiple electrons [46],

$$2H_3O^+ + 2e^- = H_2 + 2H_2O \tag{18}$$

whose equilibrium potential is $E^0{}_{H+/H2} = 0$ V. While reaction (18) may proceed by different mechanisms, all must be subject to the same thermodynamic constraints in the sense that, under equilibrium conditions, it must be fulfilled that

$$(E^0_{H^+/H_{ads}} + E^0_{H_{ads},\ H^+/H_2})/2 = E^0_{H^+/H_2} = 0 \tag{19}$$

The equilibrium condition for a step like Volmer's equation (6) must recognize the fact that it is an electrochemical process, with counterbalanced components accounting for the chemical and electrostatic contributions:

$$\Delta G^0(H_{ads}) = \Delta G(H^+ + e^-) \tag{20}$$

where G is the Gibbs free energy, G^0 refers to this energy in the standard condition, and $\Delta G^0(H_{ads})$ is the change of adsorption energy of H_{ads}. If the energy that binds the adsorbed intermediary H_{ads} to the surface does not depend heavily on the potential, i.e., the nature of the interactions is mainly chemical with a negligible electrostatic contribution, then the changes in potential should only be reflected in the energy change of electrons [46, 47]. Thus, the equilibrium potential for Volmer's step is given by:

$$E^0_{H^+/H_{ads}} = -\Delta G^0(H_{ads})/e \tag{21}$$

where e is the charge of the electron. From Eq. (21) it may be noticed that if $\Delta G^0(H_{ads}) < 0$, i.e., when adsorption on the surface is thermodynamically favorable, the potential is positive. In terms of the overall reaction, such a situation implies that

$$E^0_{H^+/H_{ads}} = -E^0_{H_{ads},H^+/H_2} < 0 \tag{22}$$

This means that the desorption reaction will have a negative potential in equilibrium (referred to as the same standard state) and therefore requires energy to occur. It is then understood that very high adsorption energies prevent the release of the product from the reaction, while very low energies lead to very slow reaction rates [44, 46, 47]. Therefore, both cases lead to poor catalytic performance.

Thus, under the conditions where $\Delta G^0(H_{ads}) < 0$ there is an energy barrier in the global reaction to be able to produce hydrogen, and this is reflected as a thermodynamic overpotential, $\eta = E - E_0$, which under strict energy considerations can be defined as the equilibrium potential of the thermodynamically less-favored step, $E^0{}_{\text{Hads,H+/H2}} - E^0{}_{\text{H+/H2}}$. Since the lowest possible thermodynamic overpotential is zero, it follows that the ideal surface should satisfy the condition:

$$\eta = E^0_{H^+/H_{ads}} = E^0_{H_{ads},H^+/H_2} = \Delta G^0(H_{ads})/e = 0 \tag{23}$$

As set out in Eq. (23), it is necessary to establish a compromise between the adsorption and desorption energies of H_{ads}. In fact, for an efficient process, a condition of thermoneutrality, $\Delta G^0(H_{ads}) = 0$, is necessary [43]. This trend of engagement between two opposing forces is what is known as the Sabatier principle [37, 38, 45, 46], which allows to drawing of linear Gibbs free energy relationships that in the case of HER typically yield volcano-like curves (volcano plots) [46, 48].

A volcano plot is a graph of a kinetic variable [$-log(j_0)$] as a function of a thermodynamic variable (ΔG or M-H bond strength) acting as a descriptor, (i.e., it provides an overview but essential description of surface properties). j_0 is the exchange current density at overpotential equal to zero ($\eta = E - E_0 = 0$). These plots were studied and reproduced extensively by Trasatti in the 1970s [41, 49], who rationalized all kinetic studies in terms of these linear relationships, and found that the analysis and predictions obtained were quite satisfactory, thus establishing a reference point for studying HER in a new family of catalysts. For example, Fig. 4 illustrates a volcano plot for HER in metals, highlighting Pt as the best electrocatalyst at the top of the volcano [41, 50].

The apparent activity of a cathode for HER is generally characterized by an overpotential at a given cathodic current density of 10, 100, or 250 mA cm^{-2} [34]. It should be noted that the exchange current densities are highly dependent on the Tafel slope and can only be compared when the Tafel slopes are equal [34, 45]. There are current reports in the literature that have made comparisons between different Earth-abundant electrocatalysts at a particular overpotential, such as -10 or -100 mV

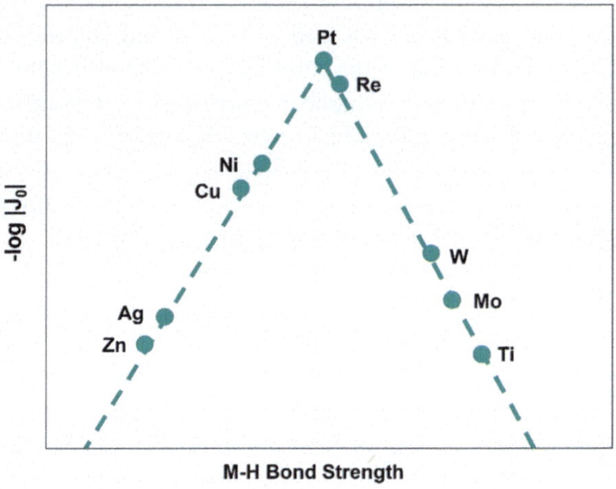

Fig. 4 Illustration of a volcano plot for HER in metals

[24]. In all cases, depending on the properties of the cathodes evaluated for HER, it is necessary to verify that the reliability of the current data *vs.* potential is not compromised by uncompensated resistances [34, 51]. It will be further commented that the arguments described here are consistent with the microkinetic aspects of the reaction mechanisms described for HER [34]. This framework established a structure-dependent descriptor: $\Delta G^0(H_{ads})$, that has a meaningful impact on the RLS of the reaction. It has served not only to understand but to anticipate activity, based on the current understanding of the reaction mechanism.

Thereafter, the rational design and screening of new (electro)catalysts have traditionally relied on the reaction energetics, using $\Delta G^0(H_{ads})$ or Tafel slope as a common benchmark of catalytic activity. Using this conceptual framework and looking at other processes in nature, researchers have been able to find and screen new non-noble alternative materials. Indeed, the search for materials from abundant sources that have interesting catalytic activity for HER, remains a very dynamic and open field of research. Metal phosphides at acidic pH [24] and Ni-Mo alloy at basic pH [52] are cathodes based on non-noble materials that continue to show representative yields of practical interest.

3.3 *Electrode Materials*

Platinum has the highest performance towards HER according to multiple reports in the literature [2, 41, 44, 46, 50]. Pt has high catalytic activity and durability under HER operating conditions since this metal can satisfy the thermoneutrality condition to the best degree. However, its high cost due to its low abundance limits its applicability, motivating the search for more affordable alternatives such as other transition metals [24]. A relatively recent trend to develop high-performance metals is the combination of metal and nonmetals on binary or multicomponent systems, the metal-nonmetal combination provides a bifunctional interface that can outperform their metal counterparts and exhibits good stability and electrical conductivity in different conditions, thus creating a suitable chemical environment for the HER. The ability of nonmetal elements to form hydrogen bonds, due to their electronegativity, can significantly affect the kinetics of HER [53]. In this sense, some researchers have looked for inspiration in nature to find better catalysts. For example, hydrogenases are biological catalysts that promote reversible proton reduction in complex systems. These enzymes often have metal-sulfur groups as active sites embedded in a complex biological cavity providing a suitable chemical environment for proton reduction [54]. By virtue of this, extensive studies have been developed to try to mimic this type of active site, reporting interesting results on MoS_2 electrodes [54–56].

The work of Hinnermann and collaborators shows that the active MoS_2 sites are very different from the sites of a conventional Mo-metal alloy. The chemical environment mimics the proton adsorption like that of enzymes [54, 55]. MoS_2 is among the most used catalysts for hydrodesulfurization [57], a process where sulfur impurities are removed from hydrocarbons. It has been shown through computational simulations that both processes, hydrodesulfurization, and hydrogen evolution, despite being chemically distinct, are in fact regulated by the binding energy of H_{ads} on the surface [54–56]. Given that the best catalysts for hydrodesulfurization should be able to satisfy the same requirement of thermoneutrality $\Delta G^0(H_{ads}) = 0$, several research groups have proposed to study these as possible candidates to find suitable catalysts for HER [54–57]. By the use of this conceptual framework and looking at other processes in nature, researchers have been able to find and evaluate new alternative non-noble materials. Some of the most interesting contributions of the last decades include the combination of metallic and non-metallic catalysts, for example, it was found that Ni_2-P mimics the properties of HER like those of [NiFe] hydrogenase [58]. The promising activity was related to an ensemble effect, in which the number of active metal sites decreases due to the presence of the non-metal, leading to a more moderate binding energy of the intermediate. Similarly, the identification of MoS_2 [59], and Mo_2C [60] as low-cost active catalysts towards HER was possible. Non-metallic sites actively participate in the reaction and can be tuned to the electronic structure of nearby metal sites [61].

In the case of multicomponent materials, the appropriate structure-dependent descriptors are not straightforward and are a current topic of discussion in electrocatalysis [62]. Interesting activities derived from the synergistic interaction between the active elements are expected, as documented in cathodes based on high entropy alloys [63]. These new materials not only provide viable alternatives but also open new questions and expose gaps in our fundamental understanding of the most studied reactions in electrochemistry. Now, due to its conceptual simplicity, the Nørskov and coworkers [64, 65] framework remains highly relevant and widely implemented in the rational design of new HER catalysts and trend studies, where the calculation of first principles is rigorously developed to obtain the energy contributions of the processes that take place in the electrode|electrolyte interface. However, the incorporation of other phenomena such as solvent rearrangement effects from a microkinetic perspective is relevant when it comes to going beyond adsorption descriptors and establishing new design guidelines [66–68].

Table 1 shows some electrode materials with outstanding performance for HER, reporting the values of the Tafel slope, the current density at overpotential equal to zero (j0) and the overpotential measured at 10 mA/cm^2 (η_{10}). Now, if the reader is interested in knowing the performance of various electrode materials, it is recommended to consult the following research and the references therein: the work of Trasatti [41] to identify materials based on iron and mild steel, nickel, interstitial compounds as carbides, sulfides, and phosphides, the work of Pletcher et al. [52] for alloy cathode considerations, the researches of Lasia [34] and Yu et al. [24] for Earth-abundant electrocatalysts, and the work of Zeng and Li [69] for cathodes based on alloys, chalcogenides, carbides, nitrides, borides, phosphides, and metal-free catalysts. As can be seen in Table 1, metal phosphides emerge as cathodes with an interesting behavior, since η_{10} is relatively low compared to noble metals. For this reason, some aspects of metal phosphide cathodes will be discussed.

Table 1 Tafel slope (b), current density at zero overpotential (j_0), and overpotential at 10 mA/cm^2 (η_{10}) for electrode materials with outstanding performance for HER

Electrode	Conditions	Tafel slope (b, mV/dec)	j_0 (mA/cm^2)	j_{10} (mV)	Refs.
Pt polycrystalline	0.5 M H_2SO_4, 23 °C	36	3.20	18*	[70]
Pt polycrystalline	0.5 M NaOH, 23 °C	75	0.31	113*	[70]
Ir	0.5 M H_2SO_4, 25 °C	35	0.25	56*	[71]
Ni polycrystalline	50% KOH, 80 °C	140	0.11	274*	[72]
Ni/NiO on Ni foam	1.0 M KOH	43	-	145	[73]
WO_3 on GCE**	0.5 M H_2SO_4	38	-	38	[74]
Fe_2O_3/Fe@CN on GCE	1.0 M KOH	114	-	330	[75]
NiMo	1 M NaOH, 25 °C	132	0.08	278*	[76]
Ni_2Mo	1 M NaOH, 25 °C	142	0.07	306*	[76]
Ni_3Mo	1 M NaOH, 25 °C	148	0.03	369*	[76]
$Ni_{60}Mo_{40}$	30% KOH, 70 °C	107	0.04	254*	[77]
$LaPO_4$ bonded Rh (3.5%) on Ni	1 M KOH, 25 °C	77	2.70	44*	[78]
MoS_2/graphene hybrid	0.5 M II_2SO_4	41	-	140	[79]
WC nanoparticles	0.5 M H_2SO_4	84	-	125	[80]
Mo_2C/graphene hybrid	0.5 M H_2SO_4	54	-	175	[81]
CoNiB on GCE	0.5 M PBS	51	-	170	[82]
Ni–P on carbon fiber	0.5 M H_2SO_4	59	-	98	[83]
Ni_xP on Ni foam	0.5 M H_2SO_4	39	-	90	[84]
Ni_3PMPs on Ti foil	0.5 M H_2SO_4	177	-	65	[85]
CoP NPs on Ti foils	0.5 M H_2SO_4	50	-	75	[86]
Co_2P NPs on Ti foils	0.5 M H_2SO_4	45	-	95	[86]
FeP on FTO***	0.5 M H_2SO_4	79	-	138	[87]
Fe_2P on FTO	0.5 M H_2SO_4	66	-	83	[87]
MoP NPs on Ti foil	0.5 M H_2SO_4	45	-	90	[88]
MoP/CNT# on carbon nanofiber	0.5 M H_2SO_4	60	-	83	[89]
FeCoP on carbon cloth	0.5 M H_2SO_4	30	-	37	[90]
NiCoP on carbon felt	1.0 M KOH	34	-	58	[91]
Oxidized carbon nanotubes	0.5 M H_2SO_4	71.3	-	220	[92]

(continued)

Table 1 (continued)

Electrode	Conditions	Tafel slope (*b*, mV/dec)	j_0 (mA/cm^2)	j_{10} (mV)	Refs.
N and S co-doped nanoporous graphene	0.5 M H_2SO_4	80.5	-	280	[93]
C_3N_4 nanoribbons on graphene nanosheets	0.5 M H_2SO_4	54	-	207	[94]

* Values of η_{10} estimated using j_0 and the Tafel equation at $j = 10$ mA/cm^2: $\eta = b \log_{10} (j/j_0)$
** GCE Glassy Carbon Electrode
*** FTO: Fluorine Tin Oxide
\# CNT: Carbon nanotubes

3.3.1 Metal Phosphides and the Search for Abundant Materials for HER

Metal phosphides (MPs), represented by the general formula M_xP_y, are solid-state compounds formed by the combination of metallic elements with phosphorus. Crystal structures depend on phosphorus-metal bonding and stoichiometric composition. Metal-rich phosphides ($x \geq y$ in M_xP_y) are often semiconductors and, in some cases metallic, depending on the presence of a significant metal–metal bonding. Appropriate incorporation of P can change the conductivity and the susceptibility of the surface to corrosion/passivation, which plays an important role in the catalytic activity and the durability of the electrode material [53, 95–98].

Theoretical studies by Liu and Rodriguez reported in 2005, based on density functional theory (DFT), suggested that the synergistic effect between proton acceptor centers (P) and hydride acceptors (Ni) exposed on the surface (001) of Ni_2P could mimic the characteristics of active hydrogenase enzyme sites to facilitate efficient HER [99, 100]. The formation of M-P bonds produces a partial redistribution of charges from metal to non-metal due to the electronegativity difference. The electrons partially transfer from metal to P atoms, generating positively charged metal sites that can act as hydride-acceptors, whereas P sites, with a partial negative charge, act as proton-acceptors. The existence of P atoms in the metallic lattice has the effect of diluting the density of metal atoms at the surface and regulates the binding energy with the reaction intermediates. This so-called "ensemble effect" provides moderate binding energies to optimize the reaction pathway and prevents the deactivation of the catalysts.

These predictions were then experimentally demonstrated with the synthesis of Ni_2P nanoparticles ten years later by Popczun and collaborators [100]. It would not be the first time that a metal phosphide was synthesized, but it would be the first time that it was promoted as a catalyst for HER. Since then, there has been a steady increase in research regarding these materials.

In general, the synthesis of metal phosphides can be classified according to the phosphorus source [98]. Regarding their application as catalysts, the main sources of phosphorus used are organic, inorganic phosphines, and other sources such as hypophosphite or phosphane gas [101]. Using organic phosphine, such as trioctyphine in an organic solvent at temperatures between 200 and 400 °C, different crystalline phases are obtained that can be adjusted depending on temperature, reaction time, or molar ratio [98, 101]. Inorganic sources, such as phosphates, allow MP to be synthesized with classical heat treatment strategies. They are mixed with a solid-state metal in a reactor at high temperatures and allowed to be incorporated into the structure, under an inert atmosphere, in what is known as the annealing method [101].

The hypophosphite (HP) route allows for expanding the range of options for metal precursors, greater flexibility, and more direct MP production free of sophisticated steps [97]. All these processes follow a similar mechanism where the phosphorus source plays the role of reducing agent, acting on the metal ion, and in turn decomposing, by the energy conditions imposed. The latter allows the phosphorus to be incorporated within the metal lattice. Different synthetic pathways give rise to different morphologies and particle sizes, which have a great impact on catalytic activity [98, 101]. However, this type of synthesis can be dangerous, since the high temperatures can lead to the formation of pyrophoric by-products such as white phosphorus or phosphine [100]. To solve these drawbacks, electrochemical synthesis is rather appropriate since the necessary energy is supplied as a potential difference instead of heat.

The use of HP as a source of phosphorus is inherited from electrolyte coating mechanisms, in which the conditions are prepared for the redox reaction to occur and the metal in solution becomes a solid film [102]. When hypophosphite has been used as a reducing agent, phosphorus impurities have been found in metal films. This leads to the idea of being able to use similar conditions and catalyze the process in an electrochemical cell with electrical potential. However, so far this approach has not been refined and tends to generate amorphous phases with a wide range of phosphorus content [102–104]. In this regard, the electrochemical synthesis of metal phosphides is understood as an open frontier to be explored with much potential due to its recent technological interest.

Electrodeposition has gained unprecedented attention as the method of choice for the synthesis of nanostructured catalysts for electrochemical energy conversion [105]. Precisely, the recent study by Torres and coworkers [106] supports the possibility of electrodeposition of a binary phase composed of metal and phosphorus, in addition to providing the kinetic description and thermodynamic consequence of the process. For the electrochemical nucleation of an MP phase with diffusion-controlled growth, and provided that the P content in the resulting phase follows the proportion of the precursor ion in the electrolytic solution, the following expression for the current density (j_{MP}) *vs.* time (t) response is reported:

$$-j_{MP}(t) = \frac{FD_w c_M}{(\pi D_a t)^{1/2}}\left(1 - exp\left\{-N_0 \pi k_{MP} D_a^{\frac{1}{2}} D_w^{\frac{1}{2}}\left[t - \frac{\left(1 - e^{-At}\right)}{A}\right]\right\}\right) \quad (24)$$

where N_0 is the density number of active sites for nucleation on the surface and A is the rate constant of their conversion into growing nuclei. By defining, $\gamma = c_M/c_P$ is the ratio of the bulk concentrations of the metal (c_M) and HP (c_P), the constant k_{MP} can be written as $k_{MP} = [8\pi c_M(x_M \overline{\upsilon}_M + x_P \overline{\upsilon}_P)/(\gamma(z_M x_M + z_P x_P)(1+\gamma))]^{1/2}$ where $\overline{\upsilon}_M$ and $\overline{\upsilon}_P$ are the molar volumes, z_M and z_P the charge transferred, and: $x_M = c_M/(c_M + c_P)$ and $x_P = c_P/(c_M + c_P)$ represents the relative proportions in solution, the subscripts M and P indicate metal, and HP, respectively. On the other hand, the diffusion coefficient can be written as $D_a = [(\gamma D_M + D_P)/(\gamma + 1)]$, where D_M and D_P are the diffusion coefficient of the metal and P precursor in solution, respectively, and $D_w = [(z_M D_M \gamma + z_P D_P)/\gamma(z_M x_M + z_P x_P)]$ is the coefficient that accounts for charge and mass transport. It should be noticed that N_0 and A are the kinetic parameters that characterize the nucleation and growth phenomena, and the experimental methodology to measure them and extract thermodynamic information using the classical theory of nucleation and growth has been reported, specifically for the electrocrystallization of copper phosphide [106]. The exploration of the electrosynthesis of MPs with HP as precursors of P but using metals other than Cu remains a subject of recent research. Extending the methodology and theory using metals that, when deposited, catalyze the reduction of protons and the HP remains an open topic.

4 Storage and Diffusion Processes

The storage of hydrogen in the form of metal hydrides has attracted attention since the discovery of the dissolution of hydrogen in palladium by Graham in 1866 [2]. The peak of active research was observed in the 1960s when intermetallic compounds capable of absorption reversible hydrogen at moderate temperatures and pressures were discovered. In addition to the development of suitable methodologies for studying the H processes into an electroactive metal, as Devanathan and Stachurski [107], Conway and Tilak [44] and Zoltowski [108].

4.1 Absorption Hydrides and Mechanism

This type of material is characterized by having the ability to break the hydrogen molecule. Separated atoms are absorbed into the metal network, and stored in the interstices of the host metal [11, 21]. Mainly this type of hydride is produced in metals such as palladium (Pd), magnesium (Mg), lanthanum (La), or their alloys with a wide variety of elements [11]. One of the factors that have aroused the most interest in this type of material is its ability to store hydrogen in a non-stoichiometric way, which implies that the amount of hydrogen absorbed depends mainly on its pressure or the applied temperature.

The non-stoichiometric storage in addition to the fact that hydrogen is stored in the "free" spaces of the metal network, suggests that there must necessarily be a change in the structure of the metal as the internal hydrogen pressure increases. This structural change is due to a phase change, from phase α (i.e. the structure of the metal's internal network before the entry of hydrogen), to phase β (i.e., the crystal structure expanded by hydrogen) [11, 21]. This expansion of the crystalline lattice does not occur immediately with the entry of hydrogen. The internal structure is progressively expanding from phase α to phase β, leading to the coexistence of both phases in the material. The phase equilibrium between α and β allows the material to be able to store hydrogen in a reversible way. This coexistence occurs for certain values of internal pressure of hydrogen and is determined by both the nature of the material and the temperature. A phase change is also observed in hydrides of the ionic type or in different alloys with the ability to absorb hydrogen [11]. The main advantages of this type of hydride are the reversibility in the process of entry and exit of hydrogen from host material and, that large amounts of energy are not required to carry it out. However, one of the relevant disadvantages of these types of materials is that they store rather low amounts of hydrogen, around 1.2%. This fact makes them unsuitable for use on mobile devices [11, 21], preferring stationary applications with the possibility of recharging.

Now, it is worth asking: *how does the accumulation of hydrogen occur?*

Figure 5a shows a series of isotherms in hydrogen pressure coordinates within a metal as a function of its concentration at various temperatures, or *P–C-T* diagram. It is observed how the coexistence of both phases is presented as a constant value of the internal pressure of hydrogen, even though the concentration of these increases. This is the typical behavior observed in a phase transition phenomenon. For example, if the pressure decreases, the α phase is favored, whereas when the pressure increases, the β phase is then favored. The plateau in the *P vs. C* curve corresponds to the amount of hydrogen that can be stored in a metal reversibly without significantly damaging its internal network [11, 21]. It is also observed that as the temperature of the system increases, this plateau becomes narrower, until reaching a critical temperature (T_c), in which the phases α and β are indistinguishable. In equilibrium conditions, the potential depends on the phase ratio and the amount of stored hydrogen, so E depends linearly on the inverse of the absolute temperature, as seen in Fig. 5b.

Figure 5c presents schematics of some interpretations commonly reported in the literature about hydrides formed by changes in pressure and/or temperature [11]. Basically, H_2 is first dissociatively adsorbed on the surface forming a surface hydride, then the growth of the new metal hydride phase occurs under diffusional control into the solid. The idealized mechanism involves diffusion-controlled hydride formation in the solid state, but it has been determined that certain physicochemical factors may play a critical role in affecting the kinetics. For example, not all particles are of the same size, and not all particles experience similar growth rates. Therefore, at least three types of scenarios can be easily visualized: (*i*) *surface-controlled chemisorption*, where the growth of the hydride phase depends on the initial hydride distribution on the surface, (*ii*) *geometric contraction shrinking core*, where there is a continuous growth of the hydride phase shell into a core–shell type structure and, (*iii*) *nucleation and growth mechanisms*, where instantaneous or progressive nucleation followed by growth with hierarchical overlapping of diffusion zones is considered to occur. To

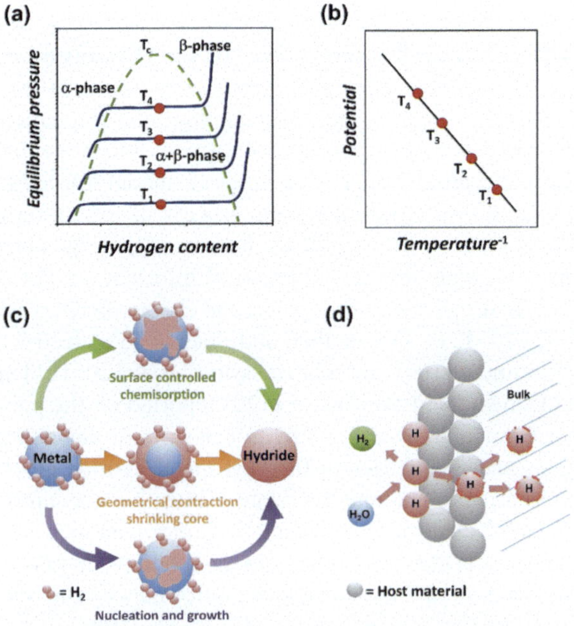

Fig. 5 **a** *P–C-T* diagram for hydrogen storage, **b** Potential *versus* temperature^{-1} plot, **c** mechanisms of metal hydride formed by changes in pressure and/or temperature (from H_2), and **d** mechanisms of metal hydride formed by polarization (from water reduction reaction)

detail the mathematical expressions of these kinetic models, as well as to obtain guidelines for their implementation and correlation of experimental data, the reader is encouraged to review the work of Puszkiel [109].

On the other hand, Fig. 5d shows a scheme in which metal hydrides are formed in electrodes immersed in an aqueous electrolyte under appropriate polarization [110]. The hydride is first formed on the surface by reduction of the water molecule, and then diffusion of hydrogen adsorbed on the electrode occurs. It should be noted that a subsurface phase can form and define the growth of the new hydride phase in the solid. In the situation described, HER can occur from adsorbed hydrogen, and for hydrogen storage purposes it must be considered as a side reaction that reduces storage efficiency.

Regarding hydrogen storage materials and methods from an electrochemical point of view, it has been recognized that the synthesis method has an essential effect on the shape and structure, which impacts the storage capacity [111]. Studies detailing the effect of the morphology of active materials as an important parameter for electrochemical energy storage have been reported [112]. On the other hand, the operating parameters of the functional cells are also important [113, 114], for example, preparation of the working electrode, type of reference electrode and counter electrode, cell with three or two electrodes, electrolyte, temperature, and current density, are variables that affects the performance of energy storage. In the literature [113], the effect of the type of electrode and the operating parameters that increase the yield of hydrogen storage has been discussed. In all cases, we recommend the reader review the suggested references to learn about the main techniques applied to the study of hydrogen storage in electroactive materials, these would be: chronopotentiometry (CP) [113], cyclic voltammetry (CV) [115], open-circuit potential decay [44], and electrochemical impedance spectroscopy (EIS) [108].

To contribute to the understanding and implementation of processes based on hydrogen technology, it is necessary to rationalize the experimental data and define the conceptual bases of a suitable model for hydrogen accumulation kinetics, which allows representing the process of formation of metal hydride under diffusional control. In fact, the evaluation of hydrogen charging and discharging cycles provides both kinetic and practical information. For example, amounts of accumulated hydrogen, lifetime cycles, and charge/discharge efficiency can be characterized [11, 116–121]. Devanathan and Stachurski's electrochemical hydrogen permeation experiment [107] and related assays [108, 120, 121] are suitable for these purposes.

Conway and Wojtowicz [120] reported on the protocols for estimating the electrochemical time scales of hydrogen sorption and desorption in relation to the dimensions and geometries of the host metal hydride electrodes. They also reported the appropriate mathematical relationships that account for diffusion behavior through the solid electrode. Understanding the dynamics of the phase transitions in these materials is also important, and the relationship between the structural properties of the material and the consequent deformations due the hydrogen permeation can be studied using impedance-based methods [108].

As a first approximation, the classical model of hydrogen diffusion in Pd is suitable to describe phenomenologically the formation of the hydride phase under diffusion control [120, 121]. Suárez and coworkers [121] evaluated the physicochemical properties of diffusion processes on Pd. The differences between a massive or semi-infinite metallic phase and a micro-scattered phase were considered, allowing the identification of kinetic limitations associated with the structural defects of the new materials. The trends of the diffusion transport coefficient as a function of the potential for the formation of the hydride phase or the electrical charge, as well as the trends in the hydrogen charge/discharge process, allowed showing no idealities due to microstructure.

Currently, other kinetic models consider the effect of limitation by the hydride formation reaction and the morphology of the host material [11, 117]. Deepening and understanding these phenomena can have a profound effect on the search for better materials for hydrogen accumulation.

4.2 Alloys for Hydrogen Storage.

The use of pure metals for the storage and generation of hydrogen has proven to be inefficient, either because of problems with the kinetics of the reaction and the strength of the Metal-H bond or in ionic hydrides, due to low storage capacity with high costs of pure metals [21]. Metal alloys combine the properties of the constituent metals, demonstrating possibilities in more durable, efficient, and economical hydrogen storage. There are two main types of metal alloys used in the hydrogen storage area: type AB_5 alloys and A_2B type alloys [11]. In these alloys, element A corresponds to the metal capable of absorbing hydrogen, while metal B is added to improve the kinetics of the process, prevent the formation of oxides, provide durability, or reduce the strength of the A-H bond.

In this group of compounds, the most studied example is the Mg_2Ni alloy, which has great advantages over AB_5-type alloys among which is the fact that Mg is a light metal [116–119]. Mg is also an element whose hydrogen absorption capacity per unit mass is greater than 7.6%, which makes it more feasible to be used on mobile devices [11, 21]. However, Mg presents several problems, such as the high stability of magnesium hydride (MgH_2), which requires 300 °C to be broken, it is a poor catalyst to break the H–H bond during the dissociative adsorption process, and Mg oxidizes very easily. Therefore, Ni is added, mainly to decrease the stability of the hydride as well as to reduce the energy required for the H release. In fact, the resulting material has more stability and catalyzes more efficiently the incorporation of the H atom [11, 116–119]. In alloy materials based on Ni and Mg, the hydrogen accumulation kinetics is improved. However, the storage capacity also decreases, maintaining a maximum value of 3.6% [21]. Finally, the reader is encouraged to review the works of Eftekhari and Fang [122], Kaur and Pal [115], and Lai et al. [11], which summarize information on several hundred known metal hydrides and intermetallic compounds for hydrogen storage applications.

5 Electrocatalysis of the Hydrogen Transfer Reaction

HTR to organic and/or organometallic molecules to obtain reduction products with higher added value is a current research challenge. The main limitation is about the reaction kinetics. It is possible to generate the adsorbed state of hydrogen on the surface of suitable cathodes (-H or H_{ads}), but hydrogen transfer to another H atom adsorbed is faster than the transfer to another adsorbed molecule. In all cases, there are systems of interest for which HTR has been successfully achieved. Some examples are reduction of CO_2 on Cu-based cathodes [123] or heteroatom-doped carbon materials [124], reduction of nitrate to ammonia on Cu-Sn surfaces [125], the reductive transformation of priority drinking water contaminants as halogenated compounds, N-nitrosamines, azo dyes, NO_3^-, NO_2^-, ClO_4^-, ClO_3^-, BrO_3^- and other oxyanions [31], and upgrading of petroleum effluents [126].

5.1 Mechanism and Alloy Electrodes

The electrochemical reduction of organic compounds through hydrogen transfer from water on alloy cathodes (M_xN_y), occurs through complex mechanisms that involve multiple steps, including transport of the organic compound (R) to the electrode surface, followed by the establishment of an adsorption-desorption equilibrium, typically by interaction with one of the metals in the active sites ([]) of alloy:

$$R + M_xN_y[] \rightleftarrows M_xN_y[R] \tag{25}$$

followed by electron transfer at the electrode–electrolyte interface,

$$M_xN_y[R] + e^- \rightarrow M_xN_y\left[R^{\bullet -}\right] \tag{26}$$

In addition, under these conditions, water decomposition into adsorbed hydrogen readily occurs according to the Volmer reaction, due to the reactivity of the second metal that composes the alloy

$$M_xN_y[] + H_2O + e^- \rightarrow M_xN_y[H] + OH^- \tag{27}$$

Hydrogen on the electrode surface may follow different competitive pathways including the highly reactive. It generally promotes H_2 generation, H storage by permeation into the solid bulk, and/or reduction of organic substrates via hydrogen transfer [127]. In the case of hydrogen transfer, both the reaction with the radical anion (Eq. 28) or with the organic molecule (Eq. 29) to form the respective stoichiometric products that can occur:

$$M_xN_y[H] + M_xN_y\left[R^{\bullet -}\right] \rightarrow M_xN_y[] + M_xN_y\left[HR^-\right] \tag{28}$$

$$M_xN_y[H] + M_xN_y[R] \rightarrow M_xN_y[] + M_xN_y[HR] \tag{29}$$

However, the reduction reaction of organic species through hydrogen transfer is invariably accompanied by hydrogen evolution, either according to the Heyrovsky (Eq. 30) or Tafel (Eq. 31) reactions on the metal alloy cathode (M_xN_y).

$$M_xN_y[H] + H_2O + e^- \rightarrow M_xN_y[] + OH^- + H_2 \tag{30}$$

$$2M_xN_y[H] \rightarrow 2M_xN_y[] + H_2 \tag{31}$$

Figure 6 schematizes the electrochemical hydrogen transfer reaction (HTR) for a typical molecule, illustrating this advanced reduction process (ARPs) with the occurrence of the inevitable and parallel hydrogen evolution reaction (HER).

Typically, in alloy electrodes, one metal must have appropriate properties to electrogenerated hydrogen adsorbed on the surface, and the second metal is added to improve the interactions with the target compound for HTR. Notably, Ni and Mo-based alloys show very promising characteristics for carrying out a hydrogen transfer reaction. Nickel is in the middle of the volcano plot of log $|j|$ *versus* M-H energy (Fig. 4), so after overcoming the division of water to form an adsorbed state of H, this material promotes the hydrogen evolution reaction (HER) quite well [128]. However, several studies have shown that hydrogen atoms enter into the nickel lattice

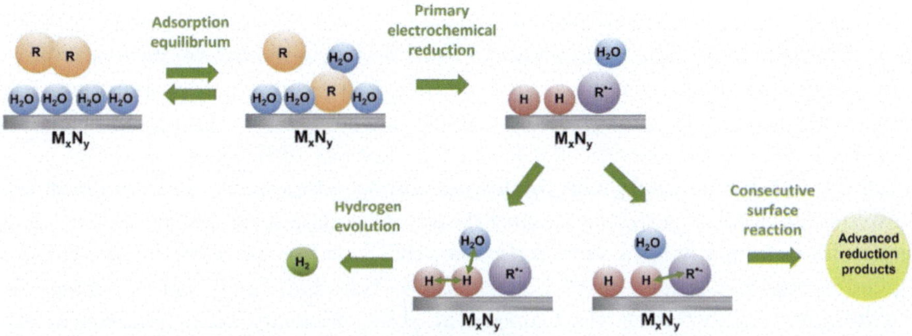

Fig. 6 Scheme of electrochemical hydrogen transfer reaction

during this process, creating a subsurface hydrogen layer that distorts the unit cell and modifies interatomic distances [110]. This fact leads to the passivation of the material as a catalyst and therefore a loss of reaction efficiency. It has been reported that before the hydrogen completely enters the Ni crystal to form the hydride, it is located very close to the surface, forming an equilibrium between the adsorbed hydrogen and a kind of subsurface. This subsurface hydrogen then establishes equilibrium with the absorbed hydrogen that forms the respective hydride [129].

Hydrogen absorption and subsequent passivation of the Ni electrode can be avoided by forming a NiMo alloy cathode. Mo can stabilize the adsorbed hydrogen species much more than Ni, thus avoiding the formation of this subsurface layer and the subsequent absorption of hydrogen. This results in greater stability in the catalyst and prevents the loss of efficiency to perform the HER [129–131]. In addition to stabilizing H on the surface, Mo has been reported to be a promoter of Mo-Metal and Mo-Heteroatom bonds in the case of organometallic compounds and organic molecules with heteroatoms [132], respectively. This means that cathodes based on Ni-Mo alloys have a double benefit: (*i*) stabilization of hydrogen on the surface, with useful life long enough for an adequate reaction with adsorbed organic compounds, and (*ii*) promotion of hydrogen transfer to adjacent chemical species, i.e., -H or -molecule. In fact, the use of NiMo as cathode is shown as a very favorable option for both HER and HTR [38, 129–133].

5.2 *High Entropy Alloys (HEAs)*

A novel ground-breaking approach used nowadays in the development of new electrode material for HER, OER, ORR, or even CO_2 reduction, is the use of multi-elemental alloys, which are characterized by the absence of a major constituent element, or multi-elemental alloys where all the components have equimolar composition. These multi-elemental materials have been studied since the 1980s without any major applications, mostly because of the presence of a wide variety of crystalline phases and defects, leading to brittleness, low resistance, and poor mechanical properties. In 2004 Cantor and collaborators [134], first reported the synthesis of alloys with multiple elements ranging from 5 to 16 and 20, with equal proportions. The alloys with the largest number of elements followed the same predicted behavior, poor mechanical properties mostly due to the presence of too many different crystallinities; but the materials with the lowest number of elements, between 6 and 9, showed a remarkable performance, mostly accommodating all the components in a single phase, typically body-centered cubic (BCC) and face-centered cubic (FCC), depending on the elements used. Later in 2004, Yeh et al. [135] reported materials like those previously reported by Cantor et al. [134] and analyzed the phase structure of various alloys with different compositions. The presence of more than 5 elements in a solid solution would increase the mixing entropy, thus increasing the stability of the material in terms of temperature and thermodynamics. This increase in the entropy of the materials led Yeh and collaborators [135] to coin the term high entropy alloys (HEAs) for the first time.

There are two major definitions when it comes to the so-called HEAs, the composition-based and the entropy-based definitions [136]. The first case states that HEAs are materials containing at least five major or dominant elements, whose compositions range between 5 and 35%, and the minor elements, if present, should have a composition smaller than 5%:

$$N_{\mathrm{major}} \geq 5, \text{ and } 5\% < C_{\mathrm{i}} < 35\% \tag{32}$$

$$N_{\mathrm{minor}} \geq 0, \text{ and } C_{\mathrm{j}} < 5\% \tag{33}$$

where N_{major} and N_{minor} are the numbers of elements in the composition, and C_{i} and C_{j} are the corresponding concentrations of the major and minor elements, respectively.

The second definition, entropy-based, takes into account all the factors comprising the entropy of mixing:

$$\Delta S_{mix} = \Delta S_{mix}^{conf} + \Delta S_{mix}^{vib} + \Delta S_{mix}^{mag} + \Delta S_{mix}^{elec} \tag{34}$$

where ΔS_{mix}^{conf} is the configurational entropy, ΔS_{mix}^{vib} the vibrational entropy, ΔS_{mix}^{mag} the magnetic dipole entropy, and ΔS_{mix}^{elec} the electronical randomness entropy. In this definition, the configurational entropy plays the dominant role, which increases with the number of elements of the alloy following the next relation [136].

$$\Delta S_{mix}^{conf} = -R\ln\frac{1}{n} = R\ln n \tag{35}$$

R being the gas constant, and n the number of elements.

Based on this definition, Yeh [137] made a distinction between low entropy ($\Delta S_{conf} < R$), medium entropy ($R \leq \Delta S_{conf} \leq 1.5R$), and high entropy ($\Delta S_{conf} >$

1.5*R*) alloys. Using this definition, the limits stated in the composition-based (5%–35%) can be used to set the boundary between major and minor elements; this is, if an element has at least 5% concentration, has 0.15*R* value for ΔS_{conf}, which is 10% of the required value. On the other hand, if the number of elements is too high ($n > 13$) the effect of each element on the configurational entropy would be too small ($\approx$2.7%). Based on this, the high entropy alloy composition is comprised of between 5 and 13 elements [136].

Now, the superior mechanical and catalytic properties of the HEAs, can be attributed to three major characteristics [136]. First, as its name implies, the high entropy mixing effect, this parameter is useful when analyzing the thermodynamics of phase formation, considering changes in Gibbs free energy (ΔG_{mix}), enthalpy (ΔH_{mix}) and entropy (ΔS_{mix}):

$$\Delta G_{mix} = \Delta H_{mix} - T\Delta S_{mix} \tag{36}$$

and considering the three more common cases: elemental phases (small ΔH_{mix} and ΔS_{mix}), intermetallic phases (large ΔH_{mix} and small ΔS_{mix}), and solid solutions (medium ΔH_{mix} and high ΔS_{mix}). This relation $\Delta H_{\text{mix}}/\Delta S_{\text{mix}}$ is what makes the materials stable and the single-phase solid-state solution becoming competitive and stable [134–137].

The second effect is the slow diffusion of atoms in the material, the highly heterogeneous environment surrounding each atom makes it more difficult for each element to move inside the matrix to different defects, giving these materials a much higher thermal stability. The severe lattice distortion is the third effect and is strongly related to the diffusivity. The high configurational entropy of the solid solution is much more important than the vibrational component, strongly limiting the mobility of atoms within the crystal, increasing the thermal and mechanical stability, but also these lattice distortions and defects lead to an abundance of highly stable active sites for catalyzing different reactions [137].

The final, and probably more interesting effect from the catalysis point of view, is the so-called "cocktail effect", the presence of multiple atoms on the surface opens the possibilities for a wide range of possibilities for a single HEAs, if chosen properly, individual atoms can be used to catalyze specific reactions, while the rest of the matrix remains either inactive or increasing the activity of the electrode. In any case, the stability of the catalyst would increase because of the previously mentioned effects. The research of these materials skyrocketed since 2004 and is currently one of the hottest topics in the catalysis field, where HER [138], OER [139], ORR [140], and H_2 absorption [141, 142] have been the most studied reactions. A recent report on a high-entropy perovskite cathode for fuel cells [143] has been published. In addition, regarding HTR by electrochemical route, a simple, controllable, and eco-friendly synthesis of FeCoNiCuZn-based HEA electrocatalysts has recently been reported [63]. In fact, it was interesting that due to the dependence of the selectivity of the reaction on the electrode potential, the surface dynamics during the hydrogenation of nitrobenzene on FeCoNiCuZn electrocatalyst can be controlled. HEAs are currently seen as a viable option for substituting the use of noble metals in most reactions, or just to reduce the amounts required, showing similar catalytic performances with apparently higher stabilities.

6 Fuel Cells (FCs)

As emphasized at the beginning of this chapter, one of the most interesting and convenient uses of hydrogen is to use it as fuel in a cell to obtain electricity. Gaseous H_2 and O_2 from the air are continuously fed to the anode and cathode compartments, respectively. Electrochemical reactions take place at the electrodes to produce an electrical current, and electroneutrality is supported by ionic flow in the electrolyte [144]. The reaction is the electrochemical oxidation of a fuel, typically H_2, and the reduction is due to an oxidant, typically oxygen from the air [12].

The basic structure of fuel cells includes an electrolyte in contact with a porous anode and cathode on each side. Generally, the anodic and cathodic compartments are separated by a Membrane Electrode Assembly (MEA) [144]. The ionic flow is established through the MEA which matches the external flow of electrons due to the redox reactions. Currently, there are different types of fuel cells. The most common are those that use a polymeric electrolyte membrane (PEM) for the permeation of protons (H^+) throughout the MEA, see Fig. 7a, as well as the so-called solid oxide fuel cell (SOFC) which contains an MEA through which O^{2-} species permeate due to transport processes activated at high temperature, see Fig. 7b. When methanol is supplied as fuel instead of H_2, the resulting system is known as a direct methanol fuel cell (DMFC) and has a similar configuration to the PEM fuel cell (PEMFC).

Table 2 summarizes the anodic, cathodic reactions, and selected characteristics of the commented fuel cells, specifically fuel cells with potential application to sustainable microgrid systems and vehicles applications [145–147].

Currently, a common way to classify fuel cells is according to the choice of electrolyte and fuel [148]. For example, under this criterion there are six main types of fuel cells, which are currently available in the market:

1. - Proton Exchange Membrane Fuel Cell (PEMFC)
a. - Direct Formic Acid Fuel Cell (DFAFC)

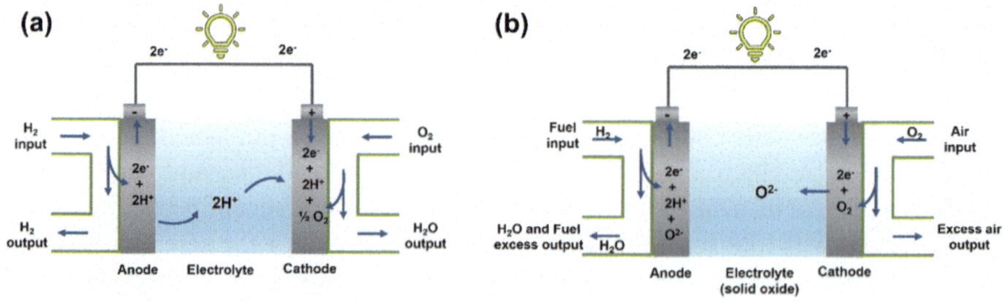

Fig. 7 Scheme of a PEMFC (**a**), and a SOFC (**b**)

Table 2 Characteristics of selected fuel cells

	Fuel Cell Technology		
	PEMFC	DMFC	SOFC
Anode	Pt supported on C	Pt supported on C	Ni- or Co- doped Yttria Stabilized Zirconia (YSZ) cermet
Cathode	Pt supported on C	Pt supported on C, Pt-Ru	Sr- doped $LaMnO_3$
Electrolyte	Nafion®	Sulfuric acid solution	Oxygen - ion conductor
Fuel	Pure Hydrogen	Methanol	Impure hydrogen, variety of hydrocarbon fuels
Operation Temperature (°C)	50–100	70–100	800–1000
Efficiency (%)	35–45	~ 35	50–70
Lifespan (h)	up to 3000	1000	1000
Volumetric Power Density (kW/m^3)	> 420	1.00–300	4.20–19.25
Energy Density (kWh/m^3)	112.2–770	29.9–274	172–462.09
Start-up time (h)	< 0.1	< 0.1	1–5

b. - Direct ethanol fuel cell (DEFC)
2. - Alkaline Fuel Cell (AFC)
a. - Direct borohydride fuel cell (DBFC)
3. - Phosphoric Acid Fuel Cell (PAFC)
4. - Molten Carbonate Fuel Cell (MCFC)
5. - Solid Oxide Fuel Cell (SOFC)
a. - Proton ceramic fuel cell (PCFC)
6. - Direct Methanol Fuel Cell (DMFC)

Furthermore, fuel cells can also be classified based on their operating temperature. The low operating temperature is in the range between 50 and 250 ºC for PEMFCs, AFCs, and PAFCs, and the high operating temperature is in the range between 650 and 1000 ºC for MCFCs and SOFCs [12]. According to Kirubakaran et al. [148], PEM has high power density, fast start-up, lower cost, long service life, and is almost suitable for all applications. MCFC and SOFC are the right choices for medium and large power density applications due to higher efficiency, internal reforming, and combined heat and power cogeneration of hybrid systems. On the other hand, it should be noted that research into fuel cells continues to bring improvements. For example, the operating temperature of SOFCs can drop to 550 ºC due to the use of ceria-based electrolytes [149], just like PCFCs, which reach a lower operating temperature due to lower activation energy (0.4 - 0.6) eV of the proton-conducting electrolyte [150]. Studies on fuel cell technology have intensified with rapid and continuous growth. For more details on the different types of fuel cells, the reader is encouraged to review the following literature [12, 144–148].

6.1 *Phenomenological Response: Reversible Voltage and Potential Losses*

Concerning the characterization of the physicochemical aspects that govern the fuel cell, it is important to notice that the simplest and most practical definition of efficiency for a fuel cell operating near room temperature is the relationship between the real voltage (E) and the reversible voltage (E_0), which is consistent with the second law of thermodynamics [151, 152]. Now, considering the difference between reversible potential and internal losses is a strategy that allows characterizing the performance of a fuel cell.

In the reversible situation, the potential difference of the fuel cell ideally corresponds to the electromotive force of the cell constituted by the redox processes of oxidation of H_2 and reduction of O_2 and it is defined by the Nernst equation [153]. Conceptually, lowering the voltage below the reversible value E_0 is a good measure of the thermodynamic imperfections of the cell, which are known as losses [151–153]. The voltage *vs.* current density curve in fuel cells is shown in Fig. 8 [151, 153]. Then, by requiring the cell to operate by delivering a certain amount of current, the loss due to polarization of the electrodes is first noticed, due to the activation of redox reactions. A subsequent increase in the current leads to greater losses due to the ohmic drop of the system. Finally, high consumption of reagents leads to higher currents, defining losses due to limitations due to mass transport or concentration. In all cases, the voltage (V_{FC}) of a fuel cell can be represented as the deviation of the ideal voltage (E_0) due to the following overpotentials or internal losses; polarization or activation (ΔV_{act}), ohmic (ΔV_{ohm}), and concentration (ΔV_{con}), respectively [153, 154], as follows:

$$V_{FC} = E_0 - (\Delta V_{act} + \Delta V_{conc})_{cathode} - (\Delta V_{act} + \Delta V_{conc})_{anode} - \Delta V_{ohm} \quad (37)$$

Using a simple kinetic analysis, Bockris and Khan [153] correlated the activation overpotential with the Butler and Volmer model for electron transfer, the system resistance with Ohm's law, and the concentration overpotential with a limiting current defined by the integrated Fick equation, yielding an analytical expression for the voltage as a function of the current density. Then, the simplest equation characterizing a fuel cell is:

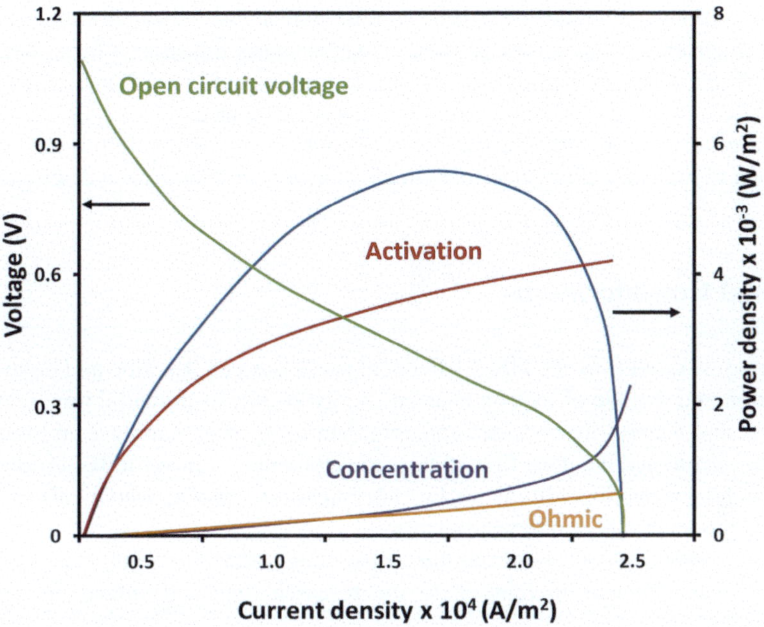

Fig. 8 Voltage and power density *versus* current density curves for a fuel cell. The contribution of activation, ohmic, and concentration losses to the voltage drop is shown

$$\begin{aligned} V_{FC} &= E_0 - (RT/\alpha_c F)\ln(j/j_{0,c}) \\ &-(RT/nF)\ln(1 - j/j_{L,c}) - \\ &(RT/\alpha_a F)\ln(j/j_{0,a}) \\ &-(RT/nF)\ln(1 - j/j_{L,a}) - jAR_i \end{aligned} \tag{38}$$

where $j_{0,c}$ and $j_{0,a}$ are exchange current densities, $j_{L,c}$ and $j_{L,a}$ are the limit current densities and α_c and α_a are the electron transfer coefficients, the subscripts "c" and "a" referring to cathode and anode, respectively. A is the cross-sectional area of the current flow through the electrodes and R_i is the internal resistance of the fuel cell. Equation (38) describes the behavior indicated in Fig. 8.

Finally, the relevance of the analysis described in the previous paragraph lies in the fact that the maximum operating power density (P_m) of the fuel cell can be defined. More precisely this is the recommended condition to obtain the maximum chemical to electrical energy conversion efficiency. The power density curve is illustrated on the right axis of Fig. 8. This parameter can be obtained by calculating the power density (P) as the product of the current density (j) and the voltage (V_{FC}). Power density calculations from experimental data show that the power limits of a fuel cell depend on the parameters of the system, e.g., the magnitude of the current, the number of mass transfer units, the polarization, the electrode's area, redox reaction rates, cell stack architecture, and others [144–154]. Currently, the literature reports several multiphysics models that involve time dependence, three-dimensional and porous electrodes, temperature gradients, reagent mass transport effects, chemical nature of the electrolytes, degradation of the components, other phenomena, and design [144, 149, 155–160], all to represent the performance of fuel cells assertively. In all cases, optimizing applications for renewable energy generation is key, so it is important to understand the electrochemical concepts that have been discussed.

7 Final Considerations

Having pointed out how the electrochemical processes can describe and control the generation and storage of hydrogen as an energy vector, its potential impact on our economy and way of life becomes apparent. But for hydrogen to have an impact on our future economy, we must first address the challenges it poses today. Addressing these challenges requires a fundamental understanding of physicochemical principles and required technological developments.

In this sense, recent contributions that could serve to continue scientific-technological advances in relation to the research topics discussed are pointed out:

(i) the use of adequate experimental methods to elucidate/deepen undisclosed kinetic effects [4, 161];
(ii) analyses based on rate laws consistent with microkinetic phenomena during HER [34] and for electrochemical synthesis of new materials with the concomitant hydrogen production [162];
(iii) design of key material activation strategies to achieve a universal pH working cathode for HER [163];
(iv) synthesis of new precious metal-free electrocatalytic materials for HER [164];
(v) delve into the photocatalytic phenomena of solar fuel production [165];
(vi) new electrodes based on abundant and/or recycled materials to seek double benefit: in environmental issues by degrading pollutants, and in energy by producing hydrogen [166, 167];
(vii) understanding and optimization of the phenomena that occur in the solid state during the sorption of hydrogen [11];
(viii) electrocatalytic systems for a more efficient hydrogen transfer [31, 63, 125];
(ix) energy optimization of fuel cells according to the approach of thermodynamics far from equilibrium and dynamical systems [168];
(x) the integration of electrochemical processes for a new technological era [4, 169, 170].

8 Summary

The advancement of scientific and technological developments in hydrogen-based technologies requires a thorough understanding of physicochemical principles. It is important that new contributions can consider the elucidation of unrevealed kinetic effects, employ rate laws consistent with microkinetic phenomena, develop precious metal-free electrocatalytic materials, optimize hydrogen sorption phenomena, improve electrocatalytic systems to transfer hydrogen efficiently, optimize fuel cells through thermodynamics and dynamic systems analysis and integrate the different electrochemical processes to drive a new technological era. Finally, by taking an interdisciplinary scientific approach, the chemical potential of hydrogen-based technologies can be unlocked and used efficiently.

Acknowledgements Research partially supported by a grant from Decanato de Investigación y Desarrollo of Universidad Simón Bolívar (S1-FEP-DID-017-2017). We would like to thank Fundación Empresas Polar (Venezuela) and CONICET (Argentina).

References

1. J. O'M. Bockris, Int. J. Hydrog. Energy **38**, 2579 (2013)
2. O. Petrii, in, *Chemistry, Electrochemistry and Electrochemical Applications | Hydrogen*, ed. by J. Garche. *Encyclopedia of Electrochemical Power Sources*, (Elsevier, 2009), p. 751
3. Y. Li, Q. Li, H. Wang, L. Zhang, D. Wilkinson, J. Zhang, Electrochem. Eng. Rev. **2**, 518 (2019)
4. B.C. Tashie-Lewis, S.G. Nnabuife, Chem. Eng. J. Adv. **8**, 100172 (2021)
5. K. Mazloomi, C. Gomez, Int. J. Hydrog. Energy **45**, 19620 (2020)
6. P. Nikolaidis, A. Poullikkas, Renew. Sust. Energ. Rev. **67**, 597 (2017)
7. A. Fujishima, K. Honda, Nature **238**, 37 (1972)
8. OPEC. Monthly Oil Market Report. Mon. Oil. Mark. Rep. 100 (2016).
9. OPEC. Monthly Oil Market Report. Mon. Oil. Mark. Rep. June 2021 (2021).
10. S. Shiva Kumar, V. Himabindu, Mater. Sci. Eng. Tech. **2**, 442 (2019)
11. Q. Lai, Y. Sun, T. Wang, P. Modi, C. Cazorla, U.B. Demirci, J.R. Ares Fernandez, F. Leardini, K-F. Aguey-Zinsou, Adv. Sus. Sys. **3**, 1900043 (2019)
12. J.M. Andújar, F. Segura, Renew. Sust. Energ. Rev. **13**, 2309 (2009)
13. Directive (EU) 2018/2001 of the European Parliament and of the Council of 11 December 2018 on the promotion of the use of energy from renewable sources. PE/48/2018/REV/1. OJ L 328, 21.12.2018, pp. 82–209 (2018).
14. Z. Yilmazer Hitit, P.C. Hallenbeck, Biomass. Bioeng. **147**, 106014 (2021).
15. P. Allulema-Pullupaxi, P.J. Espinoza-Montero, C. Sigcha-Pallo, R. Vargas, L. Fernández, J.M. Peralta-Hernández, J.L. Paz, Chemosphere **281**, 130821 (2021)
16. E.N. Aguilera González, S. Estrada Flores, A. Martínez Luévanos, in, Nanomaterials: Recent Advances for Hydrogen Production, ed. by O.V. Kharissova, L.M.T. Martínez, B.I. Kharisov, *Handbook of Nanomaterials and Nanocomposites for Energy and Environmental Applications*, (Springer, Cham, 2021), p. 1.
17. N. Dubouis, A. Grimaud, Chem. Sci. **10**, 9165 (2019)
18. K. Scott, in, Introduction to electrolysis, electrolysers and hydrogen production, ed. by K. Scott. *Electrochemical Methods for Hydrogen Production,* (RSC, 2019), p. 1.

19. P. Millet, S, Grigoriev, in, Water Electrolysis Technologies, ed. by L. M. Gandía, G. Arzamendi, P. M. Diéguez, *Renewable Hydrogen Technologies* (Elsevier, 2013), p. 19.
20. J.T.S. Irvine, D. Neagu, M.C. Verbraeken, Ch. Chatzichristodoulou, Ch. Graves, M.B. Mogensen, Nat. Energy **1**, 15014 (2016)
21. L. Schlapbach, A. Zuttel, Nature **414**, 353 (2001)
22. L.N. Kustov, A.N. Kalenchuk, V.I. Bogdan, Russ. Chem. Rev. **89**, 897 (2020)
23. X. Du, H. Zhang, K. Sullivan, P. Gogoi, Y. Deng, Chemsuschem **13**, 4318 (2020)
24. P. Yu, F. Wang, T. Ahmed Shifa, X. Zhang, X. Lou, F. Xia, J. He, Nano. Eng. **58**, 244 (2019).
25. M. Inci, M. Buyuk, M. Hakan Demir, G. Ilbey, Rene. Sus. Eng. Rev. **137**, 110648 (2021).
26. T. Roach, M. Schmitz, V. Leach, M. Miller, B. Chan, S. Kalman, J. Organomet. Chem. **873**, 8 (2018)
27. Y. Cai, F. Li, Y.-Q. Li, W.-B. Zhang, F.-H. Liu, S.-L. Shi, Tetrahedron Lett. **59**, 1079 (2018)
28. B.M. Trost, in, Atom economy: challenge for enhanced synthetic efficiency, ed. by P. Anastas. Handbook of green chemistry, (Wiley, 2010) p. 1.
29. P. Wender, V. Verma, T. Paxton, T. Pillow, Acc. Chem. Res. **41**, 40 (2008)
30. N. Burns, P. Baran, R. Hoffmann, Angew. Chemie. Int. Ed. **48**, 2854 (2009)
31. B.P. Chaplin, M. Reinhard, W.F. Schneider, Ch. Schuth, J.R. Shapley, T.J. Strathmann, Ch.J. Werth, Environ. Sci. Technol. **46**, 3655 (2012)
32. O. Yépez, B.R. Scharifker, J. Appl. Electrochem. **29**, 1185 (1999)
33. O. Yépez, B.R. Scharifker, Int. J. Hydrog. Energy **27**, 99 (2002)
34. A. Lasia, Int. J. Hydrog. Energy **44**, 19484 (2019)
35. R. de Levie, J. Electroanal. Chem. **476**, 92 (1999)
36. J. Tafel, Z. Phys, Chem. **50**, 641 (1905)
37. P. Sabatier, Ber. Deutsch. Gem. Ges. **44**, 1984 (1911)
38. H. Ooka, J. Huang, K.S. Exner, Front. Eng. Res. **9**, 1 (2021)
39. J. Heyrovsky, J. Recl. Trav. Chim. Pays-Bas. **46**, 582 (1927)
40. T. Volmer, M. Erdey-Gruz, Z. Phys, Chem. **150**, 203 (1930)
41. S. Trasatti, in, Electrocatalysis of hydrogen evolution: Progress in cathode activation, ed. by H. Gerischer, Ch.W. Tobias. Advances in *Electrochemical Science and Engineering,* (Wiley-VCH, 1992), pp. 1–85.
42. F.P. Bowden, E.K. Rideal, Proc. R. Soc. London. Ser. A. **120**, 59 (1928)
43. H. Kobosew, P. Nekrassow, Z. Elektrochem. **30**, 529 (1930)
44. B.E. Conway, B.V. Tilak, Electrochim. Acta **47**, 3571 (2002)
45. D. Pletcher, R. Greff. R. Peat, L.M. Peter, J. Robinson, 1st ed, *Instrumental Methods in Electrochemistry,* (Ellis Horwood Serie in Physical Chemistry, 2001), pp. 229–250
46. M. Koper, J. Electroanal. Chem. **660**, 254 (2011)
47. M. Duca, M. Koper, in, Fundamentals Aspects of Electrocatalysis, ed. by K. Wandelt. Surface and Interface Science: Interfacial *Electrochemistry,* (Wiley-VCH, 2020), pp. 773–890.
48. A. Appleby, J.H. Zagal, J. Solid State Electrochem. **15**, 1811 (2011)
49. S. Trasatti, J. Electroanal. Chem. **39**, 163 (1972)
50. J. Novak Hansen, H. Prats, K. Krojer Toudahl, N. Morch Secher, K. Chan, J. Kibsgaard, I. Chorkendorff, ACS. Eng. Lett. **6**, 1175 (2021).
51. E. Fachinotti, E. Guerrini, A.C. Tavares, S. Trasatti, J. Electroanal. Chem. **600**, 103 (2007)
52. D. Pletcher. X. Li, S. Wang, Int. J. Hydrog. Energy **37**, 7429 (2012).
53. J. Callejas, C. Read, Ch. Roske, N. Lewis, R. Schaak, Chem. Mater. **28**, 6017 (2016)
54. B. Hinnermann, P.G. Moses, J. Bonde, P. Jogersen, J.H. Nielsen, S. Horch, I. Chorkendorff, J.K. Norskov, J. Am. Chem. Soc. **127**, 5308 (2005)
55. B. Hinnermann, J.K. Norskov, H. Topsoe, J. Phys. Chem. B **109**, 2245 (2005)
56. F. Keivanimehr, S. Habibzadeh, A. Baghban, A. Esmaeli, A. Mohaddespour, A.H. Mashhadzadeh, M. Reza Ganjali, M. Reza Saeb, V. Fierro, A. Celzard, Sci. Rep. **11**, 3958 (2021).
57. Y. Liu, X. Xu, H. Li, Z. Si, X. Wu, R. Ran, D. Weng, Catal. Lett. (2021).
58. F. Sun, Q. Tang, D. Jiang, ACS Catal. **12**, 8404 (2022)

59. T.F. Jaramillo, K.P. Jørgensen, J. Bonde, J.H. Nielsen, S. Horch, I. Chorkendorff, Science **317**, 100 (2007)
60. G. Gao, A.P. O'Mullane, A. Du, ACS Catal. **7**, 494 (2017)
61. R.B. Wexler, J.M.P. Martirez, A.M. Rappe, ACS Catal. **7**, 7718 (2017)
62. A.R. Zeradjanin, P. Narangoda, I. Spanos, J. Masa, R. Schlögl, Electrochim. Acta **388**, 138583 (2021)
63. V. Márquez, J.S. Santos, J.G. Buijnsters, S. Praserthdam, P. Praserthdam, Electrochim. Acta **410**, 139972 (2022)
64. J.K. Nørskov, T. Bligaard, J. Rossmeisl, C.H. Christensen, Nat. Chem. **1**, 37 (2009)
65. J.K. Nørskov, T. Bligaard, L. Ashildur, J.R. Kitchin, J.G. Chen, S. Pandelov, U. Stimming, J. Electrochem. Soc. **152**, J23 (2005)
66. A.R. Zeradjanin, G. Polymeros, C. Toparli, M. Ledendecker, N. Hodnik, A. Erbe, M.
67. F.L. Rohwerder, Mantia. Phys. Chem. Chem. Phys. **22**, 8768 (2020)
68. L. Rebollar, S. Intikhab, N.J. Oliveira, Y. Yan, B. Xu, I.T. McCrum, J.D. Snyder, M.H. Tang, ACS Catal. **10**, 14747 (2020)
69. K.S. Exner, ACS Catal. **9**, 5320 (2019)
70. M. Zeng, Y. Li, J. Mater. Chem. A **3**, 14942 (2015)
71. B.E. Conway, L. Bai, J. Electroanal. Chem. **198**, 149 (1986)
72. E. Lamy-Pitara, J. Barbier, J. Electroanal. Chem. **416**, 47 (1996)
73. M.H. Miles, G. Kissel, P.W.T. Lu, S. Srinivasan, J. Electrochem. Soc. **123**, 332 (1976)
74. X. Yan, L. Tian, X. Chen, J. Power. Sources **300**, 336 (2015)
75. T. Zheng, W. Sang, Z. He, Q. Wei, B. Chen, H. Li, C. Cao, R. Huang, X. Yan, B. Pan, S. Zhou, J. Zeng, Nano Lett. **17**, 7968 (2017)
76. D. Su, J. Wang, H. Jin, Y. Gong, M. Li, Z. Pang, Y. Wang, J. Mater. Chem. A **3**, 11756 (2015)
77. J.M. Jaksic, M.V. Vojnovic, N.V. Krstajic, Electrochim. Acta **45**, 4151 (2000)
78. J.Y. Huot, M.L. Trudeau, R. Schultz, J. Electrochem. Soc. **138**, 1316 (1991)
79. H. Dumond, P. Los, A. Lasia, H. Ménard, J. Appl. Electrochem. **23**, 684 (1993)
80. Y.G. Li, H.L. Wang, L.M. Xie, Y.Y. Liang, G.S. Hong, H.J. Dai, J. Am. Chem. Soc. **133**, 7296 (2011)
81. A.T. Garcia-Esparza, D. Cha, Y.W. Ou, J. Kubota, K. Domen, K. Takanabe, Chemsuschem **6**, 168 (2013)
82. L.F. Pan, Y.H. Li, S. Yang, P.F. Liu, M.Q. Yu, H.G. Yang, Chem. Commun. **50**, 13135 (2014)
83. S. Gupta, N. Patel, R. Fernandes, R. Kadrekar, A. Dashora, A.K. Yadav, D. Bhattacharyya, S.N. Jha, A. MiotelloD, C. Kothari, Appl. Catal. B **192**, 126 (2016)
84. X. Wang, W. Li, D. Xiong, D.Y. Petrovykh, L. Liu, Adv. Funct. Mater. **26**, 4067 (2016)
85. Z. Jin, P. Li, X. Huang, G. Zeng, Y. Jin, B. Zheng, D. Xiao, J. Mater. Chem. A **2**, 18593 (2014)
86. A.B. Laursen, R.B. Wexler, M.J. Whitaker, E.J. Izett, K.U.D. Calvinho, S. Hwang, R. Rucker, H. Wang, J. Li, E. Garfunkel, M. Greenblatt, A.M. Rappe, C. Dismukes, ACS Catal. **8**, 4408 (2018)
87. J.F. Callejas, C.G. Read, E.J. Popczun, J.M. McEnaney, R.E. Schaak, Chem. Mater. **27**, 3769 (2015)
88. D.E. Schipper, Z. Zhao, H. Thirumalai, A.P. Leitner, S.L. Donaldson, A. Kumar, F. Qin, Z. Wang, L.C. Grabow, J. Bao, K.H. Whitmire, Chem. Mater. **30**, 3588 (2018)
89. J.M. McEnaney, J.C. Crompton, J.F. Callejas, E.J. Popczun, A.J. Biacchi, N.S. Lewis, R.E. Schaak, Chem. Mater. **26**, 4826 (2014)
90. X. Zhang, X. Yu, L. Zhang, F. Zhou, Y. Liang, R. Wang, Adv. Funct. Mater. **28**, 1706523 (2018)
91. C. Tang, L. Gan, R. Zhang, W. Lu, X. Jiang, A.M. Asiri, X. Sun, J. Wang, L. Chen, Nano Lett. **16**, 6617 (2016)
92. R. Zhang, X. Wang, S. Yu, T. Wen, X. Zhu, F. Yang, X. Sun, X. Wang, W. Hu, Adv. Mater. **29**, 1605502 (2016)
93. W. Cui, Q. Liu, N.Y. Cheng, A.M. Asiri, X.P. Sun, Chem. Commun. **50**, 9340 (2014)
94. Y. Ito, W. Cong, T. Fujita, Z. Tang, M. Chen, Angew. Chem. Int. Ed. **54**, 2131 (2014)

95. Y. Zhao, F. Zhao, X. Wang, C. Xu, Z. Zhang, G. Shi, L. Qu, Angew. Chem. Int. Ed. **53**, 13934 (2014)
96. H. Du, R.-M. Kong, X. Guo, F. Qu, J. Li, Nanoscale **10**, 2018 (2018)
97. A. Parra-Puerto, K. Ling Ng, K. Fahy, A.E. Goode, M.P. Ryan, A. Kucernak, ACS Catal. **9**, 11515 (2019).
98. S.-H. Li, M.-Y. Qi, Z.-R. Tang, Y-J, Xu. Chem. Soc. Rev. **50**, 7539 (2021)
99. C.-C. Weng, J.-T. Ren, Z.-Y. Yuan, Chemsuschem **13**, 3357 (2020)
100. P. Liu, J.A. Rodriguez, J. Am. Chem. Soc. **127**, 14871 (2003)
101. E. Popczun, J. McKone, C. Read, A. Biacchi, A. Wiltrout, N. Lewis, R. Schaak, J. Am. Chem. Soc. **135**, 9267 (2013)
102. Y. Shi, M. Li, Y. Yu, B. Zhang, Eng. Environ. Sci. **13**, 4564 (2020)
103. T. Harris, J. Electrochem. Soc. **140**, 81 (1993)
104. K. Sridharan, K. Sheppard, J. Appl. Electrochem. **27**, 1198 (1997)
105. Y. Pei, Y. Yang, F. Zhang, P. Dong, R. Baines, Y. Ge, H. Chu, P. Ajayan, J. Shen, M. Ye, A.C.S. Appl, Mater. Interfaces. **9**, 31887 (2017)
106. M. Bernal Lopez, J. Ustarroz, Curr. Op. Electrochem. **27**, 100688 (2021).
107. D. Torres, L. Madriz, R. Vargas, B.R. Scharifker, Electrochim. Acta **354**, 136705 (2020)
108. M.A.V. Devanathan, Z. Stachurski, Proc. Royal. Soc. London. Serie. A, Math. Phys. Sci. **270**, 90 (1962).
109. P. Zoltowski, J. Electroanal. Chem. **600**, 54 (2007)
110. J. A. Puszkiel, in, Tailoring the Kinetic Behavior of Hydride Forming Materials for Hydrogen Storage, ed. by M. Rahman, A. M. Asiri. *Gold Nanoparticles - Reaching New Heights,* (IntechOpen, 2018), pp. 1–31.
111. B. Conway, G. Jerkiewicz, J. Electroanal. Chem. **357**, 47 (1993)
112. F. Sedighi, M. Ghiyasiyan-Arani, M. Behpour, Fuel **310**, 122218 (2022)
113. M. Ghiyasiyan-Arani, M. Salavati-Niasari, A.F. Zonouz, J. Electrochem. Soc. **167**, 020544 (2020)
114. M. Baladi, M. Valian, M. Ghiyasiyan-Arani, M. Salavati-Niasari, Int. J. Hydrog. Energy **46**, 21026 (2021)
115. T. Gholami, M. Pirsaheb, Int. J. Hydrog. Energy **46**, 783 (2021)
116. M. Kaur, K. Pal, J. Energy Storage **23**, 234 (2019)
117. D. Vojtech, V. Knotek, Int. J. Hydrog. Energy **36**, 6689 (2011)
118. Q. Luo, X.-H. An, Y.-B. Pan, X. Zhang, J.-Y. Zhang, Q. Li, Int. J. Hydrog. Energy **35**, 7842 (2010)
119. W.-H. Liu, J. Alloy. Comp. **404–406**, 694 (2005)
120. E. Lass, Int. J. Hydrog. Energy **36**, 14496 (2011)
121. B. Conway, J. Wojtowicz, J. Electroanal. Chem. **326**, 277 (1992)
122. I. Suárez, C. Borrás, B.R. Scharifker, J. Mostany, in, Diffusion in solids: Hydrogen transport in massive and microdispersed palladium, ed. by M. Palomar-Pardavé, M. Romero-Romo, *Electrochemical and Materials Engineering,* (Research Signpost, 2007), p. 173.
123. A. Eftekhari, B. Fang, Int. J. Hydrog. Energy **42**, 25143 (2017)
124. D. Fermín, F. Marken, in, Introduction to the electrochemical and photo-electrochemical reduction of CO_2, ed. by F. Marken, F. Fermín. *Electrochemical reduction of carbon dioxide: Overcoming the limitations of photosynthesis,* (RSC, 2018), pp. 1–16.
125. A. Perez, M.A. Díaz-Pérez, J. Serrano, Catalysis. **10**, 1179 (2020)
126. S. García-Segura, M. Lanzarini-López, K. Hristovski, P. Westerhoff, App. Catal. B. Environ. **236**, 546 (2018)
127. C. Ovalles, I. Rojas, S. Acevedo, G. Escobar, G. Jorge, L.B. Gutierrez, A. Rincón, B.R. Scharifker, Fuel. Proc. Technol. **48**, 159 (1996)
128. J.M. Chapuzet, A. Lascia, L. Lessard, in, Electrocatalytic hydrogenation of organic compounds, ed. by J. Lipkowski, P. Ross. *Electrocatalysis,* (Wiley-VCH, 1998), pp. 155–196.
129. G. Kreysa, B. Hakansson, P. Ekdunge, Electrochim. Acta **33**, 1351 (1988)
130. B. Pauw, W. Kalisvaart, S. Tao, M. Koper, A. Jansen, P. Notten, Acta Mater. **56**, 2948 (2008)
131. J. Cermak, B. David, Int. J. Hydrog. Energy **36**, 13614 (2011)

132. C. González-Buch, I. Herraiz-Cardona, E. Ortega, J. García-Antón, V. Pérez-Herranz, J. App. Electrochem. **46**, 791 (2016)
133. L. Madriz, H. Carrero, J.R. Domínguez, R. Vargas, L. Fernández, Fuel **112**, 338 (2013)
134. J. Hwan Kim, J. Kim, H. Kim, J. Kim, S. Hyun Ahn, J. Ind. Eng. Chem. **79**, 255 (2019).
135. B. Cantor, I.T.H. Chang, P. Knight, A.J.B. Vincent, Mater. Sci. Eng. A **375–377**, 213 (2004)
136. J.W. Yeh, S.K. Chen, S.J. Lin, J.Y. Gan, T.S. Chin, T.T. Shun, C.H. Tsau, S.Y. Chang, Adv. Eng. Mater. **6**, 299 (2004)
137. X. Wang, W. Guo, Y. Fu, J. Mater. Chem. A. **9**, 663 (2021)
138. J.W. Yeh, JOM **65**, 1759 (2013)
139. S. Wang, H. Xin, Chem. **5**, 502 (2019)
140. X. Cui, B. Zhang, C. Zeng, S. Guo, MRS Commun. **8**, 1230 (2018)
141. S. Li, X. Tang, H. Jia, H. Li, G. Xie, X. Liu, X. Lin, H.-Q. Qui, J. Catal. **383**, 164 (2020)
142. D. Karlsson, G. Ek, J. Cedervall, C. Zlotea, K. Moller, T. Hanse, J. Bendnarcik, M. Paskevicius, M.H. Sorby, T,R. Jensen, U. Jansoon, M. Sahlberg, Inorg. Chem. **57**, 2103 (2018).
143. G. Ek, M. Nygard, A. Pavan, J. Montero, P. Henry, M.H. Sorby, M. Witman, V. Stavila, C. Zlotea, B. Hauback, M. Sahlberg, Inorg. Chem. **60**, 1124 (2021)
144. Q. Yang, G. Wang, H. Wu, B.A. Beshiwork, D. Tian, S. Zhu, Y. Yang, X. Lu, Y. Ding, Y. Ling, Y. Chen, B. Lin, J. Alloys Compd. **872**, 159633 (2021)
145. A.A. Kulikovsky, 2nd ed, *Analytical Modelling of Fuel Cells,* (Elsevier, 2019), pp. 1–33.
146. S. Sabihuddin, A.E. Kiprakis, M. Mueller, Energies **8**, 172 (2015)
147. D. Akinyele, E. Olabode, A. Amole, Inventions **5**, 42 (2020)
148. R.M. Dell, P.T. Moseley, D.A.J. Rand, 1st ed, *Towards Sustainable Road Transport,* (Academic Press, 2014), pp. 260–295.
149. A. Kirubakaran, Sh. Jain, R.K. Nema, Renew. Sustain. Energy Rev. **13**, 2430 (2009)
150. B. Timurkutluk, C. Timurkutluk, M.D. Mat, Y. Kaplan, Renew. Sus. Energy Rev. **56**, 1101 (2016)
151. I.T. Bello, Sh. Zhai, Q. He, Ch. Cheng, Y. Dai, B. Chen, Y. Zhang, M. Ni, Int. J. Energy Res. **46**, 2212 (2022)
152. S. Sieniutycz, Int. J. Heat. Mass. Transfer. **53**, 2864 (2010)
153. S. Sieniutycz, Int. J. Mod. Phys. B **26**, 1246001 (2012)
154. J.O'M. Bockris, S.U.M. Khan, 1st ed, *Surface Electrochemistry A Molecular Level Approach,* (Springer Science + Business Media, LLC, 1993), pp. 861–926.
155. F. Barbir, 2nd ed, *PEM Fuel Cell: Theory and Practice,* (Elsevier, 2012), pp. 33–72.
156. C.C. Boyer, R.G. Anthony, A.J. Appleby, J. App. Electrochem. **30**, 777 (2000)
157. K. Promislow, B. Wetton, SIAM. J. App. Math. **70**, 369 (2009)
158. S. Beale, M. Andersson, C. Boigues-Muñoz, H. Frandsen, Z. Lin, S. McPhail, M. Ni, B. Sundén, A. Weber, A.Z. Weber, Prog. Eng, Comb. Sci. **85**, 100902 (2021).
159. M. Arif, S.C.P. Cheung, J. Andrews, Eng. Fuels. **34**, 11897 (2020)
160. C. Pacheco, R. Barbosa, L.C. Ordoñez, J. Sierr, B. Escobar, Int. J. Hydrog. Energy **51**, 26197 (2021)
161. K. Jiao, J. Xuan, Q. Du, Z. Bao, B. Xie, B. Wang, Y. Zhao, L. Fan, H. Wang, Z. Hou, S. Hou, N.P. Brandon, Y. Yin, M.D. Guiver, Nature **595**, 361 (2021)
162. L. Botello, J. Feliú, V. Climent, App. Mater. Interfaces. **12**, 42911 (2020)
163. M. Palomar-Pardavé, B.R. Scharifker, E.M. Arce, M. Romero-Romo, Electrochim. Acta **50**, 4736 (2005)
164. Q. Wang, Z.L. Zhao, S. Dong, D. He, M.J. Lawrence, S. Han, C. Cai, S. Xiang, P. Rodriguez, B. Xiang, Z. Wang, Y. Liang, M. Gu, Nano Eng. **53**, 458 (2018)
165. S. Hadimane, S. Aralekallu, K. Prabhu, M. Hojamberdiev, L.K. Sannegowda, ACS Appl. Eng. Mater. **4**(10), 10826 (2021)
166. L. Peter, J. Electrochem. Soc. **166**, H3125 (2019)
167. L. Madriz, J. Tatá, D. Carvajal, O. Núñez, B.R. Scharifker, J. Mostany, C. Borrás, F.M. Cabrerizo, R. Vargas, Renew. Energy **152**, 974 (2020)
168. H. Rueda, M. Arenas, R. Vargas-Balda, S. Blanco, P. Delvasto, Sustain. Mater. Technol. **29**, e00296 (2021)

169. S. Sieniutycz, Eng. Conver. Manag. **68**, 293 (2013)
170. M. Ostadi, K.G. Paso, S. Rodruiguez-Fabia, L.E. Oi, F. Manenti, M. Hillestad, Energies **13**, 4859 (2020)
171. M. Yue, H. Lambert, E. Pahon, R. Roche, S. Jemei, D. Hissel, Renew. Sus. Energy. Rev. **146**, 111180 (2021)

Modification of TiO_2 as SO_4/TiO_2 Acid and CaO/TiO_2 Base Catalysts and Their Applications in Conversion of Waste Frying Oil (WFO) into Biodiesel

Karna Wijaya, Remi Ayu Pratika, Wega Trisunaryanti, and Alfrets Daniel Tikoalu

Abstract The SO_4/TiO_2 acid and CaO/TiO_2 base catalysts have been successfully prepared for the conversion of waste frying oil into biodiesel. The acid catalyst was prepared through direct sulfation of TiO_2 with H_2SO_4, and the base catalyst was prepared through a thermal method between TiO_2 and CaO in an autoclave reactor. This process has been carried out with variations in the concentration and temperature of calcination. The purpose of this study is to obtain catalysts with the highest acidity and basicity values and then apply them to the esterification and transesterification processes of waste frying oil into biodiesel. The results show that the SO_4/TiO_2-1.5–600 (where 1.5 is the concentration of H_2SO_4, and 600 is the calcination temperature in °C) was the catalyst with the highest acidity (2.49 mmol NH_3/g). Application of this catalyst in the esterification of waste frying oil was successfully achieved, with the highest reduction in free fatty acid (FFA) content of 66.21%. The transesterification reaction of the esterified oil was carried out using a catalyst with the highest basicity, CaO/TiO_2-20–700 (8.13 mmol HCl/g), (where 20 is the concentration of CaO, and 700 is the calcination temperature in °C). This catalyst succeeded in converting waste frying oil into biodiesel of 58.13% with the main composition of methyl ester being methyl oleate.

Keywords Catalysts · TiO_2 · Sulfation · Transesterification · Biodiesel

K. Wijaya (✉) · W. Trisunaryanti
Department of Chemistry, Faculty of Mathematics and Natural Sciences, Universitas Gadjah Mada, Yogyakarta, Indonesia
e-mail: karnawijaya@ugm.ac.id

R. A. Pratika
Department of Chemistry, Faculty of Mathematics and Natural Sciences, Universitas Palangka Raya, Palangka Raya, Indonesia

A. D. Tikoalu
Institute for Nanoscale Science and Technology, College of Science and Engineering, Flinders University, Sturt Road, Bedford Park, South Australia 5042, Australia

S. J. Ikhmayies (ed.), *Advances in Catalysts Research*, Advances in Material Research and Technology, https://doi.org/10.1007/978-3-031-49108-5_11

1 Introduction

Biodiesel is a fuel derived from vegetable oil or animal fat consisting of mono-alkyl esters of long-chain fatty acids. Biodiesel is an alternative fuel that can replace fossil fuels due to it being environmentally friendly, biodegradable, non-toxic, and renewable [1]. Compared to petroleum fuels, it has a fast-growing market due to its lower production costs. Its raw materials, which are renewable and available locally, make it an alternative fuel to replace petroleum fuels [2]. In addition, biodiesel is preferred because it can prevent air pollution due to its low hydrocarbon and carbon monoxide emissions [3].

The general method for converting fatty acid into biodiesel is through esterification and transesterification by combining fatty acids or triglycerides and alcohol with a catalyst. The esterification reaction occurs in the free fatty acid (FFA) oil with methanol to produce methyl esters using an acid catalyst. The purpose of this process is to reduce the levels of free fatty acids and water contained in the oil before proceeding to the transesterification stage. Pretreatment of oil with an acid catalyst in the esterification reaction can reduce the levels of free fatty acids (FFA) and water content of the oil to prevent the saponification reaction, which results in deactivation of the catalyst and a decrease in biodiesel productivity [4, 5].

The products of the esterification or transesterification reaction depend on the physical and chemical properties of the oil raw material. The use of oil with high levels of FFA can also block the separation of methyl esters and glycerol, and cause damage to the reactor or separator used [6]. The quality of the product biodiesel must comply with biodiesel standards: not contaminated by impurities such as catalyst residue, water, glycerol, and excess alcohol which can precipitate and cause corrosion which can damage vehicle engines [7]. In addition, one of the factors that influence biodiesel products is the type of alcohol. Generally, the types of alcohol used in biodiesel production are methanol and ethanol. The use of methanol is preferred because its reaction rate and purity are higher than ethanol [8].

Esterification can be carried out using a homogeneous base or heterogeneous catalyst. Generally, the type of homogeneous catalyst can reduce free fatty acid with a high product percentage and a shorter reaction time. However, it is difficult to separate the catalyst from the product because the catalyst mixes homogeneously with the product. The mechanism of the esterification reaction with a homogenous catalyst is shown in Fig. 1. The opposite of this reaction is hydrolysis, in which water will react with the ester and return to form free fatty acids. The conversion of the biodiesel process must be done carefully to prevent a hydrolysis reaction. The water content (>0.5 wt.%) of methyl esters can produce methanol proton complexes with a high moisture content that is less hydrophobic and polar. This makes protons (H^+) difficult to reach triglycerides. In addition, water can approach free fatty acids, which are also less hydrophobic, which can increase free fatty acids and produce low methyl ester [9].

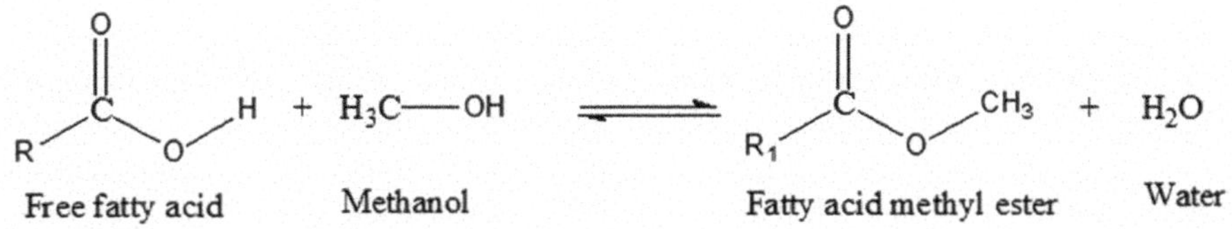

Fig. 1 Mechanism of esterification reaction [10]

Heterogeneous catalysts are generally more active than homogeneous catalysts. These catalysts are non-corrosive, non-toxic, environmentally friendly, easy to separate from the products, and can prevent soap formation. In addition, the advantages of using heterogeneous catalysts are that they do not require a washing process and produce high a percentage of biodiesel [11]. The esterification mechanism with a heterogenous catalyst is shown in Fig. 2.

Titanium dioxide (TiO_2) is a material that can be used as an acid or base heterogeneous catalyst because of its amphoteric properties. It is nontoxic, inexpensive, environmentally friendly, and widely used as a material for catalysts, photocatalysts, paint, ink, and cosmetics [13]. Its application as a photocatalyst to degrade various organic and inorganic pollutants in water based on its nature as a metal oxide semiconductor shows excellent stability and activity in photocatalytic reactions [14]. TiO_2 as a transition metal oxide has higher activity in the transesterification reaction than other acid solid catalysts. The application of those materials as catalysts in biodiesel production from vegetable oil shows that TiO_2 catalyst has high catalytic activity and can work on oil with high FFA levels of up to 15% and produce less soap [15].

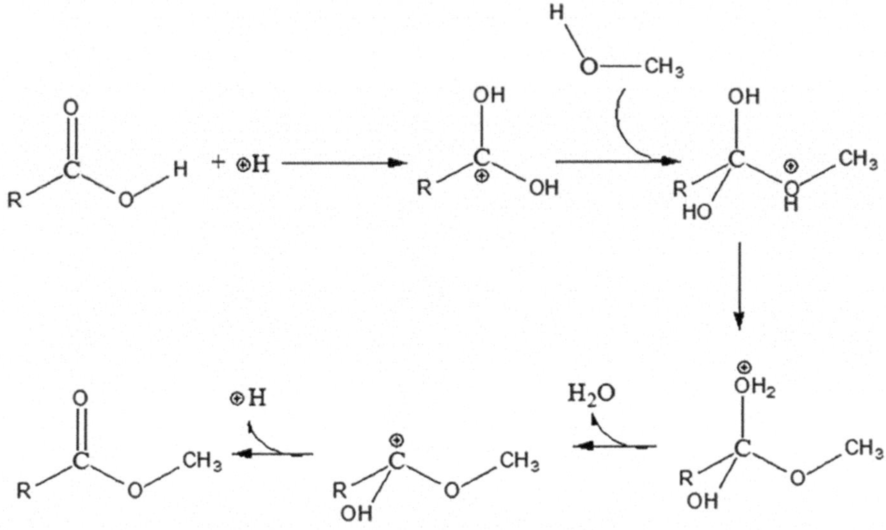

Fig. 2 Mechanism of a heterogeneous acid catalyst in the esterification of FFA [12]

Alsharifi et al. [16] in the production of biodiesel from canola oil using a catalyst Li/TiO_2 suggests that the addition of Li to the TiO_2 surface could increase the catalytic efficiency in the transesterification of canola oil to produce 98% of biodiesel with a catalyst weight of 5 wt.%, a mole ratio of methanol to oil of 24:1, and reaction temperature of 55 °C for 4 h. Gardy et al. [17] reported the use of a SO_4/Fe–Al–TiO_2 catalyst to produce biodiesel from used cooking oil at a reaction temperature of 90 °C, and a mole ratio of methanol to oil of 10:1 for 2.5 h. The yield of biodiesel was produced at up to 96%.

TiO_2 as a heterogeneous acid catalyst can be developed as sulfated titania (SO_4/TiO_2) through a sulfation process with the impregnation of sulfate ions from H_2SO_4 or $(NH_4)_2SO_4$ solutions on the surface of TiO_2. The process leads to the formation of Brønsted and Lewis acid sites, which are the major acid sites of the acid catalyst [18, 19]. Compared with other sulfate metal oxide catalysts such as SO_4/ZrO_2 and SO_4/SiO_2, the SO_4/TiO_2 catalyst exhibits the strongest acidity on its surface [20, 21]. The SO_4/TiO_2 catalyst has low toxicity, easy to handle, and inexpensive. This solid acid catalyst shows good catalytic activity, which has more than half the activity of a liquid sulfuric acid catalyst in the fatty acid esterification reaction with ethanol and the transesterification reaction of vegetable oils [22].

Sulfated Titania (SO_4/TiO_2) has a stronger acidity than pure TiO_2. The source of the acidity in the SO_4/TiO_2 catalyst is the Brønsted acid site, which is formed by the bonding of sulfate and TiO_2. The structure of the SO_4/TiO_2 catalyst is shown in Fig. 3. Sulfated titania consists of two oxygen atoms from S–O bonds bonded to Ti atoms and S=O groups coordinated with Ti atoms. The hydroxyl surface of TiO_2 induced by sulfate groups produces acidic protons that are easily released and produced at the Brønsted acid site [19].

Based on previous research, SO_4/TiO_2 catalyst was reported to have good catalytic activity in the esterification and transesterification. Esterification of benzoic acid with n-butanol using sulfated-TiO_2 was reported by Sarvari et al. [22]. Gardy et al. [23] reported the production of biodiesel from waste frying oil with a 1.5 wt.% $Ti(SO_4)O$ catalyst at a reaction temperature of 75 °C, which yielded 97.1% methyl

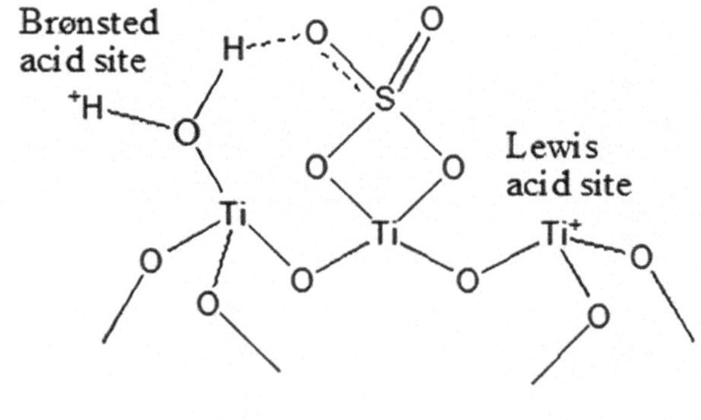

Fig. 3 Structure of Sulfated Titania catalyst [20]

ester. Ropero-Vega et al. [24] reported the use of $TiO_2 - SO_4^{2-}$ catalyst in the esterification of oleic acid at a reaction temperature of 80 °C for 3 h, showing an oleic acid conversion of 82.2 and 100% selectivity.

Transesterification is the chemical conversion of triglycerides with alcohol into alkyl ester or biodiesel with the help of a base catalyst. Transesterification with triglycerides or oil that has a low FFA level can produce good-quality biodiesel products [5]. The heterogeneous base catalysts that are often used in transesterification reactions are alkaline earth metal oxides such as CaO, MgO, BaO, and SrO which show high basicity and a high percentage of biodiesel products due to the presence of cations containing Lewis and Brønsted base sites that can supply electrons or serve as proton acceptors from the reactants [25]. Heterogeneous acid catalysts such as sulfated metal oxides, zeolite, or bentonite can also be used in the transesterification reactions. However, the use of these catalysts requires large amounts of alcohol, high reaction temperature, high concentration of catalyst acidity, long reaction time, and high pressure to keep alcohol in the liquid phase [26, 27].

Calcium oxide (CaO) is a type of solid-based catalyst that is commonly used in transesterification due to its high basicity and selectivity, environmentally friendly, and availability [2]. CaO is a type of metal oxide that has positive metal ions (cations) from Lewis sites which can act as electron acceptors and has negative ions (anions) as proton acceptors and Brønsted base sites. The presence of Brønsted and Lewis base sites as a source of basicity exhibits CaO to be a good base catalyst candidate. According to Pandit and Fulekar [28], biodiesel production from *Chlorella vulgaris* using CaO catalyst resulted in a biodiesel yield of 92.03% with an optimum temperature of 70 °C, the ratio of methanol to biomass 10:1, and reaction time of 3 h. Dias et al. [29] reported the optimization of transesterification using waste frying oil over a CaO catalyst, showed an optimum condition of 60 °C, 12:1 methanol to oil ratio, a 4 h reaction time, 3.39% catalyst, and an 87% biodiesel yield.

The calcination temperature of the CaO catalyst also affects its catalytic activity. Calcination at low temperatures (400–600 °C) results in the formation of a $CaCO_3$ calcite phase due to the absorption of CO_2 and H_2O on the CaO surface, which is very reactive to air [30]. The $CaCO_3$ phase has lower basicity than CaO. Calcination of CaO at a temperature of 700 °C results in the formation of a mixture of CaO, $Ca(OH)_2$, and $CaCO_3$, whereas it forms the majority of the crystalline phase of CaO at a temperature of 800–900 °C. The catalyst basicity of the CaO crystal phase is CaO > $Ca(OH)_2$ > $CaCO_3$. Thus, it can be concluded that the synthesis of the CaO must be carried out at high temperatures ($\geq$ 700 °C) to produce the formation of CaO, which has high basicity and catalytic activity. However, the high catalytic activity of CaO comes from the chemical properties and texture of the CaO [31].

Research on the use of different metal oxide catalysts (AB), where A = alkaline earth metals (Ca, Ba, Mg) and B = transition metals (Ti, Mn, Fe, Zr, Ce) found that $CaTiO_3$, $CaMNO_3$, $Ca_2Fe_2O_5$, $CaZrO_3$, and $CaO–CeO_2$ showed higher basicity and better biodiesel conversion ($\geq$90%) than CaO in the transesterification [11]. The synthesis of a mixed metal oxide catalyst $CaO–ZrO_2$ at different CaO concentrations was reported by Ore et al. [32]. The increase in Ca concentration in the Zr ratio

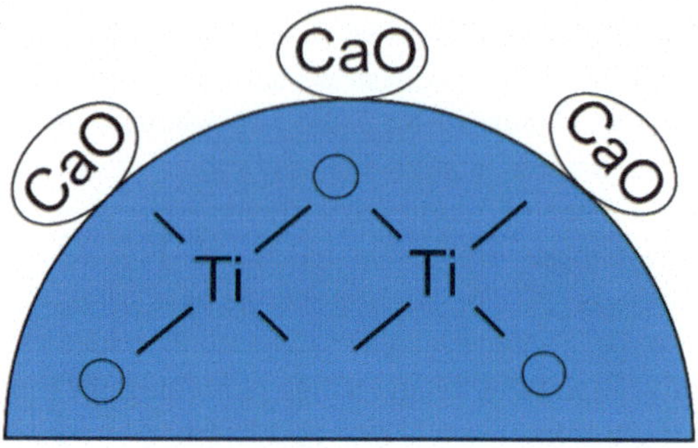

Fig. 4 Illustration of CaO/TiO_2 catalyst

results in higher basicity of the $CaO–ZrO_2$ catalyst and biodiesel conversion up to 79%.

Titanium dioxide is a material that has low basicity. Modification of TiO_2 with CaO could increase the basicity and catalytic activity of TiO_2. The illustration of the CaO/TiO_2 catalyst is shown in Fig. 4. CaO is converted to $Ca(OH)_2$ in the reaction with TiO_2 and then dissolved to produce Ca^{2+} ions before entering the TiO_2 crystal lattice to form CaO/TiO_2 [33, 34]. Biodiesel conversion using a CaO/TiO_2 catalyst has been reported. Biodiesel production from oleic acid has been studied by Sistani et al. [35] with a 3% $CaO/ZrO_2–TiO_2$ catalyst, mole ratio of oil to methanol of 10:1, temperature of 90 °C for 2.5 h indicated the presence of TiO_2 anatase as support for CaO resulted in 96% biodiesel conversion. Pratika et al. [36] reported a high conversion of Jatropha oil into biodiesel by employing a CaO/TiO_2 catalyst with a biodiesel yield of 79.68%. Yahya et al. [37] employed 0.2 wt.% CaO/TiO_2 catalyst to produce biodiesel from waste frying oil (acidity of 3.75 mg KOH g^{-1}). The optimal biodiesel yield was achieved at conditions of a ratio of oil to methanol of 3:1, a reaction time of 1 h, a temperature of 65 °C, and yielding 80% of biodiesel.

The type of oil that is usually used in biodiesel production is soybean, canola, and palm oil which contain high fatty acids. However, the use of these oils requires a higher production cost [18, 36]. Waste frying oil is sought to be a potential source for making biodiesel due to its high fatty acid content, and this oil can be easily obtained from industrial, household, and commercial disposals at a low cost [38]. Tangy et al. [26] reported the use of waste frying oil with $SrO@SiO_2$ catalyst using microwave irradiation successfully converted to biodiesel as high as 99.4%. Transesterification of WFO that employs a $TiO_2–MgO$ catalyst was studied by Wen et al. [39]. The WFO, with an initial FFA and water content of 3.6 mg/KOH and 1.9 wt.%, was successfully converted into biodiesel with a methyl ester yield of 92.3%.

The drawback of using waste frying oil is that it contains a high content of free fatty acids which promote saponification and reduce the formation of methyl esters [6]. The heating process of waste frying oil (150–200 °C) causes the formation of various types of compounds containing oxygen, water, oxidants, and antioxidants that come from fried foods or ingredients. In addition, the heating process of oil also causes oxidation, hydrolysis, polymerization, isomerization, and decomposition of oil into volatile compounds. The oxidation and hydrolysis reactions lead to the formation of polar compounds, such as free fatty acids (FFA) which will form soap during biodiesel production [40]. The high FFA and water contents in waste frying oil require a pre-treatment process to reduce them [41]. Oil with an FFA content of more than 1% will undergo saponification during transesterification [8]. Hence, esterification with an acid catalyst is necessary to reduce the FFA of WFO.

In this research, SO_4/TiO_2 acid catalyst is used to reduce the free fatty acid of WFO via esterification. Next, the biodiesel production from esterified WFO is done through transesterification using a CaO/TiO_2 base catalyst. Preparation of SO_4/TiO_2 and CaO/TiO_2 catalysts was also conducted using $NaHCO_3$ as a porogen to see its effect on the structure of the resulting catalyst. Sodium bicarbonate ($NaHCO_3$) is a raw material whose production comes from natural or synthetic soda ash. $NaHCO_3$ is precipitated from a saturated solution of sodium carbonate (Na_2CO_3) and sodium bicarbonate by injection of carbon dioxide (CO_2) in the solution. Sodium bicarbonate crystals are now widely used in the fields of medical, pharmaceutical, food, gas treatment, and other important industrial activities [42].

$NaHCO_3$ is also commonly called baking soda, which is relatively easy to use, safe, and has low toxicity and cost [43]. It has been found that $NaHCO_3$ can increase the catalytic efficiency of the catalyst for the oxidation of organic compounds in the presence of H_2O_2 as the oxidation center. Zheng et al. [44] studied the effect of $NaHCO_3$ on TiO_2 and suggested that $NaHCO_3$ can be decomposed into $NaCO_3$, H_2O, and CO_2 at low temperatures, causing increased agglomeration and size reduction of TiO_2 that increases the surface area, dispersion, and catalytic activity of the TiO_2 catalyst. Many studies related to the use of pore-forming agents to produce catalysts with uniform pore distribution have been carried out. Gu et al. [45] reported the use of $(NH_4)_2CO_3$ as a pore-forming agent in bioceramic materials $CaHPO_4 \cdot 2H_2O$, $CaCO_3$, La_2O_3, and TiO_2. The results showed the use of $(NH_4)_2CO_3$ resulted in larger pores and evenly distributed pores throughout the surface of the material. In addition, the use of $(NH_4)_2CO_3$ also increases the number of pores in the material.

2 Experimental Section

2.1 Materials

Titanium dioxide ($TiO_2 \geq 99.0\%$), calcium oxide ($CaO \geq 99.0\%$), sulfuric acid (H_2SO_4), hydrochloric acid (HCl), methanol (CH_3OH), ethanol (C_2H_5OH), potassium hydroxide (KOH), and ammonia (NH_3) were purchased from E-Merck. Waste frying oil was obtained from a crispy mushroom business at Depok District, Sleman Regency, D. I. Yogyakarta.

2.2 Methods

2.2.1 Preparation of SO_4/TiO_2 Catalyst

10 g of TiO_2 was added to 100 mL of H_2SO_4 at various concentrations: 0.7, 0.9, 1.1, 1.3, and 1.5 M. To each solution, 25 mL of 5% $NaHCO_3$ was added and stirred for 24 h. Each solid obtained from the sulfation process was dried at 100 °C for 2 h and then calcined at 500 °C for 4 h. The resulting solids were labeled as SO_4/TiO_2-0.7, SO_4/TiO_2-0.9, SO_4/TiO_2-1.1, SO_4/TiO_2-1.3, and SO_4/TiO_2-1.5. The catalyst with the highest acidity value was calcined at temperatures of 400, 500, 600, and 700 °C for 4 h and labeled as SO_4/TiO_2-*x*-400, SO_4/TiO_2-*x*-500, SO_4/TiO_2-*x*-600, and SO_4/TiO_2-*x*-700, respectively, where x is the concentration of H_2SO_4. The catalysts were characterized by Fourier transform infrared (FTIR) spectroscopy, X-ray diffraction (XRD), and acidity tests. The best acidity catalyst was characterized using scanning electron microscopy-energy dispersive X-ray analysis (SEM-EDX), transmission electron microscope (TEM), N_2 adsorption-desorption analysis, thermogravimetric analysis-differential scanning calorimetry (TGA–DSC), and then applied in esterification to reduce FFA of waste frying oil.

2.2.2 Acidity Test of SO_4/TiO_2 Catalyst

The total acidity of the SO_4/TiO_2 catalyst is determined using a thermogravimetric method with ammonia. Initially, an empty porcelain cup was dried in an oven at 105 °C for 1 h to obtain a constant mass (W_o). A total of 0.05 g of sample was put into the empty porcelain cup and weighed to constant as the total mass of the porcelain cup and sample before adsorption (W_1). The porcelain cup with the sample was placed in a desiccator, vacuumed, flowed with ammonia gas, and allowed to sit for 24 h. The porcelain cup containing the sample was removed from the desiccator and then weighed to constant as the total mass of the porcelain cup and sample after adsorption (W_2). The total acidity of the catalyst was calculated using the formula in Eq. (1) [46].

$$\text{Total acidity}\left(\text{mmolg}^{-1}\right) = \frac{W_2 - W_1}{(W_1 - W_0) \cdot MW_{NH_3}} \times 1000 \quad (1)$$

where MW is the molecular weight.

2.2.3 Preparation of CaO/TiO_2 Catalyst

10 g of TiO_2 was suspended in 100 mL of distilled water and put into an autoclave. It was treated with CaO at concentrations of 1, 5, 10, 15, and 20% (w/w) and 25 mL of 5% $NaHCO_3$, before being heated at 120 °C for 24 h. The solids obtained were dried

in an oven at 100 °C for 2 h, calcined at 500 °C for 4 h, and then labeled as CaO/TiO_2-1, CaO/TiO_2-5, CaO/TiO_2-10, CaO/TiO_2-15 and CaO/TiO_2-20. The catalysts were characterized by FTIR, and basicity test. The highest basicity catalyst was calcined at 400, 500, 600, 700, and 800 °C for 4 h, and was denoted as CaO/TiO_2-*x*-400, CaO/TiO_2-*x*-500, CaO/TiO_2-*x*-600, CaO/TiO_2-*x*-700, and CaO/TiO_2-*x*-800, where x is the concentration of CaO. The catalysts were characterized by FTIR, XRD, and basicity test. The best acidity catalyst was characterized using SEM-EDX, TEM, N_2 adsorption-desorption, and TGA-DSC, and then applied in the transesterification of the esterified oil into biodiesel. The basicity of CaO/TiO_2 catalysts was determined by the titration method. The total basicity of the catalyst was calculated using the formula in Eq. (2).

$$\text{Total basicity (mmol HCl/g)} = \frac{(V_{\text{NaOH blank}} - V_{\text{NaOH with catalyst}}) \times M_{\text{NaOH}}}{\text{mass of sampel } (g)} \tag{2}$$

where V is the volume and M is the molarity.

2.2.4 The Esterification Reaction of Waste Frying Oil

Preparation of Waste Frying Oil (WFO)

WFO was prepared by heating at 105 °C followed by filtration with the testing sieve of mesh size 200 μm to remove water content and impurities. The initial FFA level of WFO was determined by the titration method using KOH and calculated using Eq. (3).

$$\text{FFA level (\%)} = \frac{V_{\text{KOH}} \times N_{\text{KOH}} \times \text{MW}_{\text{of waste frying oil}}}{\text{mass of sampel}(g)} \times 100\% \tag{3}$$

where N is the molality.

Catalyst Weight Variation

The optimal amount of SO_4/TiO_2 catalyst in the esterification of waste frying oil was determined by varying the catalyst weight 1, 2, 3, 4, and 5% of the total weight of oil and methanol. The mole ratio of oil to methanol used in each experiment was 1:9. Prior to esterification, the catalyst was mixed with methanol and refluxed at 45 °C for 10 min. After that, 25 g of waste frying oil was added to the mixture and it was refluxed at 65 °C for another 30 min. The percentage of FFA reduction was then calculated using Eq. (4) [38].

$$\text{FFA reduction (\%)} = \frac{\text{FFA of waste frying oil} - \text{FFA of esterified oil}}{\text{FFA of waste frying oil}} \times 100\% \tag{4}$$

Mole Ratio of Waste Frying Oil to Methanol Variation

SO_4/TiO_2 catalyst, at the optimum FFA reduction by catalyst weight, was mixed with methanol at mole ratios of 1: 9, 1:12, 1:15, 1:18, and 1:21. Each variation was refluxed at 45 °C for 10 min. Next, 25 g of waste frying oil was added, and then refluxed at 65 °C for 30 min.

Reaction Time Variation

SO_4/TiO_2 catalyst, at the optimum FFA reduction by catalyst weight and methanol mole ratio, was refluxed at 45 °C for 10 min, added to 25 g of oil, and refluxed again at 65 °C with a variation of reaction time of 30, 60, 90, and 120 min.

2.2.5 Transesterification of Esterified Oil to Biodiesel

Transesterification was carried out using waste frying oil with the optimum FFA reduction in the esterification and CaO/TiO_2 catalyst. CaO/TiO_2 catalyst with a weight of 3% was mixed with methanol (ratio oil to methanol 1:18). The mixture was refluxed at 45 °C for 10 min. Subsequently, 20 g of esterified oil was added to the mixture and refluxed again for 120 min at 65 °C. The transesterification product was then centrifuged for 30 min at 4000 rpm to separate biodiesel, glycerol, and catalyst. The resulting biodiesel was characterized using FTIR, GC-MS, and ^{1}H-NMR.

2.3 *Characterization*

Characterization by Fourier transform infrared (FTIR) was performed using thermo scientific Nicolet™ iS™ 5 with thermo scientific iD5 ATR Accessory and scanned in the range 400–4000 cm^{-1}. X-ray diffraction (XRD) patterns were obtained using Bruker D2 phaser that employs a Cu line focus x-ray tube with Ni k_β absorber and K_α ratio = 0.5; λ = 1.54060 Å in the 2θ range of 5–90°. Scanning electron microscope (SEM) analysis was performed using JEOL JSM-6510, operated at 15 kV accelerating voltage along with Energy dispersive x-ray spectrometry (EDX), JED-2300. Transmission electron microscopy (TEM) was performed using a JEOL JEM 1400. Surface area analysis (SSA) was performed using a NOVA quantachrome with Brunauer-Emmet-Teller (BET) from N_2 gas adsorption isotherms measured at 77 K. Thermogravimetric-differential scanning calorimetry (TG-DSC) was performed using Linseis TGA PT1000 with a heating rate of 10 °C/min. Gas chromatography-mass spectroscopy (GC–MS) analysis was carried out using a Shimadzu QP2010S, employing the Rtx 5 MS column. Proton nuclear magnetic resonance (^{1}H-NMR) spectra were obtained using JEOL ECS-400 using chloroform as a solvent.

3 Results and Discussion

3.1 *Characterization of SO_4/TiO_2 Acid Catalyst*

FTIR spectra of TiO_2-600 and SO_4/TiO_2 catalysts are shown in Fig. 5. The absorption peaks at 596 and 662 cm^{-1} correspond to the stretching vibration of the O–Ti–O bond [47]. The absorptions at around 3300–3400 and 1643–1648 cm^{-1} correspond to the stretching of –OH and bending vibrations of H_2O molecules coordinated onto the catalyst [48]. The intensity of the –OH absorption peak was increased with increasing sulfate concentration, due to the large number of –OH groups in the catalyst. The sulfation of TiO_2 provided new peaks at 1130–1143, 1053–1079, and 947–956 cm^{-1} which were assigned to be the asymmetric vibration of S=O, asymmetric, and symmetric vibrations of S–O from the sulfate on the TiO_2 surface, respectively [20, 22, 32]. Higher SO_4^{2-} ion concentration increased the intensity of vibrations of S–O and S=O, indicating that the sulfation was successfully achieved. The highest S–O and S=O absorption peak intensity was shown by the SO_4/TiO_2-1.5.

The acidity test of the catalysts was carried out by the gravimetric method using NH_3 solution in a vacuum condition. The amount of NH_3 absorbed by the catalyst is equal to the total number of acid sites of the catalyst. The acidity value of SO_4/TiO_2 at various concentrations is shown in Table 1. The acidity result showed that TiO_2-600 had an acidity value of 1.16 mmol/g. The addition of sulfate causes an increase in the acidity of catalysts. The SO_4/TiO_2-1.5 catalyst had the highest acidity value (2.32 mmol/g). Similar studies about the synthesis of SO_4/TiO_2 catalysts with concentrations of 0.5; 1; 1.5 and 2 M were studied by Chen et al. [19]. The optimum acidity catalyst was achieved by the SO_4/TiO_2 1.5 M catalyst.

The catalyst with the highest acidity (SO_4/TiO_2-1.5) was then calcinated at temperatures of 400, 500, 600, and 700 °C. The FTIR spectra of the SO_4/TiO_2-1.5–400 catalyst (Fig. 6) show a high-intensity absorption of –OH stretching and O–H–O bending vibrations at 3370 and 1630 cm^{-1}, which indicates the presence of water from the calcination process at a low temperature. The intensity of –OH groups decreased significantly as the calcination temperature increased from 500 to 700 °C. This condition explains that the number of –OH groups in the catalyst decreases as the calcination temperature rises [47].

A new absorption at 1221 cm^{-1} also appears in the spectra, representing S=O symmetry vibration [49]. The highest intensity of S=O and S–O bonds from SO_4 absorption indicates that SO_4/TiO_2-1.5–400 exhibits a high concentration of acid sites, while the acidity value of SO_4/TiO_2-1.5 had practically decreased after calcination at 500–700 °C.

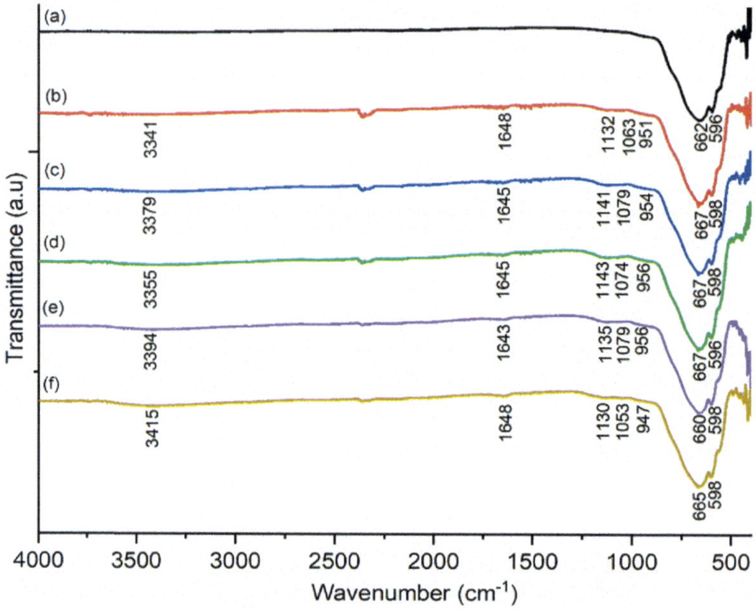

Fig. 5 FTIR spectra of **a** TiO_2, and SO_4/TiO_2 at various concentrations of H_2SO_4: **b** 0.7 M, **c** 0.9 M, **d** 1.1 M, **e** 1.3 M, and **f**. 1.5 M

Table 1 Acidity values of TiO_2 and SO_4/TiO_2 catalysts

Catalyst	Acidity (mmol NH_3/g)
TiO_2-600	1.16
SO_4/TiO_2-0.7	2.14
SO_4/TiO_2-0.9	2.17
SO_4/TiO_2-1.1	2.21
SO_4/TiO_2-1.3	2.28
SO_4/TiO_2-1.5	2.32

The catalyst was tested for its acidity. It was found that calcining SO_4/TiO_2-1.5 catalyst at 400 °C was able to increase the catalyst acidity from 1.16 to 6.00 mmol/g (Table 2) and it was found that the SO_4/TiO_2-1.5 catalyst was the catalyst with the highest acidity. However, a high number of water molecules in the SO_4/TiO_2-1.5–400 catalyst can generate saponification in the esterification reaction. The incomplete calcination process causes a large number of water molecules in the catalyst that can facilitate the formation of saponification in transesterification for biodiesel production [46]. In addition, the high number of water molecules facilitates the absorption of ammonia into the catalyst during the acid test process by gravimetric method, resulting in a high catalyst acidity value [50]. As a result, the second-highest total

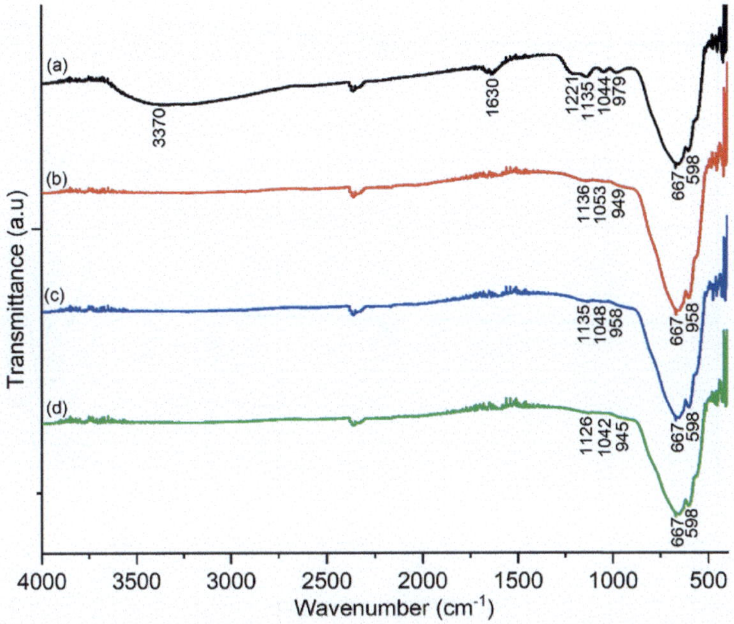

Fig. 6 FTIR spectra of SO_4/TiO_2-1.5 calcined at **a** 400 °C. **b** 500 °C. **c** 600 °C. **d** 700 °C

acidity (SO_4/TiO_2-1.5–600) was then selected for esterification with an acidity value of 2.48 mmol NH_3/g.

The SO_4/TiO_2-1.5 catalyst after the acidity test was then characterized using FTIR to show the interaction between –NH groups and the catalyst. Figure 7. shows the FTIR spectra of the catalysts from the acidity test by ammonia adsorption. The SO_4/TiO_2-1.5–400 spectra showed a specific peak at 1096 and 1443 cm^{-1} that are the symmetric stretching vibrations of NH_3 coordinated onto the Lewis acid sites and the Brønsted acid site from the interaction between conjugated acid and hydrogen [51]. Higher intensities of Lewis and the Brønsted acid site indicate that the SO_4/TiO_2-1.5–400 catalyst has higher total acidity.

The high intensity of NH_3 on SO_4/TiO_2-1.5–400 was caused by the large number –OH groups that can facilitate the chemisorption of ammonia on the surface of the

Table 2 Acidity values of SO_4/TiO_2-1.5 catalysts

Catalyst	Acidity (mmol NH_3/g)
SO_4/TiO_2-1.5–400	6.00
SO_4/TiO_2-1.5–500	2.32
SO_4/TiO_2-1.5–600	2.48
SO_4/TiO_2-1.5–700	2.34

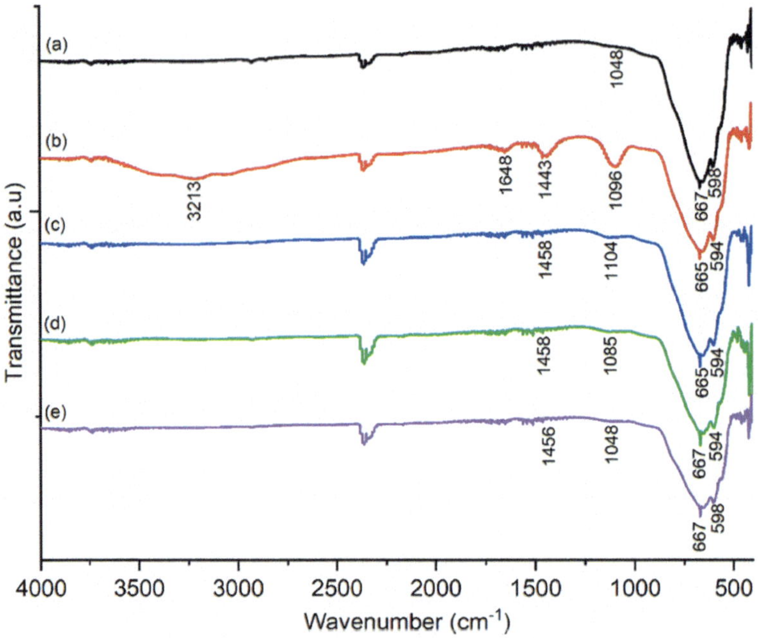

Fig. 7 FTIR spectra acid tested. **a** TiO_2 and SO_4/TiO_2-1.5 calcined at: **b** 400 °C. **c** 500 °C. **d** 600 °C. **e** 700 °C

catalyst [33]. When the calcination temperature was increased, the absorption intensity of NH_3 decreased due to the decrease in the acidity of the catalysts. A similar study about the acidity of the SO_4/TiO_2 catalyst has been carried out. Chen et al. [19] reported the acidity value of the SO_4^{2-}/TiO_2 catalyst based on the NH_3– temperature programmed desorption (TPD) analysis, which shows that TiO_2 contains weak and moderate acid sites. The addition of SO_4^{2-} increases the acidity of TiO_2 due to the presence of weak, moderate, and strong acid sites. The higher the SO_4^{2-} concentration, the stronger the acid sites on the SO_4^{2-}/TiO_2 catalyst. In addition, based on the results of the FTIR-pyridine analysis, TiO_2 has Lewis acid sites while SO_4^{2-}/TiO_2 catalyst has Lewis and Brønsted acid sites formed from the coordination between SO_4 and TiO_2 and through adsorption on SO_4 and TiO_2, respectively. The presence of Lewis and Brønsted acid sites makes the SO_4^{2-}/TiO_2 catalyst have good acidity.

Figure 8 shows the X-ray diffraction patterns of TiO_2-600 and SO_4/TiO_2-1.5 400, 500, 600, and 700 °C catalysts. The main diffraction peaks of TiO_2-600 appeared at $2\theta = 25.38°$, 37.07°, 37.81°, 38.69°, 48.17°, 52.95°, 55.13°, 62.82°, 68.91°, 70.39°, and 75.13° [JCPDS no. 21-1272], that indicates the formation of the TiO_2 anatase [50]. The TiO_2-600 diffraction peaks showed high crystallinity and confirmed the rutile and brookite phases were not formed. Nagaraj et al. [52] reported that TiO_2 anatase calcined at a temperature of 550–750 °C only formed the anatase crystalline phase. Afshar et al. [53] reported the formation of a crystalline anatase phase without the formation of rutile and brookite phases in the synthesis of the SO_4^{2-}/TiO_2-400.

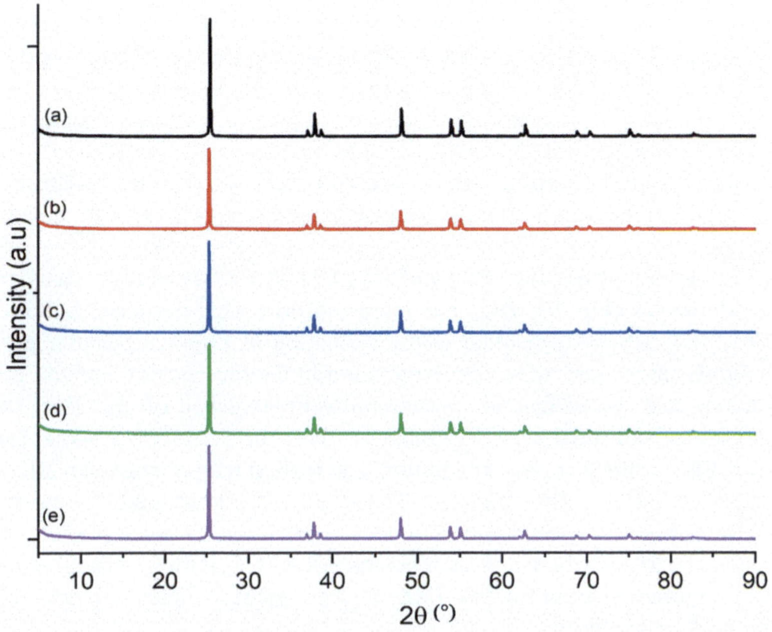

Fig. 8 XRD pattern of **a** TiO_2, and SO_4/TiO_2-1.5 calcined at: **b** 400 °C. **c** 500 °C. **d** 600 °C. **e** 700 °C

The mixture of anatase and rutile phases begins to form at 800 °C, while the rutile phase is formed at temperatures above 1000 °C [54].

The same characteristic peaks appeared in the diffractograms of TiO_2 and SO_4/TiO_2-1.5 catalysts, which indicated no change in the structure of titania after sulfation with sulfuric acid. However, a decrease in the x-ray diffraction intensity of TiO_2 was observed. The presence of SO_4^{2-} ion can cover the surface of TiO_2, resulting in the formation of amorphous Ti–O–S [55]. An increase in calcination temperature has caused the diffraction intensity of TiO_2 to increase due to the decomposition of SO_4^{2-} ion. In addition, the use of $NaHCO_3$ also affected the diffraction peaks of TiO_2. Zheng et al. [44] explained that the addition of $NaHCO_3$ results in the diffraction peak of TiO_2 anatase being weak due to the ion radius of Ti^{4+} which is smaller than that of Na^+ (74.4 and 102 pm), making Na^+ ions easily enter the TiO_2 lattice, resulting in an increase in lattice parameters. Next, the average crystal size (D) of the catalysts was determined using the Debye–Scherrer equation:

$$D = K\lambda/\beta\cos\theta \tag{5}$$

where D = average crystal diameter (nm), K = Scherrer constant (0.9), λ = x-ray wavelength = 1.5418 Å, β = full width at half maximum (FWHM), and θ = Bragg angle.

The average crystal size of TiO_2 and SO_4/TiO_2-1.5 temperature calcinated at 400, 500, 600, and 700 °C was 64.58, 45.01, 46.47, 48.53, and 47.18 nm, respectively. The crystal size of catalysts shows a decrease with the presence of SO_4^{2-} ion that covers the TiO_2 surface. Increasing calcination temperature resulted in an increase in the crystal size of the SO_4/TiO_2 catalyst. In addition, the use of $NaHCO_3$ also has an effect on decreasing the particle size of SO_4/TiO_2. It was confirmed by Zheng et al. [44] who reported the decrease of the average crystal size of TiO_2 with the addition of $NaHCO_3$.

The morphology of the TiO_2-600 and SO_4/TiO_2-1.5–600 catalysts was explored using SEM images (Fig. 9). TiO_2 has more uniform-sized particles. Sulfation on TiO_2 (SO_4/TiO_2-1.5–600) resulted in the formation of brighter granular particles with a slightly larger size due to agglomeration in the presence of sulfated groups. This signaled that the sulfate was successfully impregnated on the TiO_2 surface [49], which was confirmed via EDS (Table 3) where the SO_4/TiO_2-1.5–600 catalyst contains 1.36% sulfur by mass. In addition, calcination temperature also affects the morphology and size of TiO_2 particles. Haider et al. [54] reported that calcination of SO_4/TiO_2 catalyst at a temperature of 400 °C produces spherical particles with a particle size of <20 nm, whereas at a temperature of 600 °C, it forms an agglomeration with a larger particle size, and at 700–1000 °C, it produces particles with non-uniform sizes and high crystallinity.

TEM images of TiO_2-600 and SO_4/TiO_2-1.5–600 catalysts are shown in Fig. 10. TiO_2 particles had a particle size in the range of 70–160 nm. Sulfation exhibited a catalyst morphology that did not suffer from damage due to the presence of SO_4^{2-}

Fig. 9 SEM images of TiO_2-600 (**a**), and SO_4/TiO_2-1.5–600 (**b**)

Table 3 Composition of TiO_2 and SO_4/TiO_2-1.5–600

Catalyst	Mass percentage (%)			
	Ti	O	C	S
TiO_2-600	76.26	20.77	2.97	–
SO_4/TiO_2-1.5–600	50.29	45.54	2.82	1.36

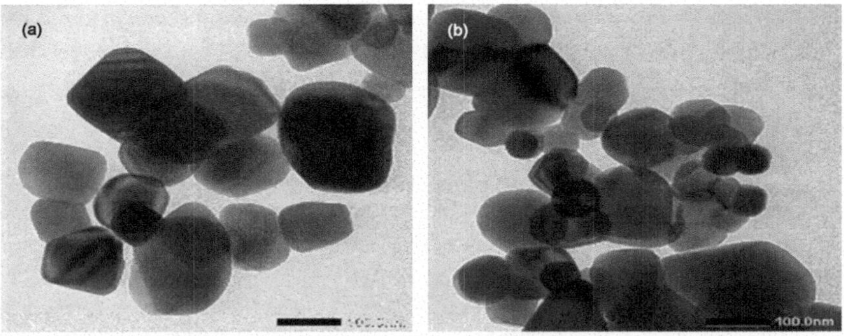

Fig. 10 TEM images of TiO_2-600 (**a**), and SO_4/TiO_2-1.5–600 (**b**)

on the TiO_2 surface. The SO_4/TiO_2-1.5–600 catalyst had a particle size in the range 40–180 nm with a gap between the catalyst particles. According to Fadhil et al. [56], the presence of gaps between the catalyst particles will facilitate the contact between the reactants and the catalyst in the esterification/transesterification, which increases the increase in the methyl ester product or the resulting biodiesel.

The textural properties and pore structure of TiO_2-600 and SO_4/TiO_2-1.5–600 catalysts were determined by N_2 adsorption-desorption. The porosity of catalysts is shown in Table 4. TiO_2 has a surface area of 7.055 m^2/g, a pore volume of 0.0624 cc/g, and an average pore diameter of 34.48 nm. The sulfation process reduces the specific surface area, total pore volume, and pore diameter of the catalyst due to the formation of aggregates that block the pore of the catalyst through the interaction between SO_4 and TiO_2 [20]. Zhao et al. [57] revealed that the specific surface area of the SO_4/TiO_2 was also influenced by the calcination temperature. Calcination at low temperature (400 °C) resulted in a catalyst with a high specific surface area, whereas calcination at high temperature (≥600 °C) resulted in catalysts with a decreased surface area.

Figure 11 depicts the profiles of a catalyst's N_2 adsorption-desorption isotherm (right) and pore size distribution (left). The N_2 adsorption-desorption isotherm curve of TiO_2-600 and SO_4/TiO_2-1.5–600 implies a type IV adsorption-desorption isotherm with a type H3 hysteresis loop. Based on the International Union of Pure and Applied Chemistry (IUPAC) isotherm classification, type IV was the characteristic of a mesoporous material. Type H3 hysteresis loop indicated the presence of slit-like pores in aggregates with a wide pore size distribution [58]. The pore distribution of TiO_2-600 and SO_4/TiO_2-1.5–600 showed the catalysts consisted of mesoporous material, and were uniformly distributed throughout the surface of the

Table 4 Porosity of TiO_2 and SO_4/TiO_2-1.5–600 catalyst

Catalyst	Surface area (m^2/g)	Total pore volume (cc/g)	Average pore diameter (nm)
TiO_2-600	7.055	0.0624	34.48
SO_4/TiO_2-1.5–600	6.559	0.0592	34.08

catalyst. TiO_2-600 showed a narrow peak and has an optimum pore distribution size of 2 nm, which explains the homogeneous pore size [41]. The SO_4/TiO_2-1.5–600 showed a wider pore distribution, which leads to a heterogeneous pore size with an optimum pore distribution of 3 nm.

Figure 12 shows the TGA-DSC curve for TiO_2-600 and SO_4/TiO_2-1.5–600 catalysts. The mass change in the TGA curve appears in the temperature range of 50–250 °C, which is the loss of water molecules adsorbed physically on the catalyst. Furthermore, calcination temperatures of 500–700 °C indicated the loss of the organic material and the decomposition of sulfate (SO_4/TiO_2-1.5–600) from the surface of TiO_2. Finally, the mass change at 750–900 °C indicated the formation of the rutile phase in the TiO_2 [13, 52]. The higher mass loss in the SO_4/TiO_2-1.5–600 catalyst indicates this catalyst has water molecules and organic material, (sulfate) higher than TiO_2 due to the catalyst synthesis process.

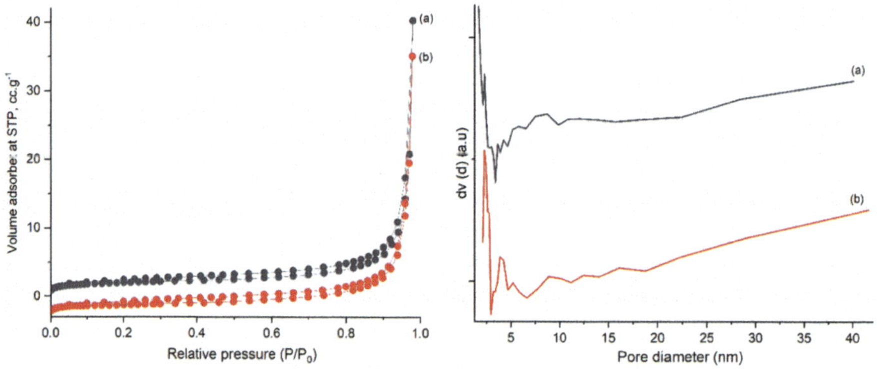

Fig. 11 Curves of N_2 adsorption-desorption isotherms (left), and Barrett-Joyner-Halenda (BJH) pore size distribution (right) of TiO_2-600 (**a**), and SO_4/TiO_2-1.5–600 (**b**)

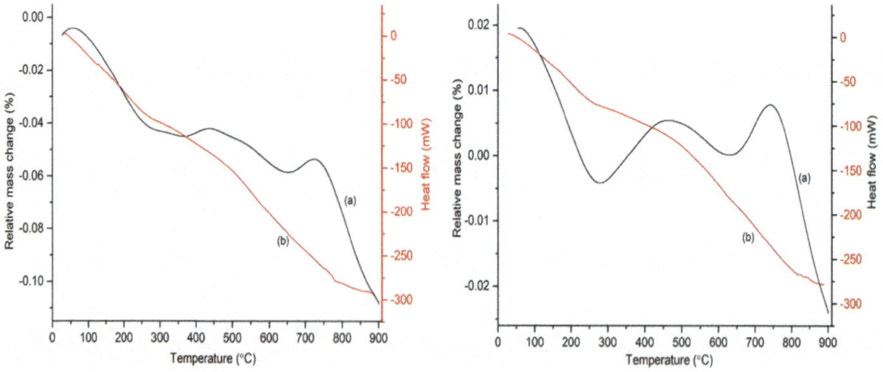

Fig. 12 TGA (**a**), and DSC (**b**) curves of TiO_2-600 (left), and SO_4/TiO_2–1.5–600 (right)

The DSC curve of the TiO_2-600 and SO_4/TiO_2-1.5–600 catalysts showed the formation of an endothermic curve at 168–322 °C, which contributes to the loss of physically adsorbed water from the catalyst. The second curve in the TiO_2-600 formed in the range of 752–854 °C, indicates the loss of organic material and the formation of the TiO_2 rutile phase. The SO_4/TiO_2-1.5–600 showed an endothermic curve at 613–835 °C, which indicated the decomposition of sulfate ions from the surface of the catalyst [52, 56].

3.2 *Characterization of CaO/TiO_2 Base Catalyst*

The FTIR spectra of TiO_2-600, CaO, and CaO/TiO_2 catalysts at various concentrations is shown in Fig. 13. The stretching vibration of the O–Ti–O bond corresponds to the absorption peaks at 596 and 662 cm^{-1} [47]. The impregnation of CaO on TiO_2 showed new absorption peaks at 1413, 872, and 710 cm^{-1}, which correspond to the stretching and bending vibrations of C–O and C=O bonds from $CaCO_3$ [2]. The higher the concentration of CaO, the larger the absorption intensity of $CaCO_3$, which indicates that CaO has been successfully impregnated onto TiO_2. CaO spectra revealed peaks at 557 and 3640 cm^{-1}, corresponding to the CaO bond and –OH stretching from $Ca(OH)_2$ [2]. The presence of these peaks refers to the fact that these bonds are formed at high calcination temperatures. According to Roschat et al. [46], a crystalline phase of CaO and $Ca(OH)_2$ will be formed when the calcination temperature is above 700 °C.

The basicity test of the CaO/TiO_2 catalyst was carried out to see the effect of CaO on the basicity of TiO_2. The total basicity of the CaO/TiO_2 catalyst was determined by the titration method using HCl. The titration is carried out on excess HCl which is neutralized by the catalyst. The amount of HCl that is neutralized by the catalyst is considered the total basicity contained in the catalyst. The basicity value of CaO/TiO_2 catalysts is shown in Table 5. It was found that the addition of 1% CaO to the TiO_2 catalyst was able to significantly increase the basicity value by around 34 times, from 0.13 to 4.45 mmol HCl/g.

The higher the concentration of CaO, the greater the basicity of the CaO/TiO_2 catalyst. The catalyst with the highest basicity value was found to be the CaO/TiO_2-20 catalyst with a total basicity of 7.82 mmol HCl/g. According to Li et al. [34], the base site of the CaO/TiO_2 particles comes from the CaO (single) base site and the $CaTiO_3$ surface. CaO/TiO_2 catalyst preparation at high temperatures causes CaO to convert to $Ca(OH)_2$, which dissolves to produce Ca^{2+} ions, and then enters the TiO_2 crystal lattice to form $CaTiO_3$. The reaction mechanism of TiO_2 and CaO to form $CaTiO_3$ is shown in Eqs. (6), (7), and (8).

$$CaO + H_2O \rightleftharpoons Ca(OH)_2 \tag{6}$$

$$Ca(OH)_2 \rightleftharpoons Ca^{2+} + 2OH^- \tag{7}$$

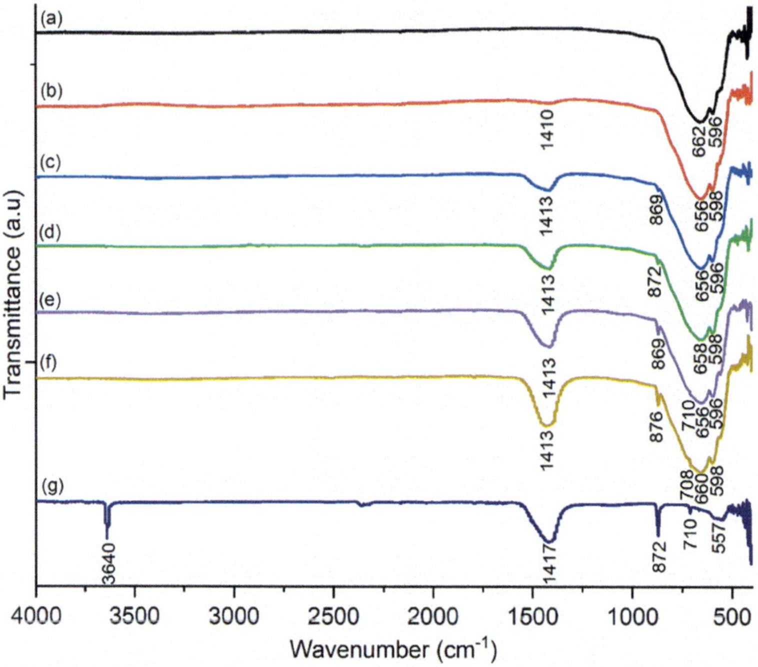

Fig.13 FTIR spectra of **a** TiO_2, and CaO/TiO_2 at different concentrations of CaO: **b** 1%, **c** 5%, **d** 10%, **e** 15%, **f** 20%, and **g** CaO

Table 5 Basicity values of CaO, TiO_2, and CaO/TiO_2 catalysts

Catalyst	Basicity (mmol HCl/g)
CaO	24.21
TiO_2-600	0.13
CaO/TiO_2-1	4.45
CaO/TiO_2-5	5.77
CaO/TiO_2-10	6.19
CaO/TiO_2-15	6.97
CaO/TiO_2-20	7.83

$$Ca^{2+} + TiO_2 + 2OH^- \rightleftharpoons CaTiO_3 + H_2O \quad (8)$$

The FTIR spectra of the CaO/TiO_2-20 catalyst calcined at various temperatures displayed in Fig. 14 show a decrease in the absorption intensity of the C–O and C=O bonds with the increase in calcination temperature. The CaO/TiO_2-20–700 catalyst spectra revealed the appearance of new peaks in the 3640 cm^{-1}, indicating the physical adsorption of water molecules (–OH stretching) from $Ca(OH)_2$. This fact implies that $CaCO_3$ would decompose at higher temperatures and begin to form

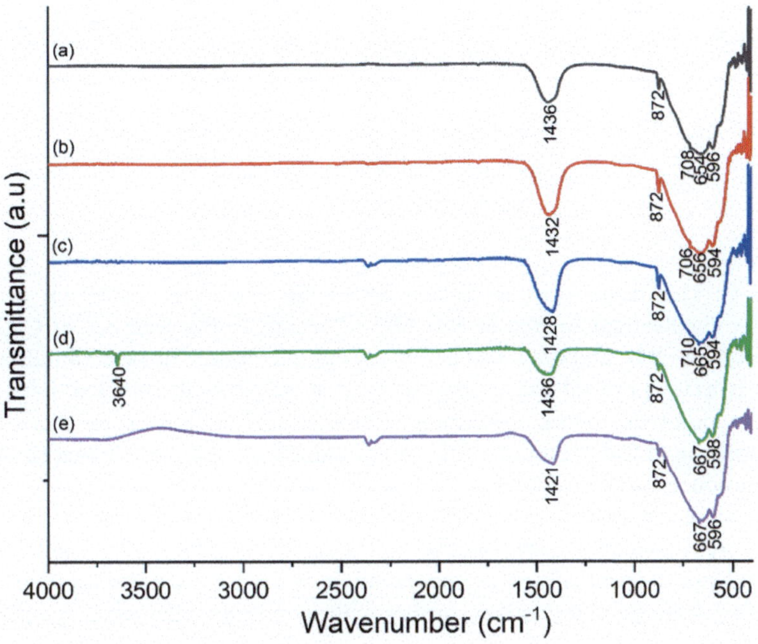

Fig.14 FTIR spectra of CaO/TiO_2-20 calcined at **a** 400 °C. **b** 500 °C. **c** 600 °C. **d** 700 °C. **e** 800 °C

$Ca(OH)_2$ at 700 °C. However, at a calcination temperature of 800 °C, the spectra of $Ca(OH)_2$ disappeared, confirming that $Ca(OH)_2$ would decompose at a temperature of 800 °C and form a CaO crystalline phase. This is consistent with the finding of Roschat et al. [46], who explained that the synthesis of CaO at 700 °C forms a mixture of $Ca(OH)_2$ and CaO phase, whereas at 800–900 °C it forms crystalline CaO.

The basicity values of the CaO/TiO_2 catalysts at different calcination temperatures are listed in Table 6. It was found that higher calcination temperatures increase the basicity of catalysts. The highest basicity catalyst was shown by CaO/TiO_2-20–700 with a basicity value of 8.13 mmol HCl/g. The interaction between CaO and TiO_2 at high calcination temperatures can lead to the formation of CaO and $CaTiO_3$, which is formed through the bond between CaO and TiO_2 [59]. The CaO and $CaTiO_3$ formed phase can form Lewis and Brønsted base sites which contribute to the increase in the basicity of the CaO/TiO_2 catalyst [34].

Figure 15 shows the XRD patterns of TiO_2-600 and CaO/TiO_2 catalysts at various calcination temperatures. The impregnation of CaO on TiO_2 at temperatures of 400–600 °C generated new peaks at $2\theta = 29.51°$, 36.03°, and 39.60°, which correspond to the lime-$CaCO_3$ and the calcite-$CaCO_3$ phase [2]. The higher the calcination temperature, the lower the absorption of $CaCO_3$. Calcination at 700 °C resulted in the loss of $CaCO_3$ peaks, and the formation of new peaks at 18.11° and 34.12°, which correspond to the portlandite-$Ca(OH)_2$, and at 33.21°, which represents the CaO crystalline phase. The portlandite-$Ca(OH)_2$ peaks then vanished at the temperatures

Table 6 Basicity values of CaO/TiO_2-20 catalysts

Catalyst	Basicity (mmol HCl g^{-1})
CaO/TiO_2-20–400	7.41
CaO/TiO_2-20–500	7.56
CaO/TiO_2-20–600	7.70
CaO/TiO_2-20–700	8.13
CaO/TiO_2-20–800	7.92

of 800 °C. This result is in agreement with FTIR spectra of CaO/TiO_2 catalysts (Fig. 14). Catalyst calcination at 400–600 °C results in the presence of $CaCO_3$, a mixture of $Ca(OH)_2$ and CaO, and a CaO crystalline phase. The peak of the calcium titanate ($CaTiO_3$) crystal phase was observed at 59.29° with very low intensity at the calcination temperature in the range 700–800 °C. This peak is produced by the decomposition of $CaCO_3$ and $Ca(OH)_2$ to produce Ca^{2+} ions, which react with Ti^{4+} ions to produce $CaTiO_3$ [34, 60].

The average crystallite size of TiO_2-600 and CaO/TiO_2-20 catalysts at calcination temperatures of 400, 500, 600, 700, and 800 °C are 64.58, 51.04, 49.48, 48.02, 49.08, and 46.82 nm, respectively. The impregnation of CaO caused the crystalline size of TiO_2 to decrease and decrease further with increasing calcination temperature.

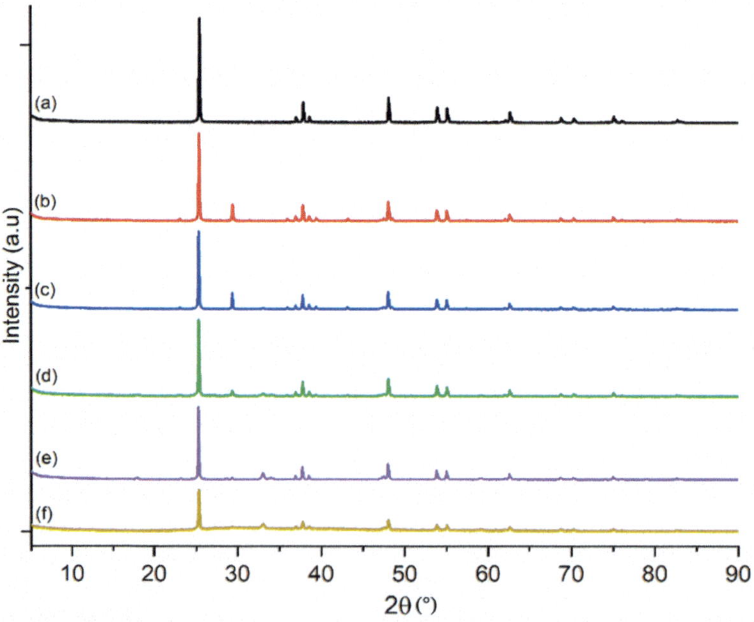

Fig. 15 XRD pattern of TiO_2 (**a**), and CaO/TiO_2-20 calcined at: **b** 400 °C. **c** 500 °C. **d** 600 °C. **e** 700 °C. **f** 800 °C

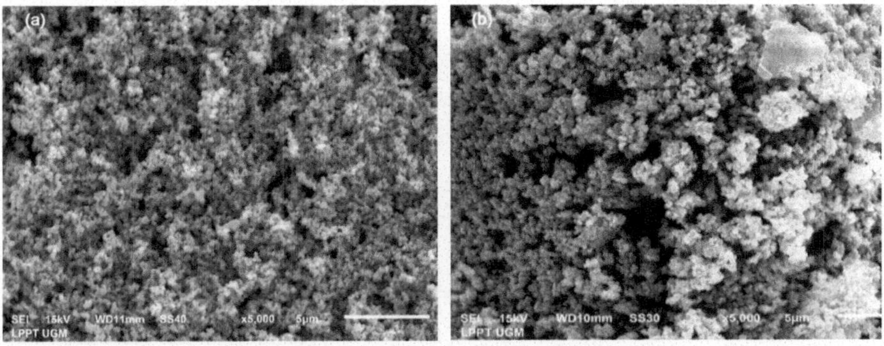

Fig. 16 SEM images of TiO_2-600 (**a**), and CaO/TiO_2-20–700 (**b**)

Pandit et al. [28] reported that CaO on TiO_2 results in the formation of aggregates with smaller particle' sizes.

The surface morphology of TiO_2-600 and CaO/TiO_2-20–700 catalysts is shown in Fig. 16. The impregnation of CaO on TiO_2 caused the morphology of the catalyst to be irregular or non-uniform in size. In addition, larger particle sizes are also formed due to the formation of aggregates [27]. Fadhil et al. [2] reported the presence of CaO particles, causing the formation of aggregates with non-uniform particles sizes. EDS analysis of CaO/TiO_2-20–700 revealed the presence of CaO with 9.96% of calcium (Table 7). This indicates that CaO has been successfully impregnated on TiO_2.

The TEM images of the TiO_2-600 and CaO/TiO_2-20–700 catalysts are shown in Fig. 17. TiO_2 particles had a spherical shape with a particle size in the range of 70–120 nm. The CaO/TiO_2-20–700 exhibited aggregate particles that were distributed equally throughout the catalyst. The presence of aggregations results in non-uniform particle sizes with a size in the range of 20–120 nm. According to Dalai et al. [3]; Aghilinategh et al. [61], calcination of CaO/TiO_2 at 700 °C will form an agglomeration of CaO particles, which have an irregular particle size and can increase the surface area and decrease the pore diameter of the TiO_2.

The N_2 adsorption-desorption analysis was performed to determine the textural properties and pore structure of TiO_2-600 and CaO/TiO_2-20–700 catalysts. The porosity of the catalyst influences the catalytic activity of heterogeneous base catalysts in the transesterification reaction [28]. Table 8 shows that the surface area of TiO_2 increased after impregnation with CaO, while the total pore volume and average pore diameter decreased. At high temperatures, the thermal decomposition

Table 7 Composition of TiO_2-600 and CaO/TiO_2-20–700

Catalyst	Mass percentage (%)			
	Ti	O	C	Ca
TiO_2-600	76.26	20.77	2.97	–
CaO/TiO_2-20–700	27.72	48.41	13.91	9.96

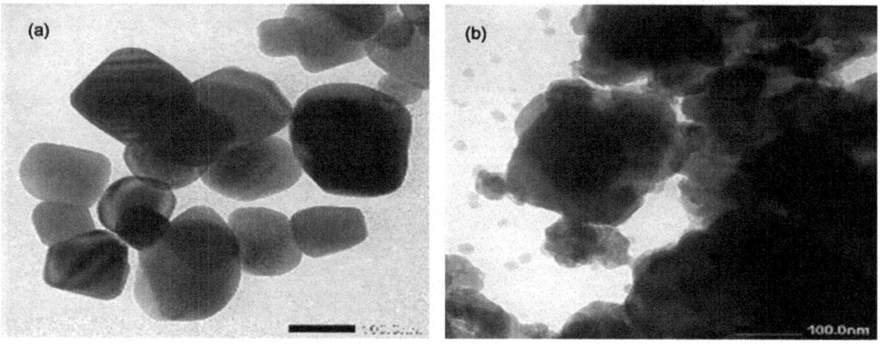

Fig. 17 TEM images of TiO_2-600 (**a**), and CaO/TiO_2-20–700 (**b**)

of $CaCO_3$ and $Ca(OH)_2$ results in the loss of CO_2 and H_2O, resulting in the formation of porous CaO. The presence of porous CaO can increase the specific surface area of the catalyst. In contrast, the total pore volume and average pore diameter of CaO/TiO_2-20–700 decreased due to the formation of CaO aggregates covering the TiO_2 surface and blocking the pores of the catalyst [43, 61].

Figure 18 shows the profiles of N_2 adsorption-desorption isotherm curves of TiO_2-600 and CaO/TiO_2-20–700 catalysts. As the same with TiO_2-600, CaO/TiO_2-20–700 shows an isotherm type IV adsorption-desorption isotherm which is the characteristic of a mesoporous material and a hysteresis pattern of type H3, confirming slit-like pores in aggregates form with a wide distribution of pore size of the catalyst [41]. The pore distribution of the CaO/TiO_2-20–700 based on BJH desorption showed wider pore distribution and an optimum pore distribution peak at 4 nm, which explains the presence of heterogenous pore size particle, while the TiO_2-600 has a narrow peak with optimum pore distribution at 2 nm. This fact explains the greater specific surface area and non-uniform particle sizes of CaO/TiO_2-20–700 [6]. The textural properties of the catalyst (Table 9) confirmed the increase in surface area of the CaO/TiO_2-20–700 catalyst.

Figure 19 shows the curves of TG-DSC for TiO_2-600 and CaO/TiO_2-20–700 catalysts. Temperatures of 50–250, 500–700, and 750–900 °C resulted in mass loss in TiO_2-600, indicating loss of water molecules, organic material or volatile substances, and anatase phase transformation, respectively. The TGA curve of CaO/TiO_2-20–700 shows stability up to a temperature of 600 °C, indicating there are no water molecules contained in the CaO/TiO_2-20–700. FTIR spectra of CaO/TiO_2 catalysts (Figs. 13 and 14) confirmed the absence of stretching vibrations of –OH and bending

Table 8 Porosity of TiO_2-600 and CaO/TiO_2-20–700 catalysts

Catalyst	Surface area(m^2/g)	Pore volume (cc/g)	Pore diameter(nm)
TiO_2-600	7.055	0.0671	34.48
CaO/TiO_2-20–700	8.292	0.0592	30.57

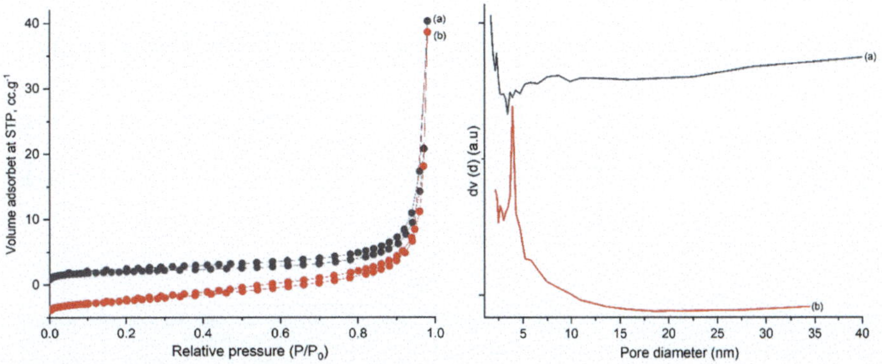

Fig. 18 N_2 adsorption-desorption (left) and BJH distribution (right) of TiO_2 (**a**), and CaO/TiO_2-20–700 (**b**)

Table 9 Specification of methyl ester compounds

Peak	Compound	Molecular formula	Molecular weight	Weight (%)
1	Methyl tetradecanoate (methyl myristic)	$C_{15}H_{30}O_2$	242	0.76
2	Methyl 9-hexadecanoate (methyl palmitoleate)	$C_{17}H_{32}O_2$	268	0.13
3	Methyl hexadecanoate (Methyl palmitate)	$C_{17}H_{34}O_2$	270	36.87
4	Methyl 9-octadecanoate (Methyl oleate)	$C_{19}H_{36}O_2$	296	58.23
5	Methyl octadecanoate (Methyl stearate)	$C_{19}H_{38}O_2$	298	4.29

vibrations of H_2O molecules in the CaO/TiO_2 catalysts. The mass loss of CaO/TiO_2-20–700 occurred in two stages. The first mass loss occurred at 510–780 °C, with a greater loss of mass than the second stage, indicating the decomposition of $CaCO_3$ to form $Ca(OH)_2$ and CaO [46, 62]. Transformation of the TiO_2's anatase phase to the rutile phase caused the second mass loss at 780–900 °C [52].

The DSC curve of TiO_2-600 showed the formation of two endothermic curves at 168–322 and 752–854 °C, which contribute to the loss of physically adsorbed water, organic material, and the formation of the TiO_2 rutile phase. The DSC curve of CaO/TiO_2-20–700 catalyst revealed the formation of an endothermic curve at 697–891 °C, indicating the loss of volatile substances bound to the catalyst [1], the decomposition of $CaCO_3$–$Ca(OH)_2$ and CaO [45], and the transition of phase anatase TiO_2 [62].

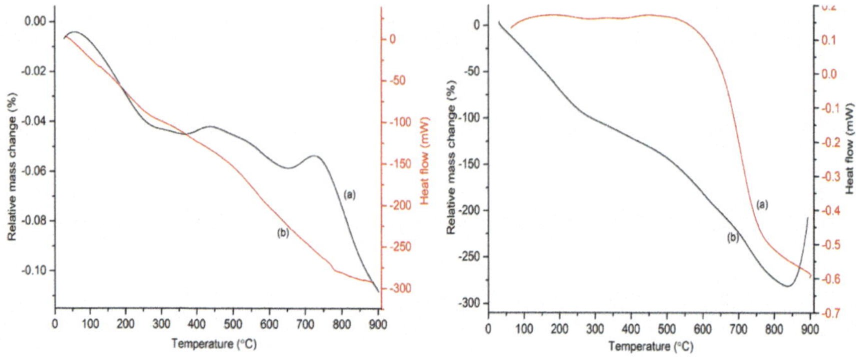

Fig. 19 DSC (**a**) and TGA (**b**) curves of TiO_2-600 (left) and CaO/TiO_2-20–700 (right)

3.3 Esterification of Waste Frying Oil

3.3.1 Waste Frying Oil Preparation

Waste frying oil was tested for initial FFA content and saponification number to determine the molecular weight of used palm oil. The results showed that the FFA content and lathering numbers of used palm oil were 1.55%, and 1.98 mg KOH g^{-1}, respectively. The high FFA content of waste frying oil requires the esterification reaction to reduce the FFA level of waste frying oil. Thangaraj et al. [8] reported the maximum value of FFA or fatty acid content in oil for biodiesel production of 1%. The reaction mechanism of free fatty acid adsorption in the esterification using an acid catalyst to produce methyl ester is shown in Fig. 20. The esterification reaction is carried out using an acid catalyst, SO_4/TiO_2-1.5–600. The esterification reaction was carried out on three parameters to see the optimal condition for reducing FFA levels of waste frying oil, which is the effect of catalyst weight (1, 2, 3, 4, and 5%), the mole ratio of oil and methanol (1: 9, 1:12, 1:15, 1:18 and 1:21) and reaction times (30, 60, 90 and 120 min).

3.3.2 Catalyst Weight Variation

The initial esterification experiment was carried out using various catalyst loadings (1–5 wt.%) to find the optimal amount of catalyst weight that could significantly reduce the FFA of waste frying oil. The use of a catalyst in the esterification stage has an effect on increasing the rate of reaction and establishing the reaction at a lower temperature. The more catalyst is used, the more active the site catalyst is, which allows equilibrium to be reached more quickly [2]. Esterification using 3 wt.% of the catalyst loading was found to optimally reduce the FFA content by 58.82% (Fig. 20). The catalyst weight of 4–5% shows a lower decrease in the FFA content. Employing

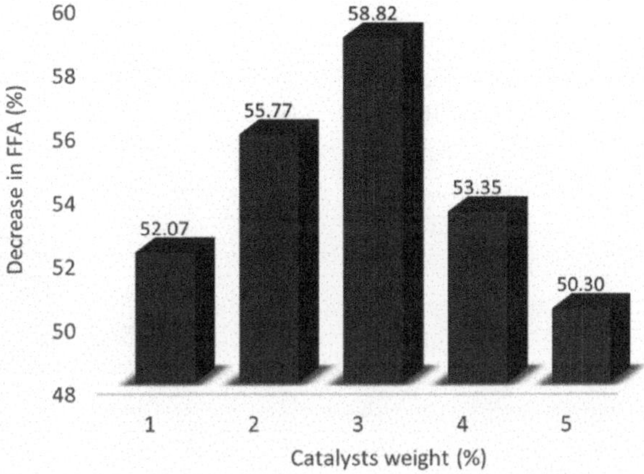

Fig. 20 A graph of the relationship between catalyst weight and FFA level reduction of waste frying oil

a large amount of catalyst can lead to the formation of soap in the reaction mixture, thereby increasing the FFA level of the oil [56].

3.3.3 Methanol to Oil Mole Ratio Variation

Next, the optimal catalyst weight obtained from the previous experiment (3%) was employed to find the optimum mole ratio of methanol to oil. Figure 21 shows the relationship between the molar ratio of oil to methanol and the decrease in the FFA of waste frying oil. Excess use of methanol can affect equilibrium by forcing the reaction to shift to the product, so the FFA of the oil will decrease [63]. The methanol to oil mole ratio of 1:18 showed the highest decrease in FFA level (61.33%). However, the use of methanol at a ratio of 1:21 caused a lower decrease in the FFA level. The excess methanol can be converted to glycerol, which is soluble in oil and produces free fatty acids [27].

3.3.4 Reaction Time Variation

With optimal catalyst weight (3%) and mole ratio of methanol to oil (1:18), the next experiment was carried out to evaluate the effect of reaction time in esterification to reduce the FFA content of waste frying oil. Long reaction times can increase the collisions between molecules that enable the equilibrium to be reached by forcing the reaction to shift to reactant (reversible reaction) [4]. Figure 22 shows the relationship between reaction time and the decrease in the FFA content in waste frying oil. The results showed that the optimum reaction time was 90 min with a decrease in FFA of 66.23%. The reaction time of 120 min caused a lower decrease in FFA level. A longer reaction time can shift the equilibrium toward the reactants, causing the –SO_3H acid site to become deactivated and thus increasing the FFA [64].

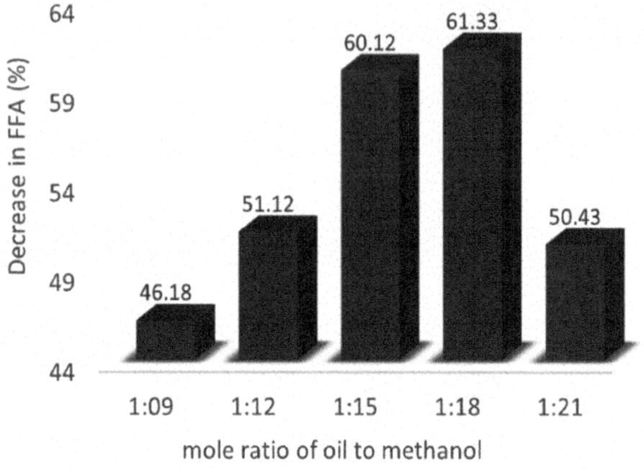

Fig. 21 A graph of the relationship between the molar ratio of oil to methanol and FFA level reduction of waste frying oil

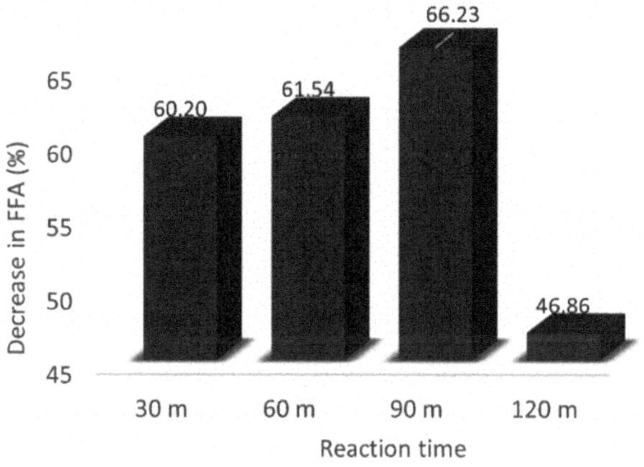

Fig. 22 A graph of the relationship between reaction time and FFA level reduction of waste frying oil

The significant property of evaluating the relationship between the acidity and catalytic activity of the catalyst was carried out by applying the TiO_2, SO_4/TiO_2-1.5–400, and SO_4/TiO_2-1.5–600 catalysts in the esterification to reduce the free fatty acid (FFA) of waste frying oil. This reaction was carried out at a catalyst weight of 3%, a ratio of oil to methanol of 1:18, and a reaction time of 90 min. SO_4/TiO_2-1.5–400 was selected due to the higher total acidity than SO_4/TiO_2-1.5–600 (Table 2) but showed the presence of high-water molecules.

The relationship between the catalysts and FFA level reduction of waste frying oil is shown in Fig. 23. TiO_2, SO_4/TiO_2-1.5–400 and SO_4/TiO_2-1.5–600 catalysts reduce the FFA level of WFO by 21.82, 45.79, 61.33%, respectively. This fact indicated that the catalytic activity of the catalyst was influenced by total acidity. TiO_2 has a low total acidity (1.16 mmol/g), while SO_4/TiO_2-1.5–400 has a total acidity of 6.00 mmol/g. Higher strength acids can facilitate the reduction of FFA levels [4]. However, SO_4/TiO_2-1.5–600 showed a higher decrease in FFA. A high number of water molecules

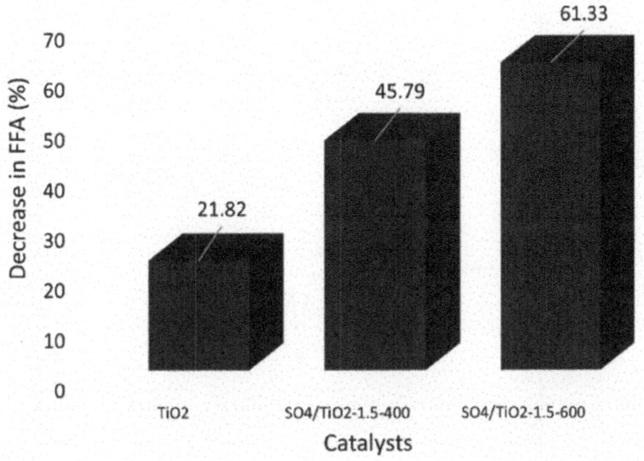

Fig. 23 The relationship between the FFA level reduction of waste frying oil and the catalysts

in the SO_4/TiO_2-1.5–400 can generate saponification, which can produce more FFA [5]. The result revealed that SO_4/TiO_2-1.5–600 was the catalyst with the optimum condition for decreasing FFA levels in WFO.

3.4 *Transesterification of Esterified Oil*

The transesterification of esterified waste frying oil was done by CaO/TiO_2-20–700 base catalyst. The reaction mechanism of transesterification using a base catalyst is shown in Fig. 24. The triglyceride was reacted using 3 mol of methanol with the help of CaO/TiO_2-20–700 catalyst to produce biodiesel or fatty acid methyl ester and glycerol as a secondary product.

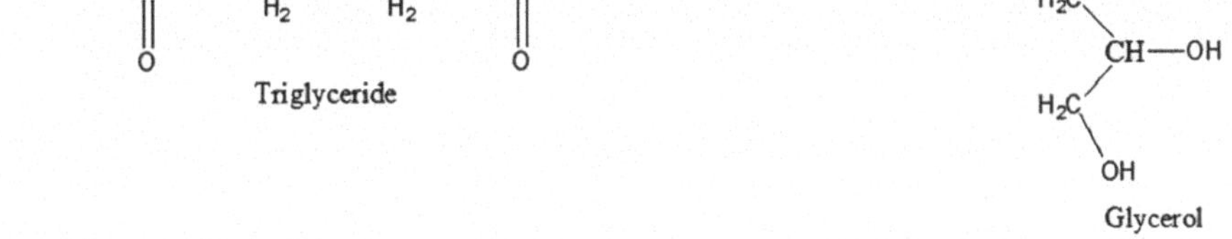

Fig. 24 The reaction mechanism of biodiesel conversion in the transesterification

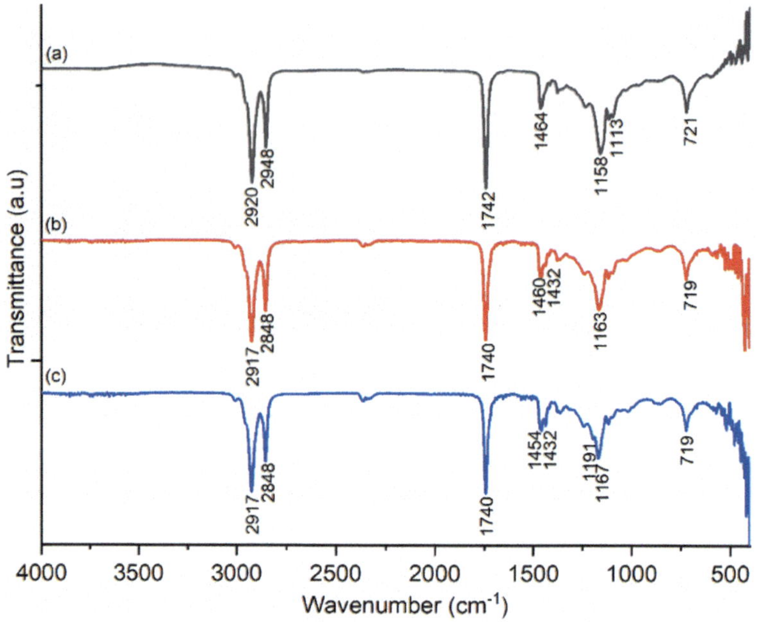

Fig. 25 FTIR spectra of waste frying oil (**a**), esterified (**b**), and transesterification product (**c**)

The FTIR spectra of waste frying oil, the esterified oil, and the transesterification product are shown in Fig. 25. All spectra show two strong absorptions at 2917–2920 and 2848 cm^{-1}, which correspond to the asymmetric and symmetric stretching vibrations of C–H sp^3. A medium absorption at 1454–1464 cm^{-1} corresponding to the C–H bending vibration, while an absorption of 719–721 cm^{-1} is assigned to be a C–H rocking vibration [56].

Bands at 1158–1167 and 1740–1742 cm^{-1} indicated the vibrations of the C–O and C=O bonds of the ester groups. These peaks will normally appear in oil and products due to the same fatty acid composition [40]. The esterification and transesterification of biodiesel showed new peaks in the 1432 cm^{-1}, which correspond to the asymmetric bending vibration $-CH_3$. The band at 1191 cm^{-1} in the transesterification is attributed to the presence of the methoxy $-OCH_3$ group in the transesterification product. The presence of vibration in the $-CH_3$ and $-OCH_3$ groups confirmed that the waste frying oil triglycerides were successfully converted to methyl ester [65].

The gas chromatogram (GC) of methyl ester compounds contained in the biodiesel from waste frying oil is shown in Fig. 26. Specifications for compounds based on the peak of the methyl ester formed (Table 7) consist of methyl myristate (0.48%), methyl palmitate (36.87), methyl palmitoleate (0.13%), methyl oleate (58.23%), and methyl stearate (4.29%). This reveals that the dominant types of methyl ester compounds are methyl oleate and methyl palmitate. A study conducted by Elkadi et al. [66] also showed that the major compound of methyl ester found in biodiesel from waste

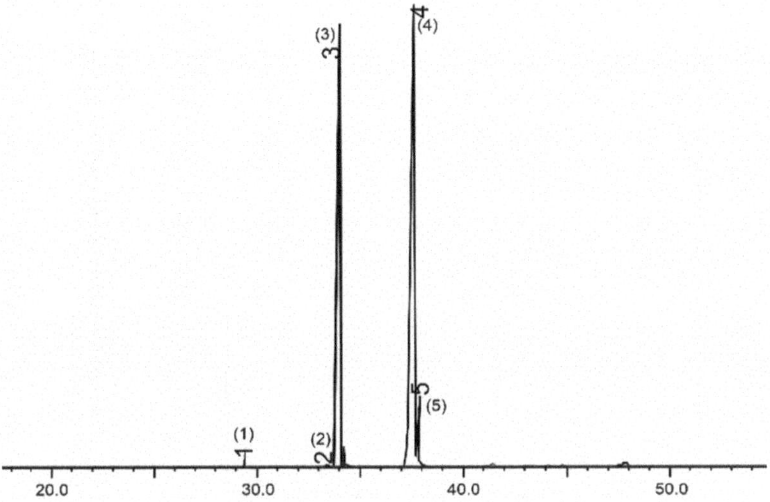

Fig. 26 Gas chromatogram (GC) of transesterification biodiesel

frying oil is methyl oleate. The formation of methyl ester suggests that waste frying oil has been successfully converted to biodiesel.

The ^{1}H-NMR spectrum data for biodiesel is presented in Fig. 27. The structural characteristics of the methyl ester were characterized by the presence of a proton peak of the methoxy group ($-OCH_3$) in the methyl ester at $\delta = 3.66$ ppm with the integration of 3.00. This is a typical peak of biodiesel. The 2.31 ppm singlet is the α-methylene ($-CH_2COOMe$) group of the methyl ester, which supports the presence of the biodiesel methyl ester. The multiple peaks at 5.33–5.34 ppm indicate the presence of an alkene group (–CH=CH–) and the peak at 1.25 ppm is a proton of the methylene group ($-(CH_2)n-$). The peak at 4.14–4.16 ppm, which is a component of glyceride (–CH–COOR) [67, 68], explains that the triglycerides in waste frying oil are not completely converted into methyl esters. The conversion of triglycerides to methyl ester is determined based on the proton integration value of the methoxy proton (3.66 ppm; 3.00) and α-methylene proton group (2.31 ppm; 3.44) using Eq. 9. The results obtained a methyl ester conversion of 58.13%.

$$\text{Biodiesel yield}\ (\%) = \frac{2 \times \text{Integration of methoxy proton(methyl ester)}}{3 \times \text{Integration of methylene proton(triglycerides)}} \times 100\% \qquad (9)$$

The structural characteristics of the biodiesel that uses CaO/TiO_2-20–700 catalyst was compared with those of the biodiesel that uses TiO_2-600 as a catalyst in the transesterification to evaluate the relationship between the basicity and catalytic activity of the catalyst. The ^{1}H-NMR spectrum of transesterification biodiesel with

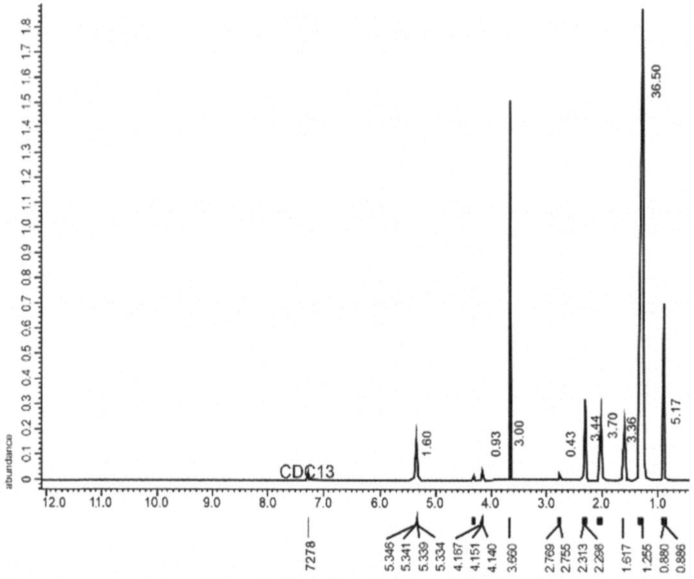

Fig. 27 ^{1}H-NMR spectrum of transesterification biodiesel with CaO/TiO_2-20–700 catalyst

a TiO_2-600 catalyst is shown in Fig. 28. The structure of biodiesel using TiO_2-600 showed the peaks of the methoxy proton ($-OCH_3$) and α-methylene proton with an integration of 2.71 and 9.59, respectively. The conversion yield of biodiesel reached 18.83%. This result indicated that the base of the catalyst was influencing the catalytic activity of the catalyst. The CaO/TiO_2-20–700 catalyst that has a total basicity of 8.13 mmol HCl/g succeeded in converting esterified waste frying oil into biodiesel with a conversion rate of 58.13%, while the TiO_2-600 with a low total basicity (0.13 mmol HCl/g) only obtained 18.83% of biodiesel.

The porosity of the catalyst has a different role in the catalytic activity of heterogeneous base catalysts [28]. The correlation structure of the surface structure of the catalyst and the catalytic activity of this study compared with other studies is shown in Table 10. The conversion of biodiesel in this work was lower compared to other studies. Conversion of palm oil into biodiesel by Pratika et al. [37] using a CaO/TiO_2 catalyst that has a surface area of 10.68 m^2/g produced biodiesel up to 79.68%. Yahya et al. [36] synthesized a CaO/TiO_2 catalyst using Titanium tetrabutoxide ($Ti(OCH_2CH_2CH_2CH_3)_4$) as a TiO_2 precursor. Due to the synthesis processes of the catalyst from precursor titanium tetrabutoxide ($Ti(OCH_2CH_2CH_2CH_3)_4$), the CaO/TiO_2 catalyst has a higher surface area (31.40 m^2/g). The use of precursor Titanium tetrabutoxide will produce a CaO/TiO_2 catalyst with a high surface area, which will accelerate the production of biodiesel with a high yield (80%). This fact indicates that the surface area of the CaO/TiO_2 catalyst is related to its catalytic activity in biodiesel production.

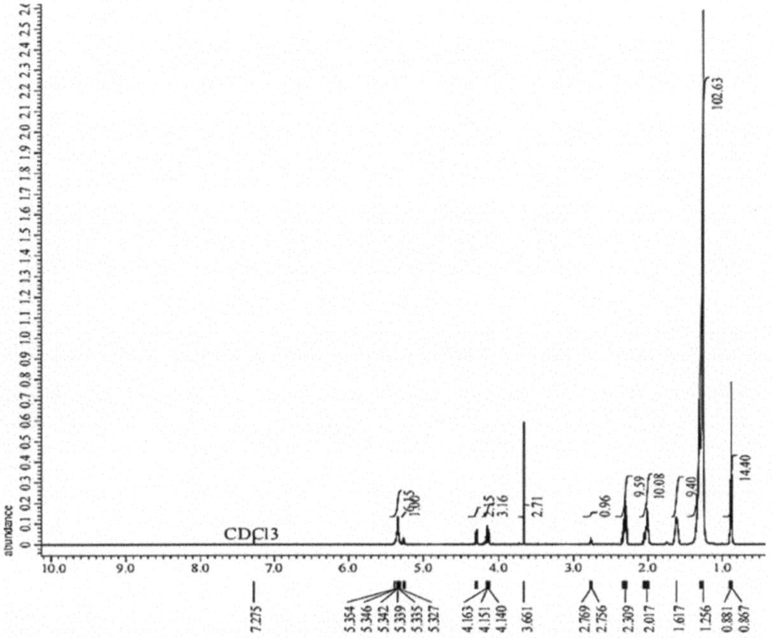

Fig. 28 ^{1}H-NMR spectrum of transesterification biodiesel with TiO_2-600 catalyst

Table 10 Catalytic activity of CaO/TiO_2 catalyst in the biodiesel production

Parameter	This study	Pratika et al. [35]	Yahya et al. [34]
Catalyst	CaO/TiO_2	CaO/TiO_2	CaO/TiO_2
Feedstock	Waste frying oil	Jatropha oil	Waste frying oil
TiO_2 source	TiO_2 anatase	TiO_2 anatase	Titanium tetrabutoxide
Surface area (m^2/g)	8.29 m^2/g	10.68 m^2/g	31.40 m^2/g
Biodiesel product (%)	58.13	79.68	80.00
Reaction condition	C: 3 wt.%, M: 18:1, T: 65 °C, R: 2 h	C: 3 wt.%, M: 1:15, T: 65 °C, R: 1.5 h	C: 0.2 wt.%, M: 3:1, T: 65 °C, R: 1 h

Where C = catalyst loading, M = molar ratio of oil to methanol, T = temperature, R: Reaction time

4 Conclusion

Modification of TiO_2 material as SO_4/TiO_2 and CaO/TiO_2 catalysts has been successfully carried out at each variation of sulfate and CaO concentration and calcination temperature. The SO_4/TiO_2-1.5–600 acid catalyst is the catalyst with the highest total acidity (2.48 mmol NH_3^{-1}). Sulfation on TiO_2 based on EDS analysis confirmed the presence of sulfate element by 1.36%, besides that it caused the surface area of SO_4/

TiO_2-1.5–600 catalyst to decrease from 7.055 to 6.559 m^2/g. The application of SO_4/TiO_2-1.5–600 catalyst in esterification successfully reduced the FFA content of used cooking oil up to 66.21% under the reaction conditions of 3% catalyst weight, 1:18 oil and methanol mole ratio, and 90 min reaction time. Modification of TiO_2 with CaO as a basic catalyst showed that the CaO/TiO_2-20–700 catalyst had the highest basicity value of 8.13 mmol/HCl. SAA analysis and SEM or TEM morphology confirmed the presence of aggregates (Ca content of 9.96%) which caused the surface area of the CaO/TiO_2-20–700 catalyst to increase from 7.055 to 8.292 m^2/g). Transesterification of used cooking oil using CaO/TiO_2-20–700 catalyst successfully produced biodiesel products up to 58.13% with the main component of methyl ester being methyl oleate (58.23%).

Acknowledgements This work was supported by the World Class Research (WCR) Project of DRPM Kemenristekdikti 2019 under Contract Number: 1980/UN1.DITLIT/DIT-LIT/LT/2019, and by The Indonesian Endowment Fund for Education (LPDP)-Ministry of Finance, Indonesia.

References

1. B. Tabah, P.N. Anjani, P. Nina, P. Aharon, Solar-heated sustainable biodiesel production from waste cooking oil using a sono chemically deposited SiO catalyst on microporous activated carbon. Energy Fuels **31**, 6228–6239 (2017). https://doi.org/10.1021/acs.energyfuels.7b00932
2. A.B. Fadhil, E.TB. Al-Tikrity, A.M. Khalaf, Transesterification of non-edible oils over potassium acetate impregnated CaO solid base catalyst. Fuel. **234**, 81–93 (2018). https://doi.org/10.1016/j.fuel.2018.06.121
3. A.K. Dalai, T. Issariyakul, C. Baroi, Biodiesel production using homogeneous and heterogeneous catalysts: a review, in *Catalysis for Alternative Energy Generation, Department of Surface and Chemistry and Catalyst*. ed. by L. Guczi, A. Erdohelyi (Springer, Heidelberg New York, 2012), pp.237–262
4. M.D. Serio, T. Riccardo, L. Pengmei, S. Elio, Heterogeneous catalyst for biodiesel production. Energy Fuels **22**, 207–217 (2017). https://doi.org/10.1021/ef700250g
5. Y. Luo, Z. Mei, N. Liu, H. Wang, C. Han, S. He, Synthesis of mesoporous sulfated zirconia nanoparticles with high surface area and their applies for biodiesel production as effective catalysts. Catal. Today **298**, 99–108 (2017). https://doi.org/10.1016/j.cattod.2017.05.047
6. G.L. Maddikeri, A.B. Pandit, P.R. Gogate, Intensification approaches for biodiesel synthesis from waste cooking oil: a review. Ind. Eng. Chem. Res. **51**, 14610–14628 (2012). https://doi.org/10.1021/ie301675j
7. J.M. Fonseca, J.G. Teleken, V.C. Almeida, C.D. Silva, Biodiesel from waste frying oils: methods of production and purification. Energy Convers. Manag. **184**, 205–218 (2019). https://doi.org/10.1016/j.enconman.2019.01.061
8. B. Thangaraj, P.R. Solomon, B. Muniyandi, S. Ranganathan, L. Lin, Catalysis in biodiesel production—A review. Clean Energy **3**(1), 2–23 (2019). https://doi.org/10.1093/ce/zky020
9. A.K. Dalai, T. Issariyakul, C. Baroi, C. Biodiesel production using homogeneous and heterogeneous catalysts: a review, in *Catalysis for Alternative Energy Generation*, eds. by L. Guczi, A. Erdôhelyi (Springer, New York, NY, 2012), pp. 237–262. https://doi.org/10.1007/978-1-4614-0344-9_6
10. S.Z. Abidin, K.F. Haigh, B. Saha, Esterification of free fatty acids in used cooking oil using ion-exchange resins as catalysts: an efficient pretreatment method for biodiesel feedstock. Ind. Eng. Chem. Res. **51**, 14653–14664 (2012). https://doi.org/10.1021/ie3007566

11. F. Guo, Z. Fang, Biodiesel production with solid catalyst, in *Biodiesel–Feedstocks and Processing Technologies.* eds. by M. Stoytcheva, G. Montero (InTech, 2011), 22 pp. 339–358. https://doi.org/10.5772/25602
12. I.M. Lokman, U. Rashid, Y.H. Taufiq-Yap, Production of biodiesel from palm fatty acid distillate using sulfonated-glucose solid acid catalyst: characterization and optimization. Chinese J. Chem. Eng. **23**, 1857–1864 (2015). https://doi.org/10.1016/j.cjche.2015.07.028
13. B. Caratão, E. Carneiro, P. Sá, B. Almeida, S. Carvalho, Properties of electrospun TiO_2 nanofibers. J. Nanotechnol. **2014**, 472132 (2014). https://doi.org/10.1155/2014/472132
14. H. Valdés, L.M. Molina, J.A. Alonso, Water adsorption and dissociation on gold catalysts supported on anatase-TiO_2 (101). Appl. Surf. Sci. **487**, 244–252 (2019). https://doi.org/10.1016/j.apsusc.2019.04.249
15. N. Kaur, A. Ali, Preparation, and application of Ce/ZrO_2–$TiO_2/SO_4{}^{2-}$ as solid catalyst for the esterification of fatty acid. Renew. Energy **81**, 421–431 (2015). https://doi.org/10.1016/j.renene.2015.03.051
16. M. Alsharifi, H. Znad, S. Hena, M. Ang, Biodiesel production from canola oil using novel Li/TiO_2 as a heterogeneous catalyst prepared via impregnation method. Renew. Energy **114**, 1077–10089 (2017). https://doi.org/10.1016/j.renene.2017.07.117
17. J. Gardy, A. Hassanpour, X. Lai, M.H. Ahmed, Synthesis of $Ti(SO_4)O$ solid acid nano-catalyst and its application for biodiesel production from used cooking oil. Appl. Catal. A-Gen. **527**, 81–95 (2016). https://doi.org/10.1016/j.apcata.2016.08.031
18. C. Carlucci, L. Degennaro, R. Luisi, Titanium dioxide as a catalyst in biodiesel production. Catalyst **9**(1), 75 (2019). https://doi.org/10.3390/catal9010075
19. C. Chen, L. Cai, X. Shangguan, L. Li, Y. Hon, G. Wu, Heterogeneous and efficient transesterification of Jatropha curcas L. seed to oil to produce biodiesel catalyzed by nano-sized $SO_4{}^{2-}/TiO_2$. R. Soc. Open Sci. **5**, 181331 (2018). https://doi.org/10.1098/rsos.181331
20. R.M. de Almeida, L.K. Noda, N.S. Gonçalves, S.M.P. Meneghetti, M.R. Meneghetti, Transesterification reaction of vegetable oils using superacid sulfated TiO_2-base catalysts. App. Catal. A-Gen. **347**, 100–105 (2008). https://doi.org/10.1016/j.apcata.2008.06.006
21. X.H. Lin, X.J. Yin, J.Y. Li, S.F.Y. Li, Elucidation of structures of surface sulfate species on sulfated titania and mechanism of improved activity. App. Catal. B-Environ. **203**, 731–739 (2017). https://doi.org/10.1016/j.apcatb.2016.10.068
22. M.H. Sarvari, E. Sodagar, Esterification of free fatty acids (biodiesel) using nano sulfated-titania as catalyst in solvent-free conditions. C. R. Chim. **16**, 229–238 (2013). https://doi.org/10.1016/j.crci.2012.10.016
23. J. Gardy, A. Hassanpour, X. Lai, M.H. Ahmed, Synthesis of $Ti(SO_4)O$ solid acid nano-catalyst and its application for biodiesel production from used cooking oil. App. Catal. A-Gen. **527**, 81–95 (2016). https://doi.org/10.1016/j.apcata.2016.08.031
24. J. Rovero-Vega, A. Pérez, R. Gómez, M.E. Gómez, Sulfated titania $\{TiO_2/SO_4{}^{2-}\}$. A very active solid acid catalyst for the esterification of free fatty acid with ethanol. App. Catal. A-Gen. **379**(1–2), 24–29 (2010). https://doi.org/10.1016/j.apcata.2010.02.020
25. B.M. Marinković, M. Stanković, A. Veličković, J. Avramović, M. Miladinović, O. Stamenković, V. Veljković, D. Jovanović, Calcium Oxide as a promising heterogeneous catalyst for biodiesel production: current state and perspectives. Renew. Sust. Energy Rev. **56**, 1387–1408 (2016). https://doi.org/10.1016/j.rser.2015.12.007
26. A. Tangy, I.N. Pulidindi, A. Gedanken, SiO beads decorated with SiO nanoparticles for biodiesel production from waste cooking oil microwave irradiation. Energy Fuels **30**, 3155–3160 (2016). https://doi.org/10.1021/acs.energyfuels.6b00256
27. E.P. Susi, K. Wijaya, Wangsa, R.A. Pratika, P.L. Hariani, Effect nickel concentration in natural zeolite as catalysts in hydrocracking process of used cooking oil. Asian J. Chem. **32**(11), 2773–2777 (2020). https://doi.org/10.142333/ajchem.2020.22708
28. P.R. Pandit, M.H. Fulekar, Biodiesel production from microalga biomass using CaO catalyst synthesized from natural waste material. Renew. Energy **136**, 837–845 (2019). https://doi.org/10.1016/j.renene.2019.01.047

29. J.M. Dias, M.C.M. Alvim-Ferraz, M.F. Almeida, J.D.M. Diaz, M.S. Polo, J.R, Comparison of the performance of different homogeneous alkali catalysts during trans-esterification of waste and virgin oils and evaluation of biodiesel quality. Fuel **87**, 3572−3578 (2008). https://doi.org/10.1016/j.fuel.2008.06.014
30. A. Shaheen, S. Sultana, H. Lu, M. Ahmad, M. Asma, T. Mahmood, Assessing the potential of different nano-composite (MgO, Al_2O_3–CaO and TiO_2) for efficient conversion of *Silybum eburneum* seed oil to liquid biodiesel. J. Mol. Liq. **249**, 511–521 (2018). https://doi.org/10.1016/j.molliq.2017.11.053
31. I. Lawan, N.Z. Garba, W. Zhou, M. Zhang, Z. Yuan, Synergies between the microwave reactor and CaO/zeolite catalyst in waste lard biodiesel production. Renew. Energy **145**, 2550–2560 (2020). https://doi.org/10.1016/j.renene.2019.08.008
32. A.A. Dabbawalla, S.M. Alhassan, D.K. Mishra, J. Jegal, J.S. Hwang, Solvent free cyclodehydration of sorbitol to isosorbide over mesoporous sulfated titania with enhanced catalytic performance. Mol. Catal. **454**, 77–86 (2018). https://doi.org/10.1016/j.mcat.2018.05.009
33. M.S. Ore, K. Wijaya, W. Trisunaryanti, W.D. Saputri, E. Heraldy, M.W. Yuwana, P.L. Hariani, A. Budiman, S. Sudiono, The Synthesis of SO_4/ZrO_2 and Zr/CaO catalysts via hydrothermal treatment and their application for conversion of low-grade coconut oil into biodiesel. J. Environ. Chem. Eng. **8**, 104205 (2020). https://doi.org/10.1016/j.jece.2020.104205
34. S. Li, J. Zhang, S. Jamil, Q. Cai, S. Zang, Conversion of eggshells into calcium titanate cuboid and its adsorption properties. Res. Chem. Intermed. **44**, 3933–3946 (2018). https://doi.org/10.1007/s11164-018-3332-1
35. A. Sistani, N. Saghatoleslaml, H. Nayebzadeh, Influence of calcination temperature on the activity of mesoporous CaO/TiO_2–ZrO_2 catalyst in the esterification reactions. J. Nanostru. Chem. **8**, 321–331 (2018)
36. M. Mohamad, N. Ngadi, Effect of TiO_2 mixed CaO catalyst in palm oil transesterification. App. Mech. Mater. **695**, 319–322 (2014). https://doi.org/10.4028/www.scientific.net/AMM.695.319
37. R.A. Pratika, K. Wijaya, W. Trisunaryanti, Hydrothermal treatment of SO_4/TiO_2 and TiO_2/CaO as heterogeneous catalysts for the conversion of Jatropha oil into biodiesel. J. Environ. Chem. Eng. **9**, 106547 (2021). https://doi.org/10.1016/j.jece.2021.106547
38. N. Kondamudi, S. Jason, M. Mano, K.M. Susanta, A green process for producing biodiesel from feather meal. J. Agric. Food Chem. **57**, 6163–6166 (2009). https://doi.org/10.1021/jf900140e
39. Z. Wen, X. Yu, S.H. Tu, J. Yan, E. Dahlquist, Biodiesel production from waste cooking oil catalyzed by TiO_2–MgO mixed oxides. Bioresour. Technol. **101**(24), 9570–9576 (2010). https://doi.org/10.1016/j.biortech.2010.07.066
40. S.H. Park, N. Khan, S. Lee, K. Zimmermann, M. Derosa, L. Hamilton, W. Hudson, S. Hyder, M. Serratos, E. Sheffield, A. Veludhandi, D.P. Pursell, Biodiesel production from locally sourced restaurant waste cooking oil and grease: synthesis, characterization, and performance evaluation. ACS Omega **4**, 7775–7784 (2019). https://doi.org/10.1021/acsomega.9b00268
41. F.U.M. Allah, G. Alexandru, Waste cooking oil as source for renewable fuel in Romania. IOP Conf. Ser. Mater. Sci. Eng. **147**, 012133 (2016). https://doi.org/10.1088/1757-899X/147/1/012133
42. A. Gerard, H. Muhr, E. Plasari, D. Jacob, C. Lefaucheur, Effect of calcium based additives on the sodium bicarbonate crystallization in a MSMPR reactor. Powder Technol. **255**, 134–140. https://doi.org/10.1016/j.powtec.2013.08.009
43. D.D. Nindarwi, A.N. Rochman, M.R.N. Tsani, V. Rachmawati, E.D. Masithah, Study of calcium hydroxide ($Ca(OH)_2$) and sodium bicarbonate ($NaHCO_3$) treatment on the dynamics of pH, Cod, N/P ratio and plankton abundance. J. Aquacult. Fish Health **8**(2), 72–79 (2019). https://doi.org/10.20473/jafh.v8i2.12428
44. H. Zheng, A. Zhang, R. Shi, Z. Zhao, Effect of $NaHCO_3$ on synthesis of anatase TiO_2 nanopowders by thermal processing of the precursor. Adv. Powder Technol. **20**, 588–592 (2009). https://doi.org/10.1016/j.apt.2009.09.007
45. H. Gu, Q. Fu, B. Sheng, E. Ca, G. Tang, Q. Liu, Effect of pore-forming agent quantity on pore structure, phase composition, micro-hardness of gradient bioceramic coating under optimal

laser process parameters. Ceram. Int. **46**(8), 11275–11281 (2020). https://doi.org/10.1016/j.ceramint.2020.01.153
46. W. Roschat, T. Siritanon, B. Yoosuk, V. Promarak, Biodiesel production from palm oil using hydrated lime derived CaO as a low-cost basic heterogeneous catalyst. Energy Convers. Manag. **108**, 459–467 (2016). https://doi.org/10.1016/j.enconman.2015.11.036
47. Y. Li, Y. Guo, Y. Liu, Synthesis of pure TiO_2 nanoparticles from $Ti(SO_4)$ in presence of EDTA as complexing agent. China Particuol. **3**(4), 240–242 (2005). https://doi.org/10.1016/S1672-2515(07)60195-X
48. S. Ardizzone, C.L. Bianchi, G. Cappelletti, F. Porta, Liquid-phase catalytic activity of sulfated zirconia from sol-gel precursors: the role of the surface feature. J. Catal. **227**, 470–478 (2004). https://doi.org/10.1016/j.jcat.2004.07.030
49. M. Utami, K. Wijaya, W. Trisunaryanti, Effect of sulfuric acid treatment and calcination on commercial zirconia nanopowder. Key Eng. Mater. **757**, 131–137 (2017). https://doi.org/10.4028/www.scientific.net/KEM.757.131
50. K. Wijaya, M.A. Kurniawan, W.D. Saputri, W. Trisunaryanti, M. Mirzan, P.L. Hariani, A.D. Tikoalu, Synthesis of nickel catalyst supported on ZrO_2/SO_4 pillared bentonite and its application for conversion of coconut oil into gasoline via hydrocracking process. J. Environ. Chem. Eng. **9**, 105399 (2021). https://doi.org/10.1016/j.jece.2021.105399
51. A. Rachmat, W. Trisunaryanti, Sutarno, K. Wijaya, Synthesis and characterization of sulfated zirconia mesopore and its application on lauric acid esterification. Mater. Renew. Sustain. Energy. **6**(13), (2017). https://doi.org/10.1007/s40243-017-0097-1
52. G. Nagaraj, R.A. Dhayal, I.A. Albert, R.L. Josephine, Tuning the optical band gap of pure TiO_2 via photon induced method. Optic **179**, 889–894 (2019). https://doi.org/10.1016/j.ijleo.2018.11.009
53. S. Afshar, M. Sadehvand, A. Azad, M.G. Dekamin, M. Jalali-Heravi, A. Mollahosseini, M. Amani, A. Tadjarodi, Optimization of catalytic activity of sulfated titania for efficient synthesis of isoamyl acetate by response surface methodology. Monatsh. Chem. **146**, 1949–1957 (2015). https://doi.org/10.1007/s00706-015-1533-5
54. A.J. Haider, R.H. Al-Anbari, G.R. Kadhim, C.T. Salame, Exploring potential environmental application of TiO_2 nanoparticles. Energy Procedia **119**, 332–345 (2017). https://doi.org/10.1016/j.egypro.2017.07.117
55. W. Yang, Y.S. Ok, X. Dou, Y. Zhang, M. Yang, D. Wei, P. Xu, Effectively remediating spiramycin from production wastewater through hydrolyzing its functional groups using solid superacid TiO_2/SO_4. Environ. Res. **175**, 393–401 (2019). https://doi.org/10.1016/j.envres.2019.05.037
56. A.B. Fadhil, A.M. Aziz, M.H. Al-Tamer, Biodiesel production from Silybum marianum L. seed oil with high FFA content using sulfonated carbon catalyst for esterification and base catalyst for transesterification. Energy Convers. Manag. **108**, 255–256 (2016). https://doi.org/10.1016/j.enconman.2015.11.013
57. X. Zhao, J. Li, Y. Zhang, H. Dong, J. Qu, T. Qi, Preparation of nanosized anatase TiO_2-coated Illite composite pigments by $Ti(SO_4)_2$ hydrolysis. Powder Technol. **271**, 262–269 (2015). https://doi.org/10.1016/j.powtec.2014.10.021
58. X. Wei, G. Zhu, J. Fang, J. Chen, Synthesis, characterization, and photocatalysis of well-dispersible phase-pure anatase TiO_2 nanoparticles. Int. J. Photoenergy **2013**, 726872 (2013). https://doi.org/10.1155/2013/726872
59. M. Kouzu, T. Kasuno, M. Tajika, Y. Sugimoto, S. Yamanaka, J. Hidaka, Calcium oxide as a solid base catalyst for transesterification of soybean oil and its application to biodiesel production. J. Fuels **87**, 2798–2806 (2008). https://doi.org/10.1016/j.fuel.2007.10.019
60. X. Lei, B. Xu, B. Yang, B. Xu, X. Guo, A novel methods of synthesis and microstructural investigation of calcium titanate powders. J. Alloys Compd. **690**, 916–922 (2017). https://doi.org/10.1016/j.jallcom.2016.08.213
61. A.A. Refaat, Biodiesel production using solid metal oxide catalysts. Int. J. Environ. Sci. Tech. **8**(1), 203–221 (2011). https://doi.org/10.1007/BF03326210

62. M. Aghilinategh, M. Barati, M. Hamadanian, Supercritical methanol for one put biodiesel production from Chlorella vulgaris microalgae in the presence of CaO/TiO_2 nano-photocatalyst and subcritical water. Biomass Bioenergy **123**, 34–40 (2019). https://doi.org/10.1016/j.biombioe.2019.02.011
63. M. Farooq, A. Ramli, A.D. Subbarao, Biodiesel production from waste cooking oil using bifunctional heterogeneous solid catalysts. J. Clean. Prod. **59**, 131–140 (2013). https://doi.org/10.1016/j.jclepro.2013.06.015
64. T.T.V. Tran, S. Kaiprommarat, S. Kongparakul, P. Reubroycharoen, G. Guan, M. Nguyen, C. Samart, Green biodiesel production from waste cooking oil using an environmentally acid catalyst. J. Waste Manag. **52**, 367–374 (2016). https://doi.org/10.1016/j.wasman.2016.03.053
65. A. Rafati, K. Tahvildari, K. Nozari, Production of biodiesel by electrolysis method from waste cooking oil using heterogeneous MgO–NaOH catalyst. Energy Sour. Part A: Recov. Utiliz. Environ. Effects **41**(9), 1062–1074 (2019). https://doi.org/10.1080/15567036.2018.1539139
66. M.F. Elkady, A. Zaatout, A.O. Balbaa, Production of biodiesel from waste vegetable oil via km micromixer. J. Chem. Hindawi Pub. Corp. **2015**, 630168 (2015). https://doi.org/10.1155/2015/630168
67. T. Roy, S. Sahani, D. Madhu, Y.C. Sharma, A clean approach of biodiesel production from waste cooking oil by using single phase $BaSnO_3$ as solid base catalyst: mechanism, kinetics, and e-study. J. Clean. Prod. **265**, 121440 (2020). https://doi.org/10.1016/j.jclepro.2020.121440
68. R. Aitbelale, R. Abala, F.E.M. Alaoui, A.S. Eddine, N.M. Rujas, F. Aguilar, Characterization, and determination of thermodynamic properties of waste cooking oil biodiesel: experimental, correlation and modeling density over a wide temperature range up to 393.15 and pressure up to 140 MPa. Fluid Phase Equilibria **497**, 87–96 (2019). https://doi.org/10.1016/j.fluid.2019.06.003

Index

S. J. Ikhmayies (ed.), *Advances in Catalysts Research*, Advances in Material Research and Technology, https://doi.org/10.1007/978-3-031-49108-5

D

E

I

J

K

L

P

Q

R

S

T

X

Y

Z

SPRINGER NATURE

GPSR Compliance

The European Union's (EU) General Product Safety Regulation (GPSR) is a set of rules that requires consumer products to be safe and our obligations to ensure this.

If you have any concerns about our products, you can contact us on ProductSafety@springernature.com

In case Publisher is established outside the EU, the EU authorized representative is:

Springer Nature Customer Service Center GmbH
Europaplatz 3
69115 Heidelberg, Germany

The manufacturer's authorised representative in the EU is Springer Nature Customer Service Centre GmbH, Europaplatz 3, 69115 Heidelberg, Germany. If you have any concerns regarding our products, please contact ProductSafety@springernature.com

Printed and bound by CPI Group (UK) Ltd, Croydon, CR0 4YY
01/10/2026
02225855-0015